AF527955

Geography of Asia

GEOGRAPHY *of* ASIA

RANJIT TIRTHA
Eastern Michigan University

Drawings by
Dale Austin
University of Michigan

RAWAT PUBLICATIONS
Jaipur • New Delhi • Bangalore • Guwahati • Kolkata

ISBN 978-81-7033-691-4 (hardcover)
ISBN 978-81-7033-692-1 (paperback)

Reprinted, 2016

Published by
Prem Rawat for **Rawat Publications**
Satyam Apts., Sector 3, Jawahar Nagar, Jaipur - 302 004 (India)
Phone: 0141 265 1748 / 7006 Fax: 0141 265 1748
E-mail : info@rawatbooks.com
Website: rawatbooks.com

New Delhi Office
4858/24, Ansari Road, Daryaganj, New Delhi 110 002
Phone: 011-23263290

Also at *Bangalore, Mumbai, Guwahati* and *Kolkata*

Typeset by Rawat Computers, Jaipur
Printed at Nice Printing Press, New Delhi

Contents

List of Maps

Preface

Although Asia has always captured the imagination of the world, it has especially forced itself on public attention during the last four decades due primarily to the dramatic developments that have taken place within it. Since the birth of Israel, the oil crisis of the Persian Gulf, and the Iran-Iraq War, the Middle East has become potentially one of the most explosive regions of the world that threatened on more than one occasion to bring the major world powers into direct confrontation. The rise of Japan as an industrial giant rivaling the West has disturbed the existing international economic and political equations. Southeast Asia's internal problems have had direct American involvement. India and China have been experimenting different models of development.

Despite a revival of interest in regional geography during the last few decades in India and the United States, textbook choices for a regional geography of Asia remain limited. Thus for courses in the geography of Asia, there is currently no general-purpose text in print. The four standard textbooks: Cressey, *Asia's Lands and People*: McGraw-Hill, 1963; East and Spate, *The Changing Map of Asia*: Methuen, 1950; Ginsburg, et al, *The Pattern of Asia*, Prentice-Hall, 1958; and Spencer, *Asia East by South*, Wiley, 1961, all published in the United States, ran into several editions during the 1950s and 60s, and have since become outdated.

The inexorable march of events, notably that of population, but also of developmental change since the publication of these texts of the 1950s and 1960s, has considerably altered the economic and political picture of Asia. To capitalize on the limitations of the old texts, in a market singularly lacking in an all-purpose, general textbook on Asian geography, a new entry is presented.

This book reflects more than thirty years of experience as a traveler, a researcher, and a teacher in Asian geography. Looking back across these years, I am conscious above all of the dramatic changes that have altered, and continue to alter, the Asian scene in a number of significant ways.

Combining the regional and systematic traditions in geography, the proposed book is of moderate size. It presents an examination of Asia's physical, cultural, economic and political patterns including a discussion of its regions. In the light of Asia's growing international importance, its geography is discussed in the international context, with a particular focus on the rise of nationalism, political independence and economic development of its several regions. Literary quality is stressed, and the illustrations used will hopefully enhance the general texture of the book. Particular attention is given to the design and execution of cartographic plates. A number of imaginatively designed maps, true to scale and content, and geared specifically to the text have been introduced.

While intended primarily for the general, all-purpose beginning course on Asian geography, this book would prove of value to the students and lay persons alike. Apparently, the book holds some promise outside the somewhat restricted academic market, mainly because of its "cultural" appeal for a variety of readers, and a still wider international market of libraries. As compared to the earlier classic texts (Cressey, Ginsburg and Spencer), the present study is kept modest in scale, focus, and content with a view to establishing better communication with the readers. The text has been influenced directly or indirectly by several scholars who evinced a special interest in the book. The colleagues in the Department of Geography and Geology, Eastern Michigan University, have been unfailingly supportive. I am grateful to all of them. I am especially indebted to many generations of students in my various geography classes in India and the United States whose insights, perceptions and questions have contributed enormously to shape my own ideas.

Thanks are due to all the reviewers who have commented on various parts of the text, and saved me from errors of omission and commission, but I must claim responsibility for the undoubtedly numerous deficiencies and errors that remain. To Dale Austin I owe special thanks who skillfully executed all the illustrations and maps. My family gave the most. Rajiva, and Abha, my children, helped as and when they could with their moral support and financial assistance. Janak, my wife, gave her encouragement and perceptive suggestions. For all of their support, I am most grateful. Finally, I must offer my gratitude to Professors Ramalingam Raja and Sundaresan Swaminathan for scrutinizing the final draft and Patsy LaFave for her cooperation in transforming a disoriented manuscript into a publishable form, and above all Pranit Rawat of Rawat Publications of Jaipur, India, for his ceaseless encouragement in bringing this project to a successful conclusion.

Ranjit Tirtha

A WORD TO THE READER

Maps are kept to a minimum. These are carefully designed to function as necessary adjuncts to the text as well as to whet readers' appetite to use a standard atlas such as *Goode's World Atlas*. Spellings usually conform to the *Goode's World Atlas*, and the *Encyclopedia Britannica*. Distances, weights, measures, and volumes generally follow the international metric system.

I

The Natural Setting

1

Introduction: The Setting and Situation of Asia

If Asia has had a consistent image, it has been that of an unchanging and traditional continent, impervious and insensitive to the impressive technological advances made and prosperity generated in the West. However, for the last half a century or so, Asia has been emerging as a continent of change, challenging its own traditions, and revising its cultural mores. This change is both physical and real, evidenced not only in Japan and South Korea, but is clearly visible in the villages of India, Thailand, Taiwan, and in the oil ports of the Persian Gulf.

On the one hand, Asia still clings to its timeless traditions; on the other, it is groping for a new order. It is engaged in a struggle to balance the pressures and influences of the western civilization. The secular and temporal influences of the West have become increasingly pitched against the Asian religious and moral traditions. Social and economic values of the West have often come in conflict with the rural and traditional Asian systems. The struggle to synthesize the "tradition" and the "new" orders has not been always happy. As Asia lurches into an uncertain future, religious, social, political and economic conflicts have arisen in most Asian nations.

In the aftermath of World War II several major developments, each with its implications, have sparked worldwide interest in Asian affairs. The collapse of the European colonial regimes over much of Asia's lands and peoples was one of the momentous events in history, which unleashed forces dramatically altering Asia's political and economic environment. The nationalist forces, which had struggled to bring down the domination of the West, had gained momentum within the countries. These took two divergent paths. One, these forces consolidated the national will and triggered border conflicts with the neighboring countries. The political boundaries, artificially created by the colonial powers, were now challenged and became arenas of conflicts. The festering Israeli-Palestinian confrontation, the Cambodian tragedy, and Vietnamese disputes are some examples. On the other hand, political consciousness among the ethnic, religious, and cultural minorities within the countries intensified the forces of divisiveness, and plunged the countries into bitter struggles between the groups and weakened the countries' internal cohesion. Since achieving independence most Asian nations seemed to have lost the unifying political force so essential for their internal cohesion.

Equally uncertain was the situation in

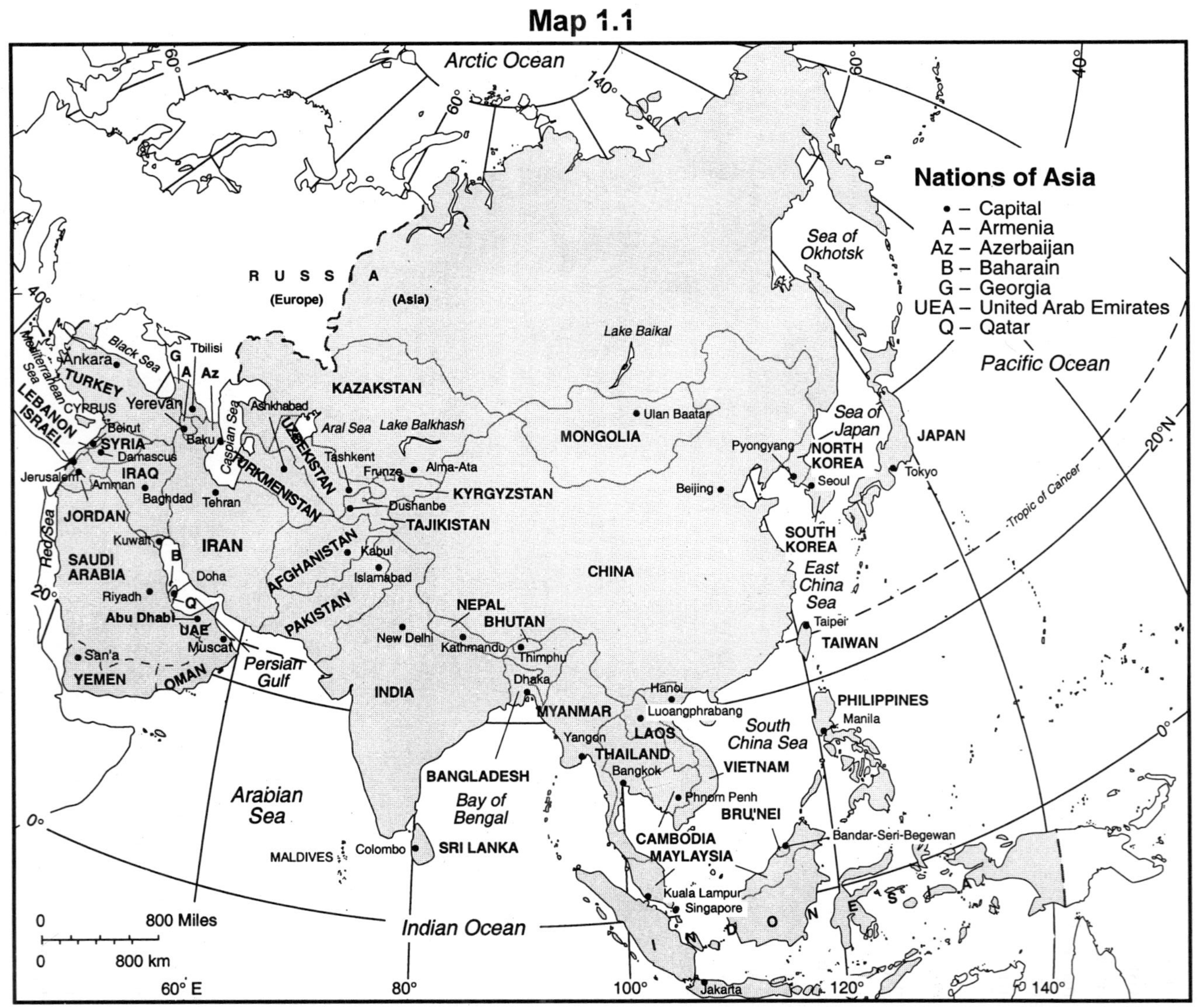
Map 1.1
Nations of Asia
• – Capital
A – Armenia
Az – Azerbaijan
B – Baharain
G – Georgia
UEA – United Arab Emirates
Q – Qatar
Arctic Ocean
Pacific Ocean
Indian Ocean
Sea of Okhotsk
Sea of Japan
East China Sea
South China Sea
Bay of Bengal
Arabian Sea
Persian Gulf
Red Sea
Mediterranean Sea
Black Sea
Caspian Sea
Aral Sea
Lake Balkhash
Lake Baikal
Tropic of Cancer
RUSSIA
(Europe)
(Asia)
KAZAKSTAN
UZBEKISTAN
TURKMENISTAN
KYRGYZSTAN
TAJIKISTAN
AFGHANISTAN
PAKISTAN
IRAN
IRAQ
SYRIA
TURKEY
CYPRUS
LEBANON
ISRAEL
JORDAN
SAUDI ARABIA
YEMEN
OMAN
UAE
MONGOLIA
CHINA
NORTH KOREA
SOUTH KOREA
JAPAN
TAIWAN
PHILIPPINES
NEPAL
BHUTAN
BANGLADESH
INDIA
SRI LANKA
MALDIVES
MYANMAR
THAILAND
LAOS
VIETNAM
CAMBODIA
MAYLAYSIA
BRU'NEI
INDONESIA
Ankara
Tbilisi
Yerevan
Baku
Beirut
Damascus
Amman
Jerusalem
Baghdad
Kuwait
Riyadh
Doha
Abu Dhabi
Muscat
San'a
Tehran
Ashkhabad
Tashkent
Dushanbe
Frunze
Alma-Ata
Kabul
Islamabad
New Delhi
Kathmandu
Thimphu
Dhaka
Colombo
Yangon
Bangkok
Hanoi
Luoangphrabang
Phnom Penh
Kuala Lampur
Singapore
Jakarta
Bandar-Seri-Begewan
Manila
Taipei
Beijing
Ulan Baatar
Pyongyang
Seoul
Tokyo
0°
20°N
20°
40°
60°
140°
60° E
80°
100°
120°
140°
0 800 Miles
0 800 km

the economic field. Although most Asian countries had received many benefits of the colonial rules by way of the establishment of an infrastructure, a system of posts and telegraphs, railroad and road networks, modernized educational facilities and an apparatus of modern western-style governments, there was little overall development, particularly in such areas as mining and industries. Most Asian nations, with the notable exception of Japan, were exporting agricultural products to Europe, where those were turned to finished products and re-exported to the Asian markets. Asian economies thus remained imbalanced and neglected, and Asia became a political and economic appendage of Europe.

THE SIGNIFICANCE OF ASIA

The single most impressive geographical fact about Asia is its enormous size, both in area and in population. It is by far the largest of the world's continents----more than four times the size of Europe, larger than North and South America put together, and nearly one-and-a-half times that of Africa. If area alone does not qualify it for importance, its large population base must surely draw our attention, for within its borders over 60 percent of the world's people live. And if size fails to interest us, what is it that makes Asia significant? What is so special about this physically vast, socially diverse, culturally complex and politically fragmented continent?

The continent has been the birthplace of all the world's great religions, including Buddhism, Christianity, Hinduism, Islam, Judaism, Sikhism, Taoism, and Zoroastrianism. Christianity moved westward to Europe. It influenced several Asian societies substantially through missionaries and the colonial rules. Now several Asian nations contain Christian minorities. The Philippines is the only Asian country that is predominantly Christian! Hinduism and Islam had more profound impacts. Buddhism is prevalent in various forms in Japan, China, Korea, Sri Lanka, and in all Southeast Asian countries. Islam fanned out of Arabia eastward to Afghanistan, Pakistan, India, Malaysia, Indonesia, and Southwest Asia and Central Asia. Hinduism, basically a non-proselytizing religion, and confined to India and Nepal, left an indelible mark on the civilizations, particularly the art and architecture of Southeast Asian societies.

Asians contributed significantly to the world's stock of cultural and technical accomplishments, many of which formed the bases for modern technology. Nearly three millennia before Christ, they knew the arts of cooking and pottery, the use of fire for smelting of ores, domestication of animals, the use of irrigation and crop rotation, the use of wheel and a form of paper for writing and had developed elaborate scripts. Early Asians were familiar with woodcarving, and the casting of metals. They built monuments in stone and metals that still evoke admiration and astonishment. They used the art of calculations including the decimal system, and employed means of measurements. Many of the indigenous systems of medicine, developed thousands of years ago in Asia, are still being practiced in many Asian nations. Intricate systems of laws and regulations, and delegating authority to institutions of administrations were developed and practiced by early Asian governments. Until the early 15th century Asia's material advancement was a source of envy by people in Europe. Asia's resources and prosperity were the main reasons that drove the adventurous from Europe to its colonization and conquest.

Even though during colonial times the economies of Asian countries were not adequately developed, Asia is fairly well-endowed with a number of agricultural produce and minerals. It is world's important producer of several foodgrains including wheat, rice, millets and corn (in rice claiming a virtual monopoly of world's production), although most of its nations are not self-sufficient in their foodgrain requirements. A leading producer of tea, natural rubber, cane-sugar, cotton, palm oil, groundnuts (peanuts) and copra with productions ranging from 40 to 80 percent of the world, it contains one-third of world's cattle, sheep, and nearly one half of world's swine. The continent produces significant amounts of tropical and subtropical fruits, which are used mostly for domestic consumption.

In the category of mining, Asia's production and reserves of several minerals, particularly of mineral fuels (such as coal and petroleum) are substantial (ranging from 20 to 40 percent of the world). It produces nearly one-fifth of world's ironore, zinc, and lead. Nearly one-half of world's tin comes from Southeast Asia. Other minerals produced in fairly large amounts (and mostly shipped to the industrialized countries of the West) are: copper, bauxite, manganese, and nickel. Asia's power resources are considerable, if not immense, due largely to its petroleum, natural gas and waterpower potentials. It produces 20 percent of world's coal, and accounts for the generation of one-sixth of world's water power (reserves and potentials are much larger). All this, at least, forms a good base for industrialization. Recently, Asians have begun to draw upon its vast human and economic resources. Infusion of capital and modern technology is slowly trickling in. Results have been particularly noteworthy where the capital formation, utilization of modern technology and well-trained labor force have generated vigorous economies, as in Japan, Taiwan, and South Korea. The Asian labor force is, generally speaking, slowly becoming skilled, and now beginning to compete with those in Europe.

Paradoxically, when we assess Asia's human resources, we are also presented with both a potential as well as a problem. The high rate of population growth in most of Asia constitutes perhaps the greatest negative factor. But several countries, including China, Japan, Taiwan, and Singapore, have recently demonstrated that progress can be achieved in the limitation of population growth.

Asia's entry into the nuclear age during the 1970s heralds its growing international importance. China became the first Asian nuclear power; India followed and tried an experimental nuclear explosion in 1974. Soon after, Israel, Pakistan, and South Korea acquired nuclear capability and very likely maintain nuclear arsenals. Before the Gulf War in 1991, Iraq was trying to develop nuclear capability. Although Japan renounced the use of atomic force as a means of settling international disputes, pressures at home have recently risen, and it might revise its policies. Given the region's political instability, Asia's nuclear proliferation has acquired an added significance.

Asia is now a continent in transition.

The European masters are no longer responsible for its political and economic destiny. Although future remains unpredictable for most Asian nations, there are clear signs that Asia is beginning to assert, relative to the rest of the world, its importance in keeping with its size, resources, and potentials.

THE LIMITS OF ASIA

In defining the limits of Asia one runs into a geographical problem. What and where is Asia? What are its boundaries? Is the gigantic landmass of Eurasia one continent or two? The problem is further compounded by the fact that Europe blends imperceptibly into Asia, with Russia sprawling between them cutting across traditional North-South trending Europe-Asia boundary along the Ural mountains. This, the so-called traditional continental boundary does not, however, follow the crest of the Ural mountains. At the southern end of the Urals, this arbitrarily defined boundary is even less convincing. It follows the Ural river and the northern side of the Caucasus mountains to the Black Sea and thence to the Mediterranean Sea.

The name 'Asia' is very ancient. The Greeks used it to designate all the lands located to the east of their homeland. It is believed that the name may have been derived from the Assyrian word *asu* (meaning the "sunrise") applied to Turkey and the lands to the east of it; whereas *ereb* ("sunset") referred to Greece and areas to the west of it. The terms "Asia" and "Europe" thus had their origins as historical expressions, and the boundary between Europe and Asia a historical-cultural concept. More recently, the division of Eurasia, largest of the continents, into Europe and Asia derives from the arbitrary decisions made by peoples in the western part of the peninsula of Europe.

Even accepting the limits of Asia along the Urals, one faces the problem of what should be included in this book. Should we include everything east of the Urals in our study? Russia, sprawls over the traditional Europe-Asia boundary, and thus spills over both Asia and Europe. Russia was developed and Europeanized into a vast European Empire by the central authority in Moscow, and thus remained culturally quite distinct from the rest of the Asian continent. It is, therefore, excluded from our discussion. The Central Asian Republics (of the former USSR or the Soviet Union) are clearly Asian in outlook, historical connections, and cultural associations. It is certainly most reasonable to include them into our discussion.

To a smaller extent, the problem of definition involves a tiny portion of Turkey and the island of Cyprus. The Ural boundary between Asia and Europe passes from the Black Sea to the Mediterranean Sea along the straits of Bosporus, the sea of Marmara, and the Dardenelles, leaving a small portion of Turkey in the European shore, where the great city of Istanbul is located. Istanbul and the island of Cyprus which lies to the south of Turkey both had Asian and European links but have sufficient Asian character as to qualify for their inclusion in our discussion.

With the limits thus defined, Asia forms a huge quadrangle covering an area of little over 17 million sq miles (44 million sq km); north-to-south extensions range from latitudes 85°N to beyond 10°S, and the west to east dimensions from 25°E

Table 1 *Area and Population Data for Asian Countries, 1999 (Estimates)*

Country	*Area (Thousand sq km/sq mi)*	*Population (millions)*	*Current Annual Rate of Growth (percent)*	*Year of Independence*
Afghanistan	647.5/250.0	25.8	2.5	1919
Baharain	0.7/0.3	0.6	1.9	1971
Bangladesh	144.0/56.9	127.1	1.8	1971
Bhutan	47.0/14.7	0.8	2.3	1910
Brunei	5.7/2.2	0.3	2.0	1984
Cambodia	181.9/70.2	10.9	2.0	1953
China	9,572.9/3,696.1	1,251.3	0.9	c. 1523 B.C.
Cyprus	9.2/3.5	0.8	0.5	1960
India	3,279.7/1,266.3	1,001.0	1.8	1947
Indonesia	1,919.3/741.0	306.1	1.4	1945
Iran	1,645.2/588.3	62.1	1.6	1906
Iraq	435.0/167.9	22.4	2.8	1932
Israel	20.3/7.8	5.1	1.4	1948
Japan	377.8/145.9	126.6	0.2	c. 660 B.C.
Jordan	88.9/34.3	4.6	3.1	1946
Kazakstan	2,724.9/1,52.1	15.3	0.4	1991
Korea, North	122.7/47.4	21.3	0.3	1948
Korea, South	99.4/38.4	46.8	0.8	1948
Kyrgyzstan	198.5/76.6	4.7	1.7	1991
Kuwait	17.8/6.9	1.8	1.8	1961
Laos	236.8/91.4	5.4	2.6	1953
Lebanon	10.2/3.9	3.5	1.6	1941
Malaysia	330.4/127.5	25.6	2.0	1957
Maldives	0.3/1.1	0.3	3.4	1965
Mongolia	1566.5/604.8	2.4	1.5	1921
Myanmar	676.5/261.2	48.0	1.6	1948
Nepal	147.1/56.8	22.4	2.5	1769
Oman	306.0/118.1	2.4	3.3	1951
Pakistan	796.1/307.3	152.2	2.4	1947
Philippines	300.0/116.8	74.7	2.1	1946
Qatar	11.4/4.4	0.6	1.3	1971
Saudi Arabia	2,248/864	21.5	3.0	1932
Singapore	0.6/0.2	3.2	0.9	1965
Sri Lanka	65.6/25.3	18.9	1.2	1948
Syria	185.1/71.5	15.7	3.2	1946
Taiwan	36.1/13.9	22.0	0.8	1945

Contd...

Contd...

Tajikistan	143.1/55.3	6.2	2.0	1991
Thailand	513.1/198.1	61.8	0.9	1350
Turkey	779.4/300.9	64.5	1.4	1923
Turkmenistan	488.1/188.5	4.7	1.7	1991
United Arab Emirates	83.6/32.2	2.8	1.6	1971
Uzbekistan	447.4/172.7	24.4	1.6	1991
Vietnam	331/127.8	77.3	1.4	1945
Yemen	472.1/182.2	16.9	3.3	1918

Source: Britannica Book of the Year, 2000.

(western margin of Turkey) to 170°W (the Bering Strait), almost halfway around the world. To the north lies the Arctic Ocean; to the northwest the plains of northern Europe, to the west the Red Sea and beyond that the Mediterranean Sea; to the south, the Indian Ocean, into which projects the peninsula of the Indian subcontinent; and to the southeast the peninsulas and the archipelagoes of Malaysia and Indonesia, separating Asia from Australia by the mingled waters of the Indian and Pacific Oceans. In the east lies the Pacific littoral broken into various seas, like the Bering Sea, the Sea of Okhotsk, the Sea of Japan, the North China Sea, and the South China Sea. Further east, off the Asian mainland, lie the chains of island groups including those of Japan, Taiwan, and the Philippines. Asia's coastlines are short for its huge size; nearly one-fifth of which remains mostly frozen and commercially useless in the Arctic Ocean.

Asia is not one, but many; several of its broad divisions or realms contain a group of countries that share broad similarities in their physical and cultural character, and in historical and political development. Such realms may be conveniently recognized: (1) Russia, the largest in area covering the entire span of northern part of the Asiatic continent; (2) East Asia which includes China, Japan, Taiwan, North Korea and South Korea; (3) Southeast Asia comprising countries that lie between India and China, as well as the various island groups that form the countries of Indonesia and the Philippines; (4) South Asia comprising the Indian subcontinent, that includes India, Pakistan, Bangladesh, Nepal, Bhutan, the islands of Sri Lanka and the Maldives; (5) Southwest Asia or the so-called Middle East which includes several Arab nations, Israel, Turkey, Iran, and Afghanistan; and (6) Central Asia that comprises the Central Asian Republics of the former USSR, namely, Kazakstan, Kyrgyzstan, Tajikistan, Turkmenistan, and Uzbekistan. In addition, world's largest nation---Russia---lies within the territorial limits and spanning the entire northern part of the Asiatic continent, but is distinctly European in its cultural orientation and historical evolution, and thus excluded from this study.

Further Readings

Atlas of the World. 2nd ed. New York: Oxford, 1995.
Goode's World Atlas. 20th ed., Chicago: Rand McNally, 2000.
Cressey, G.B. *Asia's Lands and Peoples*, New York, 1963.
Ginsburg, N.S. (ed.) *The Pattern of Asia*. Englewood Cliffs, N.J., 1963.
Spencer, J.E. and Thomas, W.L. *Asia, East by South*, New York, 1971.

2

Asia's Physical Environments

Few aspects of human activity are free from the impact of physical or natural environment. In fact, during the 1920s and 1930s geographers tended to view that all human activity was the direct consequence of physical forces. This concept ("environmental determinism") has since lost favor, as it failed to take into account those cultural, political and historical factors that also affect human distribution and activity. The environmental determinists thus minimized the role of humans in the transformation of earth's surface. Although human beings have been agents of modifications of the earth since their inception, the pace of change has constantly and more dramatically been accelerating during the last fifty years as the humans acquired greater technological sophistication. The relationship between the natural phenomena and human activities, however, is real, although it may vary in time and space; for it is difficult to understand the human landscape without having some appreciation of those factors of the physical environment that affect human occupancy. This chapter deals with the distribution and evolution of the surface features and landforms of Asia.

BROAD PATTERNS OF ASIA'S LANDFORMS

The pattern of Asia's landforms is quite complex. It contains the widest variations in relief ranging from the highest mountains and the deepest depressions in the world. Mt. Everest (Sagarmatha) is 29,028 feet (8,848 meters) high, the highest in the world, and the Dead Sea lies at about 1,300 feet (400 meters) below sea level. Nearly three-fourths of the territory of the mainland is uplands, plateaus, and mountains; the remaining in the form of river-and-coastal plains. Most of the islands that lie in Asia's southern and southeastern sections (in the Pacific and the Indian Oceans) have also mountainous cores.

In the broadest sense, the mountain systems of the mainland may be grouped in three large belts. One extends from Asia's eastern extremity through several mountains of northeastern Siberia westward to southern Siberia (the Sayan, the Altai) and to the Tien Shan and the Pamirs. The second belt lies in the western part of the continent from the western extremity in Turkey going generally eastward along Turkey's northern and southern shores, the Pontic and the Taurus Mountain respectively enclosing the Anatolian Plateau and converging in Armenian Knot, and thence passing through the Elburz Mountains of northern Iran and the Zagros along Iran's southern coast (with

Map 2.1

Konyak Highlands
Chersky Mountains
N. Siberia Lowland
West Siberia Plain
Central Siberia Plateau
Lena-Aldan Plateau
Stanovoy Mountains
Vitim Upland
Greater Khingan Mountains
Ural Mountains
Pontic Mountains
Anatolian Plateau
Caucasus
Kazakh Upland
Sayan
Altai
Gobi
Manchurian Plain
Elburz Mountains
Pamir
Tarim Basin
Nan Shan
Iranian Highland
Hindu Kush
Kunlun Mountains
Quinling
N. China Plain
Arabian Plateau
Tibetan Highlands
Himalaya
Indo-Gangetic Plain
Szechuan Basin
Deccan Plateau
Topography of Asia
Based on: Physiographic Maps by Lobeck and Raisz
Scale 1:40,000,000
200 0 200 400 600 800 1000 MILES

Iranian Plateau between these two ranges), moving eastward through the Hindu Kush in northeast direction to meet the Pamir Knot. Thus, these two belts make the Pamirs as a pivot, as does the third belt which moves eastward of the Pamirs in a form of several radiating ranges: the Tien Shan in a northeasterly direction, the Himalaya in the southeast, and the Kunlun roughly in the easterly direction. These mountain systems extend some 5,000 miles (8,000 km) and constitute the largest body of highlands in the world. Within the series of the third belt are located some of the world's highest, and most rugged mountains, elevated plateaus, low-lying basins and deserts in the "core" or the "heartland" of the continent which form a huge continental barrier difficult to penetrate.

The origin and development of these mountains and other surface features are largely puzzling. The generally accepted theory is that of plate tectonics which is invoked to explain the massive crustal rearrangements that periodically took place in the earth's geological past, eventually leading to the formation of Asia and world's other continents. The beginnings are traced to a single super-continent, known as *Pangae* which subsequently broke apart into two massive landmasses: *Laurasia* in the northern hemisphere and *Gondwanaland* in the southern hemisphere. These two contained several "plates", which later on separated at the "weak zones" and drifted in various directions. According to the plate tectonics theory of mountain-building there existed a huge geosynclinal depression or sea, known to the geologists as "Tethys", between the two vast landmasses----the Laurasian and the Gondwanaland. In the latter contained the African, the Arabian and the Indian plates, into which were poured sediments by the rivers of the two flanking plates. The colossal infilling of the sediments in the geosyncline disturbed the gravity equilibrium of the crust of the plates and along with the collision of the plates created forces of mountain-building. As the Gondwana plate moved to the north and the Laurasian plate drifted to the south, the two collided in the zone of convergence, and the sediments in the intervening geosyncline were buckled to form mountain belts. It is believed that the Himalaya and the other ranges of Central Asia were formed by such a collision of the plates.

Subsequent splintering and movement of the plates in the Gondwanaland eventually resulted in the present-day peninsulas of Arabia, Deccan (India), as also the continent of Africa, and parts of Brazil, Australia and Antarctic. This theory maintains that the evolution of the continent's present surface occurred some 50 million years ago as a result of the plates' collision, although the plate tectonic forces as far back as hundreds of millions of years ago. The evolution of plains of North India (the Indus-Ganga plains) is traced to the subsequent history of the plates. As the Himalaya was rising under the impact of the advancing plate from the north, erosional processes were removing sediments in the marine gulfs, separating the northern (Laurasian) plate from the southern, Gondwanaland one of the present-day Indus-Ganga plains resulted from the deposited infilling by the sediments in the geosyncline subsequent to the erosional history of the mountains. The erosional processes, it may be noted, are still continuing and the deposits from the

Himalaya are being accumulated in the North Indian plains. As the plains are being depressed by the sediments, the lightened Himalaya is rising to maintain equilibrium.

East of the Himalaya, the arrangement of the land forms is convoluted. Part of the Himalaya chain turns into southeastern Asia, and follows topographically into the islands of Indonesia. Mountain chains, and hill-lands in China follow, in general northeasterly direction, and are lower in elevation, although the mountains east and north of Tibet including the Nanshan, the Tienshan, and further east the Qinling trend east-westward. The plateaus and highlands east of Tibet are enclosed, surrounded by higher mountains, somewhat comparable to the plateaus of Anatolia and Iran. The topography in the northeastern section of the continent in eastern Siberia has its own sequence, with several mountain ranges maintaining a northeast-southwest orientation.

The vast "heartland" of the continent has a complex mountain "core", which acts as a gigantic wheel, the spokes of which are formed by some of the greatest rivers of the world that fan outward from the snow-capped mountains (Hindu Kush, Pamirs, Himalaya, Karakoram, Altun Shan). But some five million square miles (12.9 million sq km) of the continent's interior have only inland or landlocked drainage, that is, without any outlet to the open seas. The rivers rise within or in the margins of Asia's mountain core, flow for several hundred miles across the deserts and steppe lands and disappear into the great saline lakes, or swamps or into the Aral or the Caspian seas (both inland water bodies). Rivers Syr and Amu drain into the Aral Sea; whereas several short, swift rivers from the Elburz mountains in North Iran drain into the southern shore of Caspian Sea; and the Ili river coming from the Tien Shan and flowing through the Taklamakan desert in western China, into Lake Balkhash. The Tarim flows into lake Lop Nor. Most of these interior inland drainage systems contain relatively small, irrigated oases offering some opportunities for settlement.

A striking feature of Asia's mainland's relief is the location of its three peninsulas: the plateau of Arabia, the Deccan in South India and that of Indo-China, composed of old and stable blocks of massifs differing in age, composition, and configuration than the mountain systems in the core of the continent. The pattern of surface configuration of the outlying islands and archipelagoes, mostly located off the southeastern and eastern coasts of the continent, is complicated, consisting mostly of, with few exceptions, mountainous territory, narrow coastal plains, active volcanoes and deeply indented, inhospitable coasts.

Asia's major rivers, particularly that flow eastward and southward of the complex mountain core in the heartland, have had a profound influence on the distribution of population, and human activity. These river lowlands were settled at very early times in history and played a crucial role in the evolution of Asia's cultural and economic patterns. Prominent among the rivers are: the Indus, the Ganga and the Brahmaputra (together with their tributaries) in the Indian subcontinent; and the three major rivers of China, viz., the Huanghe, the Chang Jiang (Yangtze Kiang), and Xijiang. Within the valleys of these rivers have developed some of world's most enduring civilizations and modern nations, and it is in them that most of Asia's people live.

In Southeast Asia, the Irrawaddy, the Salween and the Mekong are the main rivers. In the North, the rivers that flow northward to the Arctic Ocean, are the Ob, the Yenseiy, and the Lena. The lesser streams rise in the mountains in southern Siberia, cross the extensive Siberian lowlands, and freeze over in winter. In spring widespread flooding occurs. The river basins are too cold for human occupation and attempts to exploit forest and mineral resources have been made by the Soviet government only during the last century or so with some success.

For its size, Asia's coast is small; the ratio of surface area to the coastline (sq miles of surface area to the linear miles of coast) is 500 to 1 compared with less than 200 to 1 in Europe. Nearly a fifth of the coast fronts the frozen Arctic Ocean and is commercially useless for practically the entire year. The waters on the Pacific coast are stormy and in part snowbound during winter. Not surprisingly, these seas were not tempting enough for the early and medieval navigators as the Mediterranean or the Norwegian waters were. If we compare Asia and Europe in terms of the climatic and commercial impact of their coasts, we find strikingly different situations. The Atlantic and the Mediterranean coasts penetrate deep into Europe's interior, extending maritime influences far into the land encouraging communications, exchange of ideas and products. In sharp contrast, the Asian waters on the Pacific Ocean are ice-blocked, storm-ridden and confining; separating peoples of different nations instead of linking them, even though the Indian Ocean and its connecting water bodies were not as constricting as the Pacific Ocean has been.

The Asian islands are mostly mountainous. Several of these contain active volcanoes, and are frequented by earthquakes. The highlands of Sri Lanka and Malaysia reach 8,279 feet (2,524 meters) and 13,451 feet (4,101 meters) respectively. The famous peak of Japan Fuji-san reaches the altitude of 12,385 feet (3,776 meters). Many of the volcanoes in Sumatra and Java (both in Indonesia), and Mindanao in the Philippines reach the altitude of 10,000 feet (3,000 meters). Other larger islands lying off the coast of East Asia---Kurile, Sakhalin, Japan, Taiwan and Hainan---are also mountainous, and several contain volcanoes. While the broad pattern of Asia's landforms is presented here, details of the surface features of the continent are given in the section dealing with the individual countries.

CLIMATES OF ASIA

The weather (day-to-day changes in atmospheric conditions) and climate (long-run atmospheric conditions) have perhaps had a more lasting impact on the course of human history than other environmental influences. Though not entirely deterministic, climate can generally provide us a good deal of understanding of the patterns---economic and cultural---over the earth. For example, climate has a direct and obvious relationship with agriculture, transport, and human life in general. Drought in Somalia, unpredictable storm in India, frost in Florida---all have direct consequences on our lives. In a more direct relationship, climatic conditions profoundly affect the development of the physical landscape---

soils, vegetation, animal life, drainage patterns, and topography.

Before attempting to describe the Asian climates, it may be useful to briefly review the essential factors that shape them. The most basic of these is the *latitude* of a given place, measured in degrees in distance from north and south of equator. The earth being almost spherical in shape, the noon sun is always at a high angle all the year near the equator than at the poles where the sun never rises far above the horizon. Broadly speaking, these latitudinal differences result in the formation of more or less consistent *atmospheric pressure belts* based on a nearly constant heating and cooling but migrating seasonally in response to seasonal changes in the receipt of solar energy on the earth's surface. Thus, the distribution of heat and cold is the first and foremost function of latitude. The formation of zones, belts of atmospheric pressure over the earth are attributed to the unequal heating and cooling in different latitudes. Thus, a low pressure zone extends roughly 5° north and south of the equator, and a polar high (pressure zone) from 80° north and south. Such pressure belts give rise to wind systems (winds blow from high to low pressure) which are either dry or moist depending on several other factors related to the latitudinal, altitudinal and other differences, and in turn are responsible for conditions producing rainfall or dryness.

If the earth were a stationary body in space, atmospheric circulation (or wind systems) would be simple: heated air would rise from the equator, flow evenly toward the poles, cool and sink and return to the earth in a vast, convectional pattern, but such is not the case. Other factors are also at work.

The earth's rotation round its inclined axis of 23½° and its yearly revolution in space around the sun, in a more or less fixed orbit, is the next most decisive factor in patterning earth's climates. These movements complicate the location, and size of the pressure belts and the resulting wind systems and other climatic phenomena, like the amount of precipitation, as also their seasonal migrations. The rotational motion of the earth causes air to rise at the equator, pile up above 30 to 40 degrees latitude in each hemisphere, and descend due to its own weight at these latitudes, and begin to blow near the surface to areas of lower pressure (e.g., the equatorial low pressure zone and the areas between 60 and 70 degrees north and south which have become low pressure areas as compared to the high pressure areas between 30 and 40 degrees latitude). Winds blow from high to low pressure zones.

Another important factor is the direction and carrying capacity of winds. Some wind systems bring moisture and some are dry. That depends on several conditions: whether these are ascending or descending airmasses (winds), or whether these are moisture-laden winds blowing from over the water bodies (onshore) or are winds blowing from land (offshore).

There are several other factors which influence the climates, like the distribution of land and water over the earth, the role of mountains (as barriers or otherwise), the local wind direction (onshore and offshore winds mentioned above), the impact of ocean currents on the adjoining land and the shape and nature of coastlines that are all important in understanding climatic patterns. Their impacts on Asia will be briefly reviewed in the discussion that follows.

Map 2.2

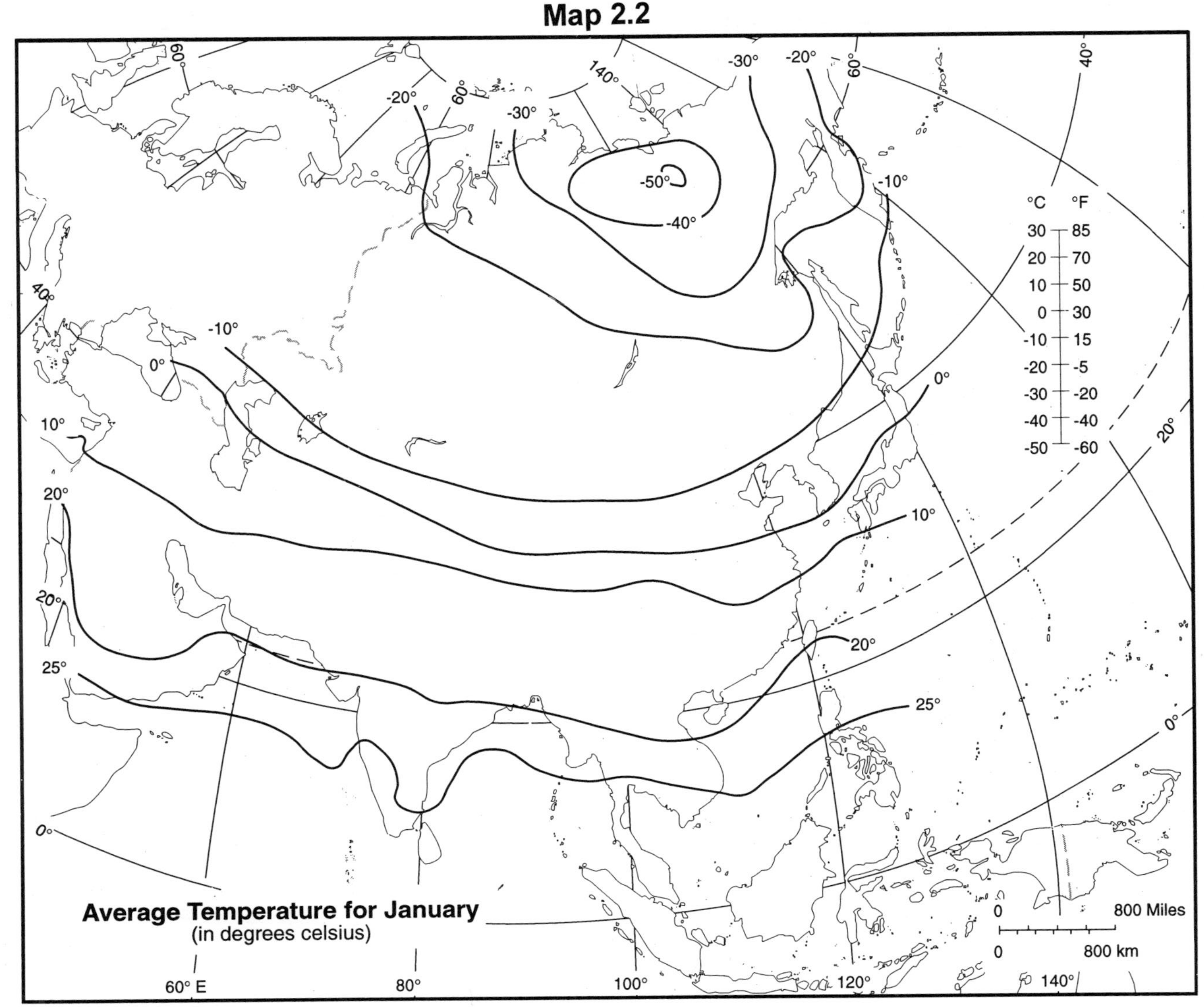

Average Temperature for January
(in degrees celsius)

Asia's enormous size spanning the entire length of the northern hemisphere, its compact shape enclosing a highland core of complex series of mountains acting as effective barriers to airmasses and several deserts tucked up in its deep interior, all interact to give most of Asia a markedly "continental" climate. Within it exist large differences in conditions of solar radiation and hence temperature distribution, atmospheric circulation, and precipitation. The outstanding characteristics of Asian climates can be described in such expressions as continentality, variety, and complexity. A continental climate is characterized by extremes of annual range of temperatures and is generally associated with large landmasses.

Seasonal Patterns: Temperature, Pressure and Wind Systems The most dominant feature that controls the Asian climates is its enormous size. The continent extends over some 6,000 miles (9,654 km) from east to west along the 40°N and well over 5,300 miles (8,527 km) from north to south along longitude of 105°E (both 40°N, and 105°E being central in the landmass). Another climatic control is the distribution of land and water. Since land and water behave very differently in their attitude toward retention of solar heat during the year, the location of places with reference to the oceans is very important. Vast interiors in the continent absorb more solar energy during summer than do water bodies, and conversely lose more energy during winter. This tendency of extreme temperature ranges (known as the "continental effect") is at a maximum on the vast Asian land surface where large part of the interiors is over a thousand miles from the oceans. As a consequence of the vastness, and remoteness of the interiors from oceans, winters tend to be colder and summers hotter in most of Asia (excepting the coastal fringes and islands) as compared to places in other and smaller continents in those latitudes. The temperature ranges (difference between the average maxima in summer and minima in winter) tends to be very large.

An examination of summer and winter isotherms (lines connecting points of equal temperature). Figures 2.2 and 2.3 illustrates Asia's marked continentality. The average January temperature over most of northeastern Siberia is below -5°F (-20°C). Over the coastal areas of the Pacific, the ocean's proximity moderates the temperatures considerably. The January isotherm of 32°F (0°C) passes through Samarkand (in the deep interior), Beijing (China), and the island of Honshu in Japan. An isotherm of 70°F (20°C) can be traced along the Tropic of Cancer and of 77°F (25°C) along the equator. In summer, the conditions are reversed. In July, most of the continent experiences hot temperatures, although the highest temperatures are found in western Asia, in Taklamakan desert (Western China) and in Thar desert in India. The 70°F (20°C) isotherm has moved as far north as 55° to 60°N, bending a bit along the comparatively cooler Pacific Ocean (the "maritime effect" as opposed to the "continental effect"). In response to the two basic controls, the latitude and the distance from the oceans, the largest ranges of temperature are experienced in the interior and in inland Siberia, particularly in Siberia's northeastern section. The annual temperature range is greatest around Verkhoyansk area in northeast Siberia (aptly named the "cold pole" due to the extremely low winter temperatures) where

Map 2.3

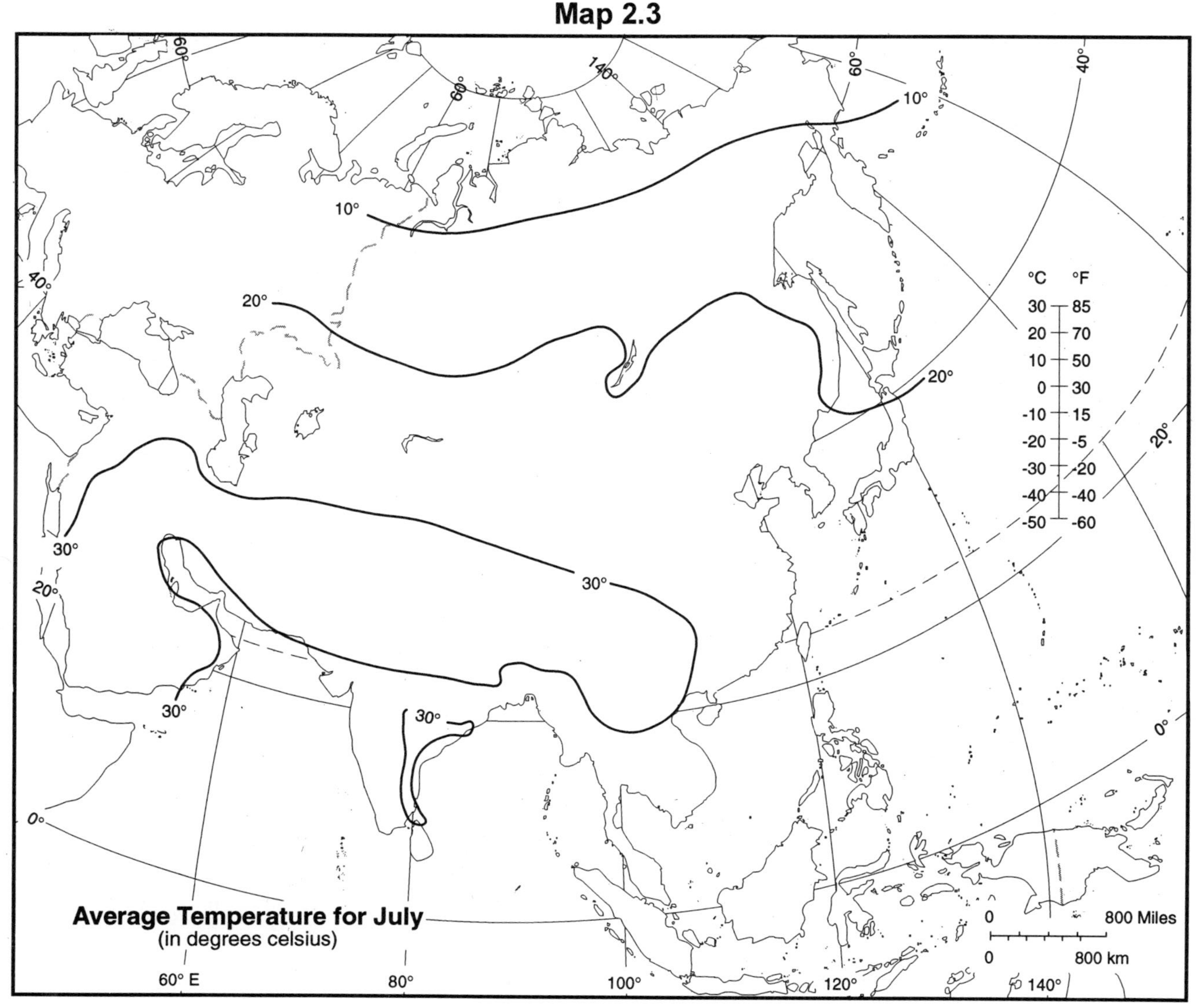

Average Temperature for July
(in degrees celsius)

the annual temperature range (the difference between January and July averages) may exceed 180°F (83°C)---the range equal to that from ice to boiling water!

The seasonal contrasts between the strong heating of the Asian landmass during the summer (May to September) and chilling in winter (January to March) are of great significance in explaining the development of Asia's pressure systems. The contrasts produce sharp seasonal variations in the atmospheric pressure and wind circulation.

During winter, the interior regions are very cold and are responsible for the formation of a high-pressure belt within East Siberia which creates conditions suitable for the outflow of strong cold winds as polar continental airmasses from this belt. The center of this belt is normally located to the south of Lake Baikal. Originating from this center, these land-oriented winds are closely packed, very cold, dry, and outflowing. The winds move eastward and southward in the continent, affecting eastern Asia and parts of southern Asia as these move over the fringes of the Pacific bringing precipitation to the lands after these have picked up moisture from the oceans over which these traverse during their long journey. These winds (in southern and eastern Asia) are known as northeast winter *monsoons* and in these latitudes may be called *temperate monsoons*.

In summer, the rapid and continuous heating of Asia's interior creates low-pressure belt centered around the basin of the Indus river and Iran which attracts hot, moist winds from the Indian Ocean. These winds are the southern or southeastern monsoons that advance along the southern edge of the Indus low bringing copious rainfall to the western Ghats and coast of peninsular India, the southern Himalaya, and as far east as mainland Southeast Asia. These monsoons may be termed as *tropical* and *subtropical monsoons*.

This phenomenon of the seasonal reversal of wind systems or the *monsoons* (derived from the Arabic word "*mausim*" meaning the "season") is most noticeable in southern and eastern Asia, but is present, to a small extent, in all other continents. In other continents, the landmass is smaller in which cooling and heating takes place, and the monsoonal impact is minimized.

The monsoon phenomenon, as noted above, is associated with a very distinctive seasonal precipitation regime, and is characterized by heavy summer rains derived from the moist onshore winds, and a pronounced dry season when the continental airmass is moving seaward. The simplistic explanation of the monsoons on the basis of unequal heating of the continent and oceans as outlined above is, however, insufficient to explain the origin, development, and behavior of the monsoons, particularly in the tropical and subtropical regions where winter is not severe enough to generate thermal conditions for the monsoonal development. The development of "low" and "high" pressure cells south of the Himalaya during the summer and winter seasons respectively and the development of seasonal wind system can be inadequately explained by the traditional theory. The traditional model, which emphasizes purely thermal factors for the timing and incidence of summer rainfall, does not explain the monsoonal "bursts," "pulls" and "breaks" of the North Indian summer monsoons. Tropical monsoon winds essentially represent large latitudinal migrations of normal trade winds and the westerly flows. Although an explanation for such extensive migrations is not clear, there is increasing evidence that it is associated with the upper-air phenomena, particularly the jet streams.

Rainfall Patterns Rainfall conditions in an extensive continent as Asia are wide-ranging. Precipitation would vary markedly from the very rainy areas of the equatorial belt to practically no rainfall in the deserts of Central Asia and Saudi Arabia. The amount and seasonality would depend upon a number of factors. In the equatorial belt (Malaysia, most of Indonesia and Sri Lanka) annual rainfall is approximately 80 inches (2,000 millimeters); between 80 to 120 inches and more on the windward maritime slopes; and less than 40 inches on the lee slopes ("rain-shadow" side) of the subequatorial areas.

In the subtropical and temperate regions which come under the monsoonal tracks, it varies between 25 and 40 inches annually according to the distance from the ocean. It is less than 10 inches in eastern Siberia, and averages less than six inches in the Central Asian and Arabian deserts.

Mountains can also act as barriers, and control the incidence of rainfall considerably. Much of the interior Asia remains in the "rain-shadow" and dry because the rain-bearing winds of maritime origin (originating mostly during the summer) cannot penetrate the mountain barriers. The "onshore" side of the mountain slopes, on the other hand, receives most of the moisture. For example, exceptionally heavy precipitation in summer characterizes the west coast of India which lies on the onshore side of the western Ghats (literally, "the mountains") but the plateau east of the Ghats receives scanty rainfall as it is located in a "rain-shadow" zone. Similarly, the North Indian plain lying on the "leeward" side of the Himalaya receives rainfall by the summer monsoon, the amount of rainfall depends on the distance the winds have traveled inland in their journey from the Bay of Bengal and the Arabian Sea; whereas Tibet which lies in the rain-shadow of the Himalaya remains dry.

Other Factors Climatic characteristics are also determined considerably by topography. Different altitudinal climatic zones are clearly represented on the southern slopes of the Himalaya, where they vary from the subequatorial and tropical climates of the foothills at the lowest levels to the snowy climate of the peaks at the highest altitudes. The degree of exposure to the sun also plays an important role. The orientation of the opposite slopes of the ridges shows contrasts in temperature conditions, and exposure to the winds. The southern slopes are sunny, windward, and receive moisture, whereas the leeward slopes in the rain-shadow are necessarily drier. In addition, the leeward slopes experience what is known as the *föehn* effect which occurs when strong wind traverses a mountain range and is deflected downhill on the lee-side, it becomes warm, dry and dusty. Such climatic contrasts and changes are observed in several mountains, like the Himalaya, the Elburz, the Tienshan, and mountains in Japan and Taiwan. The barrier and thus isolating effect of the mountains on climates has already been noted. The plateaus of Tibet, Anatolia (Turkey), Iranian plateaus, and Central Asia remain dry, enclosed, as these are, by the flanking mountains.

Cyclones and Typhoons Migrating

Map 2.4

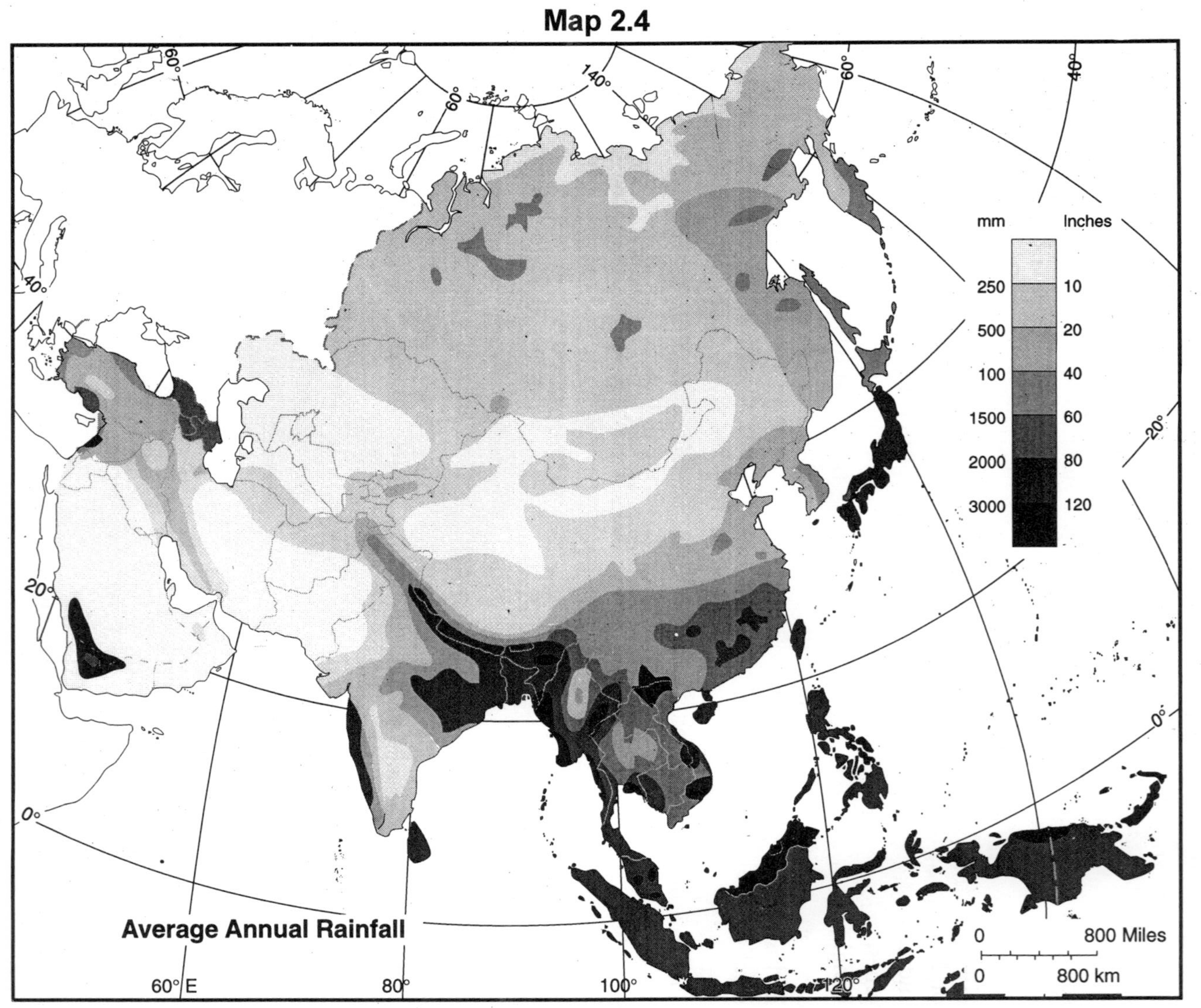

Average Annual Rainfall

cyclones or typhoons affect the climates of East Asia, and of Scuthwest Asia. Tropical cyclones (or typhoons) originate in East China Sea and South China Sea, and move swiftly westward toward the mainland, as well as curl northeast, picking up more speed, bring very heavy rainfall, particularly where these meet the mountains, or in Taiwan and the Philippines. These typhoons are like the hurricanes of the Gulf of Mexico, and bring very heavy rainfall and devastation every year to the Philippines, Taiwan, North Vietnam and Bangladesh.

Urban Climates Distinctive changes in climates also result from the economic and cultural activities of human society. The Asian cities are becoming increasingly crowded and polluted, perhaps more so than elsewhere, because these were largely unplanned. The older sections of the cities are particularly prone to environmental degradation. Recent investigations of the air quality of several Asian cities suggest severity of air pollution, due to emission of sulphur and nitrogen oxides by vehicles and acidification as many of the Asian cities have acquired a large manufacturing base. Such conditions are characteristic of northern Kyushu in Japan, and West Bengal-southern Bihar industrial region in India. Air pollution in these areas have affected locally the temperatures and wind patterns.

In the preceding paragraphs, we have discussed, at some length, factors that have shaped Asia's climates. These may now be summarized before we attempt to look at the continent's broad climatic regions. In sum, Asia's climates may be explained to a large extent by: (1) its enormous size enabling the establishment of a wide range of climatic types; (2) the location of gigantic mountain system that encloses plateaus, deserts and depressions in the heart of the continent which act as climatic barriers between the various sections of the continent, and which contain distinctive climatic patterns within them; (3) the development of monsoonal winds affecting large sections of southern and eastern Asia; (4) its compact shape with a comparatively shorter coast enabling large territory to experience "continental" climate; and (5) the oceanic influences of the Pacific and the Atlantic Oceans affect Asia's climates marginally, the coastal fringes, leaving a large territory in the interiors in Central Asia and Southwest Asia as deserts.

MAJOR CLIMATIC REGIONS

The large variety of climates that prevail in Asia may be grouped into several broad regional types. The regional division adopted here is a modification of Lyde's classification (Lyde, 1938, pp. 125-139). Unlike Köppen and Thornthwaite who apply the statistical variables in delineating the climatic boundaries, but largely ignore the important "dominant controls" in patterning climates, particularly that of relief, and pressure and wind systems, the basis of classification adopted here is "genetic" rather than statistical, and hence the climatic boundaries are generalized rather than precise.

Tundra Region It spans the entire continent from east to west along the Arctic Ocean and includes the lowlands between the Arctic circle and 70°N, the southern limits roughly coincide with the July isotherm of 50°F (10°C). Winters are long, dry and snowbound; summers are short but warm enough for some snow to melt. The snow cover, usually not thick due to scarcity of precipitation, allows the subsoil to be deeply frozen during winter; soil surface thaws are possible only in summer, when the surface becomes swampy.

East Siberia Between 50° and 70°N latitudes a large territory in eastern Siberia including the lower basin of Lena river and the area to the north of Baikal lake. It is mostly a mountainous region (Yablonoi, Stanovoi, Verkhoyansk mountains). The deep interior location and the sheltering mountains rob the region of the influences of the Pacific and the Atlantic Oceans. The winters are long, dry and among the coldest in the world. Verkhoyansk has once recorded -80°F. The summers are warm with temperatures varying between 50°F and 70°F (20°C). Given these extremes, the world's largest temperature ranges are found (up to 50°C at several stations). The region thus epitomizes the cold, continental climate. In the Kamachatka peninsula, the moderating influence of the Pacific keeps the temperature somewhat higher in winter, and lower in summer.

West Siberia The region is located west of the preceding region all the way to the Ural mountains. Temperatures are higher than in East Siberia, and there is a little more rainfall (up to 250 mm annually)----a clear evidence of the Pacific influence. The ground remains frozen in winters, and thaws during the brief summer, causing swamps and marshes.

Temperate Monsoonlands of East Asia Most of eastern China, North Korea, South Korea, and Japan are included in this region which lies entirely north of the Tropic of Cancer. Summers get rainfall, and winters are dry---a definite monsoonal trait. Only Taiwan, and southern islands of Japan including southern half of Honshu, receive rainfall in winter caused by the southeasterly winds which bring rainfall after picking up moisture on the Japan Sea, East China Sea, and the adjoining seas. The range of temperature increases in the interiors, as the winters become colder and summers warmer.

Tropical Monsoon Region It is a vast region in southern and southeastern Asia encompassing nearly all the territory north of 10° north and south of the Tropic of Cancer----from the Indus-Ganga lowlands to the Indo-Chinese peninsula (Vietnam, Cambodia, Laos) and Myanmar and Thailand. The North Indian plains are practically tropical in climate although they actually lie in the Temperate Zone. Typically monsoonal conditions----seasonal reversal of wind systems and associated rainfall regimes----prevail.

The mechanism of monsoons in India has been discussed earlier, but it is important to emphasize here that the summer monsoons in India owe their origin to the pressure and wind systems in the equatorial zone. During summer the southeast winds are sucked to the low-pressure area in western India and the moist winds from the Indian Ocean cause the onset of rain bearing summer monsoons. Most of the region, from India to China and Japan, receives rainfall during summer months. In winter, conditions are reversed. Winds

Table 2.1 *Asia: Representative Climatic Stations*

Climatic Type	*Station*	*Elevation in Feet*		*Jan*	*Feb*	*Mar*	*Apr*	*May*	*Jun*	*Jul*	*Aug*	*Sep*	*Oct*	*Nov*	*Dec*	*Annual*
East Siberia	Verkhoyansk	328	T(°F)	-57	-48	-25	5	32	54	57	48	36	5	-35	-53	
	(Russia)		P(in)	0.2	0.2	0.1	0.2	0.3	0.9	1.1	1.0	0.5	0.3	0.2	0.2	5.4
West Siberia	Tomsk	58	T(°F)	-6	-1	12	29	46	58	64	59	49	32	11	-2	
	(Russia)		P(in)	1.1	0.7	0.8	0.9	1.6	2.7	2.6	1.6	2.6	2.0	1.8	1.5	18.4
Central Asia-	Ulanbaatar	4346	T(°F)	-14	-6	9	31	42	57	61	58	47	30	9	-8	
Mongolia			P(in)	0.0	0.0	0.1	0.2	0.4	1.1	3.0	2.0	0.9	0.2	0.2	0.1	8.2
Temperate	Tokyo	20	T(°F)	37	39	45	55	63	70	77	79	73	63	52	43	
Monsoon			P(in)	1.4	2.9	4.2	5.3	5.1	6.5	5.6	6.0	9.2	8.2	3.8	2.2	61.6
Tropical	Mumbai	36	T(°F)	75	75	79	82	86	84	81	81	81	81	79	80	
Monsoon			P(in)	0.1	0.1	0.1	0.1	0.7	19.1	24.3	13.4	10.4	2.5	0.5	0.1	71.4
Equatorial	Singapore	33	T(°F)	79	81	82	82	82	82	82	81	81	81	81	81	
			P(in)	9.9	6.8	7.6	7.4	6.8	6.8	6.7	7.7	7.0	8.2	10.1	10.1	95.1
Central Asia-	Tashkent		T(°F)	29	36	45	56	67	75	78	75	66	53	44	37	
Aral-Caspian			P(in)	2.1	1.1	2.6	2.3	1.4	0.5	0.2	0.1	0.1	1.2	1.5	1.6	17.3
Iran-Sind	Tehran	4002	T(°F)	36	41	48	61	70	79	86	84	77	64	54	43	
			P(in)	1.8	1.5	1.8	1.4	0.5	0.1	0.1	0.1	0.1	0.3	0.8	1.2	9.8
Mediterranean	Beirut	112	T(°F)	79	82	84	86	84	84	82	82	82	82	79	77	
			P(in)	7.5	6.2	3.7	2.1	0.7	0.1	0.1	0.1	0.2	2.0	5.2	7.3	35.0
Tropical	Riyadh		T(°F)	58	61	69	77	86	92	93	91	86	78	75	60	
Desert			P(in)	0.1	0.8	0.9	1.0	0.4	0.0	0.0	0.0	0.0	0.0	0.0	0.0	3.3

Source: Adapted from: H.L. Nelson, *Climatic Data for Representative Stations of the World.*

Map 2.5

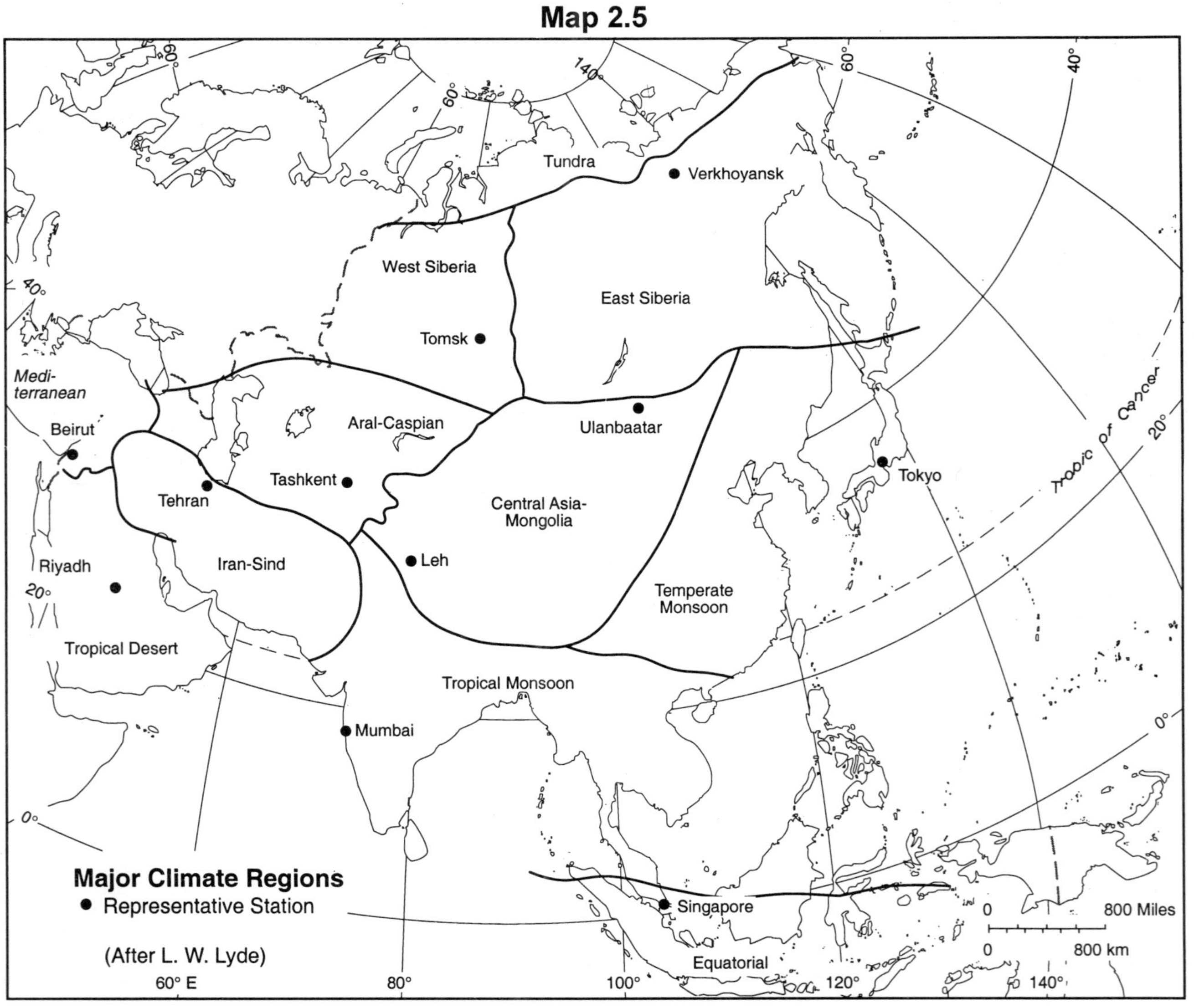
Tundra
Verkhoyansk
West Siberia
East Siberia
Tomsk
Mediterranean
Aral-Caspian
Ulanbaatar
Beirut
Tashkent
Tokyo
Tehran
Central Asia-Mongolia
Tropic of Cancer
Riyadh
Iran-Sind
Leh
Temperate Monsoon
Tropical Desert
Tropical Monsoon
Mumbai
Major Climate Regions
• Representative Station
Singapore
(After L. W. Lyde)
Equatorial
0 800 Miles
0 800 km
60° E
80°
100°
120°
140°
60°
40°
20°
0°

blow from the high-pressure region of East Siberia outward and are dry. However, during their course as they traverse the waters of East and South China seas, they pick up moisture and bring some rain to parts of east Asia and southeast Asia that lie in their tracks (southern China coast, North Vietnam, northeastern Sri Lanka, southern part of east coast of India). The countries of the Indo-Chinese peninsula (Vietnam, Laos, Cambodia) reflect monsoonal seasonality but differ from the Indian and Chinese monsoons. The range of temperature is smaller and the seasonality of rainfall depends on the precise locations with reference to the summer or winter monsoons.

Equatorial Region It includes the areas lying between 10°N and 10°S in Indonesia and Malaysia. Typical climatic conditions consist of high, seasonally well-distributed rainfall and a small range of temperature except in locations where the seasonal shifting of the trade winds leaves a part of the year relatively drier.

Central Asia: Mongolian Region
The region lies in the heart of the continent containing a vast territory of mountains, and plateaus from Tibet to Mongolia. The conspicuous climatic characteristics are extreme temperature ranges (daily and annual) and dryness. The region comes under the influence of a large body of cold, continental, and outflowing dry winds over Asia's landmass. In summer, the landmass is heated, and a low pressure system takes over, with the consequent dry, inflowing, gusty winds, the enclosing mountain system shelters this vast region from the inflow of any maritime winds.

Central Asia: Aral-Caspian Region
The region resembles the Iran-Sind region (described below) in dryness and large temperature ranges, but differs in topography, as it is mostly a lowland drained by the historic Amu and Syr rivers. The Mediterranean influences (some winter rain) reach the western section; but the eastern part gets most rainfall during the summer. Irrigation is important for all cultivation, as rainfall is scanty.

Iran-Sind Region The region includes eastern Iran, Afghanistan, and southern Pakistan (Sind province). Dryness and heat are the two outstanding characteristics of the climate. Rainfall increases somewhat from the Iran plateau with nearness to the sea. Very hot summers and cold winters account for large temperature ranges. Toward the east, monsoonal traits (summer precipitation) are becoming increasingly evident; in the western parts Mediterranean influences are evident.

Mediterranean Region Climatically, the region is transitional between the trade wind deserts (Arabia) and the zone of prevailing westerlies. It includes the territory bordering the Mediterranean from Turkey to Syria, Israel, and Iraq. Winter temperatures fall below 50°F and in summer rise above 75°F. Precipitation is mostly in winter brought by the cyclonic storms emanating in the west.

Tropical Desert Region The region lies essentially in the trade wind belt, outside the monsoonal and Mediterranean influences. Excessive temperature ranges and dryness characterize the climate.

NATURAL VEGETATION

The natural vegetation is perhaps the best summary of the physical environment, for it reflects the temperature, rainfall, drainage, elevation and soil conditions. In nature, all physical phenomena are interrelated; together, these provide us with an understanding of the physical environment. The patterns of their distribution and interactions are of crucial importance to us, as we have to depend on these for our sustenance. However, we have substantially modified the natural vegetation and the soils that were present before human intervention and that have been replaced by other forms.

It is important to note that over one-third of the continent's surface has been cleared for human settlement or given to permanent agriculture. Vegetation in such areas no longer remains "natural". In most areas of dense settlement the original vegetation has been entirely removed and no trace of undisturbed forest remains. Thus, the Indus-Ganga plains in India, lower Chao plains in Thailand, the lower basin of Huanghe, Chang Jiang, and Xi Jiang rivers in China are almost covered with cropland. Throughout these and other intensely settled areas in Java (Indonesia) and Japan, imported plants have replaced the indigenous vegetation.

Original cover can only be found in the remote and inhospitable areas, such as northern and northeastern Siberia, and parts of Southeast Asia. Reflecting the growing awareness of the consequences of deforestation and loss of natural vegetation, reforestation is becoming increasingly popular, and preservation and recovery of forest areas has become a national priority for several Asian nations, although this practice is not as widespread as in the U.S.A. or in the European nations.

The nations in the Indian subcontinent, because of their enormous populations and deficiency in timber resources, are sensitive to the issue and have launched several reforestation programmes. China, Malaysia, and Turkey have also included in their national plans aimed at increasing forest lands and in the case of Turkey to reduce timber imports.

Asia's enormous territory, immense variety of relief and a wide range of climates combine to produce most types of vegetation forms. A dominant feature of the continent's vegetation is that a larger territory on the western side has no oceanic border, and receives scanty rainfall in relation to summer temperatures that can allow appreciable forest growth.

The broad pattern of natural vegetation generally follows the climatic types; it is thus convenient and useful to describe it in association with the climatic patterns of Asia.

The Tundra The *Tundra* climatic type correlates with its namesake vegetation, *Tundra vegetation* that borders the Arctic Ocean. It is a region of cold, treeless plains with permanently frozen subsoil. Lichens, mosses, sedges and some grasses may occupy more favored locations. The Tundra belt extends to 70°N and with further south extensions on high altitudes (Chersk, Verkhoyansk and Kamachatka mountains).

The Taiga South of the Tundra region is the *Taiga*, a belt of coniferous forests, or the needle-leaf evergreen forest, with a small, transitional zone of "wooden tundra". It covers most of Siberia from the

Map 2.6

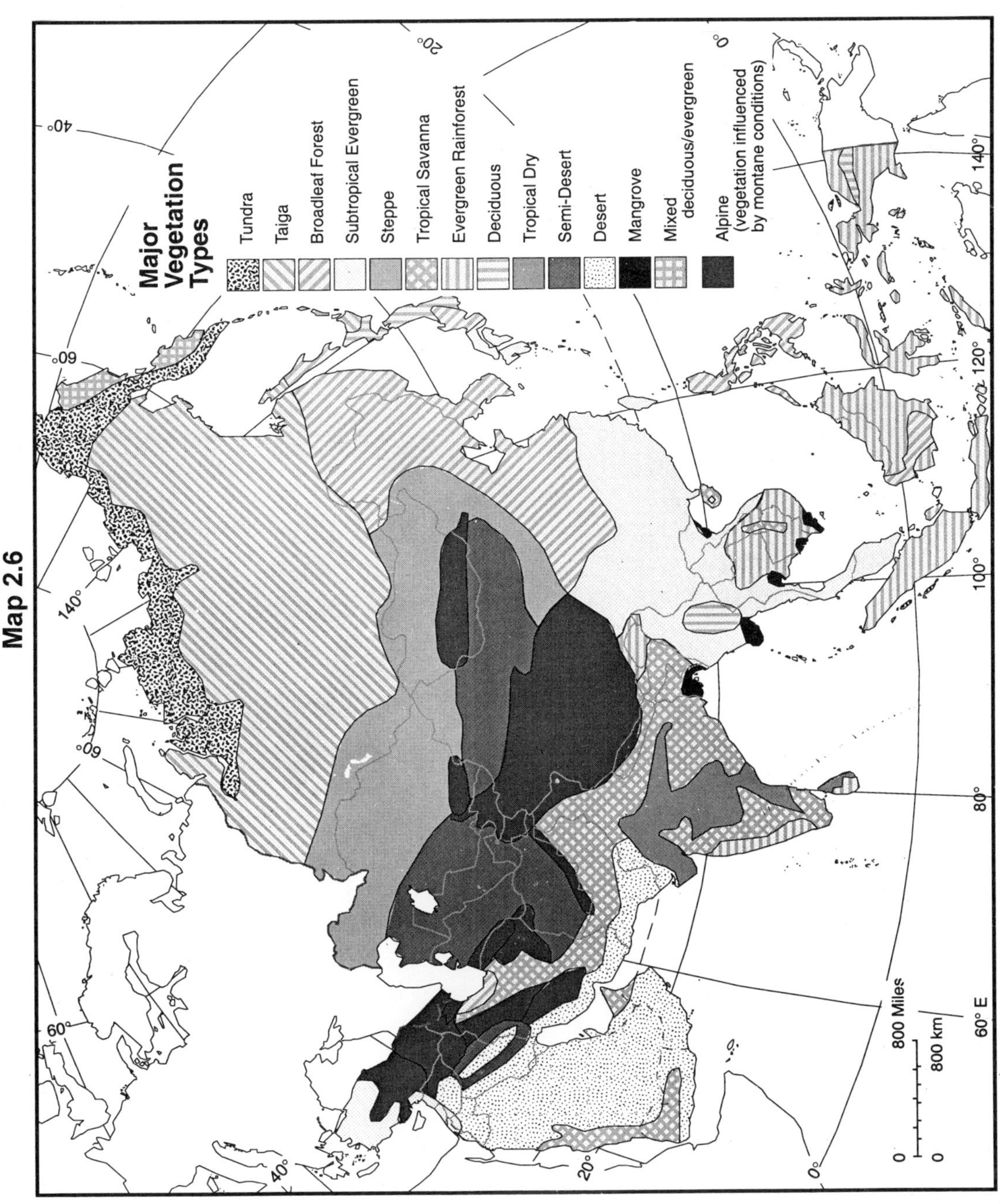
Major Vegetation Types
Tundra
Taiga
Broadleaf Forest
Subtropical Evergreen
Steppe
Tropical Savanna
Evergreen Rainforest
Deciduous
Tropical Dry
Semi-Desert
Desert
Mangrove
Mixed deciduous/evergreen
Alpine (vegetation influenced by montane conditions)
0 800 Miles
0 800 km
60° E
80°
100°
120°
140°
40°
20°
0°
60°
140°

Urals to the Pacific and northern part of Japan. The trees have small leaves, deep roots, and thick bark, and thus most species (pine, spruce, fir, etc.) are successful in cold, and dry environment. Further east and southwest of Siberia is a transitional zone of coniferous forest mixed with hardy, deciduous trees such as aspen and birch, with sections of grass and shrubs in the drier areas.

Temperate Grasslands, the Steppes
To the south of Taiga is elongated, unbroken stretch of the *Steppes* from Ukraine to Manchuria—a broad belt of several thousand miles of temperate grasslands in southern Siberia— a parkland country of open grasses. There is some precipitation although the winters are still very cold, but the extreme temperature ranges of the Taiga are softened by the warm summers. However, the higher elevations in the mountains are covered with forests, and the dry valleys are bare except for the occasional cultivated oases.

Mediterranean Scrubland and Forest
A belt of scrub forest that is particularly attuned to the Mediterranean climatic regime rims the eastern shore of the Mediterranean that includes the countries of Israel, Lebanon, Syria, Iraq, and the plateaus of Turkey and Iran. The summers are hot and dry; the winters mild and moist. Thus, the circumstances for the plant community are unusual; winters become the growing period. Plants must adapt to conditions of drought and higher temperatures in summers. They share the adaptive characteristics of small plant size, short leaves, deep roots, and thick barks to retain moisture. The most distinctive vegetation community of this region consists of a scrubland and short trees.

Desert Vegetation The desert climatic and associated vegetation types are found in the trade wind deserts of the Arabian peninsula, the deserts of Tibet, Mongolia, and the desert-like steppelands bordering the Caspian Sea. The mid-latitude deserts of Tibet and Mongolia in the interior resemble the Arabian desert only in aridity, but contain a clear altitudinal zonation of vegetation. Plant cover in the deserts is sparse, with a considerable bare ground dotted by a scattering of individual plants. Typically, the plants are moisture-combating, waxy, deep-rooted or thorny shrubs and sporadic stunted trees. The upper reaches of the highlands in Tibet or Mongolia may be bare, or contain Taiga- and Tundra-like vegetation depending on the amount of available moisture.

Monsoon Region This region is associated with monsoon climates. The natural vegetation varies with the amount of annual rainfall. Areas receiving between 40 and 80 inches of rainfall annually contain tropical deciduous (shedding leaves seasonally) forests, and those which receive less than 40 inches have *savanna* (tall grasses) and steppe-like (of short grasses) vegetation, below which semi-desert and desert vegetation prevails. The monsoon lands have been extensively modified by human settlement and put to cultivation, and little trace of the original vegetation survives.

Tropical Rainforest Tropical rain-forests are typical to the equatorial regions. In Malaysia and Indonesia rainforests cover large section of the countries, whereas in southern Sri Lanka and Java they have been almost entirely replaced by agricultural landscapes where plantations of tea, coconut palms, and rubber trees cover the mountain slopes and hills. The vegetation consists of evergreen, broad-leafed tall, dense, high-crowned trees of several species having a dense canopy above the floor because the region gets a high amount of precipitation throughout the year, and is constantly warm. The savannas and deciduous trees cover the ground, the subequatorial and the areas that lie in the rainshadow on the leeward slopes.

Vegetation in the Mountains The mountains of southern and eastern Asia show a remarkable altitudinal zonation in natural vegetation. A zone of forest followed higher up by meadows, and snow cover at the highest elevations is characteristic. On the lower slopes are the broad-leafed deciduous forests, and on higher ground the coniferous trees occur. Still higher up stunted trees, subalpine meadows and near the highest ridges perennial snow, and glaciers are found in a vertical zonation pattern. Elsewhere in the equatorial region in Malaysia and Greater Sunda Islands, (Indonesia) where the snow line is very high and moist-hot conditions predominate, the vegetation zonation is practically absent.

SOILS

Along with the climate the distribution of soils is critically important to agricultural productivity, and thus has a special significance for nations of Asia, as most Asians depend on agriculture for their livelihood. In the preceding paragraphs, wide range of interacting natural phenomena: climate, natural vegetation, parent rockmaterial and relief features that determine the development and nature of soils have been discussed for the Asian continent.

Soil classification is both complicated and subject to constant revision. In the case of Asia, the difficulty is compounded by the fact that a large part of it is physically remote, or climatically inhospitable for field surveys. Most of Asia has been surveyed unequally based on different systems with the result that comparable knowledge of soil types and their characteristics is difficult to obtain. Some part of the continent still remains unsurveyed. In addition, the most commonly used current classification at the present time—that of the U.S. Department of Agriculture known as the Seventh Approximation—has not gained much acceptance in most Asian nations.

It may be interesting and sobering to realize that a small fraction of Asia's land surface is blessed with really rich soils that have given sustenance to a large masses of humanity.

At the broad scale of our study, it may be convenient to generalize the distribution patterns of soils by recognizing that there are basically three complex processes by which soils are formed; those of the humid latitudes, of the tropical regions and of the deserts. Each of these processes are responsible for the formation of specific types of soils which can be defined in broad terms. The processes produce either the *pedocals* or calcium-bearing soils, or the *pedalfers*, the aluminum-bearing or nitrogen-bearing soils. The letters "*cal*", "*al*", "*fe*" representing three distinctive elements in soils. In simple terms, this

Map 2.7

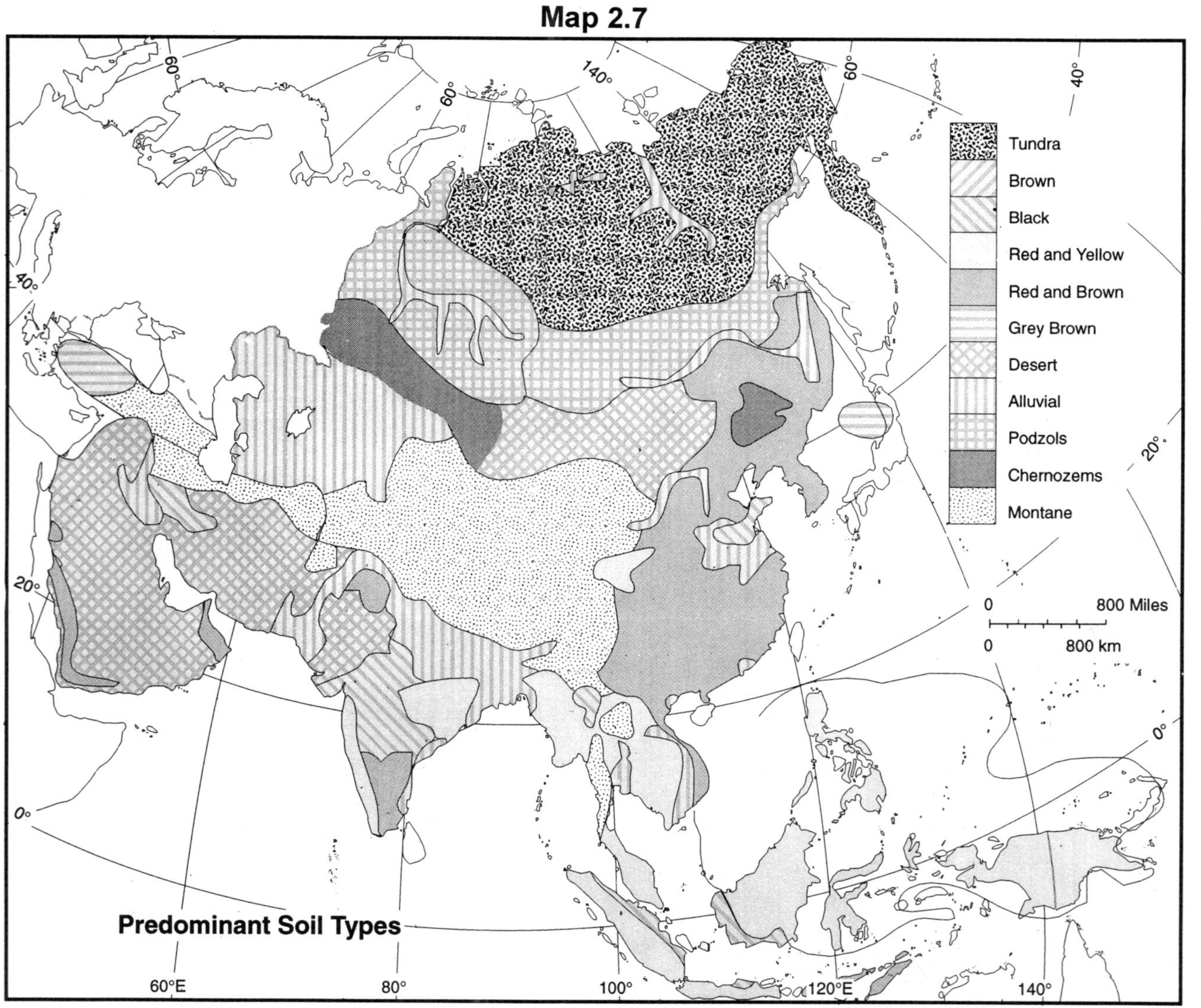

Predominant Soil Types

division separates the *basic* soils from the *acidic* soils. Soil patterns based on such classification point to some general tropical regions. Two fundamental soil-forming processes operate in the humid regions; *podzolization* in the cool, wet regions, and *laterization* in the hot, wet regions. Podzolization is thought to remove the soluble salts, iron and aluminum compounds from the surface horizons, leaving a gray silica residue; and laterization begins by removing the salts, and transferring the iron and aluminum compounds near the surface with clays accumulating just underneath them.

Soils in the tropical areas are generally *lateritic*, having varying degrees of red or black color. The soils are rich in aluminum and iron oxides, low in silica; are nearly structureless and generally of low fertility. Most soils of South and Southeast Asia belong to this category. In the Arabian desert and deserts in Central Asia the soils are usually high in their content of unleached minerals, but may be deficient in their content of nitrogen and humus (decomposed vegetative materials of the wet tropics). Where the water-table is high, evaporation salts have accumulated in excess near the surface, the soils are *alkaline*. Such red desert soils cover much of the Arabian peninsula and northwest India, while higher desert soils cover the rest of southwest Asian countries.

Large parts of northern Asia are covered by *podzolic* soils of varying, but generally low fertility. These soils cover the mid-to high-latitude locations in the region of well-distributed rainfall (in coniferous forest of a taiga forest zone) from the Urals to Kamchatka. In much of northern China and Manchuria soils belong to the gray-brown podzolic category.

In the Mediterranean lands the pattern of soils is complicated. While cool temperatures and some rainfall produce podzolic conditions during winter, hot summers complicate the conditions when surface evaporation exceeds rainfall and moisture moves upward through capillary action and the podzolization process no longer operates.

The grasslands of western Siberia, Manchuria, and most of Anatolia plateau of Turkey are underlain by *black prairie* and *chernozem* soils which have good texture and are of generally high fertility.

From the standpoint of agricultural productivity, the most important soils in Asia are the relatively *immature soils* of alluvial origin that are found in the river valleys. These soils are among the richest in plant nutrients which can maintain their fertility through periodic inundatation by silt-bearing waters from the rivers in flood. *Alluvial* soils cover the river valleys of the Tigris-Euphrates, the plains of Indus-Ganga rivers, the lower Irrawaddy basin, the deltas of the Mekong and the lower basins of Chang Jiang, Huanghe rivers, as well as the river plains of smaller rivers in India, Malaysia, the Philippines, Taiwan, Korea, and Japan. These, in a very large measure, provide substance to the growing population of Asia.

Several islands of Indonesia and southern Philippines contain active volcanoes that periodically provide a new layer of rich *basic* soils of recently deposited volcanic materials (not all volcanic soil consists of basic material; some retain their fertility for long periods as these are renewed occasionally by new deposits, and account for enormous densities of rural population in Java and some areas of the Philippines).

Soils in the hilly and mountainous regions are produced by geologic structure, altitude, and temperature conditions to a greater degree than elsewhere. These may be lumped together as *montane* soils.

A good part of the Asian surface of the uplands is too steep for the formation of deep soils. *Skeletal* soils are found in such areas, much of which are cropped by shifting cultivation, in which the cyclical use of land provides periodic regeneration of plant nutrients.

In northwest China some areas are covered by a thin layer of wind-blown material dating from the retreat of the ice sheet during the most recent Ice Age. This material is known by the term *loess*, and adds considerably to the natural fertility of the soil and thus to the agricultural productivity of the area.

Surveying the overall picture of soil distribution in Asia, it can be concluded that for its size, much of the continent offers limited opportunities for productive agricultural land. No more than 20 percent of the land surface is estimated to contain soils suitable for agriculture and a good quarter of this percentage contains soils of marginal value. Tropical and subtropical soils are generally immature and unproductive. The soils of the middle latitudes are themselves suitable, but lack of water is often a key factor. Only the alluvial soils of the river plains and the basic soils produced by the volcanic ejecta are productive soils. Several of the river basins, where the soils are rich, have been settled for hundreds of centuries and the soil cropped by hundreds of generations of farmers, and where enormous population densities are straining agricultural productivity. In such lands good soil management is an important factor as an agricultural resource. Most Asian farmers have not taken to farming practices restoring plant nutrients to the soils they farm. The diffusion of the Green Revolution is still limited to a few, selected areas in India, Pakistan, Sri Lanka, the Philippines, and Taiwan where farmers have been making use of superior seeds of hybrid plant varieties, chemical fertilizers, and mechanized farming.

Further Readings

Bridges, E.M. *World Soils.* Cambridge, 1978.

Eyre, S.R. et al, *Vegetation and Soils: A World Picture.* London, 1977.

Lyde, L.W. *The Continent of Asia*, 1938.

Spencer, J.E. and Thomas, W.L. *Asia East by South.* New York, 1971.

Trewartha, G.T. *Earth's Problem Climates.* London, 1961.

II

The Development of Asia's Human and Cultural Landscapes

3

Population Patterns and Trends in Asia

Population distribution is perhaps one of the most vital factors that affects the resource utilization of nations. The size, structure, and growth of population can make a substantial difference in how resources of nations are consumed, or exported. Many thoughtful observers have noted that whatever problems individual nations face----hunger, poverty, inflation, pollution, or political instability----rapidly increasing population usually makes conditions worse. Several Asian nations face many of those problems. In some poverty is a way of life, and their ability to save enough for economic growth is constrained by heavy dependence on foreign capital.

Comprising nearly one-third of the world's land area, Asia contains nearly three-fifths of its population, and includes the two most populous countries, China and India, which together account for more than a third of the world's people. Although its annual growth rate of 1.5 percent approximates that of the world's average, it is adding close to 65 million persons every year to its population----a figure that is higher than that of the population of Italy, France or the United Kingdom (the most populated nations of Europe). Some of the Asian nations are growing at even higher rates than the average for Asia, particularly in Southwest Asia. Only Japan, Taiwan and South Korea are the exceptions. A sobering aspect of Asia's larger population picture is that several nations in South Asia and Southeast Asia have reached a crisis stage, and only recently have been pursuing policies aiming at population control.

GROWTH OF POPULATION IN ASIA

A little over six billion people inhabit our planet today; more than 60 percent of which live in Asia. The United Nations' estimate for the year 2010 projects a staggering figure of 7 billion for Asia!

Population growth in Asia is not a new phenomenon for it has always been the world's most populated continent. Although reliable population data prior to the 20th century is lacking (and even that of the 20th century for most part is scarce, uneven, or misleading), Asia has historically been the home of large populations. Population historians estimate that large clusters of sedentary agriculturalists inhabited the Indus Valley in India and the northern river plains of China five to seven thousand years ago. In 2 A.D. China perhaps contained 60 to 75 million people, a population that exceeded that of the

Table 3.1 *World Population Summary 1650-2010 (Population in millions)* *Estimates*

Year	*World*	*Asia*[(a)]	*Europe*[(b)]	*North America*	*Latin America*	*Africa*	*Oceania*
1650	545	327	103	1	12	100	2
1750	728	475	144	1	11	95	2
1850	1,171	741	274	26	33	95	2
1930	2,008	1,069	530	134	110	155	10
1985	4,845	2,830	770	264	406	551	24
1990	5,321	3,116	791	278	447	661	32
2010	7,069	4,298	973	320	602	1,052	35

Derived from:
1. A.M. Carr-Saunders: *World Population, Past Growth and Present Trends,* 1935.
2. *Population Data Sheet, 1990 and 2000.*
(a) Excludes Russia
(b) Includes Russia

entire Roman Empire. By 1600 an estimated 150 to 180 million people lived in India and China each. During the medieval times both countries grew rather slowly and erratically responding periodically to famines, wars, and times of prosperity. China and India crossed the one-billion mark in 1999.

The population growth of the world's continents since 1650 are summarized in Table 3.1. In 1650 as North America was first being settled by Europeans there were already more people in Asia than there are in North America today. In that year, Asians accounted for 60 percent of the world's population. More than three hundred and forty-five years later, Asians still maintain that proportion in world's population. It is likely that Asia experienced high growth rates throughout historical times except during the last two decades when Africa's rate of population growth of over 3 percent a year out stripped that of Asia's. It is indeed sobering to realize that Asia's share of the world's population in 2010 will be nearly 61 percent. Most of the growth will be in South Asia where population will leap from 1.3 to 1.8 billion.

Within Asia the annual rates of population increase vary considerably. The rate in East Asia, for example, is less than half that of Southwest Asia, whereas East Asia registered lowest rates. Substantial decreases in growth rates were recorded in China, Indonesia, Singapore, and Thailand.

The annual rate of population increase is a function of the prevailing fertility and mortality patterns. The Islamic countries of Southwest Asia generally show higher rates of fertility with birth rates ranging from 35 to 45 per thousand, than the predominantly non-Muslim nations of South Asia, such as India, and Sri Lanka. There are, however, marked variations in the Muslim nations of Southeast Asia as well, for example, Indonesia or Malaysia, which have non-Muslim minorities, birth rates tend to be lower. Given the high birth rates, and generally low death rates in Southwest Asia, their rates of population increase among the highest in the world.

Table 3.2 *World Population Summary: 2000 (Estimate)*

Continent	*Population (millions)*	*Birth Rate/ Death Rate (per 1,000)*	*Annual Increase %*	*Age Structure Under 15/65+*	*GNP ($ Per Capita)*
Developed Regions:	1,184	11/10	0.1	31/7	19,480
(caution: includes some poor nations)					
Europe	729	10/11	-0.1	18/14	13,420
N. America	306	14/9	0.6	21/13	28,230
Less Developed Regions:	4,883	25/9	1.7	35/5	1,260
(includes some rich nations)					
Latin America	518	24/6	1.8	33/5	3,880
Asia	3,684	22/8	1.4	33/5	2,130
Africa	800	38/14	2.4	43/3	670
World	6,067	22/9	1.4	31/7	4,890
National Examples					
Russia	145.2	8/15	-0.6	19/13	2,260
China	1,264	15/6	0.9	25/7	750
India	1,002	27/9	1.8	36/4	470
Taiwan	22.3	13/6	0.7	21/8	13,969
Laos	5.2	41/15	2.6	44/4	320
Ethiopia	64.1	45/21	2.4	45/3	100
Brazil	170.1	21/6	1.5	30/5	3,640

Source: World Population Data Sheet, 2000.

THE DISTRIBUTION OF ASIA'S POPULATION

Within Asia there are significant variations in population distribution. As pointed out earlier, the largest two of the Asian nations---China and India---account for three-fifths of Asia's total population, and the remaining two-fifths shared by forty-six countries in Asia. Six of the Asian countries have populations greater than 100 million (Bangladesh, China, India, Indonesia, Japan and Pakistan) making each one of them larger than any European country. Twenty-one Asian countries have small populations, less than 10 million inhabitants. With a few exceptions (Cambodia, Kyrgyzstan, Mongolia, and Turkmenistan) these nations are small in area as well.

But the absolute or total population figure may also give an inadequate picture of population pressure. A better idea of population pressure is provided by a measurement, known as *population density*, which indicates population pressure in terms of the number of people per unit of land (e.g., per sq mile or per sq km). The statistical density for the land area of the world is about 110 persons a sq mile (42 per sq km). The figure for Asia rises to 270 a sq mile (104 a sq km), making the overall Asian density, the highest among all the

Map 3.1

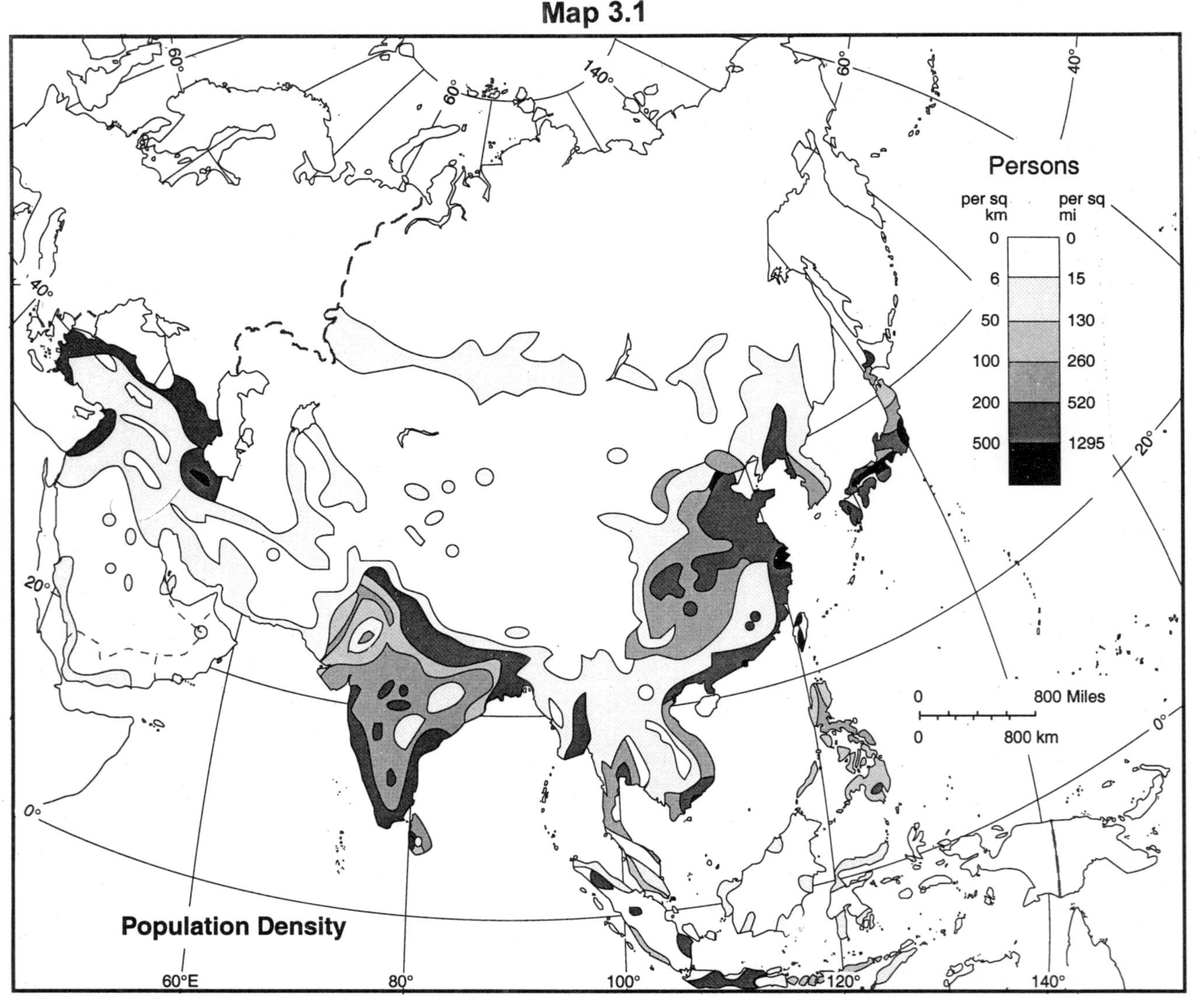

Population Density

continents. Some Asian comparisons, India as a whole averages close to 850, China's 320, and South Korea's 1,140 persons per sq mile, respectively. However, such averages have little practical value. For example, they include figures for unproductive mountains and wastelands as well. The concept of average density of population is practically unilluminating for Asia as a whole, as a large part of the continent consists of land that discourages settlement---hot deserts, high plateaus and rugged mountains. A more significant measurement that illuminates the relationship between population and development of areas relates people to the cultivated area or growth of capital and industrial production, technology and other economic and social indicators.

While looking at these figures, therefore, we should not overlook an important point: all population figures must always be considered in relation to the economic base of the area they represent. Although Asia is potentially a continent of considerable resources (see chapters on agriculture, mining and energy resources), its current overall development is inadequate to support its existing numbers at a reasonable level of well-being. Japan and a few of the Middle East countries are the exceptions. Japan can maintain a very high level of well-being by exporting manufactured items to world markets competitively and Persian Gulf countries by exporting their valuable resource of oil.

Map 3.1 shows the pattern of population density within the continent river plains in South Asia, Southeast Asia and East Asia contain among the highest densities in the world, often exceeding 2,000 persons per sq mile. The notable examples are: the Indus-Ganga Plains in north India, the lower Irrawaddy basin in Myanmar, the lower basin of Chao Phraya River in Thailand, the lower Mekong basin, and the lower basins of Huanghe, Chang Jiang, and Xijiang rivers in China. In these fertile river basins, large populations can be supported under conditions of perennial irrigation and double-cropping. Elsewhere, high densities exist in the accessible coastal plains of central and southern Japan, most of South Korea, the islands of Java and Bali in Indonesia, and the coastal plains of Vietnam and Taiwan. As a sharp contrast, large parts of Asia's interior, and of Southwest Asia are too cold, too rugged, or too dry to be attractive for settlement. Densities are as low as 5 persons per sq mile in such areas. Significantly, an important factor to note is that population densities have been, in general, on the increase everywhere in the continent, as elsewhere in the world.

THE STRUCTURE OF POPULATION

In population study, the term structure refers to a pattern of relationship, between various age groups of the nation's population. For example, if population is divided into age and sex categories, the implications of age and sex groups of a nation's population can exercise a significant impact on it because these groups represent the numbers of people at a given age and sex, which is considered a fairly good indicator of the working of society, since society usually assigns social roles on the basis of its age and sex composition. Young people and women are treated differently than the old people. It must also be remembered that the very young and the elderly are dependent on other ages for their survival. The proportions of people

Figure 3.1 *Age-Sex Distribution on India and Japan*

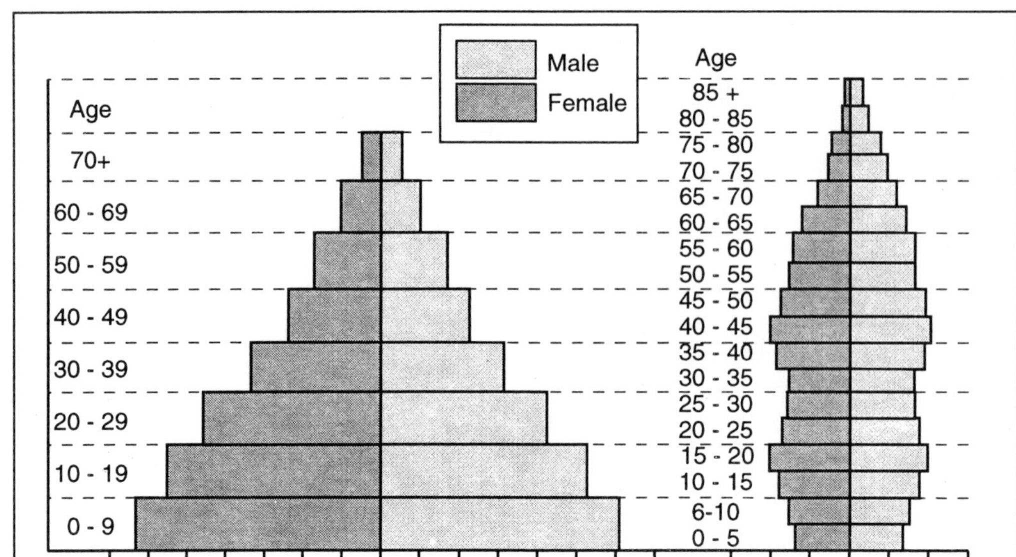

at these ages will inevitably affect the working of society.

An interesting demographic fact is that male babies outnumber the females at birth. In economically developed nations, where maternal mortality is low and where infant girls receive as much care as do the male infants, the male death rates are higher than those of the females at every stage of life. The numerical excess of male at birth is, in consequence, gradually reduced until females outnumber males in the older age-groups. In most Asian nations, however, this is not the case as females are vastly outnumbered by males. The main reason is the comparative neglect and poor treatment of the females in most of Asia. Females outnumber males in a few countries (Indonesia, Thailand and Vietnam) in Southeast Asia. In these nations women have been traditionally accorded a respectable status. There are more females in Mongolia, North Korea and Yemen as well; while the gap between the male-female ratio is narrow, the reasons for larger sex-rations of females in Mongolia, North Korea and Yemen are not clear.

Although the males predominate everywhere else, the highest proportion of males is found in many Muslim nations in Southwest Asia. In Baharain they are 59 percent of the population, and in Kuwait 56.5 percent, in Qatar and United Arab Emirates over 67 percent each, and in Saudi Arabia 54.3 percent. This excessive predominance of males in these nations is unusual, although the major reason is the presence of a large number of immigrant male workers from South and Southeast Asian countries who are employed in these oil-rich nations. In several other nations, males also outnumber the females (e.g., Afghanistan, Bangladesh, Malaysia, Brunei, Jordan, Lebanon, Turkey, all Muslim nations, and a few other nations such as India and Nepal, that contain sizable Muslim minorities.

Generally speaking, the Asian population is "young." A population with more than 35 percent of its people under age 15 is generally considered "young," and a population with more than 10 percent of its people aged 65 and over is considered "old." Applying this definition, no Asian country with the exception of Japan falls in the category of "old" population. Japan enjoys now one of the highest longevity-rates in the world. Nearly 15 percent of its population is 65 or over in age. The youthfulness of Asian countries, however, varies; while Japan is the least "youthful" and Jordan the most (a fifth of Japan's population and over half of Jordan's are below 15 years of age). Seven of the Muslim nations of the Middle East (Jordan, Iran, Iraq, Lebanon, Oman, Saudi Arabia, and Syri) rank high in "youthfulness." Over 40 percent of their population are below 15.

A useful way of depicting graphically the age-structure is by constructing *population pyramids* (Figure 3.1). These are diagrams in which the percentage of the population in each age group is shown as a bar (male and female population considered separately). Most Asian nations generally have pyramids that are very broad at the base and narrow and pointed at the top. A pyramid representing Japan's population, however, resembles that of the western European nations which generally have pyramids that are in effect more rectangular in shape. This means that Japan's population is aging, and the youthful section of its population is by comparison with most other Asian nations, a small one. Figure 3.1 depicts these two pyramids representing population structures of India and Japan. It may be noted that nearly 40 percent of India's population is under fifteen years of age as compared to 21.5 percent of Japan's. As for the old-age category, Japan has 15 percent of its population over the age of 60, as compared to only 6.5 percent of India's population in that age-group. Analyzing these figures, we can realize the enormous problems that India must face regarding the provision of education, services, and jobs for such large numbers: 350 million under 15 and 230 million of the age group of 15-29. Comparing India's situation with Hapan's, the Japanese nation has to deal with only 27 million under 15 and 26 million of the age group of 15-29. But these numerical comparisons are not strictly valid as Japan is a developed economy, and thus possesses a greater capacity to plan its population.

DEMOGRAPHIC TRANSITION IN ASIA

From the standpoint of development, nations are generally grouped into two basic categories: developed and less developed. With the exception of Japan and Singapore, all Asian nations are generally put in the less developed or underdeveloped category. It may be noted that the term development must be used with great care. Asian nations like China and India that are rich and highly developed in a cultural sense, are poor only economically; the label "underdeveloped" or "less developed" may often be inappropriate, and to an Asian mind even insulting. Therefore, the

Figure 3.2 *Classic Demographic Transition Model*

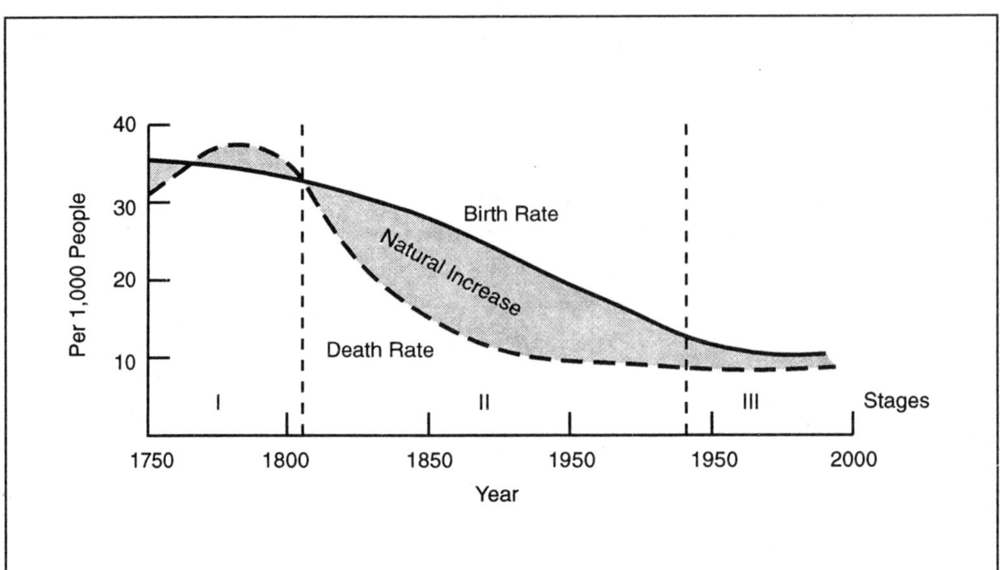

term "development" in the discussion that follows will be used in an economic sense.

In Figure 3.2, the major world regions are divided into the two groups: developed and less developed. The first three regions constitute the economically developed or "rich" world. The nations in this group share similar population traits: featuring low birth and death rates, and low rates of population growth, and relatively moderate percentage of the total population under fifteen years of age (22 to 25 percent). As regards their economic well-being (as measured by gross national product per capita) these regions clearly stand much above the world average, and far higher than the group of less-developed regions. Asia is included in the second group which contains the "have-nots" of the world, although this broad classification overlooks such relatively wealthy nations in Asia as Japan, Hong Kong, and the oil-rich states of Kuwait, Baharain, United Arab Emirates and Saudi Arabia. Here, the population characteristics of these show a broad similarity with the group, and they are much different from those of the developed regions. It may be noted that birth rates in Asia are twice as great as in Europe (the figures for some Asian nations are four times greater than Europe's average). The comparatively high birth rates of this group translate into the extraordinary rate of annual increase.

A clue to the potentially explosive situation that prevails in Asia is that most Asian countries contain over 40 percent of their population below 15 (the figure for some countries in the Middle East is between 45 and 51 percent). Finally, it is quite clear from the GNP data that a vast gap exists between the developed and less developed world in any economic sense.

It is clear from the table of World Population Summary that world population is increasing at quite a high rate, of about 1 billion people every ten years. Although the world rate of population increase has decreased form 2.0 percent to 1.5 percent over the past few decades, this still represents a disturbing rate of increase. Today nearly one-third of world's population is below fifteen; and given current trends of population growth, no matter what course the world's birth rates in the future may take, total population of the Asian countries, as indeed of the world, is likely to grow to large proportions. This prospective growth highlights the enormous problems most Asian nations (as also most other developing nations of Africa and Latin America) face where provision of services and jobs is already a massive problem.

How has population evolved in the developed and developing nations? Europe, that has gone farthest towards population stability, has been studied the most. The graphic illustration of the birth and death rates in West Europe demonstrates clearly that over the last two centuries these rates have been falling as the western European countries become economically better-off, along with the rise in educational standards, industrial growth and generally more settled conditions of living. It became clearly noticeable that there existed a correlation between population growth and economic well-being.

A consistent drop in birth rates and death rates over the past two centuries or so has been termed as *demographic transition*, depicted in Figure 3.2 which represents the European conditions. Initially, both birth and death rates were very high, at times deaths exceeding births in years of famine, wars or prevalence of plague. A quarter of Europe's population perished from bubonic plague in the 14th century. Later on, death rates dropped noticeably due to better control on diseases, or increased sanitation, and higher standards of life in general. Birthrates also began to decline gradually, as the population shifted to cities. Large families became a burden than an asset as did the expense of children's education rose. As better contraceptive devices became available, profound changes began to occur. Birth rates dropped more dramatically after the Industrial Revolution. By about the 1930's a new population equilibrium had been established in which low birth and death rates resulted in a small rate of growth.

The European demographic transition had been completed in several stages and over a relatively long period of time. If we apply the demographic transition model to Asia, we find that with the exception of Japan and Hong Kong, most other nations can be placed within the second state of the demographic transition model of Europe. In Japan and Hong Kong birth and death rates have dropped dramatically and are comparable to those of wealthy western nations indicating that these have completed their demographic transition. Some small nations, such as South Korea, Taiwan and Singapore are on the same track and likely to catch up with Japan as they become economically better-off. In other large and mid-size countries, such as Indonesia, Malaysia, Sri Lanka, Thailand and Vietnam, there are indications that birth rates are declining, although these remain high by western standards. The oil-rich nations in the Middle East fall in a category of their own. Their per capita GNP is among the highest in the world. While death rates have declined sharply in these countries and now are at the level of the West European nations, birth rates stubbornly refuse to fall. The European model thus cannot be easily applied to these situations.

MAJOR POPULATION MOVEMENTS IN ASIA

In addition to the natural factors of birth and deaths that are responsible for population change, Asian populations have been affected, if only to a small extent, by population movements as well. The numerical changes have been so striking for Asia as a whole, but for some areas like the Persian Gulf countries, the consequences of such movements have been far reaching.

The Asian population movements were basically of two types: the resettlement and refugee movements, and labor migrations. In comparison to the European massive intercontinental migrations of persons seeking greater opportunity in the United States between the mid-fifties of the 19th century and the mid 20th century, the Asian movements consisted of largely refugee movements during periods of warfare, civil war, and political turmoil. Labor movements from one country to another have been of much smaller scope. These, however, did alter the human landscapes and introduced significant social, economic and political changes even if these did not substantially alter the demographic structures of Asian societies.

Accompanying the relinquishing of British control over the Indian subcontinent in 1947, Asia experienced one of the largest of human migrations in history. It is estimated that 15-16 million persons were forced to flee across the newly-created boundary between India and Pakistan. Hundreds of thousands also perished from the tragic communal riots resulting from the partitioning of the subcontinent. In the aftermath of Vietnam War during the 1960's, and the conflicts in Cambodia, Laos and Vietnam in the 1970's, the region faced large flows of refugees from these nations after they had come under communist rule. The greatest number escaped into Thailand overland, but thousands (became known as "boat people") got out of Vietnam by sea in small boats. Those who survived the hazardous journeys, found temporary asylum in Hong Kong. Thousands also asked to be repatriated to the West or other countries. Hong Kong, already exploding with population, denied them permanent refuge. In Cambodia, where Khmer forces killed 2-3 million of people, thousands fled to the neighboring Thailand. Several thousands of Vietnamese and Cambodian refugees, however, moved to the U.S., China and the West; by 1990 nearly a million had been resettled in the United States, France, and others in the neighboring Asian nations.

Within the communist societies of China and the Soviet Union, prior to the latter's collapse in 1990 resettlement of the Russians into the Central Asian Republics of Kazakhstan, Kyrgyzstan, Turkmenistan, Tajikistan, and Uzbekistan and the Chinese into Xizang (Tibet) and Xinjiang had been going on for several decades. After 1950 the Chinese consolidated their control over Xizang (formerly Tibet). A large scale resettlement of the Chinese in Xinzang. Massive devastation of monasteries and Tibetan culture followed. This brought enormous economic and cultural changes to the landscape.

Another area greatly affected by refugee movements was Afghanistan. After its occupation by the Soviet forces in 1979 the country became the largest source of refugees in Asia. Some 3 million Afghan refugees moved to Pakistan, and another 2 million to Iran. Since the Soviet withdrawal in 1989 nearly one million refugees

----half of these from Pakistan----have been repatriated to Afghanistan. Notable among the other refugee movements across the borders of Asian countries were those of the Chinese into Taiwan, Mongolia, and North Korea. All these population shifts may be ascribed to the so-called "push" factors, i.e., people pushed out of their native lands for reasons beyond their control. However, these movements represent the sum of many of the sad human dramas: collectively they have served to shape the political and cultural map of Asia.

As a final observation, it may be noted that, for the most part, refugees in Asia, are the by-product of regional conflicts which were fought either with direct superpower involvement as in Vietnam or Afghanistan, or waged by superpower proxies. By early 1990's the superpower involvement had virtually disappeared and the refugee problems within Asia had eased *considerably*.

Another type of movement is that of labor migration, and may be solely induced by the "pull" factor. Examples of this category consist of relatively smaller numbers, and confined largely to the oil-rich nations in the Persian Gulf area. In the late 1970's and early 1980's the Arab nations of Baharain, Kuwait, Qatar, Saudi Arabia and the United Arab Emirates became the targets of rapid flows of immigrants from the lesser developed countries of South and Southeast Asia. Saudi Arabia and Kuwait became particularly attractive destinations for labor seeking employment from Indians, Pakistanis, Sir Lankans, and the Filippinos. Today, the foreign workers account for substantial numbers within the populations of the Persian Gulf nations. In 1992, approximately 10 percent of Kuwait's, one-third of Qatar's, a quarter of Baharain's and one-tenth of Saudi Arabia's population consisted of immigrant workers from India, Pakistan, Sri Lanka, and the Philippines. In 1950 the proportion of foreign workers in the populations of these countries was less than 5 percent.

While the labor migrations to the Persian Gulf countries have been comparatively recent, it may be remembered that population movements from "depressed" and "overcrowded" regions of China and India had been going on for quite some time, particularly since 1850. There are nearly 15 million "overseas" Chinese now, 12 million of which make up for the Southeast Asia's populations. Historically, the southeast-coast of China has been the feeding ground of the Chinese emigre's. Persons of Chinese descent now account for nearly 10 percent of Malaysia's, 9 percent of Indonesia's, 12 percent of Thailand's and 75 percent of Singapore's populations. In Myanmar and the Philippines they form substantial minorities. Relatively unassimilated in the native populations, these overseas Chinese have, with the exception of those in Thailand, remained an alien element.

POPULATION POLICIES IN ASIA

After World War II several Asian nations, India, China, and Indonesia among them, experienced phenomenal population growth. Some of the countries had just gained independence, and were faced with economic reconstruction at a time when mounting population pressures rendered their task difficult. As population continued to escalate, these nations became increasingly concerned with the adverse effect of demographic trends on their economic growth and social progress, and decided to adopt family planning programs.

India became the first major nation to adopt a public policy aimed at achieving reductions in birth rates in 1951 when the government formulated its first five-year plan (1951-1956). Initially the policy was designed to diffuse information on birth control among the illiterate masses, but the emphasis shifted during the succeeding plans to more tangible and effective measures, such as a large-scale male sterilization (vasectomy) program, and free supply and servicing of the intra-uterine device to females. But these measures had limited success despite government's commitment due largely to the cultural attitudes of the vast majority of the people, but also due to inadequate funding. Pakistan's population has been increasing at even a higher growth rate than India's, although both nations received good deal of funding for population control from the U.S. and international agencies.

Japan, which has had a history of consciousness of population problem even before the World War II, has experienced more dramatic success. It passed Eugenics Protection Act in 1948 legalizing abortions, and pursued a vigorous policy of population control that included the provision of contraceptive devices to the people. Its rate of population increase now approximates that of the western nations at 0.2 percent a year as compared to the over 2.0 percent annual increase before 1968.

In China, the official policy of population control included draconian methods to achieve one-child family. The goal of the administration to achieve zero-population growth by 2000, with the population stabilizing at 1.2 billion, has since been achieved. By adopting and implementing generally harsh measures, it experienced spectacular results in fertility decline during the 1970's, and 1980's. South Korea, Taiwan, Sri Lanka, Bangladesh, Southeast Asia and nations of Indonesia and Thailand also offered family-planning and birth control services to the masses using persuasive methods aimed at reduction in fertility. These policies in general, have been successful. Birth rates in these nations dropped substantially in the 1970's and 1980's. An entirely different situation prevailed in the war-ravaged nations of Cambodia and Laos where governments have been actively trying to encourage population growth that would lead to replenishing the labor force lost in the war during the 1960's and 1970's. In Laos, the government even started paying women to get married and then provided an allowance after the birth of each child. Vietnam, since the war, has had an ambivalent attitude towards birth control although a large part of its youthful labor-force had perished during the war. Malaysia has, during the last three decades, experienced fairly strong economic development, accompanied by a significant reduction in fertility that was bolstered by the presence of an active family planning program. In 1984, however, government reversed itself, declaring that Malaysia needed a more rapidly growing population to provide an adequate labor-force.

Hong Kong, Taiwan, South Korea and Singapore experienced more dramatic fertility declines during the 1960's. Singapore, on gaining independence from Great Britain in 1965, realized that its rapid rate of population growth would make it impossible to continue developing economically. The abortion laws were liberalized and measures taken by the government to offer disincentives to families with more than two children. The results were so overwhelming, that the government started considering a reversal of the policies realizing that Singapore's youthful population might soon be depleted if the government continued on its anti-natalist policies.

Generally speaking, several Southeast Asian nations lag behind in formal programs but public consciousness and some basic planning have grown universally. Among predominantly Muslim Southwest Asia this consciousness is still at a rudimentary level as the strong religious feelings militate against the institution of officially-prescribed birth control policies. In the Central Asian republics, the Soviet Union before they became independent in 1990 had pursued an ambivalent population policy; and the birth rates have remained high, comparable to those prevailing in most other Muslim nations.

Further Readings

Grigg, D. "Modern Population Growth in Historical Perspective," *Geography,* London, April 1982, pp. 97-108.

National Academy of Sciences *Population Growth and Economic Development.* Washington D.C., 1986.

Population Reference Bureau, *World Population Data Sheet, 1998.* Washington D.C., 1998 and 2000.

Wrigley, E.A. *Population and History.* New York, 1969.

4

Patterns of Asian Languages

Language and religion are the basic and vital elements of culture that identify human groups. Languages tend to promote the transmission of ideas and the functioning of political, economic and religious systems. They act as a "cultural cement" that binds one generation with the next, and one human group with the other. The distributions of languages and religious affiliations are of particular interest to geographers because they reveal the spatial interplay of political, economic and social forces that characterize human groups. This chapter examines the distributions of Asia's major languages and how these have affected Asia's cultural landscapes.

Unlike North America, which is essentially a unilingual English-speaking continent with substantial minorities of French and Spanish speakers and a largely vanished group of indigenous tongues, the continent of Asia exhibits linguistic diversity and complexity unmatched elsewhere except in the continent of Africa. According to linguistic historians, Asian languages (as well as those of Europe) originated somewhere in the present day east-central Europe and moved in two major streams: one, eastward to Iran and thence to the Indian subcontinent and the other westward to Europe between c. 2500-1000 B.C. Several migrations and branches of the migrations established subsequent patterns. Asia's present linguistic patterns reflect the continent's long and turbulent history, the historic migrations in its distant past, the conquests of the medieval rules, as well as the paths of the imperial empires established by the British, the French, the Dutch and the Portuguese during relatively recent times. The present picture, hence, bears little resemblance to the past distributions.

The present map (4.1) of Asian languages reflects patterns that had been established by the 18th century. At that time the largest language groups, by number of speakers, were the Chinese and the Indo-Aryan languages but Tungusic languages were probably used over wider areas in North Asia. Since then the Tungusic group (and many other smaller ethnic groups) have been withering away, despite the Soviet practice of publishing books and newspapers in regional languages. Mandarin Chinese has also been gaining ground at the expense of local languages and dialects. In India, however, the regional languages are not losing ground, and languages have become a political issue. Meanwhile, the Indian government continues to use English as an official language. In Indonesia, which has many local languages and dialects, Bahasa Indonesia, the national language, has failed to spread to all parts of the country. English remains the most commonly spoken language in the Philippines, despite the adoption of a national Filipino language. The marked increase in populations of China, India, Japan, and Indonesia during the 20th century means Mandarin Chinese, Hindi, and Japanese can now claim to be world's prominent languages. On the other hand, many languages have been withering as ethnic groups disappear or merge with the larger groups. Many Paleo-Siberian languages (just as most native American Indian tongues) have vanished or declined in numbers. Many ancient languages spoken in Asia have also disappeared within the last millennium and others have been greatly modified by linguistic change.

ASIA'S MAJOR LANGUAGE FAMILIES

Before examining the distribution of the Asian languages it may be essential to explain the basis of their classification. While some languages are related to each other by virtue of their origin, others may not share a common parentage. Languages grouped in a family are thought to have a shared, if a distant origin. When languages resemble in a systematic way (in grammar, syntax and structure), they are held to be genetically related and are identified with a particular family or they may be linked by contact. In the sub-family group their common traits become conspicuous. The sub-families are divided into groups consisting of individual languages. Languages that disappeared without descendants are known as extinct. These are known only through written records and comments by contemporaries. The classification for many extinct languages remains uncertain because of scarcity of data. A language that cannot be assigned to an existing family is termed "language isolate." Our knowledge of the languages of some isolated regions (e.g., New Guinea, and East Siberia) is too scanty to allow proper classification and are usually grouped as a separate entity.

Prehistoric humans perhaps used some 12,000 to 15,000 languages. Today nearly 6,000 of these are in use, although hundreds of those are spoken by merely 10,000 to 12,000 persons. Most of the smaller languages are disappearing as these are being swallowed by such languages as English, Russian, French, and Spanish. It is believed in some quarters that by the middle of the 21st century only 600 or so languages may survive.

Eight language families cover nearly all of Asia: Indo-European, Semitic, Altai-Turkic, Dravidian, Austro-Asiatic, Austronesian, Sino-Tibetan and Paleo-Siberian. The *Indo-European family* is spoken by the largest number of speakers and covers the widest extent in the world. It probably originated in the steppe region of southern Russia around 4,000 B.C., and spread into the Danube area of Europe and beyond that to other parts of Europe, and to southwest Asia as also into East and Southern Asia. There are several systematic resemblances between the European languages and Sanskrit, the oldest-attested language of Southern Asia, that point out the common ancestry of many languages of Europe and South Asia. Several well-known scholars have determined the geographical origins and lifestyle of the proto-Indo-European people by comparing the vocabulary of the extinct Indo-European languages, such as "mother," "husband," and "brother," ext. The proto-Indo-European language has words for horses, pigs, sheep, and some kind of wheeled vehicle. Words for flora and fauna

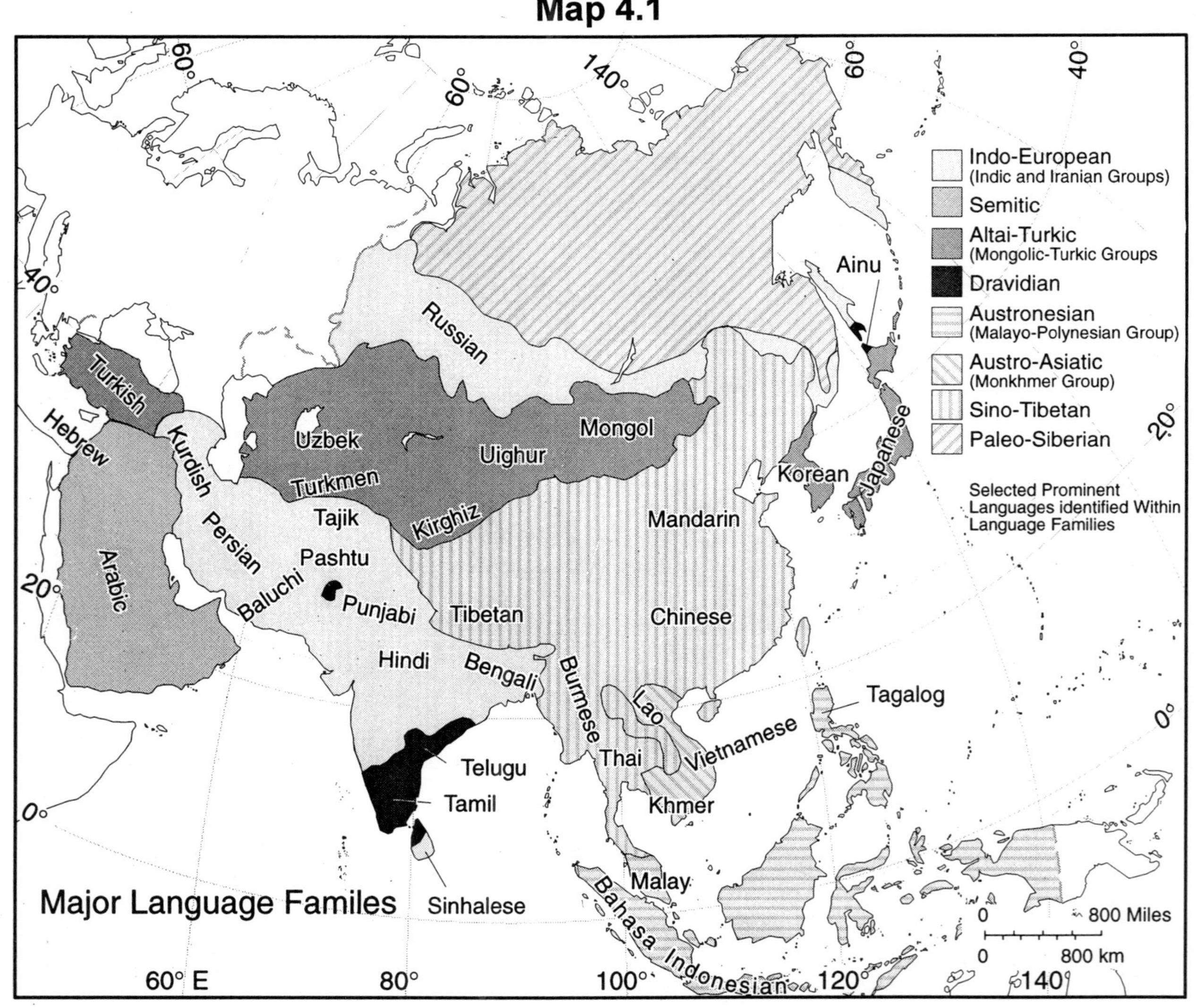
Map 4.1
Indo-European (Indic and Iranian Groups)
Semitic
Altai-Turkic (Mongolic-Turkic Groups
Dravidian
Austronesian (Malayo-Polynesian Group)
Austro-Asiatic (Monkhmer Group)
Sino-Tibetan
Paleo-Siberian
Selected Prominent Languages identified Within Language Families
Russian
Turkish
Hebrew
Kurdish
Uzbek
Turkmen
Uighur
Mongol
Kirghiz
Tajik
Persian
Pashtu
Baluchi
Punjabi
Arabic
Tibetan
Hindi
Bengali
Mandarin
Chinese
Korean
Japanese
Ainu
Burmese
Lao
Thai
Vietnamese
Khmer
Tagalog
Telugu
Tamil
Sinhalese
Malay
Bahasa Indonesian
Major Language Familes
0 800 Miles
0 800 km
60° E
80°
100°
120°
140°
0°
20°
40°
60°

provide us clues as to the place of origin of the speakers of the early (or proto-Indo-European) languages. It is now generally believed that the Indo-European family probably originated in north-central Europe, and from there one of its streams migrated to other parts of Europe, and another moved to parts of Asia.

The branch of Indo-European family that moved into Asia is divided into two large groups, known as Indo-Aryan (or Indic) and Iranian, other branches moved into Europe. The Indo-Aryan sub-family contained several subgroups, identified largely on a geographical basis: the *Midland* subgroup includes Hindi/Urdu, and related languages of northern India; the *Eastern* subgroup includes Bengali and languages of eastern India, the *western and southern* subgroups include Gujarati, Marathi, etc. in India, and Sinhalese of Sri Lanka, and *Northwest* subgroup includes Punjabi, Pahari and Dardic languages of the Indian subcontinent. The early form of Indo-Aryan dates from earlier before 1000 B.C.----the language in which the Vedas, the ancient sacred texts of the Hindus were composed. Later forms, the Prakrits, lasted 1,000 years, and were the medium of Buddhist and Jain literature. Several of the languages of western India and Pakistan such as Punjabi are related to Prakrit.

The Iranian languages---especially the Old Persian and Avestan (the sacred language of the Zoroastrians) have texts dating from the 6th century B.C. The modern Iranian languages, (Persian, Pashto, Baluchi, Kurdish, and Tajik) are spoken by over 60 million people.

The *Semitic* languages have the longest history and a large number of speakers in the world. They are located throughout Southwest Asia, including the Arabian peninsula and across the whole of North Africa. The oldest of the Semitic languages (Akkadian, Phoenician, etc.) date from the third millennium B.C. Hebrew is the best known of these whose modern version dates from 2000 B.C. or so. Its classical form was preserved as the written language of Judaism; the modern and spoken form is used by over four million in Israel, and several million elsewhere throughout the world. A form of it---Syriac---is spoken by almost a million people in the Middle East and in the U.S.A.

The major Semitic language is Arabic, spoken by over 150 million as a mother tongue. The classical form is the sacred language of Islam and used as a lingua-franca in Southwest Asia from Syria, Lebanon, Israel, the West Bank, Jordan, Iraq and the nations of the Arabian peninsula that include Saudi Arabia, Yemen, Kuwait, and the United Arab Republic.

The *Altai-Turkic family* covers a large area, from the Balkan peninsula to the northeast of Asia---an area that includes the Altai mountain region of Central Asia, from which the family receives its name. It includes nearly three dozen languages, classified into three subgroups: Turkic, Mongolian, Manchu-Tungusk. The best-known of the family is Turkish, spoken by about 45 million people of Turkey and surrounding territories. Other languages in this group are: Azerbaijani (12 million), Turkmen (around 3 million), in Central Asia, Iran, and Afghanistan, Uzbek, in Uzbekistan (6 million), Uighur in mostly China (Xinjiang province), Kazam and Tartar (both around 2 million), in the Kazakhstan.

In the Mongolian subgroup Mongol is the chief language spoken by nearly 4 million people in the Mongolian People's

Republic and adjacent ports of China. The Manchu-Tungus subgroup was spoken over a wide area in Manchuria or North-east China by nearly 3 million, but very few now speak the once-important Manchu language. Most of early Manchu texts date from the 8th century to the 17th century. In the 20th century, there has been a concerted effort to modernize the Altaic languages by injecting fresh vocabulary, and reforming the scripts as in the case of Turkish, where Latin script replaced the Arabic.

The *Dravidian* family is a group of over 20 languages, clustered chiefly in the Southern India, although one language, Brahui, is isolated. Through emigration, the speakers of Dravidian languages are today found in Southeast Asia, in eastern and southern Africa, and in several ports of the world. Tamil is the oldest language of the family, dating from the 3rd century B.C. There are several theories regarding the original home of Dravidian languages. The theory that carries a strong support among scholars is the view that these were spoken in northern India before the arrival of the Aryans, who subsequently pushed these into southern India.

The *Austro-Asiatic* family includes over 100 languages that covers a large area in Southeast Asia, in the countries between China and Indonesia, together with straggling outleis in the Nicobai Islands in the Indian Ocean and in parts of India. The origin of the family is still unclear. The largest branch of the family is the Mon-Khmer group of languages, spoken throughout Southeast Asian mainland, primarily in Vietnam, Laos, Cambodia, parts of Burma and Malaysia. Khmer is the official language of Cambodia and 5-6 million people. Vietnamese is spoken by about 50 million people of Vietnam and Laos. Some scholars relate it to the Tai language group of the Sino-Tibetan group. The Munda group is spoken in India and the Nicobar Islands. Very little is known about the origin and early movements of these languages. Possibly the various groups of this family began to split up in the second millennium B.C. but where it originated remains a puzzle of linguistic geography.

The family of *Austronesian* languages comprises close to 1,000 languages with an estimated 200 million covering a vast area extending essentially from Malaysia and Indonesia in the west, but including Madagascar in the Indian Ocean, to the Philippines and Taiwan, and from New Guinea in the east and further east in the Pacific Ocean to Hawaii----a territorial range which is reflected in an alternative name given to the family: Malayo-Polynesia. Thus, in geographical extent and the number of speakers, it is one of the largest families. The family is usually divided into two main groups, the boundary lies in New Guinea: the western group contains over 400 languages, spoken in Malaysia, the Indonesian islands, the Philippines, some parts of Cambodia, and western parts of New Guinea; the eastern group covers most of New Guinea, and throughout the 10,000 or more islands of Melanesia, Micronesia, and Polynesia. Because of many structural differences within the language, it is estimated that the family has a long history of evolution of over 4,000 years.

The classification of *Sino-Tibetan* family of language is highly controversial. The Sinitic part refers to the various Chinese languages (Cantonese, Hakka, Hunan, Kan, Mandarin, Min, and Wu), and the Tibetan part to the several languages spoken in Tibet, Myanmar, and adjacent territories. The Sinitic languages, referred to by the Chinese as Han (name derived from the Han dynasty) are spoken by over 1,100 million people, a large majority of which are in China and Taiwan, but substantial numbers reside throughout Southeast Asia, Hong Kong, Singapore, Thailand, Malaysia and in the U.S.A. There has traditionally been a single method for writing the Chinese languages, and the Chinese refer to all as "dialects." After Chinese, Burmese and Tibetan are the two main languages of the family. Burmese is spoken by over 25 million people in Myanmar as a mother tongue. The estimates for Tibetan are uncertain because of the influence of the Chinese in recent years, but a figure of 3-4 million speakers appears realistic. There are nearly 300 languages in the Tibeto-Burmen group spoken by nearly 3 million people in parts of Myanmar, Thailand, Vietnam and Laos. The Tai group (referred to by some as a separate family because of its association with both Austronesian and Sino-Tibetan families) is in Southeast Asia covering an area from Thailand to Eastern region of Burma, Laos and Vietnam.

The *Paleo-Siberian* family is the name given to several loosely-linked languages of northeastern Siberia in Russia for these old languages could not be fitted into any other category. These diverse languages currently claim several thousand speakers, though less than a century they numbered several times the present-day figure. Since the early 20th century these languages have been given a written form, based in Cyrillic alphabet.

ASIA'S LANGUAGE MAP

Of the world's ten leading languages, six, which claim over 100 million speakers each, are spoken in Asia; namely, Mandarin Chinese (850 million speakers, Hindi (340 million speakers), Arabic (185 million speakers), Bengali (180 million speakers), Indonesian (135 million speakers) and Japanese (125 million speakers). Two of these, Russian and English, each claim over 100 million speakers worldwide, and are also spoken in Asian nations. Russian speech is limited to the various republics of the former U.S.S.R., whereas English, the second largest language in the world after Chinese, is spoken widely in the Philippines, and is the *lingua franca* in several Asian nations, particularly those which formerly were a part of the British Empire. In India English retains the status of an associate official language and is used as a *lingua franca* for the country.

Perhaps over 2,000 languages are spoken in Asia; and the pattern of their distribution is complex. Map 4.1 shows the distribution of the major language families and selected languages in a highly generalized map but essentially obscures the overlapping distributions of multiple languages that have emerged in several situations. For example, around some of the Central Asian oases and in southern Siberia, migrants from Russia and China, and exiled ethnic groups have created linguistically mixed patterns. Several large cities, such as Singapore, Mumbai, Hong Kong and Manila show complex language patterns. As Chinese have moved into Lhasa and other parts of Tibet, Chinese has become the predominant language of Tibet, while the numbers of Tibetan language speakers have been shrinking. Another example is that of English speech, which is widely spoken in the Philippines and used as an associate official language of administration in India and in cities all over Asia, but is not assigned a place in our language map.

Given Asia's enormous size and cultural complexity, it may be convenient to analyze the distribution of languages within the framework of its major realms which we designated in Chapter 1.

LANGUAGES OF SOUTHWEST ASIA

The area between the Mediterranean Sea and Afghanistan is dominated by three language families: Semitic, Indo-European, and Altaic (Turkic). Linguistic diversity is greatest in the two countries of Turkey and Afghanistan that define the western and eastern peripheries of the realm. In the Semitic family, Arabic claims the largest number of speakers, and territory. It is spoken in Lebanon, Syria, Jordan, Saudi Arabia, and a small section of Southern Iran; whereas Hebrew is exclusively spoken in Israel. An ancient tongue, Sumerian, now extinct and unrelated to the Semitic or Indo-European languages, was spoken four thousand years ago in the region stretching from the Euphrates-Tigris basin to Israel. How did the splendid Sumerian civilization and language vanish remains a historical mystery?

Most of the living languages of the Indo-Iranian branch of the Indo-European language family are spoken in Iran and Afghanistan, and occasionally extend beyond into neighboring regions. Persian (or Farsi) is largely centered in Iran, Afghanistan, Tajikistan, but not entirely confined to these countries. Pashto is spoken by majority of the people in the linguistically diverse Afghanistan, and spills over to Pakistan, and Iran. Kurdish is spoken by Kurds in an area that cuts across the territories of Turkey, Syria, Iran and Iraq. Baluchi-speakers live in southern Iran, southwestern sections of Pakistan and southern Afghanistan. About half a dozen other, smaller languages classified as the Dardic sub-group of Indo-Iranian tongues, are spoken in Afghanistan.

Turkish, the principal language belonging to the Altaic family, is spoken largely in Turkey, but also spoken in the surrounding countries of Azerbaijain and Iran. Three other Turkic languages of the Altaic family: Uzbek, Turkmen and Kyrghyz are spoken in the Central Asian which carry their names.

Nearly 40 or so languages are spoken in the *Caucacus* area that lies between the Caspian Sea and the Black Sea. The region has been the major land route from south to north, and exhibits an extraordinary ethnic and cultural variety. The Caucasus languages reflect an influence of Russian, Persian and Arabic languages, and cannot be easily grouped into any family category. In all, there are around 5 million speakers of the Georgian language outnumbering all others. Armenian and Azerbaijain are two other major languages in the area, but those are classified in the Turkic family. The written form of the Caucasus languages is based on Cyrillic alphabets.

SOUTH ASIAN LANGUAGES

The Indo-European family, some indigenous families (as Dravidian) a few Sino-Tibetan languages and some "language isolates" nearly cover the territory of South Asia. The Indo-Aryan group containing about 20 languages covers the nations of India, Pakistan, Bangladesh, Nepal, Sikkim, Bhutan and most of Sri Lanka. The boundaries of these languages are difficult to demarcate as their dialects gradually grade into one another. The problem of boundary demarcation is further compounded by the degree of assimilation of Sanskritic vocabulary by these, thus making one language quite different from the other. Sanskrit, a classical Indo-Aryan language, preserved mostly in writing, enters deeply into the vocabulary of the present-day languages. Thus, Hindi of India and Urdu of Pakistan, both using the same grammatical forms, differ linguistically from one another because the former may be heavily Sanskriticized in vocabulary, while the latter draws for its vocabulary upon Persian sources.

There are over 500 Indo-Aryan languages, spoken by over 500 million people. Hindi and Hindi-related languages claim nearly 280 million speakers in India alone (there are probably another 50 million Hindi-speakers in the world). Thus Hindi is the second largest spoken language in Asia, and the third largest in the world (after Chinese and English) Among the Indo-Aryan languages, Bengali, and Urdu each claim 160 and 100 million speakers respectively. In the Iranian groups, Pashto and Baluchi are spoken in southwestern Pakistan.

The Dravidian family is almost exclusively concentrated in southern India; although one language, Brahui, is isolated in Pakistan, separated from the others by 800 miles. By virtue of their long literary traditions and by the substantial number of speakers, four Dravidian languages may be termed as the major languages of India: Telugu (75 million speakers), Tamil (78 million speakers), Kannada (38 million speakers), and Malayalam (35 million speakers). Sri Lanka, Malaysia, Myanmar, and Singapore, also contain substantial minorities of Dravidian speakers. In India, each of the four major Dravidian languages can be identified with a state.

A dozen or so Munda languages, which are possibly related to the Mon-Khmer languages of Southeast Asia and assigned to the Astro-Asiatic family are concentrated in the tribal regions of India. The various languages of the Sino-Tibetan family are distributed from Kashmir to Bhutan along India's northern borders.

LANGUAGES OF SOUTHEAST ASIA

Southeast Asia includes two subregions: mainland (Malaysia, Thailand, Vietnam, Cambodia, and Laos); and an insular section (Eastern Malaysia, all of Indonesia, and the islands of the Philippines). The languages of insular southeast Asia almost entirely belong to the Austronesian family; whereas those of the mainland belong to the Austro-Asiatic, Tai, and the Sino-Tibetan families. The Austronesian family extends to a vast and remote area in Oceania as well. Though the languages of the mainland are genetically diverse, the Mon-Khmer group (containing 50 or so languages) of the Austro-Asiatic family is most widely spoken, covering an area from Myanmar to Vietnam. In Cambodia, Khmer is the official language. Vietnamese, however, claims the largest number of speakers, nearly 60 million speakers, on the mainland, spoken primarily in Vietnam but also by smaller numbers in the neighboring countries of Cambodia, Thailand and Laos.

Sino-Tibetan and Tai languages, spoken in Southeast Asia are: Thai in Thailand; Lao in Laos, Thailand, and Cambodia; Yan in Thailand; Shan in Myanmar and a few other related languages in Vietnam. Prominent in the family of Sino-Tibetan languages is Chinese, which is spoken by over nine million people of Chinese descent distributed throughout Southeast Asian mainland with significant numbers in Thailand (7 million), and in Malaysia (4.5 million) and smaller numbers in Myanmar, Cambodia, and Laos. The Burmese language is the predominant language of Burma, where 90 percent of the population claims it as a mother tongue, and the remaining use it as a second language. Several related Sino-Tibetan tongues (like Karen, Kachin) are also used in tribal areas of Burma.

Insular Southeast Asia shows a greater linguistic diversity with 500 or so languages spoken in its hundreds of islands which may be grouped into two broad categories: western Austronesian (or Indonesian) and Eastern Austronesian (also called Oceanic). Important languages in the Austronesian family are: Malay, (includes Bahasa Indonesia, Indonesia's national language), and Tagalog (includes Pilipino, the national language of the Philippines). In addition, several non-Austronesian languages are also spoken in this vast, islandic world that stretches east and southeast of Indonesia.

EAST ASIAN LANGUAGES

Most important in terms of speakers and their influence on the other languages of East and Southeast Asia are the Chinese languages that belong to the Sino-Tibetan family. Of these, Mandarin is the most influential and is spoken by over 70 percent of China's population. It has traditionally been the language of administration. It is also the world's leading language claiming over 700 million speakers and spoken by a greater number of speakers in China than all other Chinese languages (Wu, Cantonese, Hsiang, Hakka, and Min). Prior to 1950 most Chinese were supposed to be illiterate because the Chinese languages consisted of thousands of word-images remained mostly unwritten as each generation had to learn thousands of character signs that were needed for literacy. With the introduction of simplified character, and romanization of the language by the People's Republic of China since 1950, and compulsory education in schools, most Chinese have now become literate.

Languages of the Tibeto-Burman subgroup are spoken in Tibet, parts of Nepal, Assam (India) and Bangladesh. The Altaic-family languages are spoken in western parts of China and Manchuria. Uighur, a Turkic language is spoken in Xinjiang (Sinkiang) and Kansu provinces of China. Manchu, which claims a fairly long literary tradition dating from the 16th century, is the best known of the Machu-Tungus group, is spoken in Manchuria province of China, but has been largely replaced by Mandarin and is now spoken only in scattered localities.

Korean and Japanese are two other important languages in East Asia. They are related to the Altaic family, although these have borrowed their scripts and characters from Chinese. Korean is spoken by well over 50 million people in North and South Korea and by sizable minorities in Japan and China. The language has been influenced greatly by Chinese containing more than half of its vocabulary is of Chinese origin. The Japanese subfamily includes, besides Japanese, several dialects spoken in the Ryukyu Islands by people who are bilingual in Japanese. Numerically, there are nearly 140 million Japanese speakers in the world, over 90 percent of which reside in Japan, the remaining in Taiwan, Brazil and the U.S. (particularly in Hawaii).

The Ainu language spoken only on the island of Hokkaido and in northern Japan, and the island of Sakhalin, is a linguistic puzzle as it is totally unrelated to other surrounding languages and is the speech of the Ainus who are ethnically distinct from the neighboring races.

LANGUAGES OF NORTH AND CENTRAL ASIA

The languages of this vast, sprawling region are those spoken from the Arctic Ocean to Central Asia and China, and from the Caspian Sea and the Ural Mountains in the west to the Pacific Ocean in the east. Most languages belong to the Ural-Altaic family (Turkic, Mongolian, and Manchu-Tungus) or to the Indo-European family. The Turkic languages are distributed over an extensive area in southern Siberia (Russia) through the Central Asian nations to include Turkmen in the Turkmenistan, northern Iran and Afghanistan, Uzbekistan, Kazakhstan, and Kyrgyzstan. A few Turkic languages are spoken in western Mongolia, and in northern Siberia.

The Mongolian or Mongol branch of the Ural-Altaic family is dispersed throughout Central Asia from Afghanistan to Manchuria, occupying large parts of North and East Asia. Among the Mongolian languages are Mongol, spoken in Afghanistan, and a few other languages influenced by Iranian and Turkic languages.

The Manchu-Tungus branch has speakers scattered from central interior Siberia to the shores of the seas of Japan and Okhotsk including the Sakhalin Island. Most speakers of these languages are now becoming increasingly bilingual in Russian or Chinese or altogether giving up their native languages. In northernmost Siberia several Paleo-Siberian tongues are still spoken by the indigenous tribes.

Indo-European languages in northern and Central Asia, in addition to Russian which was introduced relatively recently by the Soviets, include the Iranian languages in Tajikistan (Persian), and Baluchi in Turkmenistan. Since the occupation of these territories by the Soviet Union in the later part of the 19th century and early 20th century, Russian replaced the native languages in administration and educational institutions as the official language, and has served as the *linga franca* for the region. The Soviets encouraged the native languages in order to preserve the cultural identity of the various ethnic groups (Kazakhs, Turkmens, Tajiks, Uzbeks and Kirghis), but Russian was aggressively implanted in administration, among the elites, and in educational institutions. With the breakup of the Soviet Union in 1991 the assertion of native languages is likely to be vigorously pursued by the newly-independent nations which bear the names of their leading languages.

THE PROBLEM OF A NATIONAL LANGUAGE

Asia's language patterns outlined in the preceding paragraphs indicate that a large variety of linguistic forms are used on the continent. However, these patterns fail to bring out the complex problems that these have created in several nations.

Only seven of the Asian nations are close to being homogeneous in their linguistic composition: Bangladesh, Cambodia, China, Japan, Jordan, North Korea and South Korea, each of which claims over 90 percent of its population speaking the leading language of the country. In Afghanistan, Laos, Malaysia, Myanmar, Russia, Saudi Arabia, Sri Lanka, Thailand, Turkey, and Uzbekistan where 70 percent of their populations speak the leading language of the country but contain substantial linguistic minorities. The remaining Asian nations contain a wide-ranging linguistic diversity. In most nations no single language is spoken by a majority of the population.

The number of significantly different languages within several countries has made it necessary to establish the so-called "national" or "official" languages. The fear of neo-colonial domination has motivated the distrusts of the old languages of colonialism----English, French, Dutch, Spanish, and also of now Japanese and Russian----although these were used in administration and continue to be the *linga franca* within the countries. Language as a cohesive force for nation-states and for linguistic groups within the countries, has for long been manipulated for political ends. Language riots have occurred in India and Pakistan between rival vernacular communities. Languages have also become focus of loyalty for religious communities that consider themselves as suppressed or subjected to discrimination. Sikhs, a minority group in the state Punjab in India, successfully agitated for the formation of a linguistic state within India which was formed in 1966. Given the turbulent political climate in India and Pakistan, Hindi and Urdu were designated over the objections of other groups as the national languages although both are spoken by less than the majority of their countries' populations. India recognizes 18 languages in its constitution, several of which act as the official language for states within the country, and in the case of Hindi, for several states. Two of these, Hindi and Bengali are among the world's leading languages. In Pakistan nearly 48 percent of its population speaks Punjabi, although the status of the official language is reserved for Urdu which claims only 7.6 percent of the country's population. Sindhi, Baluchi and Pashto are some other languages in Pakistan. Both in Pakistan and in India, the decision to declare "national" or "official" languages for their countries stemmed from either historical, and political reasons, or religious consideration, rather than from reasons of the numerical strength of the languages.

Linguistic diversity is even greater in Southeast Asia. In Indonesia, where more than a dozen Malay languages are spoken, dialectical differences are so great that a national tongue, Bahasa Indonesia, has been established. In the Philippines, where more than a dozen languages are spoken, less than a third of the population speaks Tagalog, and English is universally understood and used as a *lingua franca* of commerce. However, Pilipino (based on Tagalog) is the declared national language, even though it is learned as a second language by most Filipinos. Malaysia is another multilingual nation containing substantial number of speakers of Chinese and Indian languages. The majority of the population is Malay-speaking (61 percent); the national language is Malay, but English is widely spoken and used in administration.

In Southwest Asia, Iran and Qatar are linguistically diverse where majority of the populations do not speak the leading language of the countries. Persian, the leading language of Iran claims only about 45 percent of population of Iran which contains substantial minorities of Azerbaijani, Kurdish, Baluchi, Turkmeni, and Arabic speakers. Over a third of Qatar's population consists of the speakers of South Asian languages and one-sixth that of Persian speech, while only a quarter of its population speaks Arabic. But Arabic is retained as the official language.

Among the Central Asian nations Kazakhstan, with less than 40 percent of its population speaking Kazakh language belongs to the category of linguistically diverse countries. Thirty-eight percent of its population speaks Russian, and Kazakh is the "official" language. With the newly-found independence in these nations nationalist urges are likely to find expression in the reassertion of Central Asian languages as the official languages of their nations.

Further Readings

Fromkin, V. *et al*. *An Introduction to Language*. New York, 1983.

Katzner, K. *The Languages of the World*. New York, 1975.

Thomson, D.S. *Languages*. New York, 1975.

5

Asia's Religions and Religious Landscapes

There is a tendency in the West to underrate the significance of religious values that shape our lives. This is largely due to the overwhelming role that modern technology plays in the daily life of people in the West. In the East, however, religion has been, and continues to remain, the focus of both personal and national life, where often critical decisions are made on the basis of religion. Of the diverse cultural forces influencing the social and political life of most Asians (with the notable exception of the Chinese and the Russians) religion is perhaps the most dominant one that permeates virtually all aspects of their daily personal and family lives. In Asia, religion is not only a subject of historical interest and concern, it is also a living reality. Despite increasing secularism in Europe and North America, it continues to flourish in Asian nations. Religious beliefs find direct expression on the landscape in a variety of forms like the erection of religious structures, places of worship, cemeteries and burial places. Asian landscapes are dotted with sacred altars, holy rivers, and places of pilgrimage. Hinduism, Islam, Christianity, Jainism and Buddhism have prescribed regulations regarding reproduction, life, and death which, in turn have affected the cultural landscapes of several nations such as Bangladesh, Pakistan, Saudi Arabia, Thailand and India. In India religious injunctions have stratified the Hindu society along caste divisions, which have influenced the whole range of human activities.

Association between religions and places takes several forms. The distribution of holy places and places honoring thousands of Hindu gods and Muslim and Christian saints in India and Christian saints in the Philippines may indicate the direction and history of religious movements in those countries. Hundreds of places of pilgrimages express the direct formal expression of religion on the land. River sources (Manasarowar Lake in Tibet), river confluences (e.g., Varanasi in India), mountain peaks, lakes, and caves have been sanctified by several religions. Large temples have been established in many of these locations, and these in turn, have favored urbanization. Pilgrimages as important mechanism of religious circulation afford a continual forum for the exchange of not only religious ideas but generate cultural exchanges, social mixing, and trade as in the case of the annual *hajj* pilgrimage to Mecca in Saudi Arabia.

Religions like Hinduism, Buddhism, Jainism, Islam and Judaism encourage taboos on work and food, which may interact with the cultural landscapes. For

Table 5.1 *Religious Affiliation in Asia, Mid-1999*

Faith	*Estimated Numbers (Thousands)*	*Percent of World Population*
Buddhists	331,043	9.6
Confucians	6,219	0.2
Christians	306,401	8.4
Roman Catholics	108,437	
Protestants	49,140	
Orthodox	14,161	
Anglicans	717	
Unaffiliated	11,561	
Other	153,105	
Chinese Folk Religion	380,250	10.4
Hindus	792,897	21.8
Jains	4,079	0.1
Jews	4,323	0.1
Muslims	807,034	22.2
Parsis (Zoroastrians)	2,407	0.06
Sikhs	22,015	0.6
Shintoists	2,715	0.06
Atheists and Nonreligionists	602,992	16.6

Source: Adapted from *Encyclopedia Britannica Book of the Year, 2000.*

example, the prohibition of killing of animals and meat eating among Hindus and Jains, has inhibited the development of beef and dairy industry in India. Taboos regarding the slaughter of pigs in the Islamic countries and in Israel explains the virtual absence of pork industry in these countries. Taboo on beef eating by Hindus has resulted in the accumulation of an enormous cattle population in India, a large part of which is aged, diseased and unproductive.

Religious sentiments have often dominated political behavior in several Asian nations. The partition of the Indian subcontinent in 1947 was a successful realization of Pakistan as an independent nation by a religion-inspired political party, the Indian Muslim League. Several countries in the Middle East, including Iran, and Saudi Arabia are theocratic states. Nepal and Pakistan are also theocracies. Religion-affiliated political parties often contest for elections for political gains in South Asian nations.

Asia is the birthplace of the world's major religions from the ancient Zoroastrianism and Hinduism to the relatively recent religions like Islam and Sikhism, and the home of a large number of sects and creeds. Even though within each Asian nation a single religion occupies a commanding position, there exist in each nation a number of significant religious minorities. These minorities are widely distributed and in some areas constitute local or regional majorities. Predominately Islamic Southwest Asia contains important Christian enclaves (as in Lebanon) and a Judaic nation (Israel). Within South Asia reside not only Muslims, Jains, Buddhists, Sikhs and Hindu sects, but also sizable minorities of Christians. Theravada Buddhism in Burma and Thailand contrasts with the Roman Catholicism of the Philippines, the predominant Islam of Indonesia, and the Chinese eclecticism of Vietnam. In China, the Confucius code of ethics is widely practiced along with Buddhism. In Japan State Shintoism coexists with Buddhism. In Central Asia reside large number of Muslims and Mahayana Buddhists. And, almost everywhere Christianity claims a significant minority of its adherents in larger cities. Thus, a wide variety of religious influences have shaped Asia's cultural landscapes, and defined its social and political character.

A brief review of major Asian religions given below illustrates the importance of religion in the cultural geography of the continent.

Map 5.1

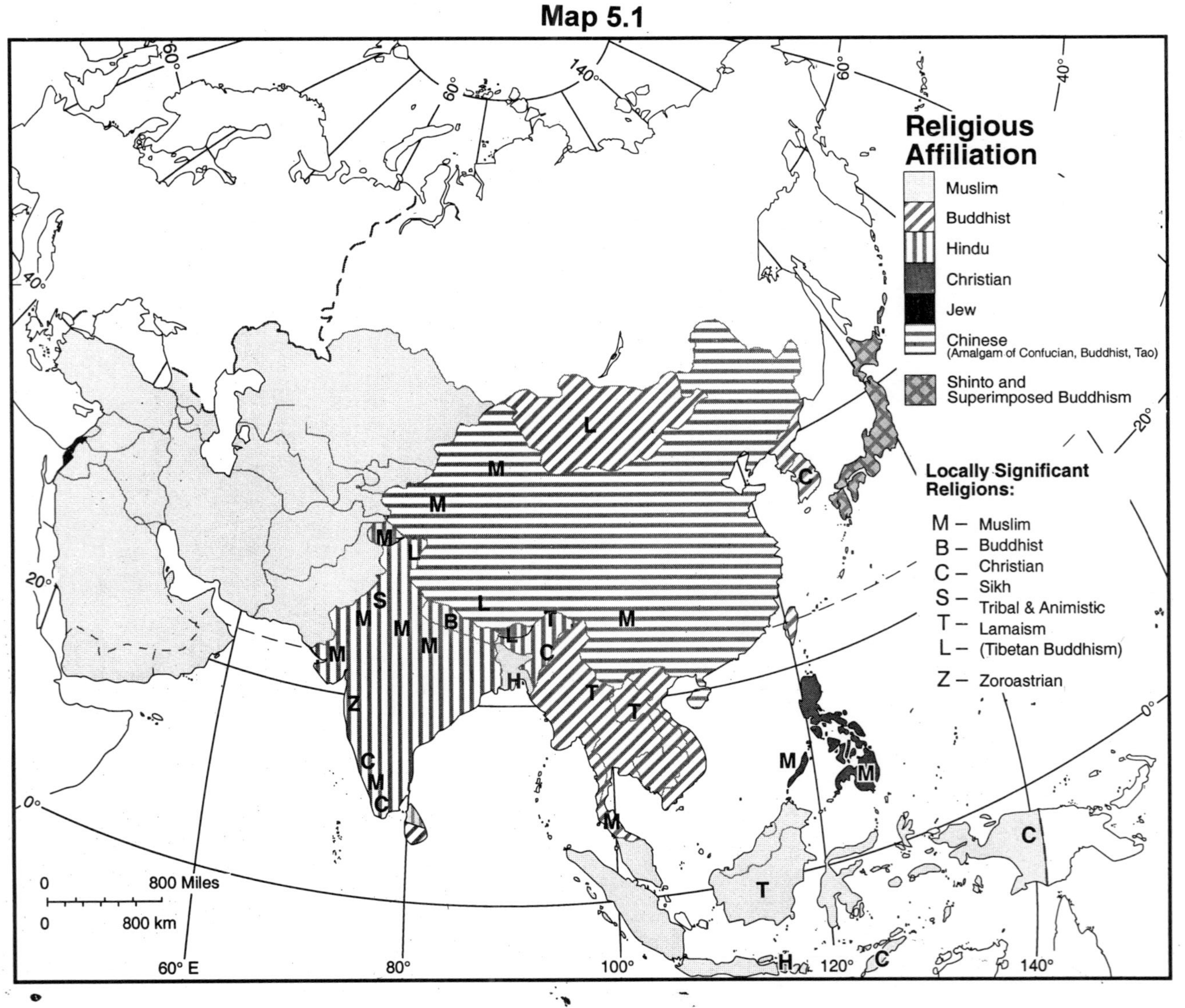
Religious Affiliation
Muslim
Buddhist
Hindu
Christian
Jew
Chinese (Amalgam of Confucian, Buddhist, Tao)
Shinto and Superimposed Buddhism
Locally Significant Religions:
M – Muslim
B – Buddhist
C – Christian
S – Sikh
T – Tribal & Animistic
L – Lamaism
L – (Tibetan Buddhism)
Z – Zoroastrian
0 800 Miles
0 800 km
60° E
80°
100°
120°
140°

HINDUISM

Unique among the great religions of the world, Hinduism has no single founder, but grew over a period of four thousand years into a religious-social system. It is more a polytheistic and ritualistic system comprising numerous cults and sects, and is the oldest, and the most durable of the religions of the East. Native to India, where over 95 percent of the world's 792 million Hindus reside, it reflects in itself the cultural history and social conditions of the Indian society from the time of the Indus Valley civilization (c. 2500-1700 B.C.) to our day. Its role has been pervasive in all aspects of the lives of its followers. It has been a unifying force in Indian culture and society. The term "Hindu" probably derives from the word "Sindhu," the Sanskrit name for the chief river in the northwest of the Indian subcontinent, used by the Muslim rulers and popularized by the Europeans in relatively modern times.

Unlike Islam and Christianity it is non-missionary and remains primarily an ethnic religion. Although 81 percent of India's and nearly 86.5 percent of Nepal's populations profess the faith, Hindus form significant minorities in the neighboring countries of Bhutan, Sri Lanka and Bangladesh. Among other Asian countries, Hindus constitute 7 percent of Malaysia's and 3.3 percent of Singapore's population. The island of Bali in the predominantly Muslim nation of Indonesia is almost entirely Hindu.

The origins of Hinduism are lost in the distant past, and its earliest documentary texts, the Vedas, relate the invasions of northwest India (the present-day Pakistan) and the settling down of the Aryan invaders in northern India. Though the Vedas provide the primary source for insight into Hindu philosophy, and are still looked upon for inspiration and guidance by Hindus, there accumulated a vast body of subsequent religious and philosophical texts within Hinduism. Over the centuries it has grown into a highly complex system, consisting of a loosely organized group of diverse religious expressions incorporating beliefs ranging from polytheism to monotheism to monism to an exalted form of mysticism. It has always been an amorphous religion, without finite creed or theology, enormously assimilative and tolerant. It has permitted the pervasiveness of the Supreme Being, the indestructibility and transmigration of the soul, non-violence, reincarnation, and preordained birth in a particular caste.

More than a religion, Hinduism embraces an entire way of life. The traditional Hindu life is arranged according to the intricate socio-religious pattern known as the caste system that mandates a hierarchical division of society. The system dictates family structure and often family occupation, status and position in life, and sanctions regarding marriage observances and indeed most aspects of life. The castes do not intermarry, and are supposed to observe an elaborate set of prohibitions with regard to interdining and other forms of personal contact. In modern India, the patently undemocratic system has been officially abolished and the system's worst manifestation, untouchability declared illegal. In the cities, and among the educated, the system is gradually eroding, but not really crumbling.

A formal expression of Hinduism on the landscape is manifested in the sacred structures. In India or elsewhere where

Hindus predominate streets, passes, bridges, trees and rivers are dotted with the statues or abstract figures of gods, which constantly receive the propitiation of the passers-by. Hindu architecture is bewilderingly complex and varied. The geography of Hindu holy places is rendered complex by the large pantheon of Hindu gods. Several river sources (Manasarowar Lake in Tibet and Badrinath in India), river confluences (Allahabad, Varanasi), mountain peaks and caves have been sanctified by religion. Large temples have been constructed in several of these locations, which in turn, have favored the growth of cities, like Varanasi, Madurai, and Ujjain in India, Borabudor in Indonesia, and Angkor Wat in Cambodia (the latter two reflect considerable Hindu-Buddhist influences). Many present-day large urban centers in India originated with a predominantly religious function as sacred places of Hindu pilgrimage.

ISLAM

Islam claims perhaps as many as 825 million followers in the world, over 76 percent of which reside in the four Asian Muslim nations of Indonesia, Pakistan, India and Bangladesh. Asia accounts for almost 90 percent of world's entire Muslim population. It is a predominant faith in Southwest Asia where it is practiced by over 95 percent of the population and is a state religion of several nations. Elsewhere in Asia, where Muslims are not in majority, they form significant minorities, with their share to the populations ranging from 2.4 percent in China to 12 percent in India. In India, a predominantly Hindu nation, Muslims number nearly 120 million, making it world's third largest Muslim country. Israel, Myanmar, Nepal, Philippines, Sri Lanka and Thailand contain sizable Muslim minorities.

Islam, the most recent of the great religions, like Christianity and Judaism, is a product of the Middle East. Muhammad, its founder, and prophet died in 632 in Medina (Saudi Arabia), and in less than a few years the Islamic armies were marching out of Arabia to both east and west, beginning campaigns of religious and military conquests, that for sheer speed remain unequaled in world history. In rapid succession they conquered Syria, Iraq, Persia, and Egypt in the west and by 712 reached India's borders in the east. The next few centuries saw the consolidation of gains made by Islam through trade, missionary activity, conquests and establishment of empires. By mid 9th century Muslim traders had reached the shores of China, and penetrated overland and deeper into Central Asia. From Central Asia Islamic conquests to India had started and by the 12th century Muslim kingdoms had been established which they ruled for over 700 years. The most glorious period of Islamic history in India was that under the Mughal rulers in the 16th and 17th centuries. Islam's penetration into Southeast Asia took place between 13th and 15th centuries when Arab and Indian Muslim traders began to dominate the maritime trade with the coastal peoples of Sumatra, Malacca, the Malay peninsula, and many of the islands eastward to Mindanao (Philippines). By the 17th century missionaries had extended Islam's sway to Indonesia, and southern Philippines.

The Islamic conquests have left deep and lasting imprints on the Asian landscapes.

Map 5.2

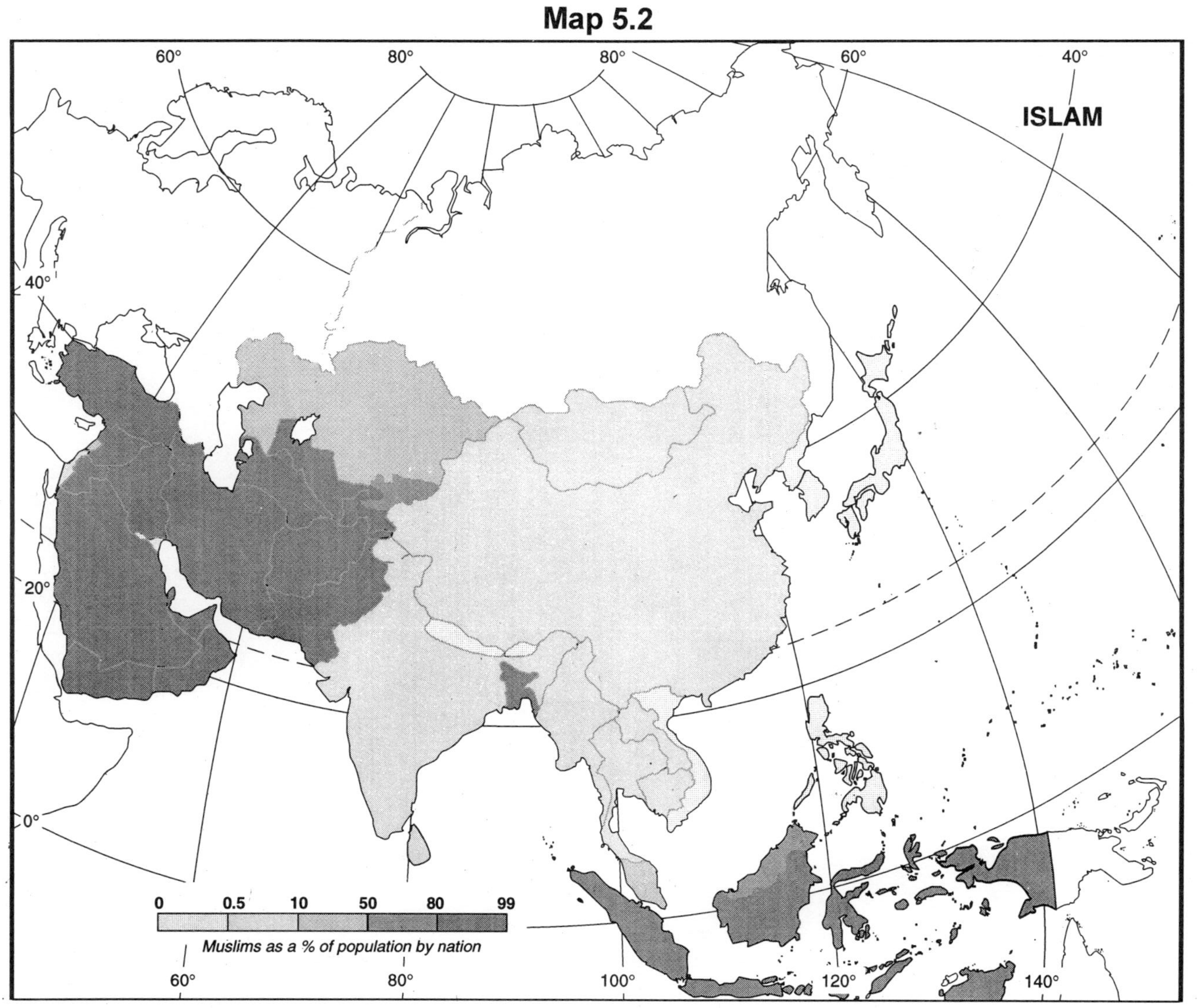
ISLAM
0
0.5
10
50
80
99
Muslims as a % of population by nation
60°
80°
80°
60°
40°
40°
20°
0°
60°
80°
100°
120°
140°

Every city of importance was required to have a mosque and a marketplace. In a typically Islamic tradition cities were generally demarcated into well-defined sections for the nobility containing palaces, gardens, marketplace and residential areas. Mosques, mausoleums, gardens and palaces adorned most imperial cities. The Taj Mahal in India is one of the most magnificent examples of Muslim architecture. Numerous examples of architecturally beautiful buildings can be seen in cities like Constantinople in Turkey, Isfahan and Shiraz in Iran, Lahore in Pakistan and Delhi and Agra in India. The Islamic rules transformed the urban landscapes by investing these with numerous universities and public-work institutions. The Islamic occupation in India laid the foundations for an administrative set-up which became the basis for the British administration later on.

Islam is a monotheistic faith, and one of the most cohesive of the religious orders. Its two major sects, Shia and Sunni have few credal differences. The Shias form only 14 percent of the world's Muslims, and are localized in Iran, neighboring Iraq and Baharain. The major difference lies in the Shiite recognition of the succession to the Califate by the descendants of Muhammad's daughter. The Shiites reject western values and accept the fundamentalist principles of *Shariat* (legal mandates as prescribed by the Koran, Muslim's holy scripture). Their religions leaders (*mullahs*) command enormous authority; because their judgments carry legal sanction of the Islamic law. The 1978-79 revolutionary movement in Iran demonstrated that Shiites have a great emotional and nationalist appeal among their followers.

Despite credal cohesion, the Islamic world contains a tremendous diversity of regional cultures. A good part of Islamic propagation was conducted not by the conquering armies or the Muslim administrations, but by the Sufi missionaries, who compromised with local customs and belief, leaving a great deal of pre-Islamic legacy in every region intact. Thus, among the Central Asian Turks, Shamanistic practices were incorporated into the Islamic fold. In India, in areas distant from the Muslim politico-religious center of power, Muslim population acquire several Hindu customs and practices. In Indonesia, the Hindu epics of the Ramayana and the Mahabharta are held in high position in national life.

The Islamic faith is now making great inroads into the West because of its simple, monotheistic beliefs and by its democratic principles of brotherhood. Historically, however, the Muslim rule was based on the use of strident emotional appeal and the sword. Spencer, the cultural geographer, ascribes the present-day political turmoil in the Muslim nations to Islam's strident and emotional policies. "In time, . . . therefore, the superficial acceptance of the Islamic faith often has become a full and ardent thing, with the result that wherever the Islamic faith is practiced today, latest political turbulence is found in which Moslems never have refrained from militant action when it seemed to promise result" (Spencer, 50).

BUDDHISM

Founded in North India in the 6th century B.C. by Gautama Buddha as a reaction to the prevailing Hindu practices, it became a leading faith of the East covering an extensive area from Central Asia to Sri Lanka and Burma to Korea by the 12th century A.D. Today, the area of its greatest concentration is in Southeast Asia, where the strength of its adherents ranges from 95 percent of the population of Thailand, nearly 90 percent of Myanmar and Cambodia, 67 percent of Laos, 70 percent of Sri Lanka, and over 50 percent that of Vietnam.

Ashoka (c. 275 B.C.), the greatest Buddhist monarch that reigned in India accorded Buddhism the status of a state religion and carried Buddha's message to the various parts of India, and through its cultural and diplomatic missions to distant lands of Sri Lanka, Syria, Egypt, and Central Asia.

Buddhism entered China along the trade routes by the 1st century and became a part of the Chinese society through the activities of exchange of scholars, missionaries and pilgrims from India, where it often heavily interblended with native Taoism. By the 7th century it had spread to southern and eastern Asia, from Sri Lanka to Myanmar, Thailand, Cambodia, Malaya, Indonesia, Vietnam, China, Korea and Japan through a vigorous missionary drive and colonization of the Buddhist monks. Tibet also became Buddhist by the 7th century, from where its distinctive form----Tibetan Lamaism----was carried to Mongolia in the 16th century. Within India, however, the destruction of the Buddhist centers by the Huns in the 6th century and by Muslims by the 11th century led to its virtual extinction and finally its absorption into the Hindu fold. Despite a surge of revival during the 1950s and 1960s in western and central India, Buddhism remains numerically an insignificant faith in India, the land of its origin, where less than 1 percent of the population professes the faith, although it has left an indelible imprint on India's civilization.

Buddhism remains essentially a religion of detachment or non-activity, and of non-violence, but retains the Hindu belief in *Karma* (the inevitability of cause and effect) and transmigration of soul. As it moved to other regions, it acquired native beliefs, and produced in each country distinctive variations of Buddhist faith, and expressions of architecture and arts. In China it was fused with the native ethnic religions such as Confucianism and Taoism, and in Japan with Shintoism to form composite faiths.

Basically, there developed two major schools within Buddhism: the Theravada or Hinayana variety and the Mahayana school. Theravada, a more conservative of the two, and closer to Buddha's teachings, is practiced in Sri Lanka, Myanmar, Thailand, Laos, and Cambodia. It holds that each person must work out one's own salvation, perform religious acts, and chart out one's personal road to Nirvana (escape from the worldly bondage or existence). Theravadins often spend a part of their lives as monks and nuns in order to gain personal merit. Mahayana Buddhism, mainly practiced in Vietnam, China, Korea, and Japan, believes that merit can be acquired from superhuman sources or from the great Buddhas (Buddhist teachers) through contemplation and meditation. In due course, Buddha became deified, and began to be worshiped, architectural expressions of which witnessed the creation of fine sculptures and statues of Buddha. A variation of Buddhism known as Tibetan Buddhism or Lamaism, which is heavily laced with mysticism, magic, metaphysics, and folklore, prevails in Tibet and

Map 5.3

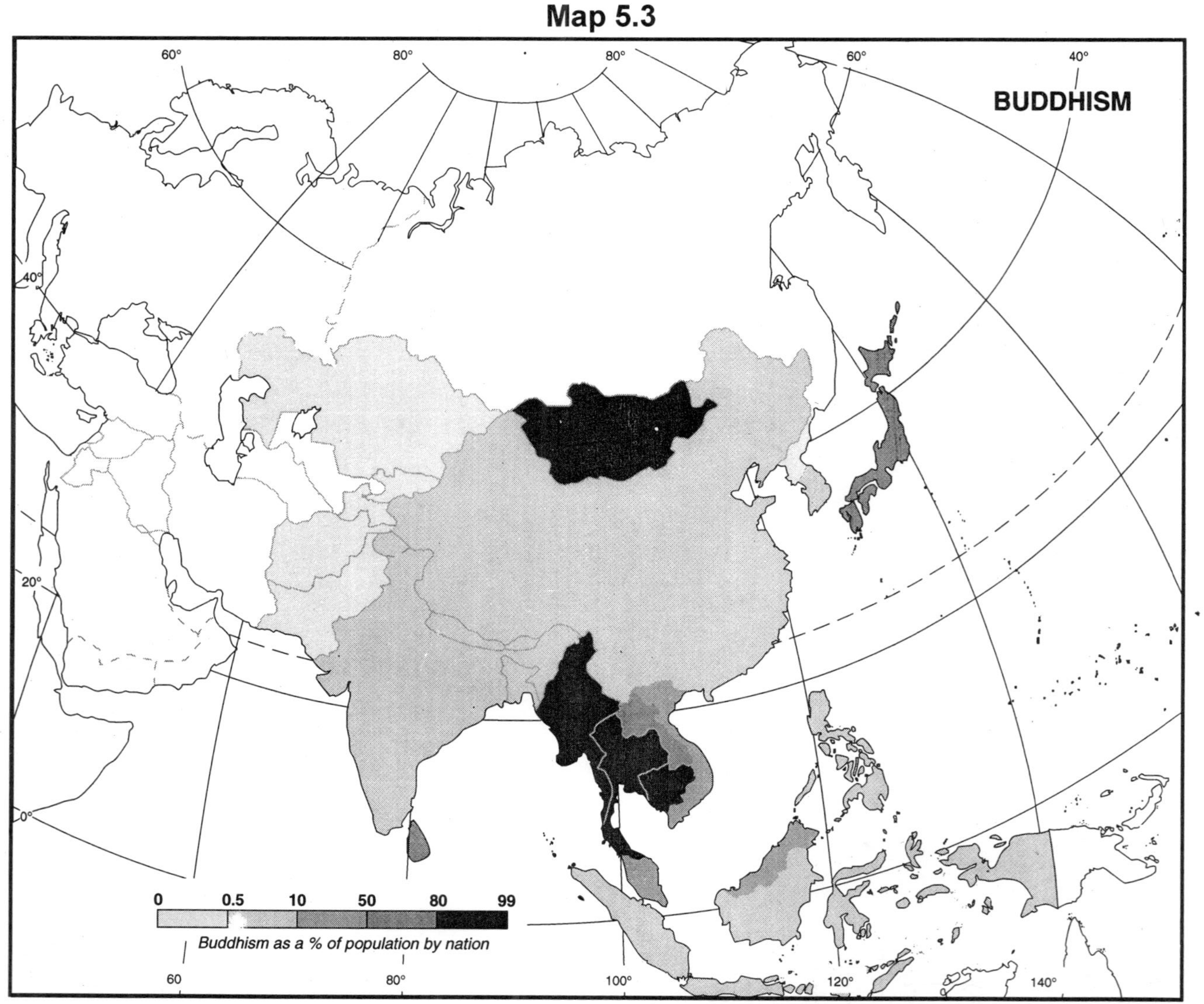
BUDDHISM
0
0.5
10
50
80
99
Buddhism as a % of population by nation
60°
80°
80°
60°
40°
40°
20°
0°
60
80°
100°
120°
140°

Mongolia. After China's occupation of Tibet in 1950, many Tibetan Buddhist monasteries were destroyed and Lamaism was suppressed. Within China, where the state actively discourages the practice of religion, only 6 percent of the population is classified as Buddhist, and 20 percent belonging to Chinese folk-religions, the latter profess Buddhism interblended with Tao and Confucian beliefs. Seventy-two percent declare themselves as atheist or non-religious in China.

CHRISTIANITY

Only one country in Asia, the Philippines, is a predominantly Christian land, where nearly 90 percent of the population professes the Christian faiths (of which 83 percent are Roman Catholics), resulting from a long and strenuous missionary activity during the three hundred years of the Spanish rule. Elsewhere, Christianity had varying degrees of success. Christian missionary activity in general, followed the imperial road of the European colonial conquests of South and Southeast Asia between the 17th century and the WWII. Today Christians account for 37.6 percent of Lebanon's and 26.5 percent of South Korea's populations. In the rest of the continent their share to the population in various countries is much smaller, from locally or regionally important to insignificant levels. The predominantly Muslim nations contain a smaller number of Christians, often less than 5 percent of their populations, with the exception of Indonesia where they account for nearly 10 percent of the population.

From its origins in the Middle East, the faith spread rapidly first throughout nearly all of Europe during the first few centuries A.D. and much later, essentially after the 16th century, to Asia through the persistent missionary activity that followed the European rules. The powerful appeal of a pragmatic and optimistic religion of the colonial masters made inroads into the lower strata of Asian societies, especially when the evangelical preaching was accompanied by the medical missionary, the agricultural experimenter, and the teacher in democracy. The greatest success was achieved in tribal and backward areas where animistic faiths prevailed or in cities where government patronage in jobs was too alluring for the underprivileged castes and classes, or competition with the existing powerful, resistant and time-honored religions was at the minimum.

Despite relatively small numerical gains, Christianity left visible imprints on the Asian landscapes and societies in the form of religious buildings, churches, and cathedrals, parochial schools, and Christian libraries in urban areas. In the Philippines, churches and cathedrals were the focal points around which cities developed, and became the centers of the communities. In numerous other ways, Christianity affected the culture and economy of nearly every Asian country.

SIKHISM

A relatively new religion of India, Sikhism dates to the 15th century. It began essentially as a protest against the corrupt practices in the Hindu society, and as a bridge between Hinduism and Islam. Its founder Guru Nanak preached against the

Map 5.4

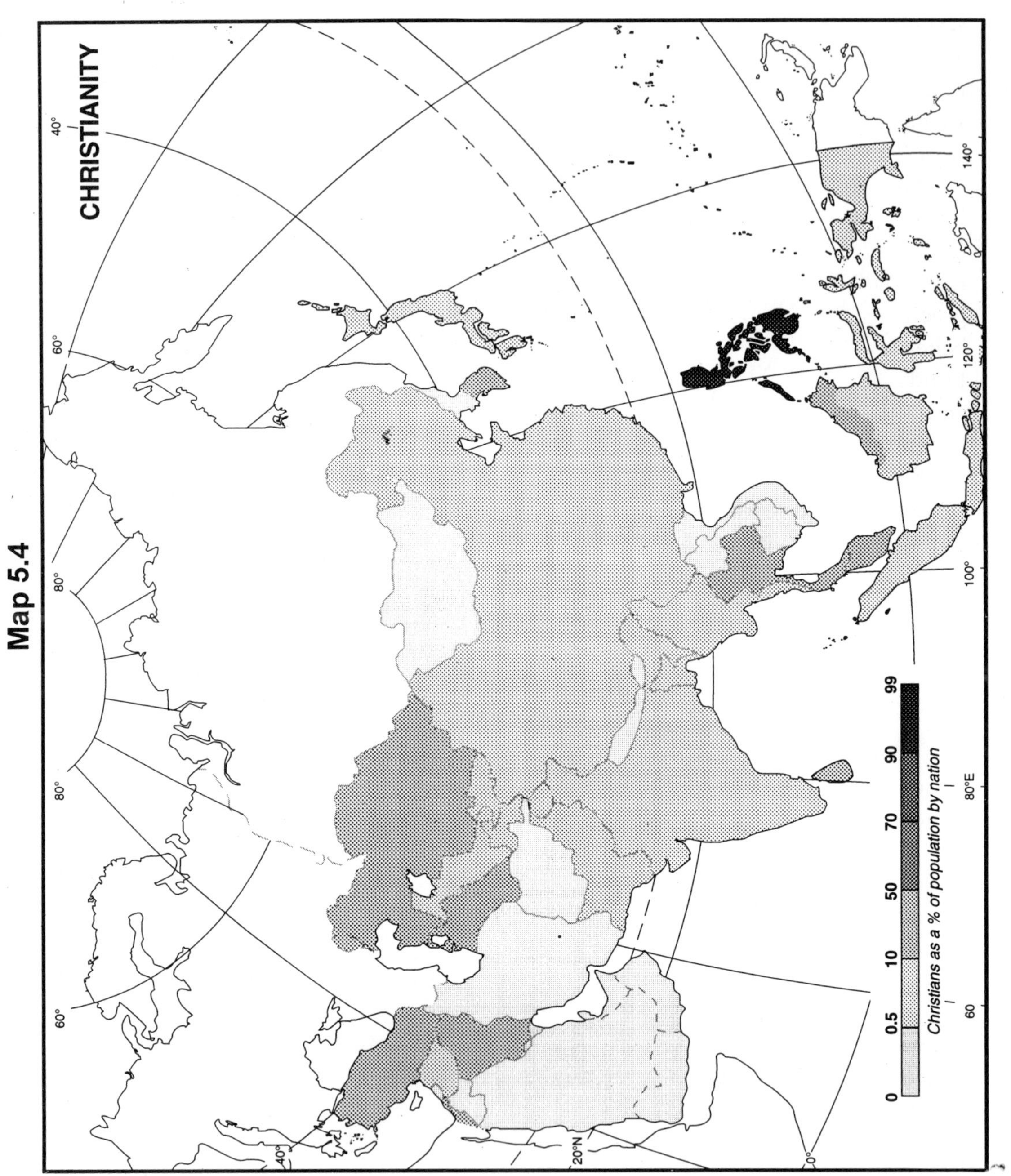
CHRISTIANITY
Christians as a % of population by nation
0
0.5
10
50
70
90
99
40°
60°
80°
80°
60°
40°
20°N
0°
60
80°E
100°
120°
140°

prevailing Hindu ritual practices and the caste system. During the occasional turbulent history of the succeeding centuries several of the Sikh leaders were persecuted and executed by the Muslim rulers. As a reaction, Sikhs were galvanized into a strong, fighting force (*Khalsa*) by their last *guru* (leader) in the 18th century.

Sikhism seemingly appears closer to Islam in its basic tenet, that is, belief in "One God" in the abstract form, but incorporated in its fold several Hindu rites and practices, including the caste system and belief in the transmigration of the soul. Hindus and Sikhs occasionally intermarry and there exists a mutual tolerance between them.

Sikhism is a congregational religion; its shrines contain spaces reserved for community prayers. A large number of their shrines are located in different parts of India (and Pakistan). The central and holiest among these stands at Amritsar known as the Golden Temple, a structure known for its simplicity and beauty.

Sikhs are mostly concentrated in a small section of India, in the state of Punjab; their original homeland borders Pakistan. Nearly 90 percent of all the Sikh population of about 23 million lives in this troubled border area of India, the remaining are concentrated in the metropolitan Delhi, or scattered in large cities. India's partition in 1947 cut directly through the Sikh territory. Nearly 1.5 million Sikhs, who found themselves in Pakistani territory, moved to adjoining India. Since then, they have been among the most active religious groups which successfully agitated in favor of a Sikh-dominated state (Punjab) within India, which was created in 1966 in which they now maintain a majority status. A more radical group among them continued a violent struggle to form an independent Sikh nation causing large-scale civil strife and bloodshed in the border state of Punjab which now appears to be subsided.

THE MINOR CREEDS

Among the numerous other religions which have regional or local importance are: Zoroastrianism and Jainism in India; Judaism in Israel; Shintoism in Japan; and Confucianism and Taoism in China.

Zoroastrianism is an ancient religion that originated in northeast Iran around 6th century B.C. Within a century it had spread over a vast area between Turkey and Iran, but its adherents moved out in 932 A.D. to settle along India's west coast in order to avoid persecution by the Muslims in Iran. There are still some 30,000 of Zoroastrians living in Iran but their numerical focus is now a small area between Mumbai and Surat in India where they are known as the Parsis. India's Parsi population is barely 100,000, but has played significant entrepreneurial roles in India's commerce and business, besides providing outstanding leaders in numerous professions.

Jains are a commercial-minded community of a little over 4 million. They are highly literate and are mostly concentrated in the urban areas in western and central India. Jainism originated in India around the same time as Buddhism and remained a highly ethnic creed, although the Jains intermarry with the Hindus, and absorbed much of the Hindu rituals and religious mores.

Judaism, the world's oldest monotheism and a parent religion of Christianity, is numerically a relatively minor religion in Asia. Despite the overwhelming political and commercial importance of its adherents, less than 5 million of the world's 14.5 million Jews reside in Asia, primarily located in the state of Israel, created specifically for them in 1950. The Jews account for 80.5 percent of Israel's population.

Confucianism, and Taoism, the two major folk-religions, professed by millions in China, lack specific doctrines but consist of codes of ethics. Confucianism emphasized that the true purpose of government was to ensure the welfare and happiness of its citizens. It was essentially a human-centered political-ethical state philosophy. Although like Confucianism, Taoism was a non-militant faith that originated in 6th century B.C. in China, but had a mystical and religious streak. As it evolved it borrowed heavily from Confucianism and Buddhism. It stresses unity of nature and emphasizes reverence and harmony between man and nature. The pragmatic Chinese took to their faiths, blending these with ancestor worship and love of nature. In modern China's politically agnostic or atheistic society, they switched to Communism as a state cult. Nearly three-fourths of China's population now declares itself as non-religious or atheist and only 20 percent professes folk-religions (Taoism or Confucianism).

Both Daoism (Taoism) and Confucianism venerated the earth as the expression of the great forces of nature. Most ancient Chinese cities were thus designed as symbolizing the square shaped earth. All important capitals in Chinese history except Nanjing and Hangzhou represent this Taoist belief in the earth's shape. The cities as well as the dwellings within them were oriented exactly to the cardinal directions so that gates and windows might receive the earth's "magic currents."

Shintoism, Japan's major faith like the Tao and the Confucian faiths of China, is not a religion but a code of ethics and points of view, interwoven with ancestor and emperor worship. Prior to World War II religion was accorded state patronage which demanded unswerving loyalty to the emperor. Since then the state has discarded special recognition to it, and the administration of Shinto shrines were separated from state control. Nearly 40 percent of Japan's population professes the faith.

In addition to the religions described above, numerous localized religious practices are found throughout Asia. Tribal animism, for example, is common in parts of South and Southeast Asia. Mystical Shamanism remains characteristic of North and Central Asian peoples, and shamanistic cults are also found in South Korea and Japan.

COMMUNISM AND RELIGION

After the Communist occupation of China in 1950, the state first declared for religious freedom misleading the people, but gradually the government tightened restrictions against all religious organizations. Initially all religious bodies were asked to further the cause of Communist propaganda, but having failed to obtain results, the government (essentially the Communist party) quickly took stern measures to dismantle all religious bodies. After the control of Tibet, its monastic order was virtually destroyed, causing large-scale social disruption and political upheaval. Dalai Lama, the Tibetan priest-ruler together with thousands of followers fled and took refuge in India. A partial liberalization of policy has recently allowed the opening of Christian churches, and selected temples of other faiths in China.

Privately, many people continued to practice religion within China, as also in the former Soviet republics in Central Asia, where Islam was not only practiced at home, but in public as well, and in the mosques, where conditions were more liberal. In the Central Asian republics of Kazakstan, Kyrgystan, Tajikistan, Turkmenistan and Uzbekistan, religion is not only reasserting itself publicly, but gaining momentum.

Further Readings

Hinnells, J.R. (ed.) *A Handbook of Living Religions.* New York, 1984.

James, W. *The Varieties of Religious Experience.* New York, 1975.

Nigasian, S.A. *World Faiths.* New York, 1990.

Sopher, D.E. *The Geography of Religions.* Englewood Cliffs, N.J., 1967.

6

Rural Settlement Patterns in Asia

There is no standard definition of what constitutes a village or a town or a city, although some basic differences among the settlements of the world can easily be observed that distinguishes the various types. In general, villages or rural settlements perform characteristically agricultural or farming functions, whereas the towns or cities contain non-agricultural service functions. Most nations usually define the settlements on the basis of their size as a cut-off limit for the rural and urban settlements. The cities or urban settlements have invariably a greater concentration of people and a much larger complexity of functions around non-agricultural activities. Thus, size alone is not the only criterion for the determination of distinction between a city and a village. For example, some of the rural settlements in India have populations of over 8,000 people and contain high population densities, and many of the urban settlements (towns) contain much smaller number of people. An important ingredient of urban settlement is the intensity of economic and social life around non-agricultural activities that clearly distinguishes it from the rural settlements. Thus, a farming village containing 5,000 or over people should not be called urban, where as a tourist center of 2,500 people may appropriately be designated as an urban place.

Nucleated and Dispersed Settlements

Given the definitions provided by the Asian governments, most Asians live in villages that accommodate nearly two-thirds of the continent's population. There are probably close to 1.5 million villages in Asia. As these villages assume a variety of shapes and forms, and their size varies from two-three house clusters to villages of several thousand houses, we can conveniently make distinction between two basic types of rural settlements: *dispersed* or isolated, and *nucleated* or compact. The dispersed settlement, in which each farm or a small group of farms is located at some distance from its neighbors, is increasingly becoming uncommon, but is widespread in hunting-gathering and fishing societies. The people live in single family homes in small, dispersed clusters, or are seasonally dispersed in the cropped lands. In some areas shifting cultivation tends to foster dispersion, as a contrast to the compact, permanent clustering of society in villages or towns. Although the dispersed settlements are mostly widespread in the remote and inaccessible hilly areas of mainland Southeast Asia, but these are also found in such areas as the Ganga delta of Bangladesh, and the Kerala coast where there is no scarcity of water for cropping and the farmsteads tend to disperse. A variation of the dispersed type common in Kerala coast of India is the linear or elongated arrangement of residences where residences are generally strung along a street backed by fields of paddy and other crops.

The Sichuan Basin in China and the nearby hilly area is another region of dispersed settlements. The compact villages are less practical in the hilly environment. With the growth of population over the surrounding hills the dispersed pattern has taken root. In northern Honshu and in Hokkaido island of Japan, large farms are possible as compared to the crowded parts of central and southern Japan. Farmers in these areas reside in homes which are close to their neighbors, and dispersed settlements are common. Such is the case in the mountainous Philippines, where the Spanish policy of concentrating the population in nucleated villages was reversed during the period of American control. The highways were improved and travel became easy, and rural settlements were strung along the road in large numbers. As a general rule, dispersed settlements are a common occurrence in the hilly areas in most of Asia as the nucleation of settlements is inhibited by the disposition of physical features, such as slope, soil condition and exposure to sun and wind. In addition, the growing pressure of population and farmlands on lower ground and the easy access of travel has made the neglected higher area targets of movement by farmers who are constantly moving and settling on higher terraces.

Much more widespread than the dispersed settlements is the pattern of compact or nucleated settlements, with populations ranging from a hundred or less to several thousands across most of Asia. Large groups of residences are usually nucleated to form villages containing virtually their entire populations. The nucleated or compact farm villages are a very characteristic feature of the rural scene in nearly all Asian countries. But the sheer compactness of the villages in the Indus-Ganga plains of northern India and Pakistan, and the North China Plain is particularly striking. The residences are placed so close to one another that there is a virtual absence of space between them. The Westerners are often impressed by the cheek-by-jowl huddling of residences, lack of gardens, streets, and other types of "living space" in such villages.

A combination of several factors has undoubtedly contributed to the nucleation of rural settlements. In areas located in dry environments and uniform relief imposed conservation of water for efficient distribution. Although uniformity of relief, soil-fertility, and depth of watertable were all conducive to the origins of nucleated settlements, grouping together of settlements also provided psychological sense of security during long centuries when lawless and dangerous conditions prevailed, especially for the weak and isolated although farm villages had little defense against large, well-equipped invaders, but could put a good front against marauding bands of casual banditry, common down to recent times.

In a different setting, where conditions of soil-fertility, or watertable, or flat topography were the prime considerations, defense sites were available on high ground such as a hilltop, or an island, or any physical location that would face invaders to fight at a disadvantage. Structures, such as a castle, fortified settlements surrounded by a moat, naturally supplemented the physical advantages of the site, where a degree of defense could be assured. Thus, numerous Asian villages (and cities) as in other parts of the world, are located around fortified hills or islands, in addition to the innumerable villages that dot the countryside.

Although the role that common protection and defense played in explaining the location of compact villages is obvious, the nucleated villages have been historically common among the Indians, Chinese and Japanese peoples, as an evolving cultural practice among them. These people had developed closely-knit social systems since the dawn of history that persist till the present day. Many of their villages are family villages, often bearing the same surname, and tracing their descent from a single ancestor. Each family has its own lands, possess certain rights and responsibilities connected with the upkeep of the ancestral land and temple. The village is thus a physical and cultural feature. Houses are close together, similar to those occupied by ancestors. The village is a compact entity, where there are a few shops, and trading is done at a nearby market town. In Japan, for example, the village is a social unit. It generally has a few tiny shops, each specializing in a single class of goods such as fish, tobacco, cakes, cloth that occupy the front rooms of a few residences. In the Indian subcontinent, on the other hand, there may be a few retail stores in the village that sell selected day-to-day items but may not specialize in a single item. In Southeast Asia, the habit of some groups of commercial living in a structure that house dozens of families, is an integral part of their social system. The village is thus a group of closely-knit dwelling units. In many regions of Asia, walled villages are also common that reflect the protective need of earlier times, but during the good policing provided during the European colonial and subsequent, native administrations, the protective walls have tended to disappear, without substantially altering the basic compactness of the settlement.

ASIAN VILLAGES: FUNCTIONS AND FORMS

The primary function of a typical Asian village is to provide housing to the local population, its household animals, crops, and agricultural tools. The structures are simple and utilitarian, built of mud, or bamboos, that serve the basic needs of shelter, storage, and privacy. Most villages generally lack the amenities, related to drinking water, waste disposal, street cleaning, fly and insect control, lighting, and entertainment, although many of these modern conveniences are slowly creeping into the village life. (See description of Sugao village in Central India given in this chapter. "Sugao" literally means "good village." Sugao is one of the several villages in transition across Asia.)

An overwhelming, but not the sole characteristic of an Asian village, is a virtual lack of secondary functions. Only a few of these have acquired service activities in recent times, especially if a railroad or a paved road has been built that links these villages with a larger market. The residential village, in general, does not contain stores, shops, fabricating plants, or government buildings. A large number of villages in India until a few years ago had no electricity or clean, drinking water.

Most villages in the continent are small. Probably there are more villages of under 50 houses than any other size category, but it is equally likely that villages of 300-500 houses contain most of Asia's rural population. Only a few of the villages in India and Pakistan have more than 5,000 inhabitants. Generally, small or large residences are densely packed in nucleated villages. High population densities per unit of land are quite characteristic of a nucleated village.

Asian villages are arranged in a variety of forms and shapes. The compact villages have generally no particular shape or pattern and may be labeled as "the cluster village." Most villages of northern China, northern and northwestern India, and southwestern Asia belong to this category. They are closely-built with a maze of alleys and access points, although the villages in China contain somewhat less tightly-packed residences. Villages in southern China, though cluster-shaped, are concentrated at river-fork, a bridge-head, or any other available space, revealing their hodgepodge arrangement. Most villages in India and in southwest Asia are merely clusters of houses arranged in a disorderly fashion. Tibetan villages are frequently elongated in cramped sites.

Elongated or string-like clusters are other common shapes of Asian villages. Such is the case with villages along a seashore, a riverbank, a natural levee or terrace, a paved road, or some other linear site that induces the arrangement of houses in a single file or a double-file. Twenty or thirty houses may be grouped in a double row in less than 100 yards, or sometimes as many as a thousand houses may extend out over a larger space but linearly arranged. The modern highway may also attract a large number of linearly-shaped pattern.

The *grid-pattern* within the nucleated villages is common in most parts of Asia, although it is characteristic of northwestern India, which probably is the origin of this form. It is also widespread in China and parts of Southern Japan.

There are also many other shapes of Asian villages that do not fit into the patterns outlined above. A traditional Malay village has a loose-straggly shape, built along the lower section of a stream. Lack of continuity, and scattered dispersion of the houses in garden patches make these neither a string nor a cluster village. A circular string of houses around a cultivable upper portion of a volcano is a characteristic shape of several Javanese villages, but mountain-ridge or mountain-top villages of most Southeastern Asian peoples are generally disorderly, adhering to no particular pattern. Some societies in Southeast Asia use a communal long-house, where all the people of the village live in one large, subdivided structure.

Generalizations regarding the precise siting of villages are difficult to make. They are located everywhere, on a variety of sites. In the floodplains, they are obviously located on natural levees, dike lines, raised platforms, or mounds created gradually by discarded rubbish during a millennium of human occupance. In Java protected mountain sites were generally preferred before the Dutch occupation. The Dutch, however, required the villages to be located near the main roads in more open sites.

A COMPACT VILLAGE IN TRANSITION

Sugao, a village of nearly 2,500 residents, is located in the State of Maharashtra, nearly 250 km (155 miles) southeast of Mumbai, in India. It is a village in transition, a microcosm of many such villages where economy is being increasingly interlinked

Map 6.1

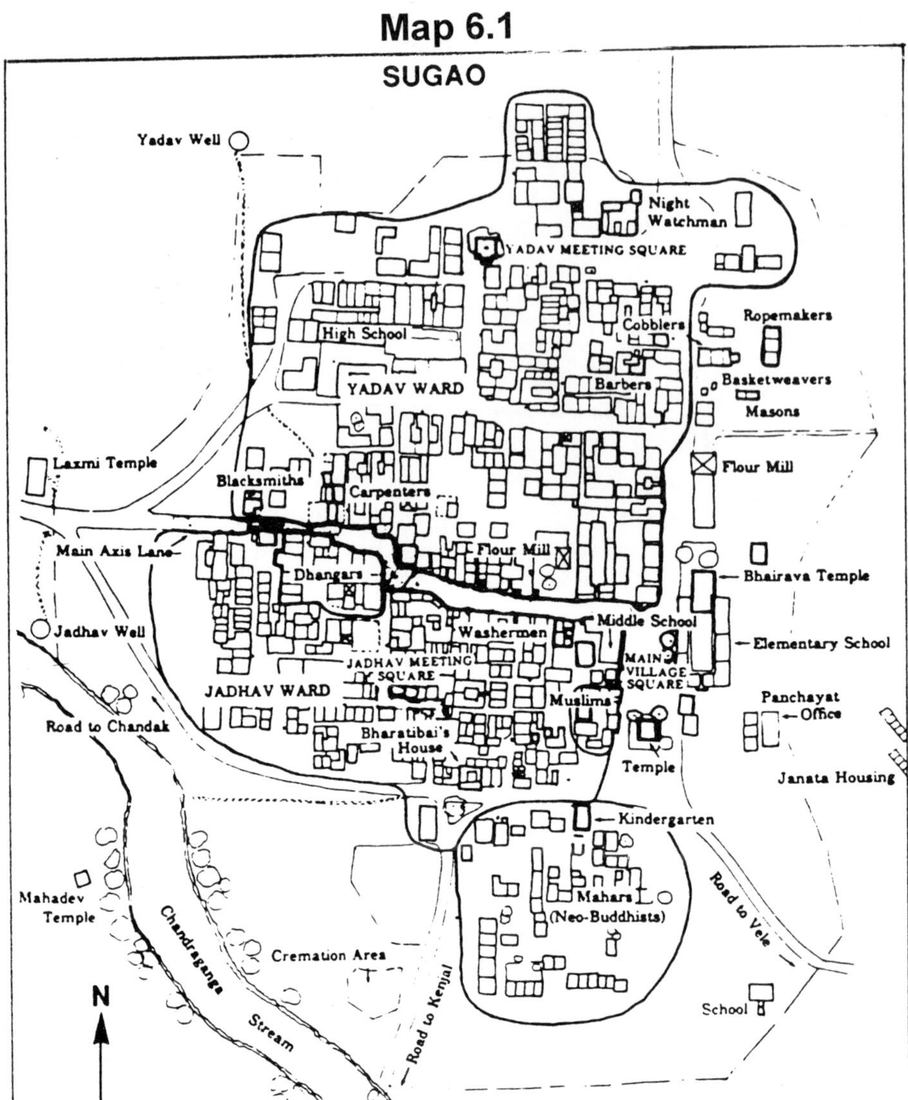

Reprinted with permission.

with the region around them. Population has outgrown the carrying capacity of land that provide cheap labor to the cities' industries. In return, villages receive remittances in cash and kind from the cities.

Sugao is a nucleated village. A railroad passes within a distance of 5 km of a larger, neighboring village of Wai that provides its most shopping needs by holding a weekly market on Mondays for the surrounding area. Electricity arrived in Wai in 1964, and in Sugao in 1968. Most residents of the two villages now make use of radio and television sets.

The village can be divided roughly into two unequal parts by a crooked, east-west axial lane of unequal width with Yadav households occupying most of the northern section, and Jadhav groups predominating in the southern section. These two groups belong to the upper caste Maratha families, and own most of the village lands and farmland surrounding the village settlement. Other groups----rope makers, carpenters, masons, basket weavers, cobblers, blacksmiths---occupying the village periphery, while the untouchables, mostly the Mahars, live outside the village. There are a few Muslim families that reside to the south of main square of the village. The village contains several temples, a few schools, a flour mill, and an office for the village elders (the Panchayat).

Nearly three-fourths of the households live in 1 to 2-room houses, the remaining in 2 or more-room houses. Households with 1 to 2-room houses contain 1 to 14 persons. The size of households accommodating 2 or more rooms range from 1 to 18 persons.

The village lands completely surround the settlement for several kilometers on all sides. The two upper castes own most of the land. Nearly one-third of the land is owned by big farmers. The availability of modern medicine and the introduction of electricity during the 1950s and 60s could only be afforded by the well-to-do upper castes. The lot of the lower castes has actually deteriorated as the new industries displaced the traditional artisans such as weavers, cobblers, carpenters, etc., who cannot afford the new cash-crop economy. Moreover, the village electrification has displaced village handicrafts to give way to machine-made goods. Thus, a large surplus labor of artisans had become unemployed since the late 1950s, who began to migrate to the cities, particularly to Mumbai, which could absorb cheap labor from the village. Usually men began to migrate about the age of 17, while most of women stayed at home. These migrants generally would return to the village after working for 15-20 years in Mumbai.

The consequences of the circular movement of migrant returnees have, however, not been entirely beneficial. But those who moved to Mumbai or other cities had a hard time surviving even in Mumbai and other cities as their earnings of the migrant laborers are consumed by both the earners in the city as well as by their families left behind in the village, who have to depend upon the remittances of the migrants to the cities.

Meanwhile, city jobs have become increasingly scarce with large influx of villagers into the cities. Simultaneously the situation in the villages has increasingly worsened as the unattended village lands became over-grazed and soil erosion became a serious problem. Moreover, the returnees to the village from the cities have been usually disinclined to work on lands when they return home. A vicious circle of village neglect has thus set in. The village of Sugao exemplifies the conditions of such villages that are in transition in most of South Asia (Dandekar, 1986).

Further Readings

Dandekar, H.L. *Men to Bombay, Women at Home.* Ann Arbor, 1986.

Johnson, E.A.J. *The Organization of Space in Developing Countries.* Cambridge, 1970.

United Nations Department of Economic and Social Affairs, *Global View of Human Settlements.* New York. 1976.

7

Urban Settlements and the Urbanization of Asia

In the seemingly long space of human history urbanization is a comparatively recent phenomenon, although we are now living in a rapidly urbanizing world. In fact, the city has become a dominant factor of modern civilization. But during most of human history, from the hunting-gathering stage of perhaps 6,000 years ago to the beginning of the 19th century, a very small percentage of world's inhabitants lived in settlements which may be defined as urban places. It is estimated that less than 3 percent of the world's population lived in urban places of 5,000 persons or more around the year 1800 or nearly 200 years ago. By 1900 the proportion had risen to over 13 percent. Currently the percentage is close to 45. The projections for the beginning of the 21st century point out to a world in which more than a half of its population will be living in urban places.

A distinction between villages and urban settlements has been made earlier. A town or a city undoubtedly is a more complex settlement than the village, performing several types of non-agricultural or secondary functions, like marketing, warehousing, processing, wholesaling and retail merchandising of goods brought from the farming communities. Such activities tend to increase the size of urban settlements. Within the towns specific areas are usually given to specialized economic functions, for example several Chinese towns have a "cotton street," or "copperbeaters" alley and Indian towns have a "silversmith's bazaar," although a single main street may usually possess such functions in branching alleys, while the major street contains most of the retail merchandising stores. Far more than American towns, the Asian towns may combine economic activities and residence in the same building and shopkeepers may live over, and behind their stores.

There is usually no clear demarcation between the town and the city except for the size and complexity of functions. The size consideration also varies with nations. Several nations do not officially employ the term "town" at all. In India places to qualify for an urban status (town or city) is that it must contain 5,000 or more inhabitants, except in certain cases, and three-fourths of its labor-force is engaged in non-agricultural activities. Such urban places that contain more than 100,000 people are designated "cities" in many nations, although the precise definitions of a city may vary from country to country.

The city is generally a much more complex affair than the town. Not only it accommodates a much larger population, it performs a complex range of economic functions---from warehousing, and retail services, to production of goods, banking, trading, recreational, and a variety of commercial, administrative, and other services. Its settlement pattern is more comprehensive, and in some cases more complex. Asian cities may not clearly possess the functional zonation of American towns, although the range of economic activities may be quite large. Population densities tend to be much higher, particularly in their old section. The "new" or Westernized section tends to possess a greater clarity of functions, is more modern in appearance, and may contain "high rise" structures, open spaces and well-regulated traffic avenues. During the colonial period of their history, Asian cities acquired some of the European-style commercial, residential and administrative functions, what may be called the "new" sections, while the "old" sections of the city present a striking contrast to these "newer" Westernized sections.

The Indigenous "Old" Sections The older, indigenous sections of the cities in Asia, known generally as the "old city" (in China known as the "Inner City"), are packed into a small place, usually a few kilometers in circumference, but house a majority of the city's population. Except for East Asian cities of Japan and China, the street pattern of Asian cities is generally confused and irregular, containing narrow streets, made narrower by protruding open shop-fronts on the ground floor. The "old" city contains mostly one or two-story brick structures, and is generally surrounded by an outer moat, broken by walks and a few gates. The street pattern reflects the original defense consideration of containing residences, administrative offices, and religious shrines within the wall.

Though essentially residential in character, the "old city" shares its limited space with commercial and manufacturing uses. Thus a jumble of confusing streets, cul-de-sacs, alleys, and byways gives access to residences, to commercial uses, and to smaller manufacturing units which often encroach on the public right of way. Outside the wall the "old city," pattern is more regular. This pattern is repeated in countless towns and cities of Southern Asia and Southwest Asia. Old Delhi (Shahjahanabad of the 17th century), Old Jerusalem, Saana (Yemen), old sections of Istanbul, the older walled sections of Tehran, Damashq (Damascus) are some notable examples of such a pattern. The older sections of the Chinese and Japanese cities, and cities in Southeast Asia (and to a lesser extent in Southern India) are, however, laid out on a more regular and clearly-defined gridiron pattern. Thus, the older sections of Tokyo, Beijing, and Jakarta show a coherent, and mostly symmetrical street patterns. There is lesser intermixing of functions, the crowding is less intense than in the cities of Southern Asia and Southwest Asia and the streets are generally wider. Although there is a combination of commercial shop-front story with residences above them on some streets, there is generally a separation of the commercial and the residential zones in cities of East Asia. The residences are made of wood in Japan and of brick and wood in China, and bamboo and wood in Indonesia.

An important element in the design of medieval cities that has much to determine the character of modern cities, was the surrounding wall or a series of walls as they repeatedly outgrew their "shell," built primarily for defense. Perhaps some port cities also possessed walls, on rivers or on the sea. Gates, bastions, battlements and encircling moats were a part of the wall-complex. Within the surrounding wall, the more affluent classes first built adobe residences or kiln-burnt brick or stone residences though the use of burnt brick is older in India than in China. In parts of Southeast Asia wood or bamboo palisades were common until the arrival of the Europeans. Most old towns still retain their walls, though most have crumbled and breached in modern times to permit modern transport to pass through. Generally speaking, most old cities now contain several newly-built suburbs outside the walls. Although colonial towns were founded in the 18th and 19th centuries by the Europeans and had walls built around them, some had walls constructed around the port and garrison cities. Fort St. George of Madras (now Chennai), the Intramunos at Manila are examples of this type. Although the concept of walled cities finally succumbed to the realities of modern times, its legacies remain. The cities have now expanded their earlier confines and gradually stretched out to newer suburbs to include European Westernized extensions.

Population densities in the "old" section of Asian cities have, over the years become very high. Parts of Kolkata, and Tokyo contain nearly half a million persons per sq mile. The original landuses have often yielded to newer functions over the ever-increasing human pressure. The dwelling units are often converted into stores or small factories, or subdivided to house more families. Thus, the high density of residences has been compounded by a mixture of landuses: residential, commercial and industrial. In general, however, the residential areas of cities lack in shopping facilities, open spaces, and such amenities as adequate water supply and sewage.

Within the cities communities are usually segregated by subdivisions or neighborhoods (known in South Asia as *mohallas*) by caste, religion, occupation and ethnicity, resulting in an essential lack of social cohesion. While this is particularly true in Southern Asia, most cities of Southeast Asia also contain neighborhoods where the Chinese are concentrated in the "Chinatowns". Indians, Japanese, Chinese, Persians, Filipinos, Turks, and Arabs usually segregate themselves in their exclusive enclaves within the Asian cities.

The hub of the retail merchandise is always the centrally-located *bazaar*, the *souk* in Southwest Asia or the *chauk* in India and Pakistan. It is crowded with countless small retail stores dealing in food, cloth, hardware, jewelry, and other consumer goods. Bankers, moneylenders, health practitioners, dentists, public letter writers all congregate here. Upper rooms of buildings are used as dwelling units. In larger cities, specialized services and goods tend to be concentrated in specific areas around the main bazaar, or in the side alleys and develop into shopping centers or small bazaars such as cloth merchants' shopping area or a grain exchange, a street of brassware stores, or a gold and silversmiths' alley or a pottery makers' shops. In these small bazaars craftsmen perform their work and display their wares in a open-front shop. There are also the ubiquitous cart peddlers, and sidewalk hawkers who do their work (ranging from corn roasting to fixing eye glasses and tooth extracting) in the customers' presence, giving them a chance to supervise the activity. In most bazaars, particularly in China, and India indigenous-style restaurants display their foods and do the cooking next to the street passers-by who may make choice not generally available to Americans or Europeans in their hometowns. An Asian bazaar offers quite a different look than the retail and merchandising section of an American market town. The colorful, if chaotic intermixture of transport, commercial, and residential functions, the intensity of human drama is a fascinating contrast to the spacious, well-laid out pattern of an American town.

The density of streets and the functional mix in the old streets are further compounded by the diversity of transportation in the larger towns and cities, though this may not be the same in smaller towns. Pedestrian traffic in larger cities is generally dense, although it fluctuates at different times of the day. Modes of transportation are many and varied----human, animal and vehicular. Between the sluggish donkey and the rapid commuter train (as in Tokyo, Mumbai, and other metropolises on the continent) are: tonga, pedicab, motorcycle rickshaw, bicycle, ox cart, taxi, private automobile, hand cart, trucks of several types, bus, and street car as the important modes of transportation of men and goods. The chaotic mixture of lan-duses produced by such diverse modes of transportation creates a melange of facilities, sharing the right-of-way in generally an uncontrolled manner in the old sections of the cities, especially in the cities of South Asia and Southwest Asia. This is generally not the case in the newer sections of the cities. In Chinese cities there is an overwhelming concentration of bicycles, although pedicabs and automobiles are becoming increasingly common in most of Asia's larger towns and cities. In the cities of Central Asian Republics, where cities are wider and better laid-out, older sections have mostly disappeared, traffic flows are smoother and more regulated. In general, however, the scene of old parts of Asian cities is confusing and chaotic.

Modern Westernized Sections In contrast to the hodgepodge settlement patterns of the "old city" are the modern, well laid-out sections of the Asian cities. These were built during the 20th century to house the colonial rulers, the local civil administrators and the growing elite. Such sections are usually set apart from the old city as in Jakarta (Batavia of colonial days) and New Delhi, and contain such settlements as civil lines, railway colonies, military or garrison towns in South Asia, and other newer colonies. These sections present an aspect of comparative spaciousness, comfort, greenery, peace and functional zonation. The tree-lined buildings are broad and surfaced, and normally follow a gridiron pattern. Buildings are usually constructed of bricks and are frequently surrounded by large fences and landscaped lawns. House lots are distinctly marked and land use zonation is generally enforced.

Modernization has been slowly changing the urban landscapes everywhere in Asia, although much of the transformation has occurred during the last half a century. Modern developments in civil and military administration, transport, agricultural processing, mineral exploitation, and petroleum refining have created new settlements that have acquired the status of towns. These activities gather around them service functions and small populations.

This is particularly true of railroad centers where service functions tend to congregate. These modern settlements are located outside of , but close to the old city which have no room for these proliferating, specialized functions within the old, crowded parts of the cities. In India and China some of the modern highways have increasingly penetrated the old walled cities, and railway stations are normally located nearby but outside the old cities. In the oil-rich nations of the Persian Gulf the modern elements of planned, well laid-out street patterns have taken place during the last four-five decades, due largely to the settlements of the American and European workers in oil refineries and the modernization schemes adopted by the native rulers. These new, modern patterns are particularly noticeable in cities like Kuwait, Dhahran, Doha, Dubai, Riyadh, Oatar, and Bahrain. Much of modernization has occurred outside the "old cities."

An important facet of modernization in Asian cities has been the growth of high-rise buildings during the last half a century, which are usually located in the "new" sections of the cities. A few notable examples of cities containing high-rise buildings used mainly as apartments and offices are: Tokyo, and Osaka, Beijing, Shanghai, Jakarta, Bangkok, Mumbai, New Delhi, Tehran, Singapore, and Hong Kong, although most other larger cities also contain areas of high-rise buildings. In Beijing, and other larger cities in China, and in Central Asian Republics most civilian employees (and factory workers) are housed mostly in high-rise buildings. In Hong Kong administration had to build several high-rise housing colonies to accommodate the influx of refugees from China mainland during the 1950s and 1960s on limited space that was available in the British colony.

A noticeable feature of several cities has been the mushrooming growth of shantytowns (*bustees* or *jhuggies* in India and Pakistan) during the last forty years. These are composed of cells of makeshift hovels which house thousands of squatters created by the large and continuing stream of migrants from the countryside. Unable to afford housing in the "old city," the migrants spread out in shantytowns, which are usually strung along the edges of the built up areas, but also in the municipal areas of the city. These resident squatters (illegal occupants) generally choose vacant land of the city, such as parks, abandoned sites, and railroad plazas and sidings. Squatters' housing usually consists of galvanized iron-roofed tenements, or one-roomed structures of mud walls. These appear in clusters of a few to several thousand units each housing one to several families, thereby producing very high density areas with few public amenities. Shantytowns are more common in southern, and Southeast Asian cities. They have added to the existing urban problems of congestion, transportation, public services, and proper functional zonation, and often impede official efforts toward rational urban planning. In Jakarta and Tehran the administrations have been trying to accommodate the inhabitants of shanty towns and the homeless in government housing but have not met with much success. Municipal authorities have attempted to contain or eliminate the shantytowns by dismantling these in India and Pakistan, only to discover their reappearance soon after.

URBANIZATION PROCESS AND TRENDS

There appears to be a general agreement that there has been a continuous history of urban development since 4000 B.C. when early cities probably started in Mesopotamia (present day Iraq), although individual, incipient cities existed at Jericho perhaps two thousand years prior to the Mesopotamian cities of Babylon, Akkad, Ur, and Nineveh, and in Egypt. Numerous excavations in the western parts of Indian subcontinent particularly in the lower Indus Basin of Pakistan indicate that a flourishing and widespread urbanization existed as early as 2500 B.C. which had perhaps connections with the Mesopotamian cities. Some of these cities like Mohenjodaro had a population of over 40,000, and the earlier Mesopotamian cities contained 50,000 to 70,000 persons, and were laid-out on a grid system like the present day American cities.

Our records of city development during the ancient times are fragmentary and discontinuous. Despite Asia's current and historic low urbanization levels, the continent holds the distinction of containing some of the oldest and largest cities in the world, such as Beijing, Delhi and Jerusalem. Several splendid and great cities such as Varanasi, Patliputra (India), Beijing (China), Jerusalem, and Baghdad (Iraq) flourished long before the Christian era and survive until the present day. A contemporary Chinese traveler reported that Patliputra in North India had a population of over half a million in the 4th century B.C. The process of urbanization was perhaps accelerated during the medieval times with the paving of roads, and ease of sea travel throughout most of China, India, Japan and the Middle East, and some of the truly magnificent cities were established such as Istanbul, Baghdad, Tehran, Makkah, Jerusalem, Damashq in the Middle East, and Lahore, Delhi, Allahabad, Agra, and Ujjain in the Indian subcontinent.

Several cities continued to flourish through the centuries to the medieval times. Changan, the capital of the Tang dynasty in China had probably a population of 800,000 in 750 A.D. Baghdad had become the world's largest city during the 9th and 10th centuries and claimed over one million inhabitants at that time. Kaifeng, the capital of most of China during the 13th century, was another large city which had a population of nearly half a million when Marco Polo, the Italian traveler visited China in the 13th century. Beijing in the 15th century probably had a population of over half a million, and was the largest city in the world for several centuries from 16th to 18th century, while Nanking and Hanchow were other great cities. By the late 18th century Yedo in Japan precursor of the present-day Tokyo, had emerged as the largest city in the world. Both Tokyo and Beijing remained world's leading cities until the present day.

Over the centuries several types of towns, particularly administrative, and religious centers increased in numbers, as the total population of Asia's nations and regions grew, although many of the ancient past towns disappeared, with the setting up of harbors, and their changed political fortunes.

The Asian cities are generally laid-out on some form of grid-pattern, prescribed by the town builders long time ago. Most of the street designers of early towns in India adhered to the regulations advocated by *Shilpashastra* (an ancient architecture treatise) and Chinese towns adopted the

Table 7.1 *Asia—Urban Growth City, Size (estimate)*

Date	*50,000*	*100,000*	*500,000*	*1,000,000*	*Largest City*
800	15	2	---	---	Baghdad (700,000)
1000	25	6	---	---	Kaifeng (China) (400,000)
1200	22	6	---	---	Hang Chow (China) (255,000)
1300	23	6	---	---	Hang Chow (China) (432,000)
1400	29	1	5	---	Nanking (China) (487,000)
1500	25	4	2	---	Peking (China) (672,000)
1600	44	4	2	---	Peking (China) (706,000)
1700	40	4	2	---	Yedo (Japan) (688,800)
1800	56	3	2	1	Peking (1,100,000)
1900	*	*	3	3	Tokyo (Japan) (1,400,000)
1950	*	*	*	6	Tokyo (7,000,000)

Source: Chandler and Fox: *3000 Years of Urban Growth*
* Numbers not calculated

rectangular grid pattern of which Beijing is a fine example. Although these basic patterns preset, at least in the Indian subcontinent, later developments during the medieval times, resulting from the accretion of intertwining streets, crowded and confused the initial patterns. The design of the towns in parts of southeast was simpler, as their growth occurred mostly during the period of European contact, although the Europeans did not exercise much control over the actual pattern of settlement except for the plantations and newer sections of the towns.

Initially, the towns were normally built on spacious sites required for the grid pattern, modified on irregular topography. The river basins thus became the most suitable sites for the towns to be located, and for them to grow, and establish commercial links on land thorough water-borne transportation. Waterways were the basic source of water supply, and initially the only means of transport access.

As most towns and cities had originated as settlement not primarily engaged in agricultural activities, the main function of the early towns was that of marketing. Beyond the purely commercial functions they were also transport terminals or way stations for both land and water transport, mining settlements, military garrisons. In due course, they acquired administrative functions. Some became places of religious pilgrimages, hill resorts, and manufacturing towns. With the passage of time, however, cities were becoming multifunctional, and began to acquire increasingly a greater functional zonation in their internal morphology. For example, grain ships clustered together or dealers in construction materials often occupied a particular section of the city, transport operators and their equipment were usually located on the city's outskirts whereas retailers of textile, jewelry, food, and other consumer items were normally found in the main bazaar of the old sections of the city. Such functional zonation established within the cities during the medieval and later the colonial times persisted to the present day. The Chinese and Japanese cities have more formal, and intricate pattern of zoning, just as most cities in Southwest and South Asia have a confusing street design and absence of formal controls, essentially a carryover of the medieval times. The newer sections of the cities, particularly set up during the colonial period, reflect the more formal and regulated zoning controls.

The rapid growth of market towns during the colonial history of several Asian nations was undoubtedly facilitated by the paving of roads and the introduction of railroads when the Industrial Revolution arrived. Urbanization process was thus pushed forward by the concentration of markets, labor and capital. The most spectacular growth of urbanization occurred after World War I, and since the World War II, as the handicraft industries of the countryside withered and the impoverished artisans and agricultural labor of the countryside flocked to the developing cities.

THE CURRENT STATUS OF ASIA'S URBANIZATION

Before we attempt to understand the patterns of urbanization in Asia, it will be useful to make a distinction between "urbanization" and "urban growth," the terms that we commonly use. The former refers to the rise in proportion (i.e., percentage share) of the total population living in urban places within the countries; it connotes the changing balance between rural and urban populations. It should not be confused with urban population growth---a measure of change in urban areas only, having no reference to rural population. However, it is quite obvious that the two terms are interrelated. Two other terms that we commonly employ---"level of urbanization" and "rate of urbanization" also need an explanation. The former is the proportion of urban population at point of time, while the latter refers to the pace of change in the level of urban population over a period of time. It is also useful to understand that the term "urban" is used by different nations according to widely different concepts. Hence, the classification of places defined as urban differs widely among nations. Thus, the comparisons of urbanization and urban growth cannot be precisely made.

By world standards, Asia is not highly urbanized. In mid-1999 a little over 45 percent of Asia's population lived in places defined as urban. This figure is far below that for Latin America, Europe, the Americas, and Australia. Within Asia, the South Asian nations have the lowest levels, with India, the largest of the South Asian countries, close to the average for the region (26 percent). In general, nearly a quarter of South Asia's population live in the cities. Basically, South Asia has been a land of villages containing one-half of world's total number of villages, although several well-known cities have existed for a long time. The land-locked nations of Nepal, Bhutan, and Afghanistan, remained the least developed and have the lowest levels of urbanization.

On the average, urbanization in Southeast Asia is higher, but there is a wider range from 21 percent for Cambodia to over 57 percent for Malaysia, and for a 100 percent for the city-state of Singapore. The regional average for East Asia runs close to that of South Asia and Southeast Asia (nearly one-quarter of its people live in the cities). Japan, the most industrialized Asian nation is an exception; 79 percent of its

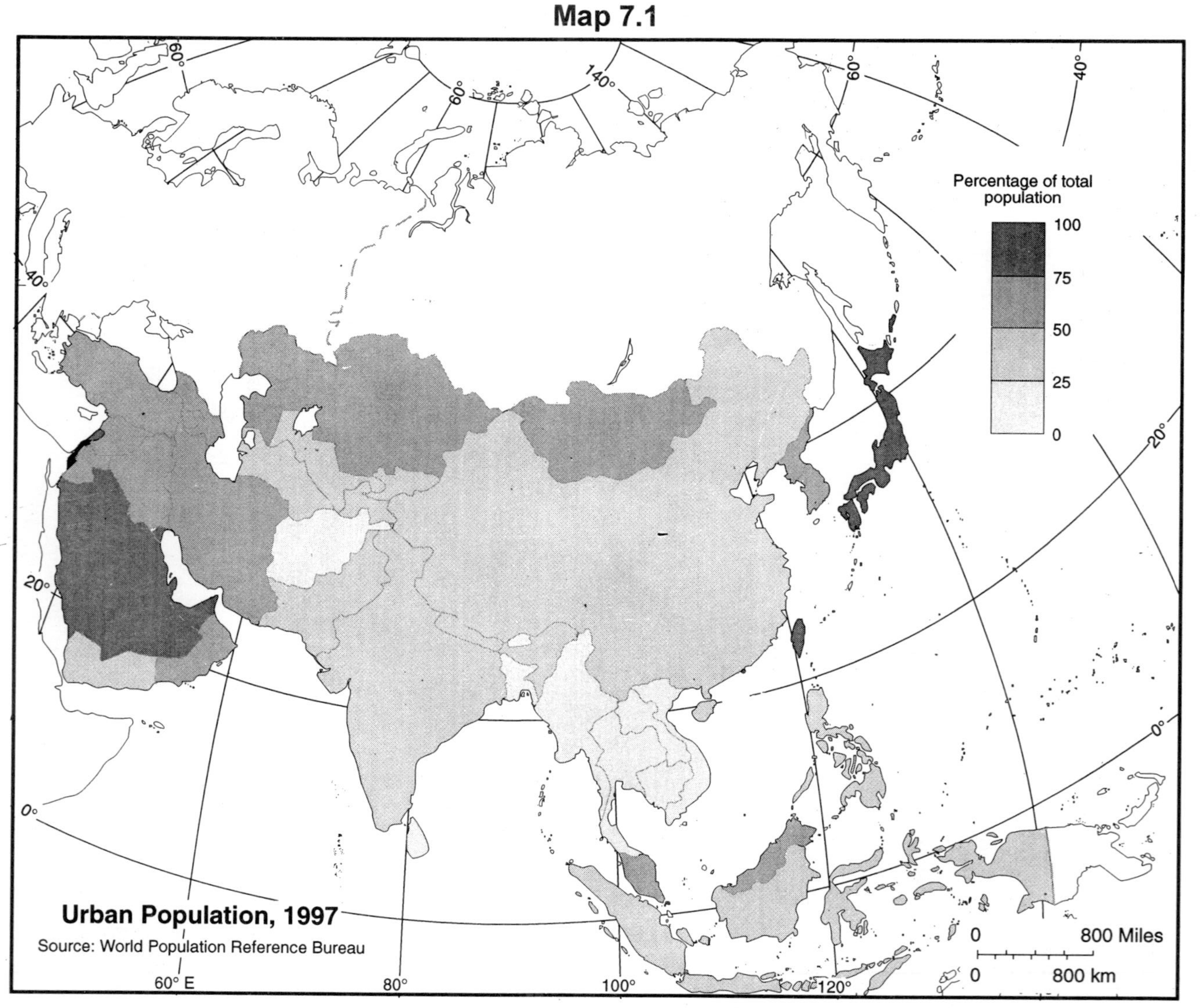
Map 7.1
Percentage of total population
100
75
50
25
0
Urban Population, 1997
Source: World Population Reference Bureau
0 800 Miles
0 800 km
60° E
80°
100°
120°
140°
60°
40°
20°
0°

Map 7.2

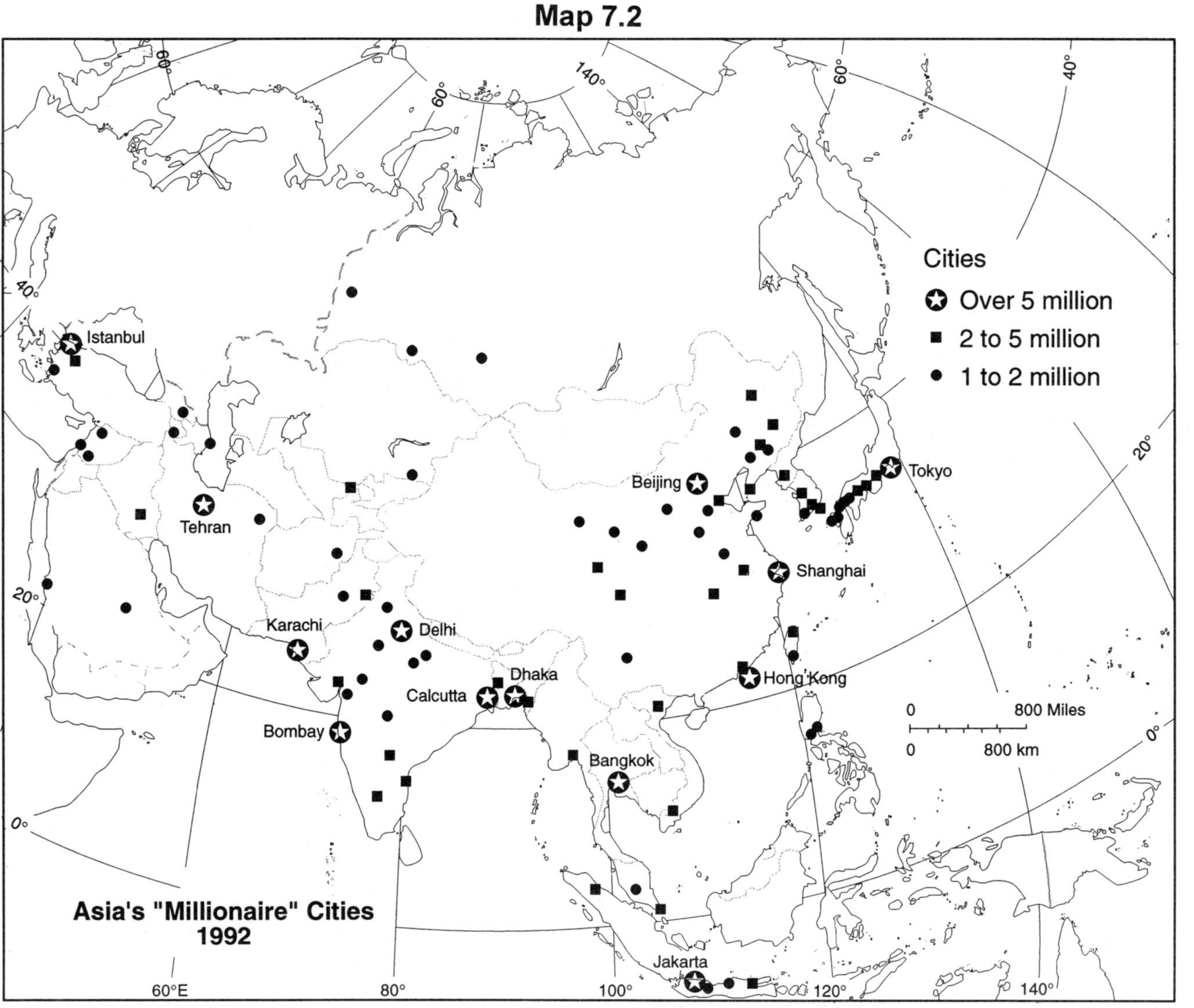

Asia's "Millionaire" Cities 1992

Table 7.2 *Urban Population, by Region, 1950, 1975, 2000, 2025 (estimates)*

Major Region	*1950*		*1975*		*2000*		*2025*	
	Millions	*% of Population*	*Millions*	*% of Population*	*Millions*	*% of Population*	*Millions*	*% of Population*
World Total	730	29	1,590	39	2,398	51	5,345	65
Europe	217	55	323	68	401	78	447	86
Oceania	8	61	16	73	25	83	32	88
Latin America	67	41	198	62	428	76	724	84
North America (excluding Mexico)	106	64	175	75	248	83	306	89
East Asia	113	17	332	30	668	45	1,066	62
South Asia	116	16	286	23	770	37	1,573	56
Africa	33	15	104	26	362	42	903	66
U.S.S.R.	71	39	152	60	230	74	296	83

Adapted from: *U.N., Estimates and Projections of Urban, Rural and City Populations, 1950-2025*

population lives in urban population. Urbanization levels in Southwest Asian nations are much higher than for the rest of the continent, resulting largely from the definition of the term "urban" applied in these nations, but also due to the recent economic progress made possible by the oil revenues. The regional average is over 60 percent. Saudi Arabia and Israel registered 79 and 90 percent of the population as city-dwellers. The land locked, and least developed country of Afghanistan is the exception. Urbanization levels for the Central Asian nations range between 28.5 percent for Tajikistan to 57 percent for Kazakstan reflecting the stages of their development.

Although Asia's urbanization levels are considerably lower in comparison to other continents, its rate of urban growth has gone up significantly during the last several decades. Between 1970 and 1980, Asia's urban growth was 4 percent annually, nearly four times that of Europe and the U.S.A., and nearly twice the rate of its population as a whole. This trend continued through the 1990s. While these figures are averages for the continent, the growth of the individual cities, particularly the larger ones, was exceptionally higher. Cities like Bangkok, Beijing, Delhi, Jakarta, and Tehran, for example, averaged between 6 and 7 percent annual growth during the 1980s. This situation is a striking contrast to mostly European cities and several American metropolises, which experienced population stability or in some instances decline during the last two decades.

Urban growth was exceptionally higher during the 1980's in two of Asia's largest nations, China and India, where it was close to 7 and 5 percent respectively. As an overall summary, some significant trends in Asia's urbanization picture may be noted. First, in general the larger cities and metropolitan areas have been growing much faster than the smaller cities for the last three-four decades. Second, despite a slight deceleration of urban growth during the 1990s, the numerical gains in urban population have been large, due largely to significant population increases in the "millionaire" cities.

Third, a notable phenomenon has been the extraordinary concentration of large number of people and economic activity in the leading city of the nations. This is generally referred to as "urban primacy." A commonly employed measure of a country's urban primacy is the percentage of a country's urban population (or total population) residing in the largest city, which is also known as its "primate city," and typically the national capital. Primate cities are generally most common in smaller or developing nations. Bangkok, Dhaka, Jakarta, Karachi, Manila, Seoul and Yangon (Rangoon) are good examples of urban primacy. Nearly one-half of Thais live in Bangkok, one-third of Bangladesh's and Myanmar's populations reside in their leading cities of Dhaka and Yangon respectively. A quarter of South Korea's population reside in the capital city of Seoul. Manila, Jakarta, and Karachi show a similar pattern. Although the degree of urban primacy usually is associated with a nation's physical size or the state of its development, other factors like system of government and policies, geographical location and historical evolution may also affect the size of the leading city. To an extent it reflects the degree of development of a nation's regions. The more developed the regions are, the lower will be the urban primacy of its leading city. Asia has now 13 cities of over 5 million inhabitants each, and over 100 "millionaire" cities. With the exception of Hong Kong all of these "mega cities" with population of over 5 million are the primate cities.

Table 7.3 *Urbanization of Selected Asian Countries, 1999*

Country	*Number of Cities Over 1,000,000*	*Percent of Population Termed Urban*
China	22	34
India	19	8
Indonesia	5	39
Iran	2	61.3
Japan	11	77.6
Malaysia	1	57
Pakistan	3	33
Thailand	1	34

Source: Derived from: *Encyclopedia Britannica, Year Book 2000.*

Urban Problems Although urbanization in Asia, as elsewhere has been an instrument of economic and social progress, it has also been accompanied by serious socio-economic problems. The physical expansion of the cities has not kept pace with the rapid increase in their population. This has resulted in excessively high densities, substandard living conditions, and the growth of slums. The average densities of population in large metropolises like Karachi, Ahmedabad, Dhaka, Hong Kong, Jakarta, Pusan, Tanjin, Bangkok, are over 100,000 persons per sq mile, and those of Manila, Kolkata, Mumbai, Tehran, Surabaja are close to this figure. Parts of these cities, particularly the older sections of the cities are incredibly congested, where the densities rise to over twice or three times the average for the cities. The housing shortage in large cities like Mumbai, Hong Kong, and Tokyo is so acute that a tiny two-room apartment with a kitchen, and dining area costs a quarter of a million dollars. Recent surveys in India and Indonesia indicate that over two-thirds of the families in the larger cities of these nations occupy one room or less for living. Shantytowns or slums have developed in nearly all major cities of

South Asia and parts of Southeast Asia. So acute is the housing problem that tens of thousands of the poor, unable to afford housing in cities like Mumbai, Kolkata, and Jakarta must literally sleep in the street or railroad sidings.

The high population densities in the cities have resulted from a large and continuing stream of rural migrants into the cities. The rising densities are usually at the root of several urban problems, such as rise in unemployment, substandard living and growth of slums and shantytowns. Problems of urban unemployment and destitution are so serious that urban unemployment is estimated at 15 to 25 percent of the urban populations in South Asian countries and in Indonesia.

The Asian cities have begun to feel the strain of population numbers. The breakdown of infrastructural facilities is being experienced in such large cities of South Asia and Southeast Asia as Bangkok, Mumbai, Kolkata, Delhi, Jakarta, Karachi, Manila and Surabaja. Electrical failures, and inadequate pressure of municipal water supply are commonplace. Increase in the numbers of automobiles without adequate legal constraint on their exhaust emissions, and the proliferation of industrial plants have exacerbated the air pollution of large cities. Parts of Delhi, for example, are perpetually enveloped in yellow-colored haze resulting from the automobile exhaust emissions, making it the fourth most polluted city in the world. Pollution problems resulting from air pollution in several other Asian cities such as Beijing, Tokyo-Yokohama, and Karaganda are also quite severe.

Further Readings

Brunn, S.D. et al, *Cities of the World: World Regional Urban Development*. New York, 1983.

Chandler, T. and Fox, G. *Four Thousand Years of Urban Growth*. New York, 1987.

Pernia, E.M. *Urbanization, Population Distribution and Economic Development in Asia*. Manila, 1993.

III

Patterns of Economic Activities

8

Agriculture and Fishing in Asia

Agriculture has played a vital role in sustaining and nurturing humankind. The initially gradual but accelerating growth of dense populations that inhabit the earth today and the development of modern civilization have been made possible by the developments and sophistications in agriculture.

Agriculture has no single or simple origin. As soon as humans began to form permanent settlements, and gave up wandering in search of food, the practice of cultivating of soil and harvesting of crops began. For a long time it was believed that agriculture began at a single center about 9000-7000 B.C. in Mesopotamia (the present-day Iraq and Syria) when human beings passed from hunting and gathering stage to that of food producing. Recent dating techniques indicate that the cultivation of crops was carried on in several regions of the world around 7000 B.C. There is considerable evidence that food grains such as wheat, rice, millets and lentils were cultivated in Mesopotamia, China, and India in the 4th millennium B.C.

Asia has long been a predominantly agricultural continent. It supports more than a half of world's population on less than one-seventh of earth's surface. Most Asian nations still adhere to their agricultural heritage, although there has been a growing tendency in the 20th century toward industrialization.

Asia's contribution to the supply of world's major food crops, beverages, fruits, nuts, fiber crops, natural rubber, cattle, sheep and fish catch is substantial. Among the grain crops it has a virtual monopoly of rice production over 90 percent of world's output. It produces one-third of world's wheat and millets; and one-fifth of maize (corn). Nearly one-half of world's cotton and practically all of the jute fibers are grown in the continent. Over 90 percent of world's natural rubber, three-fourths of palm oil (and copra), two-thirds of ground nuts (peanuts) are grown in Asia. Nearly one-third of world's cane-sugar, one-fifth of citrus fruit and over two-fifths of fish catch are also obtained from the continent. Nearly one-fourth of world's woodcut is obtained from forests in Asia. Despite these substantial productions, agriculture does not contribute significantly to the gross national product of most nations within the continent (Japan and the nations surrounding the Persian Gulf are exceptions). But in two of Asia's largest countries: China and India, two-thirds to three-fourths of the population are engaged in agricultural activities.

Forms of Cultivation Among the forms of cultivation, the most widely practiced form is peasant farming, also known as traditional farming which consists of farming by plowing the soil and planting crops in permanently occupied plots using non-mechanical tools, human and animal muscle, for growing essentially subsistence crops. A greater part of the population works on the land and depends directly on it for a living. A wide variety of food crops are usually grown for family subsistence, and occasionally a few commercial crops for marketing. Growing several different food (and other) crops within a small area tends to maintain soil nutrients better than in agricultural systems specializing in monoculture.

In a traditional agricultural system cultivation is largely carried on by the use of human and animal muscles, often all members of the family participate in farming activities, young and old, men and women. Agricultural implements are generally simple and easy to operate by hand or by animals such as a wooden and iron plow, a thresher, a simple weeder, etc. Specific tasks of planting grains may be regarded as work appropriate for males in some cultures, but for females in another. Males generally do the plowing of the fields, while the women do many other tasks, such as transplanting of rice (common in area of intensive rice cultivation), food processing, or supplying water to the field.

Often the majority of the population owns little or no land. Peasants rent land or work as laborers for the landowners who have large holdings. Even though most families depend on agriculture for their livelihood, they do not usually own resources beyond a minimum subsistence.

Farmers usually tend to be conservative and resistant to change. This has led to the perpetuation of the socioeconomic order that has developed over a long period of time, representing a time-honored adaptation to the physical environment. There is a general lack of social mobility. Possibilities of life are determined to a large extent by one's circumstances of birth, which includes gender, birth order within the family, and family's status within the community. While these observations may be true of the social and economic orders prevailing in most Asian nations, the traditional systems are particularly well-entrenched in South Asia and Southeast Asia. It may, however, be noted that the traditional agricultural systems are almost everywhere being challenged, and modified by increasing pressures of modernization. The introduction of the so-called "Green Revolution" in parts of South Asia and Southeast Asia during the 1970s and its subsequent diffusion represents one of such pressures. Among the traditional systems, *intensive subsistence agriculture* in which rice cultivation is the most predominant crop prevails largely in Southeast Asia and East Asia. Subsistence agriculture, in general, consists of growing food crops primarily of rice for domestic consumption, which occupies the largest territory in Asia, although some commercial crops such as jute, sugar cane, or cotton are also grown intensively.

Although the predominant agricultural patterns in Asia consist of subsistence cropping or traditional farming, commercial production of selected commodities on large plantations or estates occupy large sections of Southeast Asia, and to a lesser extent of South Asia. The plantations were introduced and controlled by the Europeans prior to the independence of these nations, and became a unique feature of the rural occupancy patterns in parts of Indonesia, Malaysia, the Philippines, India,

Sri Lanka, and to a smaller degree, of Thailand, and Myanmar. They are of little significance in East Asia and Southwest Asia.

The establishment of large, well-managed estates required clearing of forests, construction of railroads, roads, harbors, and building of hospitals to treat malaria and other tropical diseases. This necessitated the importation of labor from India and China for estates in Malaysia, or from India and Sri Lanka or the integration and channelization of native labor into the newly-introduced plantation system as in the case of Indonesia. The establishment of political control of India, Sri Lanka and Malaysia by the British, and of Indonesia by the Dutch gave European trading companies unrestricted power to have access to the imported or domestic labor supply for the plantations. The introduction of plantation system by the injection of European capital and management skills, proved to be an exceedingly lucrative undertaking for it utilized cheap labor, but yielding commercial agricultural products commanding built-in European markets. While the European masters were amassing large fortunes through the sale of commercial crops, the plantations brought peace, prosperity, and economic growth to these areas. However, but population growth consequent upon the reclamation of land, and disease control began exerting new pressure on land.

The most disturbing consequences of the plantation system were, however, political. Despite all the dislocations it created for the native economies or the stability it brought to the economic order, the real shortcoming of the plantation system is not economic at all; it introduced into these nations a persistent, nagging cultural intrusion by the importation of diverse ethnic groups, which became a permanent feature of these countries, and which has since proved inimical to the creation of national cohesion. The Malay-Chinese-Indian rifts in the Malaysian society and the Tamil-Sinhalese problem are grim reminders of the political consequences of the forces, initiated and generated by the ethnic diversity created within the societies of these nations.

Estate or plantation agriculture no longer holds the importance it once did and since independence in these nations not all estates are owned by the foreigners; most have passed into the hands of native administrations. Nearly a third of the plantations in rubber in Malaysia, two-thirds of plantation land in Java (Indonesia), most of tea and coffee plantations in Assam (India) and southern India and a good portion of rubber, and coffee plantations, have been distributed to smaller farmers. There has also been substantial recent expansion of smaller holdings in such crops as maize, abaca, coffee, coconuts, oil palms, pineapples, and rubber with the production of crops less oriented to export. Crop diversification on these lands has been increasing recently, rendering overall land-use patterns more complex.

Nearly 10 percent of Asia's territory is given to *shifting cultivation or agriculture*, a remnant of pre-sedentary cultivation times. It is essentially a farming technique among the cultivators of the tropical, upland forests, shifting agriculture or slash-and-burn method of cultivation and primarily practiced in Southeast Asia, although it is practiced in several other parts of the continent. In the remote, forested regions of the Outer Islands of Indonesia (Kalimantan, Sulawesi, Halmehera, Seram)

90 percent of the farmers practice some form of it. Among the upland Laotian farmers, in southern islands of the Philippines and in northern and eastern uplands of Myanmar shifting cultivation is also widely practiced. The practice is given a variety of names: *caingin* in the Philippines, *tamrai* in Thailand, *taung-ya* in Myanmar, *ladang* in Malaya and Indonesia, *jhum* in India, and *chena* in Sri Lanka. The term "swidden" has been suggested as a generic name for this type of slash-and-burn cultivation.

The most commonly used method is to make forest clearings on well-drained upland soils by cutting the brush and branches at the beginning of the dry season. When the clearings are dry, the area is burnt, letting the ash fertilize the poor tropical soil. A land is cropped for two or three years until the soil has been temporarily depleted and weed-control becomes difficult. The village then moves to a new nearby location. This relocation, or rotation, ends after ten or fifteen years as the village returns to the original site where the secondary forest cover has reestablished itself and soils have regained some fertility. A variety of crops such as maize, root crops (tapioca, yams, sweet potatoes) are the predominant ones, generally interspersed with perennials (pineapples, bananas, or fruit trees). Legumes, green vegetables are now growing in importance as farmers are becoming exposed to new crops.

Shifting cultivation on the scale practiced in these nations has led to several destructive ecological results. The periodic movement of people to new locations, lengthy fallowing of land, exposure to soil fertility, destruction of valuable forest lands and flooding of permanent agricultural lands in valley bottoms downstream from forest clearings are some ecologically destructive consequences. Modern nations tend to discourage shifting cultivation practices, as they make census and taxation difficult and destroy timber.

In the modern world, shifting cultivation presents two major problems: one, concerns with modernization, and the other with the growth of population. The first problem lies in the nature of shifting cultivation itself. Most operations in the system are conducted manually, for example, clearings, sowing, weeding, are done by hand, and occasionally with the use of mechanical handsaws. Mechanical clearing is expensive and beyond the means of most peasant farmers. The second problem is related to the capacity of shifting cultivation to feed an expanding population. Increase in population implies a progressive shortening of the fallow period, which may hinder the recuperation of soil fertility and result in the eventual decline in yields. As the proportion of fallow land progressively decreases, settlements tend to become more permanent resulting in the blurring of distinction between the sedentary and shifting systems.

Food Crops *Rice* is Asia's leading grain crop in terms of production, total area of cultivation, and social significance. Probably native to the deltas of the great Asian rivers---the Ganga, the Chang Jiang, and the Euphrates---it has been cultivated in China and India since ancient times. Rice was already the staple when early languages were taking shape. The words for rice and food are synonyms in several Asian languages.

Although nearly all Asian nations grow the crop, rice occupies an overwhelmingly important position in the monsoon lands of southern and eastern Asia. The continent as a whole produces about 91-92 percent of world's annual production; and five countries----China, India, Indonesia, Bangladesh and Thailand---account for close to 80 percent of the international production. Essentially a subsistence crop consumed largely in area of its growth, its world trade is very small, amounting to less than 3 percent of the production. In comparison, about 20 percent of wheat and maize grown enter the world trade.

Environmental and human conditions best suited for the cultivation are: high temperatures averaging 75°F (24°C) or more for the growing period; ample rainfall (60 to 80 inches/150 to 200 cms) annually, most of which falling during the growing period; clayey loams or clay with silt associated with the deltas and river basins; a level, floodable land; and a plentiful, labor force of farmers skilled in the labor-intensive techniques. Rice cultivation is an intensely demanding crop requiring much more attention than generally required for other grain crops. Farming activities such as weeding, transplantation, application of green manures, detailed water control, maintenance of irrigation channels, construction of water-retaining mud-banks, and terraced slopes are all immensely laborious tasks that do not lend themselves easily to modern, mechanized techniques. All these conditions are plentifully available in the densely populated humid and subtropical Asia where farmers are skilled in the labor-intensive farming techniques developed over the centuries.

Primarily a "wet crop," it is most suited to the flat, flooded deltas and river basins where water requirements are met by the rains in the fields themselves or by natural floods. In the uplands, water is supplied to the fields by terracing the sloping lands. The uses of irrigation techniques are the time-honored practices. The hilly areas of Japan, China and the Philippines contain some spectacular examples of elaborately terraced rice fields.

World production has grown nearly three times since 1950, from 170 million tons to 520 million tons; the chronically food deficit nations of China and India have increased their output nearly four times during the period. Nine Asian nations: China, India, Indonesia, Bangladesh, Thailand, Vietnam, Myanmar, Japan and the Philippines account for over three-fourths of world's total output. Other important producers are North Korea, South Korea, Pakistan, Sri Lanka and Malaysia. Except for Japan, Asian crop yields are low as compared to those in the U.S. (The other important rice producer and exporter). Per hectare average yield for India, China, Philippines, Indonesia, Japan and the U.S. in kilograms are: 1200, 2400, 1500, 2000, 5000, and 3600 respectively.

China is the world's leader, producing little over one-third of world's output. Rice is the dominant crop south of the Qinlin Mountain range which is the most significant line of demarcation in China's agricultural landscape. Major area of the production are the lower Huanghe basin, the area between the Qinlin Mountains and the Huanghe basin, and the Sunghua and Liao river basins in Manchuria where topographic, soil and climate conditions are most suitable. India and Bangladesh rank second and fourth in the world, but both chronically deficient in their food grain requirements prior to the mid-1980s, have attained self-sufficiency, thanks to the introduction of high-yielding strains, and techniques associated with the Green Revolution. In Bangladesh, where conditions for growth are ideal (level, flooded plains and delta of the Ganga and Brahmaputra rivers) farmers manage to obtain two to three crops a year. India is the second largest producer, producing a little over one-fifth of the world. The major areas of production are the well-watered and alluvial Ganga plains and the West and East Coastal plains.

In East Asia rice maintains supremacy as a staple food grain. The primary production areas in Japan are along the coastal plains in Honshu, generally south of 38°N and the coastal lands around the Inland Sea in southern Japan. Production has remained steady between 10-15 million metric tons since 1950; a slight decrease since 1970 reflected the dietary shifts of the Japanese toward beef and dairy products. Japan has been an importer of rice, normally importing 7 to 10 million metric tons annually, from Southeast Asian nations. Rice is also a dominant crop in the Korean Peninsula. The cultivation was greatly expanded during the Japanese control during the 1930s and 1940s by the reclamation of new areas and the use of better seeds and fertilizers with a view to export it to meet the growing demand.

Rice is the basic food crop of Southeast Asia. Producing between one-fifth to one-fourth of world production, the region accounts for nearly one-half of world's rice exports. Most of the export goes to Asia's chronically food deficit nations like China, Bangladesh, India, and Japan. Grown universally in Southeast Asia, the principal producers are Indonesia, Thailand, Vietnam, Myanmar and the Philippines. Cambodia, Laos and Malaysia are also significant producers. Production has increased substantially during the last twenty years particularly in Indonesia and Thailand. Indonesia has been an important importer of rice from Thailand, but dependence on imports has steadily decreased as it stepped up its production during the 1980s.

Wheat As a cereal plant, wheat is probably the oldest of the crops. Native to the Middle East, it was cultivated in the Euphrates valley by 4000 B.C., and in China and India by 2500 B.C. Wheat is as important a part of the diet and civilizations of the Middle East and that of the European peoples, as rice has been to the peoples of East Asia and Southeast Asia. It is a major food crop in the world, and an important commodity on the world grain market.

Wheat is characteristically different from rice in its climatic and soil requirements. Best suited to the comparatively dry, cool, irrigated lands of the temperate latitudes, the plant offers a striking contrast to that of rice which favors hot and moist areas.

Asia produces nearly 40 percent of the world's output. China leads the production which amounts to a little over one-sixth of the world. India produces nearly ten percent of the world's production. Most nations in the Middle East are also producers, among them Turkey and Pakistan are the important ones. China's production generally ranges between 100-110 million metric tons a year. China's wheat lands lie in the drier, cool, mid-latitude areas of north and northeastern part mostly north of the Qinlin. Production has more than tripled, from 31 million metric tons to 110 million metric tons, between 1970 and 1999. But China has been an important importer (importing normally between 10-15 million metric tons from the U.S.A., Canada, and Australia), although it has been able to reduce its dependence since 1985 as it stepped up its production.

After China, India is the most important producer in Asia, producing nearly one-half the output of China, but commands a third position in the world (the U.S.A. holds the second rank). The cultivation is confined to the cooler, and drier lands in the western and central parts of the country, particularly in the irrigated lands of the Indus basin (basin shared with Pakistan), the middle Ganga basin, and parts of central India The proportion of cropland declines eastward.

The area under wheat cultivation more than doubled between 1950 and 1990, but production increased ninefold. Crop yields are low by world comparison, one-half those in the U.S.A., but have been steadily improving since the 1960s following the introduction of the high-yielding varieties associated with the Green Revolution.

Wheat is a staple diet of the people of southwestern Asia and is raised everywhere provided suitable climatic conditions for its growth exist, or where man has made arrangements to provide water in these dry lands. Turkey, Pakistan and Iran are the major regional producers.

Maize (corn) Native of the New World corn was introduced by 17th century explorers to all parts of the world, where it became an important grain crop, and used both as human food and as an animal fodder. A deep-rooted leafy crop, it grows best on well-drained, deep, heavy loams in areas of assured rainfall or irrigation. It has the status of a coarse, secondary cereal crop in areas where wheat is primarily consumed.

Asia produces nearly a quarter of world's production. China is the world's foremost producer accounting for about one-fifth of its output. India, Indonesia, Philippines, North Korea and Thailand are the other significant producers; modest amounts are also produced in Turkey, Nepal and Pakistan.

In China, the major corn region lies to the north of the Qinlin Mountain range extending from Sichuan basin to southern Manchuria, although it is also raised in Central China between and in the Chan Jiang and the Huang he basins. In India most of the crop grows in the Indus-Ganga Plains. In Indonesia, corn is concentrated primarily in the island of Java. In the Philippines, crop is raised on the western coastal region of Luzon, in Panay and Mindanao islands. Corn is an important crop of North Korea as well. World trade in corn is much smaller than in other grain crops. Japan, China, and South Korea are the principal importers, and the United States, which is a world leader (producing about 40 percent of the world's crop), is the major exporter to these nations.

Other Cereals Millets as cereal crops are intermediate between rice and wheat lands in terms of location and environmental requirements. These are more drought-resistant, tolerate poorer soils, and mature in a shorter time than rice or wheat. Although there are several types of millets, chief varieties grown are *jowar, bajra* and *ragi* in India, and *kaoliang* in China. A long-standing prejudice against millets as cereal crops exists in several countries, as these are considered as inferior cereals consumed only by the poor. This prejudice stems from the fact that millets are a coarse grain, although in nutritional content, they are superior to either wheat, rice or maize.

Asia provides one-half of the world's total output, most of which comes from India and China. Producing 31 and 15 percent respectively, India and China are the world's biggest producers. The crop is raised extensively in the two countries as a secondary crop in areas where wheat is the principal crop. In India, the Deccan peninsula, where wheat is not the premier grain crop, millets such as *ragi* is the first ranking crop, as also in Central India where millet crops such as *jowar* and *bajra* are the principal grain crops. In China millets are a secondary crop almost everywhere, although the major area of kaoliang production is China's northern and northeastern sections especially the Huanghe basin, and the lower Sungari basin in Manchuria.

Fiber Crops Cotton holds a premier position among fibers. Clothing made of the cotton fiber have been used since prehistoric times. In India textiles were woven with great skill several hundred years before the Christian era. Cotton is a demanding crop: moisture restricted to a limited growing period within 60 to 86 cm rainfall zone, rich clayey soils and availability of abundant, cheap labor. Growth of the crop is, thus restricted to certain regions, usually named as the "cotton belt."

Over 50 percent of the world's cotton (lint) is obtained from four Asian nations (China, Pakistan, India and Turkey), the remaining comes from the United States and the rest of the world. Until the turn of the century the United States was the only important producer in the world. Production increased rapidly elsewhere since the 1930s particularly in the Indian subcontinent, due largely to the establishment of canal colonies in the Indus basin (area now in Pakistan) and the northern Indian plains during the British rule. More recently, since 1950, Pakistan, India and China stepped up their production considerably. By 1970 China had emerged as the largest producer, outpacing both the U.S.A. and the USSR Meanwhile, production in Pakistan had been rapidly increasing. In 1970 its production was nearly one-half that of India; but in 1991 the production of cotton lint in Pakistan were 2.1 million metric tons as compared to 1.9 million metric tons produced in India. The principal area of production in Pakistan is the "canal colonies" region in the Middle Indus basin, and in India the Deccan Plateau, particularly the northwestern section, where rich moisture-retentive lava soils, and warm temperatures combine to make it the leading production area. The Middle Ganga basin also possesses characteristics suitable to cotton growing—warm temperatures, alluvial soils, irrigation facilities, and plentiful labor supply and is another important region of production. Cotton is an important cash crop for India and Pakistan.

Cotton is China's major fiber crop, its production is more than a quarter of world's total. The country holds the first position among world producers; before World War II it held the third place, after the U.S.A. and the USSR. The principal area of production lies in the lower basins of the Huanghe and the Chang Jiang, and the area between the two basins. Most of the cotton is produced in the drier parts north of the Qinlin Mountains.

Turkey, Uzbekistan, Turkmenistan are the other cotton producers in Asia. Turkey produces nearly 3.0 percent of the world, the main areas in the valleys along the southern and western coasts. Production in Uzbekistan, Turkmenistan, Kazakstan, Tajikistan in Central Asia is based on the irrigation system introduced by the USSR in the 1930s and 1940s on the Syr and Amu rivers.

Jute Jute is a minor fiber crop in terms of its cultivated area but is an important cash crop and a foreign-exchange earner to the producing areas. It grows best under hot, moist conditions in the river deltas where land is enriched by fresh silt year after year, for it is a soil-exhausting crop. The Ganga Delta in India and Bangladesh holds a world monopoly. India is largest producer with a production of 52 percent, followed by Bangladesh with a production of over one-quarter of the world's total. Partition of the Indian subcontinent in 1947 deprived India of its major source area, and rapid expansion followed within India. India is now self-sufficient in its requirements. China and Thailand, the other major world producers also stepped up their production since the 1950s. Currently China produces one-fifth of the world's production. Most of the production enters the world trade, the major producers export to the U.S.A., Japan, and the other industrialized nations.

Plantation Crops Tea, coffee, rubber, abaca, coconuts, and palm oil are the main plantation crops grown in Asia. *Tea* is an evergreen plant indigenous to Assam (India), probably China and Japan. The plant was known in China in prehistoric times, although the commercial cultivation there dates to the 18th century. It was introduced into Europe by the Dutch in the early 17th century. The tea bush grows best in areas of abundant rainfall, gravelly soils, well-drained hill-slopes (usually between 3,000 to 4,000 feet/910-1,218 meters of altitude) in a warm climate. The leaves are handpicked involving a good deal of painstaking and methodical effort. The availability of an abundant supply of labor is essential for harvesting.

Nearly 80 percent of the world's total production comes from half a dozen Asian countries. India is the largest producer, followed by China and Sri Lanka, producing 29, 22 and 10 percent of the world's total respectively. Indonesia, Turkey and Japan are the other important producers. Except in China, where the crop is raised on small family farms, it is mostly grown on large plantations, established and managed by the European companies until the 1950s, and since then increasingly by the native enterprises. Most of the production enters world market; consumption in areas of growth is small.

Unlike tea, which is primarily an Asian crop, *coffee* plantations are limited to a few countries in Asia. Chief among these are: Indonesia, Vietnam, and India, which produced 7, 5, and 3 percent respectively of the world's output in the mid 1990s.

Native to Ethiopia coffee plant was introduced into Arabia in the 15th century. The plant requires a hot, moist climate with a rainfall of at least 50 inches (27 cms), a rich soil and well-drained slopping topography at elevations from sea level to 6,000 feet (820 meters), and thrives best on volcanic soils on elevations between 1,500 feet to 6,000 feet. Such conditions are ideally available in the islands of Java, and Sumatra (Indonesia) where it grows extensively on plantations that were established by the Dutch in the highlands utilizing cheap, and plentiful local labor. After independence most of the plantations were distributed among the local farmers. In Vietnam the French set up estates (plantations) in the Central Annamese Mountains under environmental conditions somewhat similar to those prevailing in Indonesia. In India coffee plantations are located exclusively in the warmer, tropical regions, particularly on or near the Nilgiri Hills.

Rubber (natural) is the major plantation crop of Southeast Asia. The crop has been an important foreign exchange earner for the region even after the severe competition with the synthetic rubber since 1940s. Asia has a virtual monopoly in production of natural rubber producing nearly 95 percent of the world's output. The three Southeast Asian nations----Malaysia, Indonesia and Thailand---account for three-fourths of the world's production.

Native to the Amazon basin in South America, the rubber plant was introduced into Sri Lanka, and the tropical regions of Southeast Asia (Malaya, Java and Sumatra) after 1880, thus beginning the establishment of large estates (plantations), and the development of rubber industry by the Europeans. During WWII when the supplies were cut off to Europe and the U.S., the United States began large scale manufacturing of synthetic rubber, and the production of natural rubber fell dramatically. The current production is nearly half that of synthetic rubber. Even with the diminished usage, natural rubber retains much of its importance in modern industry as it is much cheaper, and easier to work with.

The rubber tree grows best in equatorial climates---(characterizing high, constant temperatures with small daily variations, ample rainfall, well-distributed throughout the year) well-drained lateric soils, and above all the availability of cheap and abundant labor. These conditions are ideally met in Malaysia, Indonesia, and Thailand; each of these nations currently produce about 1.5 million metric tons annually. Until mid-1980s Malaysia held the foremost position, since then Indonesia and Thailand substantially increased their output and emerged as serious competitors to Malaysia. The western coastal zone in Malaya where stands of rubber tree appear in an almost continuous belt from the Thailand's border to the island of Singapore; the islands of Java and Sumatra in Indonesia; and the northern portion of the peninsular section of Thailand are the important rubber growing regions.

The British established large rubber estates in Malaya by using the labor force mostly imported from India during the 1920s and 1930s as the Chinese and the Malays were usually disinclined to work in the plantations. In Malaysia there has been a marked shift from the domination of the British-controlled estates to the smaller estates, owned chiefly by Malaysians and Chinese since the 1960s. In Indonesia, the Dutch set up large estates and built a network of roads primarily in the heavily forested islands of Java and Sumatra by utilizing the locally available cheap and plentiful labor. India, China, the Philippines and Sri Lanka are other producers, where the conditions for the growth of rubber tree are also quite suitable.

As a plantation crop, the distribution of *coconuts* is more widespread. Indonesia, the Philippines, Sri Lanka and India are the major producers, each producing between 15 and 20 percent of the world's production. Thailand and Vietnam, are smaller producers.

Sugarcane is an important cash crop of several nations in Asia, grown both for domestic use and for export. The plant is a tall, tropical perennial. Native probably to India, it has been cultivated since prehistoric times in Asia. Raw cane grows ideally in tropical climates, although a good deal of the crop is now raised in North India Plains outside the tropics. The crop is highly demanding, requiring rich soils and abundant moisture during the growing season.

Asia produces nearly one-half of the world's total output. India, Indonesia, China, Pakistan, Thailand, and the Philippines are the major producers. Production has remained static in these nations since the 1970s, although during the same period spectacular gains were recorded outside Asia, principally in Brazil. India is the world's largest producer, currently producing between 20-24 percent of the world's production. The major area of production is the middle Ganga Plains in north India, where the subtropical climate is less than ideal, but the climatic disadvantage has been offset by a longer growing season, the availability of alluvial soils, and canal-irrigation system, and above all by the cheap and plentiful labor. Yields, however, remain low, by world comparisons. Other major producers are China, Thailand, Pakistan and Indonesia (producing 7, 7, and 3 percent respectively of the world). The Pakistani production area lies in the mid-Indus basin which is an extension of the North India Plains in terms of physical environment. The crop is raised under irrigation, and in conditions similar to those of north India. Elsewhere, the crop is raised mostly within the rainy tropics, particularly in Taiwan, lower Chao Phraya basin in Thailand, the island of Java, and the western coastal areas of the islands of the Philippines.

Sugar-beet is essentially a plant of the temperate latitudes, and can tolerate comparatively smaller amounts of rainfall, lower temperatures, and can be grown on a greater range of soil conditions. China and Turkey are the important producers, each producing over 4 percent of world's output. The southern Manchurian plains and the Huanghe basin in China, and northern coastal plains in Turkey provide suitable environmental conditions for raising the crop in these nations. Iran, Japan, Tajikistan and Kyrgyzstan are other producers in Asia (each producing less than one percent of the world's production).

FISHING

The fishery resource of Asia's coastal and inland waters is probably the richest in the world, both as to the variety of species and to its economic usefulness. The aquatic environments of Southeastern Asia contain a good deal of shallow water swamps that provide inviting and productive home to several varieties of fish. The cool, and shallow shores of the northeast also present conditions particularly useful for several types of fish; whereas the southern and southeastern shores contain warm, shallow, marshlands for some other types of species. In all, perhaps there are 2,000 species in Southeast Asian waters, as compared to nearly 1,000 species off the Japanese shores (northeastern waters).

The shoreline territory of Asia may conveniently be divided into three large regional units, each presenting different climatic, topographic and geologic settings. The first unit, usually referred to as the

Indo-Pacific region, extends from the Arabian Sea around to China Sea. It is a single major tropical zone of warm climate and aquatic life that contains more families and species found anywhere else. The enormous fish wealth relates to the common history of the drainage of the old Sundaland shelf, long a part of the continent of Asia. The second unit includes the Sea of Japan, and the Northwest Pacific region containing cool to cold waters that undergo temperature variations from year to year as major warm and cold ocean currents meet and fluctuate in volume and relative location affecting fish life, as compared to the relatively uniform, warm waters of the Indo-Pacific region. The third region is that of the southern waters extending from East Indies to Australia, which contains environmental uniformity that spawns species native to the region. For example, the species here possess breathing equipment enabling them to sustain themselves under poorer ecological conditions.

These three broad regional patterns have resulted from the shifts of sea levels experienced during the past periods of glaciation, deglaciation and mountain building that rearranged land and sea features, river systems, coastal seas, and ocean currents of southern and eastern Asia. From time to time such changes produced marine environments that created the differences among the three major zones.

The importance of fishing as an economic activity in Asia can be judged from the fact that two-fifths of the world's nearly 100 million metric tons of fish catch is derived from the Asian shores and inland water bodies. Although China, Japan, India, Indonesia, South Korea and Thailand are the prominent fishing nations, nearly all of the Asian countries having marine environment report fishing as one of their major industries.

China is the world's leading fishing nation today, normally accounting for 22 million metric tons of catch which is nearly 21 percent of the world's total. As the fish forms an integral part of the Chinese diet, all rivers and coastal waters are fished extensively. Fresh-water fisheries are important in the lower Chang Jiang Valley, and southeast China, but are much less significant than the sea fisheries. Fish are also raised on one million hectares of reservoirs, ponds, and ricelands. The most important marine fishing area is along the embayed coast from Shanghai to Guangzou. China's fishing fleet is mostly lacking in new, and modern equipment, although motorized fishing is becoming increasingly important.

Japan for a long time had been the world's leading fishing nation. Her fish catch ranged between 6-11 million metric tons annually between 1970-1999. As she industrialized herself since the Second World War, China and India caught up with her and by 1990 outranked her in fish catch. Most of the Japanese fishing is marine although pond fishing is also important. The Japanese coasts lie close to the mixture of cold and warm ocean currents and are rich in aquatic life. Thousands of small fishing villages are dotted along the coastline. Nearly 80 percent of the fishing boats are motorized and the equipment is modern.

The Korean peninsula has been important for fishing. Under the Japanese occupation, fishing became an important economic activity during the 1930s and 1940s. Fish catch in North Korea and South Korea currently amounts to between

2-3 million metric tons annually for each of the two nations.

Second only to China, India today holds a prominent position in fish catch in the world. Production has rapidly increased since 1950 and amounts to about 5 million metric tons. The most important fishing area is along the West Coast, particularly that of Malabar and Konkan. Inland fisheries are limited, where there exists a vast room for expansion, but the prevailing dietary constraints among the predominantly Hindu society is an inhibiting factor.

Although fishing is carried on in nearly all the nations in Southeast, Asia, Indonesia, Thailand, and the Philippines are particularly important. Production in each of these nations ranges from 2.0 and 2.5 million metric tons annually.

AGRICULTURAL MODERNIZATION IN ASIA

Prior to and since gaining independence several countries in Asia, particularly on the Indian subcontinent, had been experiencing critical shortages in food grains as their population pressures kept on increasing. The colonial policy had primarily emphasized the production of export-oriented crops, such as cotton, jute, tea, rubber, etc., with the result that these nations' capacity to feed themselves had steadily deteriorated. The historically rice-exporting countries, Thailand, Myanmar, and Cambodia were the few exceptions. The nations in South Asia were particularly concerned as their populations began to mount, and they were faced with the task of increasing their food grain production.

The traditional approach to agricultural development had relied heavily on such methods as expansion of the cultivated area and efficient utilization of land and farm labor. It was, however, quite apparent that nearly all of the available farmland had been put under plough in these countries. Agricultural modernization that could enhance crop-yield was perhaps the answer. The successful route to increased food grain production lay in the creation and introduction of improved strains of cereal crops. A major breakthrough occurred in the early 1960s through the discoveries of Norman Borlaug who had developed higher yielding varieties of rice, wheat and maize (corn) in Mexico and tried these in the Philippines. The results of initial experiments were so overwhelmingly successful that soon the new strains, the HYVs (the high-yielding varieties) of rice and wheat (and later on some other grains) were adopted in India, Pakistan, Sri Lanka, the Philippines, and later on in several other areas (including Thailand, Taiwan, parts of China).

The modernization process of the diffusion of the high-yielding varieties of new strains of crops is based on several factors if satisfactory results were to be obtained. First, the new seeds increase yields only if they are to be supported by large quantities of fertilizers and pesticides. Second, ample and careful application of irrigation is essential for the new "miracle strains" to develop properly during the growing season. Third, the new strains respond best to mechanized farming and modern agricultural techniques. The last factor is important, if not crucial, for the capital-scarce, poor farmers in countries where farming implements and techniques are traditional and lack modern sophistication.

The "core area" of the diffusion of the new HYV of crops was India and Pakistan (the Indus-Ganga Plains) where the production of wheat and rice increased over 90 percent between 1966 and 1971, attributed largely to the use of these "miracle strains," although promising results were obtained in Sri Lanka, Thailand, and the Philippines as well. The success of the "food grain revolution" or the so-called "Green Revolution" depended largely on the grant of subsidies and loans to the farmers, and the development of irrigation facilities by the state in these nations where the farmers in many instances were largely poor, and short of capital. The last item was important for the purchase of fertilizers, and agricultural implements. The bigger landlords and farmers could spare capital for the purchase of fertilizers, pesticides, and introduce mechanization and make use of irrigation facilities. They were the largest beneficiaries of the Green Revolution. Some studies indicate that the Green Revolution created wider social gaps between the big farmers and the smaller cultivators in many instances.

Clearly, the short-run prospect of the modernization process in agricultural development in several Asian countries has been that of averting food famines and other serious problems in food-deficit regions, for example, India and Pakistan are now self-sufficient in their food requirements, and several other countries have been able to increase their foodgrain supply. But the long-term impact still remains uncertain for several reasons. Firstly, the diffusion of new varieties is spotty in space and among farmers. Those areas have benefited most where the states could provide facilities such as irrigation facilities, and provide capital subsidies to the farmers among the smaller and poorer farmers. Secondly, much depends on the rate of population growth. A rapid increase in population growth means that population can outpace the gains made through greater production of foodgrains. And thirdly, much of the success of the Green Revolution would depend upon the behavior of the economic and political conditions within the countries. For a variety of reasons the implementation and diffusion of the high-yielding varieties of crops in areas which become politically and economically unstable will be difficult as the Green Revolution programs could be put in jeopardy.

Further Readings

Barlowe, R. *Land Resource Economics.* Englewood Cliffs, N.J., 1978.

Bray, C., *The Rice Economics: Technology and Development in Asian Societies.* New York, 1986.

Tarrant, J. *Agricultural Geography.* New York, 1974.

9

Asia's Minerals and Energy Resources

Minerals have been used since prehistoric times. Although modern mining came of age with the beginning of the Industrial Revolution in the 16th and 17th centuries, the names "Stone Age," "Bronze Age," and "Iron Age" identify the major stages of our cultural and economic development. In fact, the history of human culture and the rise of civilization have been intimately related to the history of mining.

Nearly 3000 minerals are currently known to us, of which iron-ore, coal, bauxite, copper, molybdenum, and phosphates are extensively used in modern industry. Given Asia's vast size, and its long and complex geologic history, practically all known minerals are found within its borders. Its reserves of coal, oil (petroleum), natural gas, iron, bauxite, and tungsten are particularly extensive, although exploitation has been limited except that of oil, coal and iron ore. Much of China and Southeast Asia remain unexploited in terms of their mineral wealth. Inaccessibility of the fields, lack of funds, and political instability have hindered exploitation in several Asian nations. Therefore, production has been much smaller than its potential.

Nearly one-fifth of the world's coal, one-sixth each of iron-ore, lead, chromite, and nickel, and one-seventh of phosphates are produced in the continent. One-third of the world's oil comes from the nations of Asia and reserves are much larger. More than 50 percent of the world's tin and tungsten is produced within Asia.

ENERGY RESOURCES: MINERALS AND FUELS

Coal World's coal reserves are quite large, estimated at 4.5 trillion metric tons, and probably will last a long time as we increasingly utilize newer forms of energy. Asia's share of reserves is about one-eight of the world, which is more than adequate for its current needs. The distribution of the fields is, however, uneven. China, Russia and the Central Asian republics contain the largest reserves, while those of India, Indonesia, North Korea, Taiwan and Japan are more modest.

China and India are the major producers. China contributes 23-24 percent of the world's production and its coal is of high-grade quality. The largest fields are in Shanxi, Shensi, Ssuchuan and Shandong provinces. Most of the production is utilized in heavy metallurgical industry in the industrial plants in the northeast region (Fushun, Anshan and Liaoning) and in Central Anhwei area, primarily north of the Chang Jiang River, although the country's growing industry is increasingly coal

Map 9.1

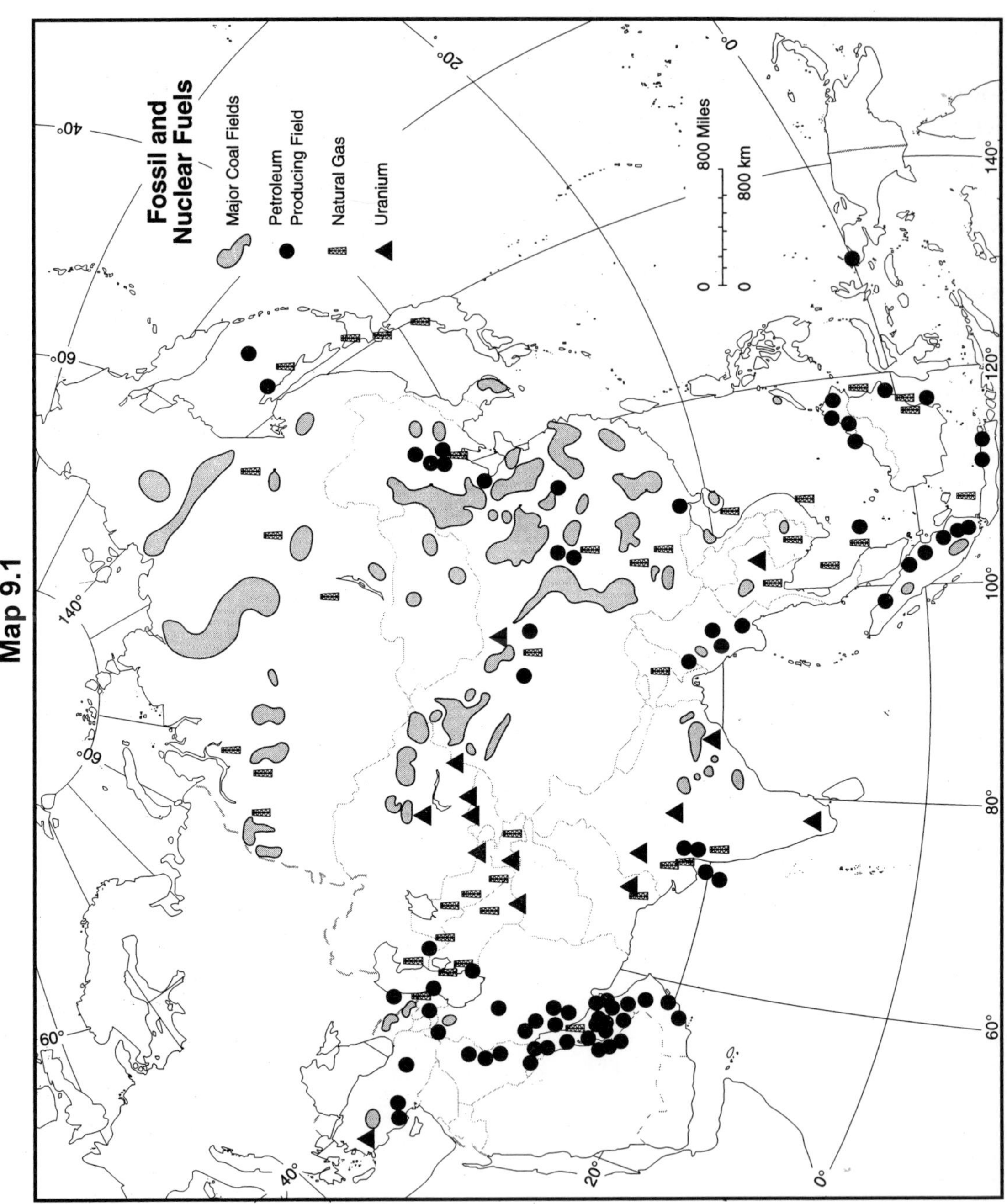

Fossil and Nuclear Fuels
Major Coal Fields
Petroleum Producing Field
Natural Gas
Uranium
0 800 Miles
0 800 km

in other industrial areas as well.

The former USSR was the world's third largest producer, after China and the United States. Its known reserves and production fields lie in Siberia, the Urals, the Soviet Far East and the Central Asian republics. Kazakhstan, Uzbekistan, Kyrgyzstan and Tajikistan contain fair amounts, the largest producer is the Karaganda field in Kazakhstan. Transcaucasia has an abundant reserve of coal and hydrocarbons and is an important producer.

India is Asia's third largest coal producer. Major mining areas are in the northeastern part of the Deccan Plateau (Chota Nagpur region) close to the industrial regions of South Bihar and Kolkata. Indian coal is generally of low-grade. Although the domestic demand is large, a part of the production is exported to Japan.

Petroleum and Natural Gas Two-thirds of the world's proven oil and natural gas reserves lie in Asia. This proportion is likely to go up as new fields are discovered, particularly on islands bordering East Asia, and island-chains in Southeast Asia, many of which have geologic formations that indicate good prospects of petroleum reserves. Several of the fields explored during the last thirty years or so have already begun to yield oil, e.g., in Sumatra, Kalimantan, Java, and islands in South China Sea. The continent's largest known reserves lie in Southwest Asia, particularly in the area bordering the Persian Gulf. Saudi Arabia alone claims nearly one-quarter of the world's total proven reserves. Significant amounts also exist in Kuwait, Iran, Iraq, and the United Arab Emirates.

In East Asia, China's reserves are still being explored; there are several areas which have bright prospects. The known proven reserves of the Indian subcontinent are considerably small. The Philippines has little prospects and Japan, North Korea, and South Korea virtually none. In Southeast Asia, Vietnam has some off-shore prospects.

Russia, and the Central Asian republics have large reserves. In the late 1980s the USSR overtook the countries of Southwest Asia in the production of oil and natural gas, although in known reserves, Southwest Asia retains a commanding lead.

Asia produces a little over one-third of the world's crude oil. Saudi Arabia alone contributes one-quarter of Asia's production; Iran, Iraq, Kuwait and the United Arab Emirates a little less than the remaining half, leaving the rest of Asian nations collectively to share one-quarter of the Asian total. During the Gulf War in 1991, production suffered greatly particularly that of Iraq and Kuwait. During the 1970s and 1980s, western nations had started to conserve oil by reducing mileage driven, encouraging public transportation, and by promoting the development of smaller and more fuel-efficient vehicles.

Since oil consumption in the industrial, automobile-oriented countries had dramatically increased since WW II, the threat of an embargo by the OPEC (Organization of Petroleum Exporting Countries) during the 1970s had only heightened the concern for, and interests in Southwest Asian oil. In the absence of domestic reserves in the European nations and Japan, and the shrinking reserves in the U.S.A. (which was the world's leader in oil production before the 1950s and now remains its leading consumer), the industrialized nations have little recourse but

Map 9.2

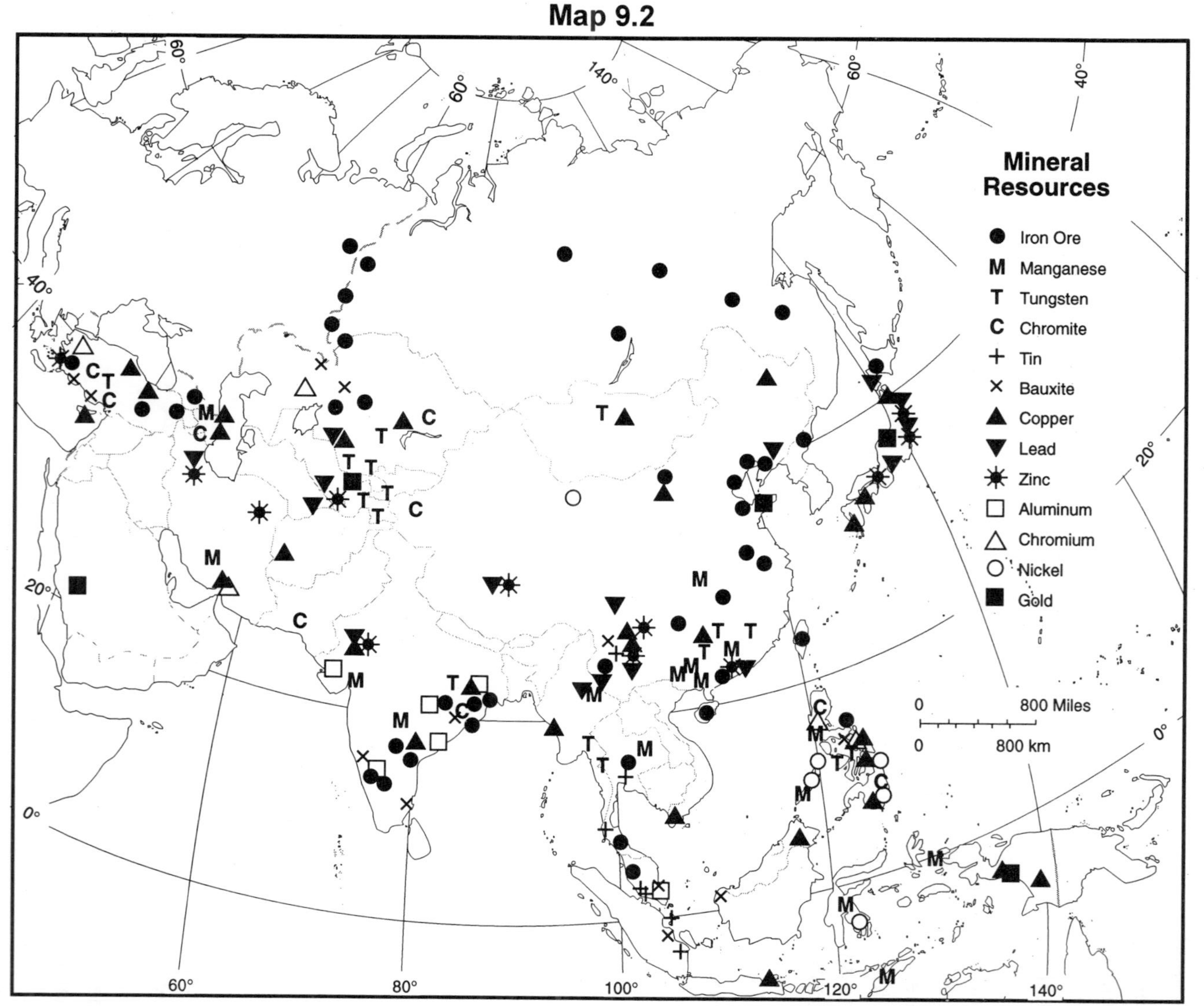
Mineral Resources
Iron Ore
Manganese
Tungsten
Chromite
Tin
Bauxite
Copper
Lead
Zinc
Aluminum
Chromium
Nickel
Gold
0 800 Miles
0 800 km

to turn to the oil producing nations, particularly the Persian Gulf countries. In a dollar amount, the world's large commodity flow is the movement of oil from the Persian Gulf to Europe and Japan. Thus, the Persian Gulf nations have assumed an unprecedented geopolitical significance as the major supplier of oil to the industrial nations.

China, Indonesia, Malaysia, Brunei, and the Central Asian republics of Turkmenistan, Uzbekistan, and Kazakhstan are the other major Asian oil producers. China's production rose substantially during the late 1980s and 1990s and currently stands at 5 percent of the world's production. The major fields are in the provinces of Ssuchuan, Kansu, Xinjiang, and in the northeastern provinces. Recently, oil companies from the U.S. and the West European countries have jointly contracted with the Chinese government to explore and expedite the development of oil in the offshore waters of East and South China Seas.

The fields of Transcancasia particularly of Azerbaijan and Georgia are rich and lie in a belt between the Caspian and Black Sea, and are important producers.

The fields of Turkmenistan are on the eastern shore of the Caspian Sea, and that of Uzbekistan and Kazakhstan along the shores of Aral Sea. There exist several untapped reserves of oil and natural gas in several other regions in Central Asia as well.

Indonesia's major fields are located in the islands of Sumatra, Kalimantan (formerly Borneo) and Java. The country utilizes most of its output, and has now begun to export some crude oil and crude-oil products. Malaysia's production is still very small by world comparisons. Brunei, a small British protectorate, is a regionally important producer, and an exporter. Production, by world standards, is small but enough to raise Brunei's per capita income to the level of prosperous industrialized nations.

Related to the production of oil is the extraction of natural gas because the two usually share the same geologic structures. Level of production varies with the capacity of individual nations to extract, pipe and deliver the natural gas to the demand areas. Asia accounts for nearly thirty percent of the world's total reserves, but production is only ten percent, indicating the variable nature of demand-supply and cost of transportation over-views. Russia, the United States, and the European nations are the main producers.

Asia's major fields of natural gas lie close to the oil fields. The Persian Gulf countries produce nearly 6 percent of the world's total. Another 7 percent is contributed by the remaining Asian nations. The production is steadily improving in Southeast Asia (Malaysia, Indonesia, Cambodia and Vietnam) with the technological assistance provided by the West.

Uranium Asia's reserves of this important mineral critical to the generation of nuclear energy for commercial and defense needs, is small and limited mostly to China and India. The production of nuclear energy and its consumption is largely in the hands of a comparatively few industrialized countries such as the U.S.A., Russia, Japan, Canada, the United Kingdom, France, Germany, Italy and Australia. Among Asian nations, Japan is the largest consumer, which utilizes about 10 percent of all the nuclear energy produced in the world. Besides Japan, a few other Asian nations are now engaged in the production of nuclear energy such as China, Israel, India, Pakistan, North Korea and South Korea. Among these, China and India have already been able to develop technology capable of producing nuclear weapons while North Korea, Pakistan, and Israel are probably close to doing so.

Nuclear proliferation has become a major issue with the West, even as several nations in Asia are developing nuclear energy.

Japan and the newly industrialized countries (NICs) now face a crucial challenge: How to bridge the gap between the cheap but declining conventional resources of oil, gas, and coal, and the yet to come (mostly experimental and possibly expensive) energy sources of the future such as nuclear, solar or geothermal? Japan's dependence on imported oil for as much as 90 percent of its total energy needs illustrates this dilemma which many of the developing nations in Asia will have to face. Although Asia's resources of coal are abundant; these are located chiefly in China and India, and are becoming increasingly of lesser importance as new, clean and efficient sources of energy are being developed. As we have already noticed, Ural oil as a source of energy is limited. Some depressing prognosticators already remind us that if the world's proven oil reserves (most of which lie in Asia) might run dry in 40 years at world's current rates of consumption.

METALLIC MINERALS

Iron Ore Iron-ore reserves of the world are very extensive, estimated to be over 65 trillion metric tons, of which less than 1 percent is annually mined. On account of the increasing use of plastics, and recycling of scrap metals, the world's supply of this useful metal is going to last for a long time. Asia's reserves amount to nearly one-eighth of the world's. Nearly all Asian nations contain some deposits of the ore; China and India contain most of it. South Korea, Taiwan, Malaysia, Sri Lanka, Kazakhstan, and Turkey have also fairly large deposits, although production in these countries is limited.

China and India, together, account for one-fifth of the world's annual output. China's reserves are widely distributed, and contain varying grades of ore. Major mining, smelting and iron manufacturing centers are located in the northeast (Anshan, Fushun); near Beijing; area west of Shanghai; in Central Changjiang basin near Wuhan; in Inner Mongolia, north of Baotou; and in the island of Hainan off the southern coast. Most of China's manufacturing regions are developed close to these coal fields.

In India, fields are located in the Chota Nagpur Plateau in southern Bihar and Orissa, in Andhra Pradesh, and in several areas of southern and western Deccan Plateau (near Goa and Kerala). Generally speaking, larger fields exist in close proximity of the coal fields and iron and steel manufacturing centers. India's iron-ore is of high grade (metal content). Some of it is exported (principally to Japan) but most of it is consumed domestically.

Indonesia, Malaysia, Myanmar, the Philippines, Taiwan, Thailand and Turkey produce modest amounts, most of which is exported. Pakistan and Vietnam produce only limited quantities. In Central Asia, major fields lie in north Kazakhstan, east of Aral Sea. With the exception of China, which utilizes nearly all of its production. Nearly all the Asian nations export the mined ore to Japan and Europe. Japan is the world's biggest importer, consuming nearly one-third of the world's total imports. Its principal suppliers are Australia, India, Brazil, South Africa, Malaysia and the Philippines.

Ferro Alloy Metals With the exception of manganese, tungsten, chromite and vanadium, Asia's reserves of ferroalloys are limited, but the production is fairly extensive amounting to almost one-half of the world's tungsten, one-eight of manganese, and one-sixth of the world's chromite. China, South Korea, Mongolia, North Korea, Myanmar, Kazakhstan, and Tajikistan are the principal producers of tungsten. Turkey, India and the Philippines are the major chromite producers. China and India are important producers of manganese.

Nonferrous Metals Reserves of tin and bauxite are abundant in Asia, while those of copper are moderate. Four Asian countries: Malaysia, Indonesia, Thailand and China account for nearly one-half of the world's tin production. The reserves of bauxite are fairly large, about 12 percent of the world. India, China, the Philippines, and Malaysia are the chief producers but production is limited (about 6 percent of the world's total output). Reserves and productions of copper both range between 9-11 percent of the world. China, the Philippines and Kazakhstan are the major producers. Turkey, Israel, India, and North Korea produce modest amounts. It may be pointed out that, in general, China is particularly rich in most of the nonferrous base metals such as mercury, magnesite, antimony, lead and zinc. Asia's production of zinc is about 15 percent of the world's production; of lead 12-13 percent; and of nickel 8-9 percent. China and North Korea are the chief producers of lead; China, Japan and North Korea of zinc; and Indonesia, China and the Philippines of nickel. The Transcaucasia area is also rich in several nonferrous minerals.

Precious Metals Traditionally, gold was obtained mainly from the alluvial stream deposits and many Asian countries (Myanmar, Indonesia, Cambodia, and China) were its important producers. The current production is insignificant like other previous metals such as silver and platinum. South Africa, Russia, U.S.A., Canada and China are the world's major producers. China and the Philippines are Asia's major gold producers. Japan, North Korea, South Korea, Taiwan and Tajikistan also have some gold ore reserves. Asia's production of silver and platinum is negligible. Japan is the only country of some consequence that produces silver (about 2 percent of the world's production).

Nonmetallic Minerals Reserves of asbestos, gypsum, mica, phosphates, rock salt and sulphur are relatively large. Russia and India are the principal producers of mica; China, Jordan and Israel of rock phosphates; Israel of potash and Iraq of sulphur.

ENERGY RESOURCES: WATERPOWER

A large portion of Asia's surface is blessed with circumstances favorable for the development of waterpower and for irrigation in terms of the nature of its terrain, volume of water in rivers, existence of waterfalls, etc. Nearly a quarter of the world's potential is estimated to exist in the continent, but only one-fifth of the world's total capacity is developed, and thus a good deal remains undeveloped. Although water resources of China, India, Japan, Pakistan, Bangladesh, Myanmar, Sri Lanka, Indonesia, the Philippines, Laos and Vietnam are potentially abundant, only Japan has been able to fully exploit its waterpower resources. Japan accounts for about 6 percent of the entire world's, and a little less than one-half of Asia's total developed capacity of waterpower. Potentials in Russia are enormous, but exist primarily in Siberia, where rivers have an excellent hydroelectric potential, but extreme cold, low winter levels and inaccessibility hinder the exploitation. (Russia is considered a European nation for this analysis. It ranks below the United States and Canada as a producer of hydroelectricity in the world.)

China leads in potential waterpower resources in Asia. With its extensive river network, and large area of mountainous terrain, China leads in potentials of waterpower resources in Asia, but it has not yet fully tapped its vast potential. It has concentrated chiefly on thermal sources, and more recently to a smaller extent on nuclear sources for the development of energy, particularly on account of the easy availability of coal of which it has enormous supplies. Most of its hydroelectric development has taken place in the southern region (i.e., south of the Qin Ling Mountains) where coal supplies are poor, but the demand for energy is slowly on the increase. Sites in North China, in the basin of river Chang Jiang are expensive to operate. Development has taken place near I-chang and the industrialized regions lie along the coast, especially around Guangzhou (Canton) and Hong Kong, and the lower Chang Jiang basin around Shanghai. Northeast China is rich in coal and oil and Central China has vast reserve of coal. Therefore, neither the hydro development nor the conditions suitable for energy production exist in these regions. In the lower Huangho basin, where the consumption of electricity is greatest, the topography is flat, the river flow given to seasonal and erratic fluctuations and flooding; and hydro development is much below the capacity.

In the Indian subcontinent several areas are well-endowed for the development of hydropower. A broad belt along the Himalaya in Pakistan, India and Nepal; the slopes of the Western Ghats in India; and the central mountainous region in Sri Lanka are particularly suitable but development is hampered by the seasonal fluctuations in rainfall. In India, the Western Ghats which slope abruptly to the western coastal plain permit high waterfalls, but contain little water during winter. Rivers on the eastern part of the Deccan Plateau also dry up in winters, and the dam construction in hard rocks of the plateau present additional problems. In the Himalayan region on the other hand, where terrain, rainfall and water storage conditions are suitable, problems of inaccessibility and distance from demand areas have caused the hydropower development to lag.

Hydropower development, therefore, has taken place at sites where the Himalayan rivers enter the Indo-Gangetic Plains and their intersection, and a few sites on the eastern slopes of the Deccan Plateau.

The other noteworthy region where hydroelectric development has taken place is the D.V.C. (Damodar Valley Corporation) region, which is a multipurpose project (besides electricity generation, river waters are harnessed for irrigation, and river transport purposes) modeled after the Tennessee Valley Authority of the United States. The DVC region is located close to the mineral and industrial region of South Bihar and Kolkata which utilize its electric power. Nepal has immense potentials for hydroelectric development. If developed and utilized within the country and exported to India (which is currently the principal market for most of the power generated in Nepal), it could spur the development of Nepal's economy.

Waterpower resources of Southeast Asia are abundant particularly in Indonesia, Myanmar, Vietnam and the Philippines, but not much development has taken place. The Indonesian resources are particularly rich on account of the country's rugged terrain, plenty of water in the rivers, and the existence of waterfalls, but inaccessibility to the densely-forested interior highlands of Maluku, West Irian, Kalimantan and Sumatra pose formidable problems for development. In the Philippines most power plants have traditionally utilized imported oil, but hydroelectric power generation has increased since the construction of several multipurpose dams in Luzon and Mindanao islands in the 1980s. These projects have also benefited flood control and irrigation and reduced dependence on foreign oil.

In East Asia, Japan, North Korea and Taiwan, where water resources are fairly abundant, have been extensively utilized. Japan has made full use of its water resources but the demand on power is larger than the supply necessitating the import of oil, and the exploitation of thermal sources. Most of the country is mountainous, and the resources are unevenly distributed. Development is concentrated mostly in the largest and central island of Honshu, and parts of the southern island of Kyushu close to the important industrial regions. The best sites are already utilized and further extension is becoming cost prohibitive as compared to that of the imported oil and nuclear energy. Most hydro plants operate for a few months of the year because of the seasonal variations of rainfall, and partly due to problems of adequate storage.

Energy resources of North Korea consist primarily of hydroelectric sources and that of South Korea of coal and imported oil. Hydroelectric power stations were built during the Japanese control of the country along the Yalu River bordering China which now supply most of North Korea's energy requirements, but thermal and nuclear energy is being increasingly utilized, partly to offset the deficiency of hydroelectric power during the dry season. In Taiwan power resources and development have taken place in its northern and central mountainous regions. During the 1960's and 70's dependence for power generation was shifted to the imported oil and nuclear sources as the costs of imported oil escalated. Currently nuclear plants provide nearly one-half of Taiwan's total power production.

With the exception of Transcaucasia waterpower resources of Southwest Asia are very limited. Only Turkey, Afghanistan and Iran contain mountainous areas where rainfall and volume of water in the rivers are adequate for the development of hydroelectricity. Iran is rich in oil, and has little dependence on hydropower. The Pontus and Taurus Mountains girdling the vast interior plateau of Turkey; the Elburz Mountains in Iran, and the mountainous region of northern Afghanistan offer suitable sites for hydroelectric development. Development is, however, marginal. Afghanistan is probably richer in this resource, but the demand for electricity is limited except near the capital. Hydroelectric power has, however, been well developed and intensively used in the southern part of Transcaucasia, particularly in Armenia, Azerbaijan, and Georgia. A number of electric stations in the region feed the several industrial centers.

As an important footnote to the energy production, it may be stated that hydroelectric power as a source of energy production has a limited future although it is widely appreciated because of its environmentally "clean" character, and its ability to revitalize the economy of remote mountainous regions. High cost of construction and maintenance of plants particularly in the remote, mountainous areas, where most sites are available, and the increased prospects of availability of sources like nuclear and solar power are likely to pose increasing competition to hydro energy in the future.

Further Readings

Eyre, S. *The Real Wealth of Nations*. New York, 1978.

Glassner, M.I. (ed.), *Global Resources: Challenges of Interdependence*. New York, 1985.

Sawyer, S.W. *Renewable Energy: Progress, Prospects*. Washington, 1986.

World Resources Institute, *World Resources*. New York, 1987.

10

Industrial Development in Asia

We live in an age of mass production, and measure our economic prosperity by the number of goods and services provided by industry. Modern industry aims at, and has been able to achieve to a remarkable extent, the production of large quantities of goods within the reach of the largest number of consumers. Industrial development has thus become an index of the prosperity of nations.

Despite its enormous size, relatively extensive mineral and human resources, Asia's contribution to the world's industrial output is small. Although spectacular gains in industrial production have been made by several nations such as a South Korea, China, India, and Israel, the production of manufactured goods in the continent remains limited. There has been, however, a discernible general trend during the last few decades or so toward a transition from the traditional, light and small-scale industries to heavy, metallurgical and sophisticated, high-tech industry. A most striking feature of these changes is visible in the "Pacific Rim" nations from South Korea to Indonesia. However, only three of the Asian nations: Japan, India and China produce and consume large quantities of steel, which is usually considered an important indicator of modern industrial development.

FACTORS OF INDUSTRIAL LOCATION

The pattern of industrial location in Asia, as indeed elsewhere, can be better understood if we analyze the factors that are responsible for the distribution of manufacturing activity. It is perhaps possible to recognize several of these individually, such as market, transportation, supply of materials, labor, capital cost, or the role of government, and above all the "human factor" that finally decides the precise location of individual manufacturing plants. But all these factors are interactive, and almost never operate singly or separately, although any one may be more important than the others in different situations. Taken together, they enable us to understand the complex web of operations that defines the spatial patterns of industries.

Market The triad of close-related factors: market, availability and supply of raw and semi-finished materials, and transport are, of course, of greatest importance in explaining the location of industrial activity anywhere in the world. Most critical of these generally is the market factor. Barring other compelling circumstances, industry will tend to locate closest or accessible to market outlets, thus reducing the shipping costs, as well as being in a position to adapt quickly to changes in the market.

"Market" can also imply, in some instances, proximity to large populations, presumably with some purchasing power, as for example, the cities. Industries producing consumer goods, have thus been located in, or near urban areas. As these urban areas have expanded, so have their usefulness to still more industry. Cities thus attract industry which in turn attracts additional labor, and services creating a larger city which becomes a bigger market for the expanding and proliferating industries. This concept in modern economic development is now being applied in the sense of revitalization of backward regions. Several new industrial towns and townships in India (near Delhi) and China (near Beijing) and Indonesia were developed on this economic theory.

Cities are also likely to have a greater variety of specialized markets and to represent greater wealth (i.e., purchasing power) than smaller communities or rural areas. Cities thus generate an exclusive and a greater range of manufacturing activity. The concentration of high-fashion clothing and luxury goods as manufactured products in Tokyo, Mumbai, Hong Kong, and Singapore is an example of the relationship between cities and manufacturing. It is thus no surprise to learn that the growth of industry and the expansion of cities have been contemporary phenomena.

Even though accessibility to market is a crucial factor in industrial location, markets may not always be domestic, but may be thousands of miles away. In such cases manufacturing tends to be concentrated in or close to the parts that are well connected to the outside world. The parts usually contain banking and trading facilities enabling the manufacturer to reduce shipping costs in comparison to the inland locations for the industry. Many of Asia's manufacturing regions are located close to its major ports such as Tokyo-Yokohama, Osaka, Shanghai, Mumbai, Kolkata, Jakarta, Bangkok, and Hong Kong.

In some instances, "market" represents something entirely different from simple concentration of people. In cases where manufacturers may have little appeal to vast majority of urban population, and manufacturing plant may have nothing to do with urban location, as for example, the basic production of pig iron or the production of several types of chemicals. Such manufacturing plants will locate to market for reasons other than the proximity to urban consumers. For these manufacturers, "market" assumes a different concept than for the maker of consumer goods.

Materials Supply While accessibility to materials used in manufacturing process is of vital interest to the manufacturing plant, sometimes markets may be half a world away from the source of material supply. The factor of "materials supply" is sometimes identified as "raw materials," but this term is misleading as it implies strictly unprocessed materials (coal, iron ore, wood, etc.) whereas in reality, materials used by a majority of manufacturing plants are already at least semi-finished (steel plates, cables, refined chemicals, electrical wires or semiconductor elements, etc., the last one used in the manufacture of electronic equipment).

Generally speaking, if the materials used in manufacturing process are bulky, heavy or low in value, it is advantageous to locate the industry as near as possible to the source of material, to reduce the excessive costs to transport the items to the markets. A good example of this type of location is the manufacturing of cement, which is nearly always located near the quarry from which the product is made. In fact, it is often cheaper to build a new plant near the site where material supply is available than to transport the material an extra few hundred miles. An industrial nation like Japan for example must import a large share of its unprocessed materials (oil, iron ore, food grains and wood). A major share of Japanese industry dealing with these items has become concentrated in urban areas around ports along much of the southern coast and the Inland Sea from Tokyo to Shimonoseki.

However, the source of supply may not always be the determining factor in the location of manufacturing activity. For example, if the manufacturing activity deals with the utilization of materials that are small, light-weight, or of a high value, there may not be any compelling reason to consider location on the source of supply, since any shipping costs are likely to be a minor item in the total price structure of the finished product. Many high-tech industries that might use such "raw" materials as electronic components or silicon chips by having those transported to the manufacturing sites by airfreight. Thus, unshackled from basic locational constraints such industries will locate to other locally desirable facilities. Several Japanese firms have located their branches all over the world utilizing the locally available facilities. On the other hand, several European and American companies have their manufacturing plants in Thailand, Malaysia, Indonesia, Pakistan (and more recently in India) where locally available skilled labor, and "local market" more than offset the costs of transporting the "raw materials" or the finished products to other parts of the world.

Transportation In determining industrial location manufacturing plants tend to be sited where a variety of readily accessible transport choices are generally available. Transport facilities thus are important links that bridge material supply and market. Railroads, highways, and airports close at hand are important considerations for the location of certain industries. Often of great importance in the decision regarding the type of transport most desirable for location of manufacturing plant is the cost involved in the infrastructure and cost of moving goods to and from manufacturing site. Those areas where transport network is most dense have thus a built-in advantage in attracting new industries provided suitable conditions of "material supply" and "market" exist. In general, riverways and railroad transport are of special usefulness in hauling low-value and high-weight materials as these are usually the cheapest means of moving goods around, particularly where time pressure is not a weighty consideration. Secondly, the relatively flat areas (the plains) have an advantage over mountainous regions, because the mountainous areas generally lack in transport infrastructure. The railroad and road transport usually offer greater flexibility and faster service. Railroad is generally preferred for long-distance hauling, and road transport for shorter distances. Air freight is generally the most expensive means of transport and used mostly for high-value, low-weight products, where the materials transported are very valuable and time demands are critical.

Labor There was a time when labor supply was an important factor in industrial location. Increased industrialization meant a greater demand for workers, and industrial location in cities was a necessity. The city was at once a market and a source of labor, and a close relationship between urban growth and industrial expansion existed. This was particularly true when industrial location did not depend on high-tech infra-structural facilities. During the past half a century or so, therefore, the role of labor supply as a factor of industrial location has significantly declined, but by no means disappeared, at least in most Asian nations. Moreover, labor has become migratory. Through the media (radio, television, newspapers) the news of job opportunities quickly reaches qualified workers where they may be located. Labor will then tend to move to industry rather than forcing industry to seek out labor. In yet another setting, in centrally planned economies such as China, North Korea, and Myanmar, government will move workers to industrial location. As a net result of these present-day happenings, there has been an almost universal decline of the "labor factor" in industrial location.

Capital The importance of capital as a factor of industrial location has considerably diminished as vast amounts of capital have now accumulated in the developed countries. But capital remains scarce in developing countries of Asia where capital may be provided either by the government or a few affluent entrepreneurial classes. For much of modern Asian history capital has been a rare commodity, concentrated in the hands of a few banking families, all having special interests. The classic example is that of Japan where most industries are in the hands of a few banking families, the Zaibatsu cartels, which have historically given priority to their family interests in determining the location of Japanese industries. Much of modern Japan's industrial rise has been due to the rapid growth of capital investment of the Zaibatsu clans and the administration. An important aspect of the Japanese financial system is the high degree of independence between the Central Bank (the Government's bank), the commercial banks (owned mostly by the few banking families) and industry.

In centrally planned economies, all locational decisions of manufacturing plants are determined by the governments as in China, and Myanmar, because capital is entirely provided by the state. In nations like India and Indonesia where there is a mix of private and public enterprise, industrial locations may often be distributed over several regions of the nation in order to bring out a "balanced" regional development. The government usually makes decisions regarding the location of heavy industry like metallurgical installations or industries of strategic importance like the petrochemical industry or the manufacturing of railroads, aircraft, defense-related material, and leave out light industries like textile manufacturing and food processing, etc. for private enterprise. Public funding may also be provided in the form of subsidies or imposition of import restrictions to spur industrialization. India is a good example of nations that have followed this policy. Nonetheless, the final selection of a manufacturing site owes its location at least partially to where the needed money could be obtained.

Government Planning The role of government in determining the location of modern industry is of vital significance in nearly all Asian nations. This is in marked contrast with the United States, for example, where this factor plays almost no part in determining plant locations. China and the Soviet Union are, of course, the classic cases, where all means of production have been controlled (liberalized in the Soviet Union since 1990) by the state, which makes all locational decisions. But even in several parliamentary democracies such as India, Indonesia, and Pakistan governments' influence is surprisingly pervasive. In these nations, as in several other Asian nations, all public utilities are nationalized. Power production (coal mining, electricity generation, nuclear plants) is mostly a government monopoly, as is the manufacture of iron and steel for railroad transportation, aircraft production and shipping. Financial institutions (banks, etc.) that provide capital to private industry are also nationalized by nearly all nations in Asia, although there has been a trend toward denationalizing (or "privatizing") them recently.

What this clearly means is that by controlling much of the necessary infrastructure, Asian governments, like most of the European ones, are in a strong position to influence the location of new industries. Industrial policy has also become a tool that governments can employ to fight unemployment or equalize regional imbalance. Governments also influence location of manufacturing plants in the private sector by providing incentives for privately-controlled industry to locate in selected parts of the country that can be identified as lagging behind national norms. Such "development zones" receive special privileges such as tax abatement, and the provision of infra-structural facilities (utilities, transport low cost or free land, low interest rates, etc.). India, Pakistan, Indonesia, Singapore, Saudi Arabia, Kuwait, Israel, Taiwan, South Korea are among the several nations that follow this policy. But it must also be noted that no amount of government encouragement can make an industry profitable unless its location is economically and geographically sound.

Other Considerations: Power & Fuel Costs remain important factors for industrial location for most nations. For example, in certain industries such as aluminum refining, power costs are of greater significance than some other factors because fuel costs form a major share of the price-structure of the finished product. An abundant supply of hydroelectric or thermal power has become an important factor in modern industry in general. Countries with good supply of cheap power have a decided advantage over others in their industrial output. Modern Japan's industrialization, for example, owes its development significantly to the abundant, cheap supply of power.

Space as a factor in industrial location has lately been a factor of some importance. Modern industrial plants require much larger space in order to accommodate the growth of assembly-line operations than the traditional ones which were located in sites highly impacted by dense urban growth. Many industries have moved away from their locations within the heart of the city to the countryside where land costs are much lower. In the relocation process, however, they remain close to the city in order to make use of the urban market and the infra-structural facilities like transport connections, utilities, etc. that the city generally provides. Many of the new industrial ownerships (industrial estates) have been established by the Indian government near the urban places. This has generated further industrial activity and the growth of smaller cities around such townships in the private sector as well. A string of small industrial cities around Delhi belongs to this category. A purely government controlled industrial development utilizing vacant space near Beijing (China) is another example of modern industry fanning out in the countryside. The need for more efficient, and spacious design has thus become an important factor in industrial expansion.

Human Factor A final statement with regard to industrial location may now be made. While the foregoing discussion helps us to understand the broad patterns of industrial location, clearly the individual locations of industrial sites cannot always be explained by the factors described above. Many of the manufacturing plants may be sited in locations which in principle will defy the considerations enunciated above. These operations owe their existence to nothing more than a personal decision taken earlier by the founder under circumstances much different than prevailing now. In many cases such plants may once have been suitably located. Thousands of such plants survive, based on inertia, and a sense of family loyalty. The *human factor*, however, irrational it may appear in the present context, is responsible for such decisions.

DISTRIBUTION OF INDUSTRY IN ASIA

As a broad generalization the level of industrialization for most Asian nations is low. Only a few nations such as Japan, Hong Kong, and Singapore show percentages of non-agricultural employment that approaches those of such high-tech countries of Europe as Germany or France, and the United States, although China and India rank high in the total volume of industrial output. Taking most indicators of industrialization, Japan ranks among the leading manufacturing nations. Map 10.1 identifies the main industrial centers of Asia that resemble the manufacturing regions of the United States or western Europe in scale and degree of manufacturing activity and the impact on landscape. A brief account of these is given below.

Japan Most manufacturing activity is confined to an industrial-urban belt running from the Tokyo metropolitan region southward to northern part of Kyushu Island within which are located four of Japan's largest industrial nodes: Tokyo-Yokohama, Nagoya, Kobe-Osaka, and Kitakyushu. In addition, an almost continuous series of manufacturing towns flank the Inland Sea. A few secondary manufacturing districts are located in both coasts of Central Honshu, and in southwest Hokkaido. Of these, *Tokyo-Yokohama* region accounts for nearly one-third of Japan's total industrial production. Manufacturing is highly diversified in this region, ranging from the production of iron and steel machinery to publishing, shipbuilding, electrical and electronic materials. The port of Yokohama has an eastward trade orientation toward North America. *Nagoya industrial complex* is less diversified. Textiles and ceramic pottery are the two important products, although a few other industries are also represented. *Kobe-Osaka* region emphasizes on heavier industries such as shipbuilding, diesel plants, but also contains Japan's major textile, chemical and cement industries. Kobe is an important overseas port with westward orientation towards Southeast Asia. The fourth region is a sprawling belt of a continuous string of manufacturing towns all along the margins of the *Inland Sea with Kitakyushu* at the western tip. The region produces such diverse goods as cement, glass, chemicals, sugar, beer, steel fabrications and ships. The region is an important funnel for a large number of Japanese imports because of its proximity to other Asian nations which are Japan's chief exporters (the Middle East, South and Southeast Asia, and China).

Japan is now a world's manufacturing giant, ranking second only to the United States. It is a shipbuilder, a major auto maker, world's foremost producer of crude steel, synthetic rubber, aluminum, plastics, cement, paper, cotton yarn, chemicals, precision goods (notably cameras), electric and electronic equipment, (including computers and telecommunications materials) and petrochemicals, and has some of the world's largest and most advanced industrial plants.

China Prior to the Communist takeover, manufacturing activity in China was very limited, consisting in the manufacture of textiles, metal goods, and of goods produced by its handicraft (cottage) industry. Since 1950 the Communist government has paid particular attention to the development of modernized industry, with the result that the overall industrial output has grown at a rate averaging 10 percent a year during the last four decades. Industry has, in fact, surpassed all other sectors of economic growth and in the degree of modernization. The highest priority was given to the metallurgical and machine building industries, which now account for nearly one-third of the total industrial output, although much of the specialized items must be imported for the industry. In the chemical industry, the main thrust has been to expand the production of fertilizers, plastics, and synthetic fibers. Textiles, chiefly synthetics, account for nearly one-seventh of the gross industrial output.

Since the discovery of offshore oil fields in the 1950s, the industries based on oil such as oil refining, and petrochemical products, previously located in a few inland centers, are now being developed at centers not too far from the east China coast.

Despite serious efforts from the mid-1950s to build new centers in the interior, the industries are unevenly distributed and remains concentrated near the major cities, or close to the east coast and a few inland locations. Thus Shanghai, and Guangzhou (Canton) areas on the east coast produce over 10 percent of country's gross value of industrial production. The Shanghai, and close-by centers of Huangzhan and Nanjing specialize in textile industry and the production of consumer goods. In the Northeast the manufacturing belt of *Shenyang*, *Fushun,* and *Anshan* contains the country's largest heavy metallurgical, shipbuilding, cement, and chemical industries.

The other prominent industrial areas in the country are: *Tanjkin-Tangshan-Beijing* also dealing with heavy iron and steel industry; *Guangzhou* (*Canton*) where the nation's major textile industry is located; and the inland industrial belt of *Wuhan-Wuchang* and *Hanyang* specializing in heavy iron-steel manufacturing. In addition to these major industrial regions a number of old, handicraft centers of cottage industry are scattered in the country producing primarily textile items and consumer goods.

India As in many other developing nations, industrial development was limited during the colonial rule. It consisted of the traditional, rural craft industry dealing with cloth weaving, woodwork, and making of sugar, brass and leather products, except for the two modern industries of textiles and iron and steel. Since 1950 government has attempted to expand and modernize the industry, especially through foreign aid, emphasizing in heavy metallurgical industries, and the manufacture of chemicals, and fertilizers in the public sector. Textiles and other light industries remained in the private sector. In 1993, a reversal of the industrial policy aiming at a more liberal participation of foreign investment was introduced.

The present distribution of industry shows a four-zone regional concentration, with major manufacturing regions located in northeastern India (west of Kolkata), the western coast near Mumbai-Pune area, and in the far south near *Bangalore*. The Kolkata-Jamshedpur area is the primary manufacturing region. Jamshedpur is the oldest and largest of iron and steel manufacturing center in the country. In Kolkata region several manufacturing plants are located in a sprawling belt along the tributary of Ganga River. Jute manufacture is the main industry, followed by engineering, chemical, food processing, rubber, automotive parts, paper, glass, plastics and fertilizer industries, and the manufacture of a wide range of consumer goods. The region is well served by a network of transport connections, a large urban market, port facilities of Kolkata, and draws upon the proximity to India's main mineral belt in the Chota-Nagpur Plateau.

Mumbai-Pune industrial area is the country's second ranking manufacturing region that had its early start in textiles. Since 1995, growth of several industries particularly those of engineering, chemical, petrochemical, pharmaceutical, electrical, automobile, film and food processing led to industrial diversification. Power is provided by the nearby hydroelectric plants on the western Ghats. Mumbai's easy access to the Middle East oil is an additional advantage. India's first major nuclear plant is also located nearby at Trombay. *Ahmedabad-Baroda* area dealing primarily with cotton textile manufacturing lies to the north of the Mumbai region. Petrochemical and oil refining are some newer industries in the area. In south India lies India's fourth major industrial region of *Bangalore-Coimbatore*. In addition to the traditional industry of textile, several newer industries have come up during the last two decades such as aircraft, automobile, locomotives, and telephone manufacture, mostly of which are in the public sector.

Industrialization in Other Asian Nations Nearly all the nations in Asia retain the traditional, cottage-type, small-scale handicraft industries, while a modernized manufacturing sector is slowly emerging. Only a few nations have been able to build a solid industrial base during the last two-three decades. Prominent among such nations are: South Korea, Taiwan, Hong Kong, Israel, Turkey, and Kazakhstan. The economies of most other nations are geared principally to the primary activities (agriculture, forestry, mineral extraction, etc.). Japan, India and China account for a large volume of Asia's industrial output. They have received a larger share of attention in our account. A very brief account of the manufacturing activities of the rest of the continent is provided in the following paragraphs.

In *East Asia* South Korea has recently taken great strides towards industrialization, and is rapidly emerging as an important producer of automobiles, chemicals, and electronic equipment, although the traditional textile industry remains the single most important one. North Korea is an important manufacturer of iron and steel, chemical products and defense-related heavy, metallurgical products (tanks, aircraft, etc.) In Taiwan, the labor-intensive light industry of consumer items (processed foodstuffs, textiles) predominated before 1970. Since then, investment shifted to more capital-intensive heavy industry producing steel, petrochemicals, machinery, and ships. In addition, as the economy prospered and the nation accelerated efforts to develop its high-volume, high-tech industries such as the manufacture of precision instruments, and electronic materials. Since WWII Hong Kong has emerged as an important producer of consumer goods, textiles, ships, chemical products, capitalizing mainly on its liberal economic policy.

Industrial development in the Southeast Asian nations has generally lagged, resulting in part from the prevailing disturbed political conditions in Vietnam, Cambodia, Laos and Myanmar but several nations such as Singapore, Thailand and Malaysia have been able to make significant progress in expanding their manufacturing base and modernizing it. Singapore offered a particularly liberal climate for foreign investors. Its manufactured items include rubber goods, chemical products, textiles, high-skill and capital-intensive materials, ships and virtually all types of consumer goods. Beginnings of modern industrialization are visible in Indonesia, Thailand, Malaysia, and the Philippines as well. While agriculture remains prominent in the economies of these nations, recent attempts to manufacture rubber goods, petrochemicals, steel products and textiles have been successful. The industry is primarily owned and controlled by the governments of these nations. Much of the small-scale industry of textiles and consumer items is owned by the Chinese community.

In the Indian subcontinent, there was very little manufacturing activity in Pakistan, Bangladesh, and Sri Lanka prior to gaining independence from the British control in the late 1940s. Manufacturing activity since then has also been slow. In Bangladesh jute manufacturing located in the capital city of Dhaka remains the chief industry. Pakistan is attempting to modernize and expand its traditional industries of textiles and food processing and since the mid-1950s has been producing chemical fertilizers, petrochemical and engineering products. Manufacturing remains of minor importance in Sri Lanka. However, like Pakistan, it is attempting to develop oil refining, chemical and textile industries.

Predictably industrial activity in the *Gulf States* (Persian or Arabian Gulf) is essentially associated with oil refining and the manufacture of petrochemical products and localized in the port cities of Daharan (Saudi Arabia), Abadan (Iran), Kuwait, and Baharain, while the traditional, small-scale industries like textiles, carpet-making and food processing continue to flourish in nearly all the countries, but more specifically in Iran, and Iraq. Among the newer industries developed during the last few decades the manufacturing of iron and steel, automobiles and defense-related products in Iran and Iraq are noteworthy.

In the Middle East the greatest development in manufacturing has taken place in Israel since its inception, but considerable gains have been made in Turkey as well. Israel produces defense-related materials (aircraft, tanks), sophisticated precision goods (Electronic and electrical equipment), fertilizers, chemicals and pharmaceuticals. Turkey's manufactured items include iron and steel, chemicals, oil refining and textiles.

In *Central Asia*, manufacturing development has largely taken place during the Soviet rule. The major industrial center is in Kazakhstan where iron-steel, cement and fertilizer industries were established in the mineral-rich Karaganda region. Plants producing heavy metallurgical machinery were set up in Uzbekistan as well. The nations of Kyrgystan, Tajikistan, and Turkmenistan contain such light industries as textiles of woolen and silk materials, food processing and the manufacture of carpets and leather products.

From the standpoint of industrial development, Transcaucasia is a better-developed region of Southwest Asia. It is rich in resources such as coal, hydroelectric power, and nonferrous metals. A number of industrial centers such as Baku, Batumi and Tbilisi producing cast iron and steel goods, locomotives, rubber and petrochemical products, cement, textiles, and processed food.

Prospects for Industrialization It is widely believed that Asia's future economic development depends largely upon the ability of its nations to industrialize themselves, but only a few of those countries have had success in increasing and modernizing their industries. Japan was the first Asian nation to modernize its industry so that the country has become a world leader in industrial production and a major exporter of manufactured goods. Israel, South Korea and Taiwan have recently made significant progress in industrial development. Before its breakup, the Soviet government made efforts to industrialize the Asian part of the country. China and India have large factories, but lack modernization of the plants.

Several Asian governments have recently attempted to improve the economies of their nations by creating new industries and expanding old industries, and have taken steps towards industrialization, which include offering low tax rates to foreign investors, and by opening up their markets to non-Asian nations. Indonesia, Israel, Hong Kong, Pakistan, Malaysia, Singapore, Taiwan, Thailand, Turkey and South Korea belong to this category. In 1993 India liberalized rules regarding foreign investment to bolster and modernize its industry. Most Gulf States countries have been developing their petrochemical industries with foreign participation. China has also been lately (since 1992) inviting foreign participation, and loosened its administration's noose over some important industrial undertakings. While all this is well and good, in general, most Asian nations have so far achieved a limited degree of industrial development.

On the other hand, industrialization unquestionably has a great scope for expansion, given Asia's potentially extensive resource base, but poor markets, little capital, strict and narrow government policies regarding foreign participation, and the usual emphasis on "big industry" to the neglect of an overall industrial development, have impeded real advances in industrial growth.

Further Readings

Chisholm, M. *Modern World Development.* New York, 1982.

Hamilton, F.E.I. (ed.) *Contemporary Industrialization.* New York, 1978.

Smith, M. et al. *Asia's New Industrial World.* New York, 1985.

Taylor, M. "Industrial Geography," *Geography*, Vol. 8. 1984.

IV

Nations of Asia

11

Southwest Asia: An Introduction

Lying at the southwestern corner of the huge Asian landmass is a large stretch of territory from Turkey to Afghanistan that occupies over two and a half million sq miles in area, and contains nearly 250 million people and over a dozen and a half sovereign states, two neutral zones, and bits and pieces of territory (Gaza, West Bank, and Jerusalem). Although geographers have generally settled on the term "Southwest Asia" to identify this section of the earth, the label the "Middle East" or the "Near East" have also been popularly applied to denote this region. However, these terms, on closer inspection, represent an inaccurate reflection of colonial chauvinism. People in London or Paris viewed China and Japan as the "Far East," while Iran and Saudi Arabia became the "Middle East" or the "Mideast"; Greece and Turkey became the "Near East." But what do the terms such as "Middle East" and "Near East" really mean? Middle of what? East of where? All places are central to somewhere. Thus, places can be usefully named from where they are *per se*, not merely in relation to somewhere else where the viewer may be.

If there is a justification of the term Middle East for the region, it lies in its centrality and its strategic location at the crossroads between Europe, Africa, and the bulk of Asia. Since earliest times the region has served as a route for people, trade, armies, and ideas moving from Europe eastward to Africa and Asia, and westward from Asia to Africa and Europe. North of Beirut are nineteen inscriptions in eight languages left by passing conquerors, beginning with Ramses II (c. 1220 B.C.) through Nebuchadnezzar, Marcus Aurelius, and Napolean (1769-1821). Today, the world's major sea routes from Europe to Asia pass through the Suez and Bosphorus straits and airroutes converge on Beirut, Kuwait, and Istanbul.

Cultural Patterns Southwest Asia is rich in history and culture. Few other parts of the world have provided a comparable homeland for great empires, civilizations, and religions. Several of the civilizations described by Arnold Toynbee in his survey of world history had their cradle here. Many of the earliest recorded events took place in towns like Jericho, going back to 6800 B.C. Three of the world's great monotheistic religions, Judaism, Christianity, and Islam, originated here and diffused to other parts of the world; their conflicts and interactions shaped the course of modern times. Great empires such as the Babylonian and the Assyrian, dating back to the fourth millennium B.C., the Persian Empire that reached its maximum extent in the 6th century B.C., the Arab empires, the Abbassid and the Ommaid, centered on Baghdad and Damascus,

Table 11.1 *Southwest Asia: Basic Data, 1999*

Country	*Area Thousand sq km/sq mi*	*Population (Millions)*	*GNP $/capita*	*Capital*	*Urban-populatio n Percent (est.)*
Baharain	0.7/0.3	0.6	7,840	Manama	88.4
Cyprus	9.2/3.5	0.8	13,702	Nicosia	68.0
Iraq	435.0/167.9	21.7	1,250	Baghdad	74.6
Iran	1,648.0/631.6	61.5	2,150	Tehran	57.3
Israel	20.3/7.8	5.7	15,870	Jerusalem	86.9
Jordan	89.3/34.4	4.6	1,650	Amman	78.6
Lebanon	10.4/4.0	3.5	2,970	Beirut	60.1
Kuwait	17.8/6.9	2.3	20,470	Kuwait City	97.0
Oman	309.5/119.5	2.2	5,950	Muscat	71.7
Qatar	11.4/4.4	0.6	11,600	Doha	88.0
Saudi Arabia	2,248/864	20.8	6,660	Riyadh	77.3
Syria	185.1/71.5	15.3	1,160	Damascus	52.4
Turkey	779.4/300.9	64.5	2,450	Ankara	59.0
United Arab Emirates	83.6/32.2	2.7	22,470	Abu Dhabi	84.0
Yemen	555/214.3	16.3	280	Sana	23.5

Source: Encyclopedia Britannica, Book of the Year, 2000.

(7th to 12th centuries) and that of the Turkish Ottomans (14th to mid 20th centuries) were established here. The eastern shores of the Mediterranean provided many avenues of conquest for the invaders: Assyrians, Persians, Greeks, Romans, the Crusaders, Arabs, Ottomans, and finally the French and the British. All these have left their indelible imprints on the region.

Few regions of the earth exhibit a greater cultural variety and complexity. Despite the fact that Islam is one of the few pervasive threads that tie the region together, cultural unity is far from being the case. Although four major cultures-Arab, Persian, Turkish, and Jewish-characterize the region, large numbers of diverse ethnic minorities in virtually all the nations of Southwest Asia have been the source of prevailing social unrest and political instability. Many of these groups number over a million each. Nearly five million Kurds are spread over Turkey, Iraq, Iran, Syria, Transcaucasia, and Afghanistan. Over two million Azerbaijanis are located in northern Iran. Nearly half of the populations of Iran and Afghanistan are composed of a dozen or so minority groups. Turkey once had Armenians and Greeks, numbering over a million each; Armenians have been liquidated or forced to flee and Greeks repatriated to Greece in exchange for Turks living there. Russians form significant minorities in Transcaucasia. The Gulf nations and Oman have attracted large numbers of Indians, Pakistanis, Sri Lankans, and Filipinos primarily to work in the service industries.

Islam is the predominant religion overall; Christianity and Judaism are numerically much smaller, the latter being nearly exclusive to Israel, but exceedingly significant from the political standpoint.

Much of the instability of the region has resulted from the interaction between the religious distributions. Perhaps nowhere else does religion play such a significant role in the daily lives of the people or influences the decisions and actions of nations to such a remarkable degree. A tangible evidence of conflicts arising out of religious discord can be found in the ancient city of Jerusalem, the most sacred of all places to Jews and Christians, and after Makkah and Medina to Muslims. In the old section of the city, venerated sites of all three religions are closely interlocked, a situation that has occasionally caused serious friction among the various sects. Additionally, a number of Christian and Muslim sects, and remnants of an ancient religion, Zoroastrianism, that inhabit the region further compound the already sectarian politics of the region.

Judaism is the oldest of the major modern religions native to the region. The Jewish people had established a small state in the highlands of Judea, the present-day "West Bank," during the second millennium. Since their conquest by the Assyrian kings in the 8th century B.C. and later dispersal by the Roman overlords in the 1st century, Jews became established in many parts of Europe and Asia, where they absorbed much of the culture (and even some of the racial traits) of the peoples of lands of their settlement. Yemenite, Persian, Georgian, and even Abyssinian Jews are distinguishable at the present day, and are in sharp contrast to the Jews of Europe who settled in Spain or Poland.

The creation of the state of Israel in 1948 has again provided the Jews a homeland. The variety of people who have emigrated to the new state is impressive, and each group has brought with it ideas and customs characteristic of those parts of the worlds in which their ancestors had settled during the centuries of the "Diaspora." Although most immigrants feel themselves a part of rebuilding an ancient homeland, there is diversity and disagreement, unrest and even terrorism in their methods of making their adopted land and nation secure. European Jews, and the Middle Eastern "Sephardic" Jews, and the Reformed and Orthodox versions of the faiths are often engaged in bitter doctrinal opposition. Moreover, perhaps nowhere else except in Iran is religious fundamentalism as strong as it is here.

The second major older religion in South-west Asia is Christianity, which began as a sect of Judaism. In 313, Christianity was adopted as an official religion of the Roman Empire. By 330 Constantinople (now Istanbul) had supplanted Rome as the capital of the Roman Empire. Despite Constantinople's position in the Christian world the predominantly Christian nation had failed to establish itself in the succeeding centuries despite the massive efforts of medieval Europe. This is perhaps a reflection of the vivisection in the Christian church which experienced a growing divergence in doctrine and rituals in areas so far from the Roman authority. From Constantinople arose the Greek Orthodox Church; from Alexandria the Coptic Church; and from Antioch the Syrian Church, all of which became in due course completely independent of Rome. Today, Christian religious communities-largely Catholic-remain active. Syrian, Armenian, Chaldean or Assyrian, and Maronite are a few of the communities that are distributed in varying degrees of strength. In Lebanon, where the Christians form more than a quarter of the population, they remain a vital political force in the nation's parliament.

Of all the religions that have taken root in the region, however, the most spectacular has clearly been Islam. In terms both of speed and depth of impact on the culture, language, and politics, the Prophet's faith had substantial implications for most of the known world of his times, and lasting effects on the region. Muhammad died in Medina in 632 and succeeded in bringing the vision of a monotheistic religion to the nomadic tribes. Following his death, his fiercely zealous followers invaded the long-established kingdoms of the times in all directions and succeeded in establishing the faith of the Prophet and an extensive empire. By 750 their control extended from Spain in the west to what is now Pakistan in the east.

With their conquests, the Arabs brought to a large part of the world Islam, the Arabic language, and a vast range of art, literature, science, and technology. Arabic, which represented the embodiment of power and authority, became influential wherever the conquest spread, finally to be established as the most important medium of communication in areas between the eastern shores of the Mediterranean and the Persian Gulf. It now ranks as one of the world's most widely spoken tongues.

The unity of the Islamic church was, however, short-lived. Muhammad died without naming a successor as the leader of the faithful. A majority of the followers felt that the leader should not be hereditary, but elective among the senior leaders qualified to rule. Thus, the first "Caliphs" ("successors") held elective positions, and continued to govern from Medina; in their time most of the impressive expansion of the faith occurred, and their line of authority became the dominant "Sunni" form of Islam. A few of Muhammad's followers, who believed that leadership should be based on kinship, chose Ali, Muhammad's nephew, as the rightful Caliph, but he was murdered. His followers were known as Shiites ("Partisans"), and became the troubled conscience of Islam.

To this day, the faith is divided between a Sunni majority and a sizable Shiite minority. Today ninety percent of Iran's and seventy percent of the adjacent nation of Azerbaijan's population belongs to the Shia faith. Nearly two-thirds of Iraq's, a quarter of Afghanistan's, and a fifth of the United Arab Emirates' populations profess the Shia faith, while Yemen's population is almost equally divided between the followers of the Sunni and Shia faiths. The Shiites have historically resented the "second class" status that they occupy in the Muslim world, and a large part of the social and political tension and conflict in the Islamic world is derived from the Sunni-Shia split. Apart from the Sunni-Shia division, there are in fact several other sects in Islam, some incorporating Christian and other elements, derived primarily from the Shia line of belief. The best known and most influential among these are the Ismailis, whose head, the Agha Khan, traces ancestry to the Prophet himself; and the Druzes, who have found refuge in the hill country of Lebanon, and play an active role in the affairs of the country.

Economic Activity Climatically, the most conspicuous feature of Southwest Asia that has profoundly affected its landscape and the daily lives of the people is the extreme aridity. Excepting a few favored coastal sections of Turkey and the mountainous Trans-caucasia, a major part of the region receives less than 10 inches (25 centimeters) of precipitation a year, large sections less than a half of it, and a few under two inches. Quite possibly the interior parts of Saudi Arabia and Iran are as rainless as any in the world. The dry climate has historically posed obvious problems for economic activity. Man has been able only to use the natural environment with relentless effort, for example by the introduction of camels and the resultant caravan trading. Agricultural activity has been severely restricted to relatively few locations of reliable sources of permanent water at oases. Thus, a pattern of dense population and economic activity is concentrated in a few places separated by vast, empty spaces.

By and large, the region is agriculturally deficient, particularly in basic food grains. Except in Turkey and Yemen, less than a third of the labor force in these nations is engaged in agriculture, and almost all the nations are importers in food grains. Only Turkey could reasonably be considered to possess a balanced agricultural base. This despite the fact that historically the Tigris-Euphrates area has been the agricultural "heartland" for some 9,000 years and from Mesopotamia (mainly modern Iraq) agriculture is supposed to have moved to other parts of the world. The climate was probably Steppe rather than Desert during the past glacial times when the region was a little cooler and substantially wetter than it is today. With the climatic change, the region and the resulting rise in pastoral economies, the region experienced a long period of agricultural decline.

The Importance of Oil Few regions of the world have experienced so dramatic a surge in economic importance in modern times derived from the exploitation of a single commodity as Southwest Asia. But for the presence of an abundant supply of oil, one of the modern world's most sought-after commodities, this entire region would have remained an insignificant international backwater from the economic standpoint. Oil has changed the economic relationship of the region with the rest of the world. Southwest Asia is now more important for its intrinsic wealth than for its historic roadways; dependability of oil supplies from the region is of greater concern than the region's historic transit functions for goods and passengers between Europe and the Orient. Oil has not only transformed much of the character of the region; it has made the area a focus of aggressive interests of major external powers, and has enormously exacerbated tensions within the region.

Although oil and gas seeps have been known in many locations in Southwest Asia for millennia, enormous amounts of petroleum have been discovered and exploited only since the beginning of the 20th century. The world's biggest reserves are located around the Persian Gulf and in the Tigris-Euphrates lowland. Production in commercial quantities was started as early as 1912 in Iran, by British companies that led to a long history of European investment as the demand arose in oil-deficient Europe. The Iraqi production dates from 1927 in Kirkuk field, but later production was concentrated in the region of Shatt-al-Arab, the waterway of the joint Tigris-Euphrates waters that drains into the Gulf, the area which became the focus of the 1990 "Gulf War." Discoveries in Bahrain, the tiny island Sheikhdom in 1938, and near Damman, which proved the existence of the richest field so far, put in course of a rapidly growing number of oil fields, finds, and exploitations, in the Gulf by the American companies (earlier concessions were held by the Arabian-American oil company, the Aramco). Nearly 60 percent of the world's oil is now estimated to be concentrated in the Gulf region, fully 30 percent in Saudi Arabia alone. Presently the Gulf region (that includes Iran, Iraq, Bahrain, Saudi Arabia, Kuwait, the United Arab Emirates, and Qatar) produces close to one-third of the world's total production.

As the peoples of Southwest Asia were technologically unable to produce the oil themselves and had no industrial or commercial need for it, the exploration was given to the Westerners. Currently the U.S. is responsible for much of the exploration; the British, Dutch, German, and Japanese are also getting increasingly involved. With the legal rights for the production and sale of petroleum came the political influence of the external powers. Political sympathies of the Gulf states were increasingly sought after by the western powers as the need for oil escalated since World War II in order to gain concessions in areas likely to yield oil, as well as to ensure flow of oil on favorable terms.

Most of the production is moved by tankers, although considerable amounts flow through and out of Southwest Asia by pipelines. However, political turmoil in the region has made oil flow by pipelines risky. Pipelines cross the territories of Syria, Iraq, Jordan, Lebanon, and Saudi Arabia and were built to carry oil to the Mediterranean shore. These were destroyed during the 1964 war between Israel and the neighboring Arab states, and remained closed until the late 1970s. Large tankers are now the major means of oil transportation; the sea lanes of the Gulf are among the busiest in the world, while the narrow Strait of Hormuz at the eastern end of the Persian Gulf has acquired substantial strategic significance. Oil is among the most precious types of merchandise that now moves on the oceans, and much of it originates in the Gulf.

It is easy to understand the enormous importance of oil to Southwest Asia, where small countries with otherwise limited resources have come to rely almost entirely upon oil revenues as a means of economic support. However, this fact has also made this area particularly sensitive to fluctuations in the world economy, such as the collapse of oil prices in the face of increasing production in 1982-83. Estimates suggest that each dollar reduction per barrel costs Saudi Arabia nearly half a billion in lost revenues, and Iran nearly half a billion, and Libya over $350 million. In fact, oil prices fell by about $10 per barrel at that time, and the total losses to these oil-exporting nations are truly staggering.

To counteract the price fluctuations that could have such devastating effect, most Gulf nations had helped form the Organization of Petroleum Exporting Countries (OPEC), an international cartel that monitors and controls the price and production levels of oil of the member nations (Nigeria, Gabon, Indonesia, Algeria, Libya, Ecuador, and Venezuela are some other members in addition to the Gulf states).

During recent years, the price of oil has tended to be more than double the pre-1982 levels in the free markets, reflecting both the growing world shortages of oil as demand rises in the rapidly industrializing nations, and the determination of the oil exporting nations to use their new bargaining strength to ensure that they obtain adequate benefits from their product. It may be noted that OPEC has invested in these nations new political and economic powers in relation to the industrial countries. As the buyers of the critical commodity, international corporations and importers of oil are obliged to deal with the OPEC organization. Besides being a purely administrative body, it has acquired an important political and economic role.

As a final note, the pattern of oil distribution in the Gulf region has its bitter and ironic aspect. The tiny state of the United Arab Emirates, smaller than New Jersey and with a population of 2.3 million and oil production at a little over one million barrels a day, is literally floating on oil. It has a per capita GNP of over $18,000—over five times the world's average. Close by, at the other end of the Arabian peninsula, Yemen with a population little over 17 million and a very small amount of oil production has a per capita GNP of only $270, nearly one-seventh of the world's average!

Prospects Despite the heightened awareness of the prospect of a long-term decline or shortage of oil, a vital resource on which much of the future development of the world depends, demand for it has been rising since the 1950s. Increased demand, especially from the rapidly industrializing Asian nations, makes the Persian Gulf region a globally strategic area and a source of Gulf conflict. The world is also facing a radical military technology and the proliferation of weapons of mass destruction in the future which present some ominous scenarios. On the other hand, the Middle East has the capacity to move toward a more constructive and peaceful future, if the nations in the area work toward it. Intra-regional proposals for joint infrastructural projects, shared oil and gas pipelines, and improved transportation lines linking the region to Europe already exist. If enacted, these could usher in a new era of prosperity and cooperation.

Further Readings

Al Faraqui, et al. *The Cultural Atlas of Islam*. New York, 1986.

Bullard, R. (ed.) *The Middle East: A Political and Economic Survey*. London, 1958.

Cressey, G.B. *Crossroads: Lands and Life in Southwest Asia*. Chicago, 1960.

Rahman, M. (ed.) *Muslim World: Geography and Development*. 1987.

12

Asia's Mediterranean Nations

TURKEY

Turkey is markedly unique among the nations of Southwest Asia, although it shares the Muslim faith with most of them. Distinctive in language and culture, it is a progressive nation, fairly Westernized and modern in outlook. Since World War II, several revolutionary changes have taken place in the country, which quickly pushed it along the course of social, economic, and political transformation that is still underway.

The Physical Setting A country that lies partly in Europe and partly in Asia, its location in two continents has been a key factor in its history, culture, and politics; Turkey has often been called a bridge between East and West. The Bosphorus, Sea of Mar-mara, and Dardanelles (collectively the Turkish "Straits" that conventionally mark the dividing line between Asia and Europe) separate the western three percent of its territory, the "European" part, from the much larger "Asian" section. The "Straits" act as a connecting link between the Black Sea in the north and the Mediterranean in the south. With the smaller appendage west of the Straits, Turkey is the only country in the region to share a common boundary with the European nations of Greece and Bulgaria. Historically, its location has been of crucial importance for it has given the country a strategic position in the eastern Mediterranean. The control of the Straits connecting the Aegean and Black Seas has enhanced the importance of the nation from a geopolitical standpoint. Turkey's foreign policy reflects this uncomfortable situation: the nation must align itself with the powers that are able to guarantee it the control of the Straits.

Turkey's land borders with Russia in the northeastern Transcaucasia and Iran on the east are also of critical significance as a passage for the Russian advance in modern times, and as a historical land route (despite difficult terrain) to the East via Iran.

The physical landscape of the country consists primarily of an inner plateau (the Central Anatolian Plateau) ringed by mountain ranges that for the most part drop steeply on their outer margins, either to the sea or to the lowlands. In the extreme east, however, there is an unbroken mass of mountainous territory of eastern Anatolia merging in the northeast into the Caucasus, and in the southeast into the Zagros Mountains. Less than 9 percent of Turkey's land is level or gently sloping. Its lowlands primarily fringe the coasts.

Relief is dominated in the north and south by ranges that rise to more than 10,000 feet (c. 3,025 meters) that fringe the

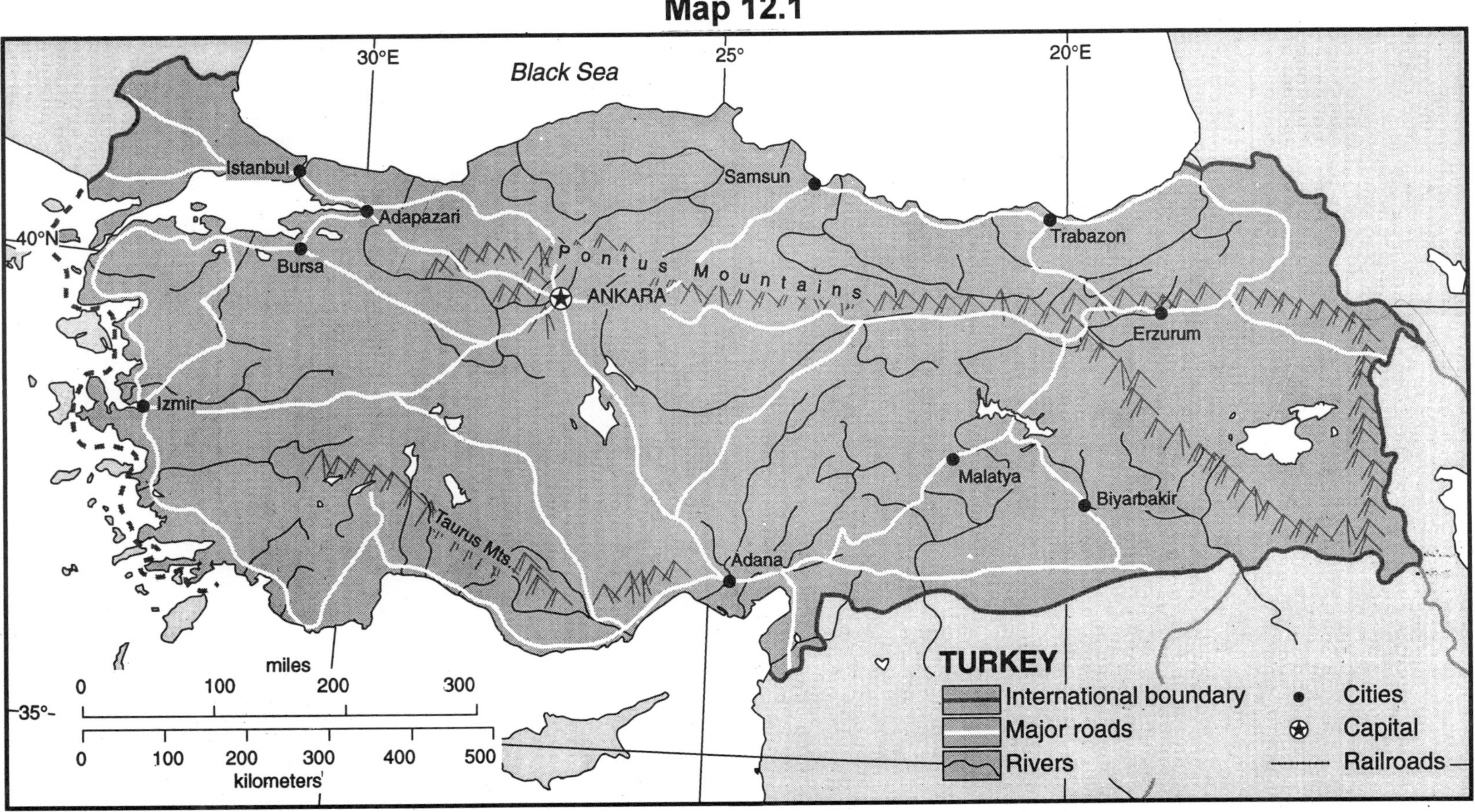
Map 12.1
Black Sea
30°E
25°
20°E
40°N
35°
Istanbul
Adapazari
Bursa
Izmir
ANKARA
Pontus Mountains
Samsun
Trabazon
Erzurum
Malatya
Biyarbakir
Adana
Taurus Mts.
miles
0 100 200 300
0 100 200 300 400 500
kilometers
TURKEY
International boundary
Major roads
Rivers
Cities
Capital
Railroads

vast Central Anatolian Plateau, separating it from the narrow coastal plains. In the north, the Pontic Mountains follow the contour of the southern shore of the Black Sea, their altitude increasing to the east to 12,900 feet (3,932 meters); a good part of the higher peaks remain snowbound. South of the central plateau are the Taurus Mountains, with several branches along the Mediterranean coast. Their highest point also reaches the elevation of nearly 13,000 feet in the east.

Enclosed within these mountain systems is Central Anatolia, much of which is a large plateau; in the eastern part of this region are high mountains, composed of a confused mass of lofty ranges and several volcanic peaks such as Mt. Ararat (altitude 16,853 feet or 5,137 meters). These mountains of eastern Turkey rise much higher than the Central Anatolian Plateau, and are interrupted by several basins, one of which is occupied by Lake Van, a salt lake formed by an extinct volcano.

Central Anatolia is a semi-arid and irregular plateau of mostly 3,000 to 6,000 feet (912--1,824 meters) in altitude fringed by mountain ranges on all sides: on the north, east, and south, mountains are 2,000 to 4,000 feet (608--1,216 meters) higher than the plateau; but on the west, towards the Aegean coast, hills are much lower in altitude. Much of the plateau region and the mountains consist of lava flows and lie in a zone of crustal instability frequented by devastating earthquakes.

Towards the western coast is the Aegean region of plains, valleys, and hills. The ridges run transverse to the coast and are separated by several valleys, such as, from north to south, the valley of Bakir Cayi (ancient Caicus), Gediz (ancient Hermis), and Mende-res, which contain the ruins of ancient cities. The landforms of northwestern Turkey around the Sea of Marmara consist of a large basin bounded by low hills and plateau-like surfaces. The European part (Thrace) is a rolling plain of recent deposits. The area around Istanbul is a deeply dissected valley. The Bosphorus and Dardanelles, which are the channels that link the Sea of Marmara with the Black Sea and the Aegean Sea respectively, are the submerged river valleys, formed by the rising sea level after the melting of the last glaciers.

Climatically, there is a great variety of conditions that reflect the geographic differences within the country. The coastal areas are mostly humid, while the vast interior, shut off from the rain-bearing winds by the encircling mountains, remains dry. Most of the coastal areas receive more than 30 inches (750 centimeters) of rainfall annually, and as much as 96 inches in the eastern Black Sea, whereas in most of Central Anatolia, it varies between 10 and 14 inches. The Marmara and Aegean coasts, and southeastern Anatolia, get moderate amounts (15 to 25 inches). The low coastal areas have warm to high temperatures in summer and warm in winter. Frosts along the Mediterranean are virtually unknown. Winters are freezing in the interiors and the mountains; the snow cover lasts up to 40 days in Central Anatolia and 120 days in the northeastern mountainous region. The south and western coasts have a Mediterranean climate characterized by hot, dry summers, and mild, rainy winters. Winters are mild and summers warm along the coasts, while the interiors have severe winters and hot summers.

Parts of the south, west, and northwest are covered by a Mediterranean vegetation, consisting mainly of thick scrubland in the lowlands and deciduous and coniferous forests at higher elevations. The humid north and northeast parts of the country are most densely wooded. The Anatolian interior is a region of sparse steppeland of short grasses.

Cultural and Historical Development

Although Turkey is an ancient land dating back to the Hittites (2000 to 1200 B.C.) and the Greeks (4th to 2nd centuries B.C.) and has been the scene of innumerable wars and the seat of several empires, the present day nation owes most of its character to the revolutionary changes introduced by a group of western-minded reformers under the leadership of Kemal Ataturk in the early 1920s. Before that, the country had been the centerpiece of a vast Byzantine Empire and the bulwark of Eastern Orthodox Christianity with its great capital Constantinople that lasted for several centuries as a successor to the Roman Empire until it succumbed to the Ottoman Turks in 1453. The Ottomans established an impressive empire across much of the region for more than four centuries, but by the early 20th century it had become a major figure on the European chessboard. Finally, the Ottoman Empire collapsed at the end of World War I.

The new government under the dynamic leadership of Ataturk ("father of the Turks") quickly set up a strong, responsible, and democratic administration in 1923 and set in motion a host of legislative measures that pushed the nation on a path of westernization and modernization. The administration reduced the dominant cultural influence of Islam where it hampered economic and political progress. The custom of *purdah* where the women were forbidden to show their faces in public and permitted only to appear heavily veiled outside the home, and denied education or social intercourse, was outlawed. Men were required to shave off their beards, and were prohibited from wearing a fez. Emphasis was placed on a secular state and society. In 1928, a modified Latin alphabet replaced the Arabic script. The old Muslim law was replaced by a western legal code. The new western identity was demonstrated by the admittance of women to the national parliament. The capital of the nation was moved from Constantinople (present-day Istanbul) to a more secure, interior location at Ankara, symbolic of the nation's new identity.

The modern republic, however, retains much of the Muslim heritage, but has been more successful than many other nations in keeping religious fundamentalism separated from the political process. With all the attempts to be modern, there is still interest in the ancient glories of the Hittites as the founders of the nation. The moving of the nation's capital from Constantinople to Ankara represents the national pride in a symbolic break-away from the past.

Emphasizing its own distinctive identity apart from the Arab culture, Turkey has reduced its formerly close ties with the Arab world, and strengthened its cultural, political, and economic ties with the West.

Culturally, the country is perhaps more European than Asian. It is a member of NATO and an associate member (and eventual candidate for full membership) of the European Community. The country continues to retain geopolitical significance because of its control of the Straits, although free passage through the Straits is recognized by international agreements (the Straits are closed to warships).

Culturally and ethnically Turkey is now quite homogeneous, and the Turkish language is dominant. Most of the minorities----Greeks, Kurds, Armenians, Assyrians (Nestarians), and Circassians---left the country following the century-old local fighting or were liquidated. Following the Greco-Turkish war of 1922, most Greeks migrated to Greece; most of the Armenians were eliminated from Turkish Armenia (adjacent to the Armenian state, formerly a part of the USSR); and the Kurds are being assimilated. Turks now form nearly 86 percent of the population. The Kurds are the largest minority representing about 10 percent of the people, but residing primarily in Kurdistan, the mountainous region in the southeastern corner of the country adjacent to Iraq and Iran. They are essentially nomadic people, distinct from the Turks in speech and culture. Their identity is being eroded by legal sanctions despite their fierce resistance. The Arabs account for 16 percent of the population and live mostly along the Syrian border. The Aegean coast was predominantly Greek before 1923, but most Greeks moved back to Greece in exchange for Turkish minorities transferred from Greece to Turkey. The Carcassian, Turkmen, and Georgians are other smaller minorities.

Economic Activity Nearly forty percent of the labor force is engaged in agriculture, accounting for about one-sixth of the nation's overall economic output. But much of the land is unsuited to cultivation; only a third of it is classed as arable. Considerable progress in the expansion of cultivable land has been made since the promulgation of land reforms by the revolutionary government after World War II, yet the agricultural sector did not keep pace with other sectors of the economy. Agriculture was no longer the dominant occupation that traditionally it had been. Most of the country is mountainous, and even on the interior plateaus neither climate nor soil are conducive to large scale agricultural activity.

In keeping with the large range of climatics, soils, and landscapes of the country, there is considerable variety in crops grown. Cereals, however, occupy a predominant position, accounting for over three-fourths of the cultivated land. Wheat is by far the most important crop grown on most of the cultivable land, particularly on the drier, interior plateau. The productivity is generally low, but the total production volume harvested is substantial: in average years, Turkey ranks seventh or eighth in the world. Barley, oats, rye, and maize, are other cereal crops grown in sizable amounts. The country is a world leader in production of barley (eighth). Rice cultivation is gaining in importance on the Aegean and Mediterranean coastal lowlands where precipitation is greater and soil and topographic conditions more favorable. Like other nations of Southwest Asia, Turkey was a wheat importer until the 1930s, but now is not only self-sufficient but has a modest surplus for exports. Beside food grain crops, Turkey is an important producer of several commercial crops, such as tobacco, cotton, oil seeds, hazel nuts (world's leader), and a variety of Mediterranean crops such as olives, figs, and grapes. Turkey is particularly famous for its fruits (apples and citrus fruits) which find a ready export market. Livestock also forms an important part of the country's agricultural economy. Sheep and goats are the most numerous and widely distributed. The country ranks high (fifth) in the world in its livestock population. Animal products are essentially consumed on farms or in nearby villages, and some are available for export.

Agriculturally, the most important areas are the coastal plains and valleys along and near the Aegean Sea, Sea of Marmara, and the Black Sea where Mediterranean crops such as hazel nuts, grapes for Sultana raisins (Turkey is the world's leader), citrus fruits, figs, and tobacco are grown in addition to the cereal crops of wheat and barley. The Black Sea coastland is, however, regionally important for corn, legumes, and sugar beets as well, for it receives heavier precipitation, a condition more conducive to the growth of these crops. The Plateau produces considerable wheat and barley, hides and skins, and wool, including *mohair* from the Angora goat which is used in the making of high-quality rugs and fabrics.

Mineral resources of Turkey are potentially significant but not exploited on a large scale. It has limited oil, mostly along the Iraq border inhabited largely by the troublesome Kurds, and exploitation is hampered by difficult terrain, and most of country's needs are met by imports. Since oil accounts for nearly 11 percent of the nation's imports, it is a major drain on the national treasury. The steep price increases of the 1970s had a devastating effect on the Turkish economy as the government had to borrow heavily to finance the fast developing industrial sector. Modest amounts of coal and iron are mined, although reserves are sufficient to meet Turkey's needs for several decades. Coal is a major mineral resource, and its deposits lie on the Black Sea coast between Eregli and Zonguldak. The iron ore deposits located at Divrigi are modest but of fairly high-grade quality. In addition, extensive fields of chrome and some of manganese are also found. Hydroelectric potentials are considerable and receiving attention by the government in view of the rapidly expanding industrial sector.

Most of the country's modern industry is in the early stages of development. Under the Ottomans much of it was in the form of handicrafts, such as the making of world-famous rugs, brassware, and woolen and cotton fabrics. Since the early 1930s the republic has attempted to industrialize the country in a hurry, without laying the basic foundations. Development has been hampered by a lack of capital, competition with foreign goods, and underdeveloped technology, and the country's overall poor infrastructure. The government's policy of state control of several industries, and repression of the experienced and skilled labor of minority groups (most of whom left for the western European countries) further aggravated the problem.

The predominant industries include textiles, processing of agricultural goods, manufacture of cement, consumer goods, and assembly of imported components for vehicles. Iron and steel, electric and electronic equipment and transport equipment account for one-third of all imports, and these must be imported to feed the growing domestic industry. The iron and steel industry is largely state owned; the major plant is located at Karabuk in the northwest. Coal comes from the nearby Black Sea fields, but the ore has to travel 500 miles by rail. Cement plants are located in the Istanbul area, and near Ankara.

The textile industry is more widespread, manufacturing a wide variety of cotton textiles, woolen and silk fabrics, and internationally known Turkish carpets. The country has a sizable sugar industry based entirely on the domestic production of beets. It has grown impressively since the 1930s, when the country had to import most of its needs. By 1946 the nation had become self-sufficient, and since then has had a small surplus for export.

Industrial development has been partly hampered by inadequate transport systems, although since the 1930s the republican administration has embarked upon a plan of extending the roads and railroads, but most of the mineral producing areas still remain poorly linked with the industrializing centers. There is only a little over 5,200 miles (8,372 km) of railroad track, and the length of the paved roads is close to 270,000 miles (434,400 km).

Predictably, the major exports of the country are: textiles which account for one-third of all exports; agricultural products (wheat, olives, cotton, fruits, and processed foods), and iron and steel. Chief imports consist of machinery (electrical, electronic equipment, transport equipment) and oil. The major trading partners are Germany, U.S.A., Italy, Saudi Arabia, Britain, France, U.S.S.R., and Japan.

Turkey's current population, estimated to be close to 65 million, has grown nearly three times during the last four decades. A little over 64 percent of it is classified as urban, of which 80 percent is located in the coastal lowlands, particularly along the Aegean and Black Seas, and the Sea of Marmara. Parts of the dry interior of the Anatolia Plateau and the mountainous area in the east are largely empty (average density for the Anatolia Plateau is only 25 people to a sq mile).

The cities, such as Ankara, Izmir, and Istanbul, have grown rapidly since World War II with the greatest massive rural outmigration to Istanbul, which is by far the largest of these. **Istanbul** (8 million) has been a major trading port since the 7th century B.C. Founded by the Greeks in 658 B.C., it was selected as a capital of the eastern Roman Empire in the 4th century A.D. by Emperor Constantine, who gave the city his name. For several centuries it was the capital of such impressive empires as the Byzantine and the Ottoman, and later of the Turkish Republic until a new capital at Ankara was founded. As an imperial capital of multinational empires, Constantinople became a cosmopolitan city of Latinos, Slavs, Greeks, Turks, Armenians, and Jews. In 1923 the revolutionary government moved the capital to Ankara, a more central location in the Anatolia Plateau, as Istanbul was considered too cosmopolitan and too vulnerable militarily. The city's location commands the Turkish Straits, the passageway that connects the Mediterranean with the Black Sea, and has historically acted as a bridgehead for land routes between Europe and Asia.

Istanbul has long been the cultural and commercial center of the country. With most of its commercial activity focused on the Golden Horn, a deep water and natural harbor, the city handles three-fourths of Turkey's external trade. The city has acquired a sizable industrial component, specializing in the manufacture of textiles, metal and glassware, and leather goods. Capitalizing on its agreeable climate, a vast array of cultural and historic sites, its commercial and industrial importance and easy accessibility, Istanbul has been one of the leading tourist attractions in Southwest Asia. **Ankara** (3 million), the national capital and the second largest city, has grown rapidly since acquiring administrative functions, from a small, interior town with its 1923 population of only 50,000. The city is a major railroad and transport center in the heart of the nation, and has attracted a few industries. **Izmir** (2 million), is the third largest city and a major port of western Turkey on the Aegean coast. It has an excellent natural harbor, and handles nearly a sixth of the country's foreign trade.

Prospects As Turkey moves towards a new century, the nation faces severe economic problems, although the over-all economic prospects appear reasonably good. Unlike some of it more fortunate neighbors, the country possesses only limited reserves of oil, but is well supplied with coal, lignite, chrome, iron, copper, and some other metals, most of which remain under-utilized. Except for steel, heavy industry has had little development, concentrated only in a few locations. But an extensive, still largely untapped, resource for the development of hydroelectric power, exists. Despite the nation's preeminence in several agricultural products, productivity, in general, remains low. Economic development is hampered, not entirely by a deficiency in basic resources, but by a weak industrial base inherited from the decadent Ottoman period, a mounting foreign debt (running currently around $48 billion in the mid-1980s) to finance the expanding industrial sector, and infrastructural deficiency.

In measuring Turkey's assets, however, one cannot overlook several of the nation's assets. Ethnically, it has a relatively homogeneous population, and politically, the democratic and secular traditions have been more or less securely laid (there have been feeble stirrings of fundamentalism in the country recently). But perhaps the greatest asset of the country is its location. Turkey's command over the Straits has made it one of the key powers in the East Mediterranean. As an ally of the European nations, it has been able to enhance its economic and military position substantially.

CYPRUS

The island of Cyprus (area: 3,572 sq miles or 925 sq km) and population 760,000) is located in the northeastern corner of the Mediterranean Sea, about 40 miles (60 km) south of Turkey, and 60 miles west of Syria. The coastline is rocky and indented with long sandy beaches. The highest altitude of 6,400 feet (c. 2,000 meters) is in the southern interior and the northern coast has a low range leaving a plain area in the middle and along the remaining coast. The climate is typical Mediterranean with mild rainy winters and hot summers, and annual precipitation ranging between 10 inches (25 cm) in dry lowlands to over 40 inches in the highest mountains.

Although the Republic of Cyprus was constituted as an independent state in 1960, after years of controversy and civil disorder between the Greek and Turkish populations, two *de facto* states currently exist on the island: the Republic of Cyprus

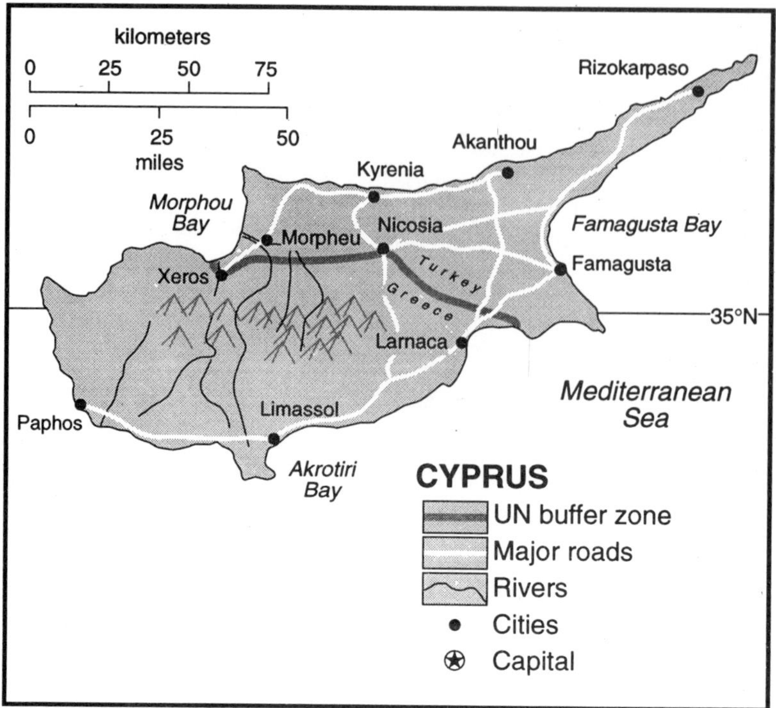

(ROC), predominantly Greek in character, occupying the southern two-thirds of the island which is still the internationally recognized *de jure* government of the entire island; and the Turkish Republic of Northern Cyprus (TRNC) proclaimed unilaterally in 1983 on territory secured by the Turkish Cypriot population by the 1974 intervention of Turkey, but the two ethnic communities have failed to establish a single state.

Predominantly Greek in its population since ancient times, the island has had a sizable Turkish minority ever since Turks conquered it in 1571 under whose control it remained until 1878. At the end of World War II Britain acquired rights to rule over it.

Of the two main ethnic groups, Greeks and Turkish, the Greeks constitute the majority, forming close to eighty percent of the population. They are descendants of a mixture of the aboriginal inhabitants and the largely Greek colonizers from ancient times. The Turkish Cypriots are primarily the descendants of the soldiers of the Ottoman army that conquered the island in 1571 and of later immigrants from Turkey. In 1960, when Britain gave independence to the island, the Constitution granted the Turkish minority a share of power in the new administration, but the power-sharing arrangement broke down soon after, and the Turkish-Greek fighting for the control of the island began that quickly engulfed the country.

The Greek element, backed by the strong Greek Orthodox Church, demanded "eno-sis", that is, union of the island with Greece. The Turkish minority strongly opposed any move for incorporation within Greece, and pressed for partition of Cyprus into Turkish and Greek sectors. After much civil war and bloodshed, the Turkish minority succeeded in controlling the northern, nearly 40 percent of the island, with the armed intervention of the government of Turkey in 1974. The remaining island became the *de facto* state of Cyprus (officially "Republic of Cyprus"). In the Civil War most Greeks, living in the northern part, moved to the south, and the Turks of the southern part moved to the northern, Turkish sector secured by the Turkish army. In 1983, the Turkish community declared the northern territory as independent Turkish Republic of Northern Cyprus (TRNC). The two sectors are now more or less ethnically homogeneous (minorities amount to less than 5 percent of the population in the two territories). The Turkish-Greek conflict, however, is far from over. Many of the Turk Cypriots' desire reintegration, while there are several factions within the Greek community that resent the strong influence of Greece within the Republic of Cyprus.

The economic effects of partitioning of the island have been disastrous for the North, the TRNC. Before partition, the island had achieved a standard of living higher than most of its neighbors, with the exception of Israel. Based primarily on agriculture and trade, and assisted substantially by the various agencies of the United Nations, the free market economy had made notable progress. Between 1960 and 1973 the domestic product grew at an annual rate of over 7 percent. Agricultural production nearly doubled and industrial production rose nearly three times. The civil war in 1973, and the Turkish occupation of the northern section dealt a serious blow to economic progress of the island. Particularly hit was the Turkish-occupied sector, from where nearly 20,000 Greek Cypriots, mostly farmers, entrepreneurs, and skilled workers, moved to the south as refugees. A much smaller return flow of Turkish Cypriots to the north was grossly inadequate to compensate for the loss of the Greek Cypriots. Following the unequal transfer of people, the Turkish-occupied area of the north, which enjoyed a flourishing economy before partition, stagnated. Trade between the two sectors ceased and their economies became almost entirely independent. By contrast, the southern Greek-occupied territory (Republic of Cyprus) continued to prosper with economic aid from Greece, the United States, and the various agencies of the United Nations. Particular gains were made in tourist-related businesses, shipping, transit trades, international banking and insurance, and miscellaneous light industries. The Republic of Cyprus, capitalizing on its excellent location in the Mediterranean and a duty-free zone, has made a remarkable recovery since 1974. Most of the refugees have been settled. Economic prosperity is reflected in the per capita income of about $16,550 that is probably three times that of the economically depressed north.

ISRAEL

As the first Jewish state to be established in nearly 2,000 years, Israel represents a unique phenomenon in world history. Created by the United Nations Commission in 1948, it became the fulfillment of the dream of the Jewish people that stemmed from a long tradition of Jewish aspirations of getting a land of Israel for their people. Surrounded by hostile Arab nations who resented the creation of a Jewish state from the territory inhabited for centuries by Arabs, the young nation since its inception has been engaged in unbroken conflict with its Arab neighbors. Despite seemingly insurmountable obstacles, it has forged a strong nation, and successfully built a solid economic and cultural base.

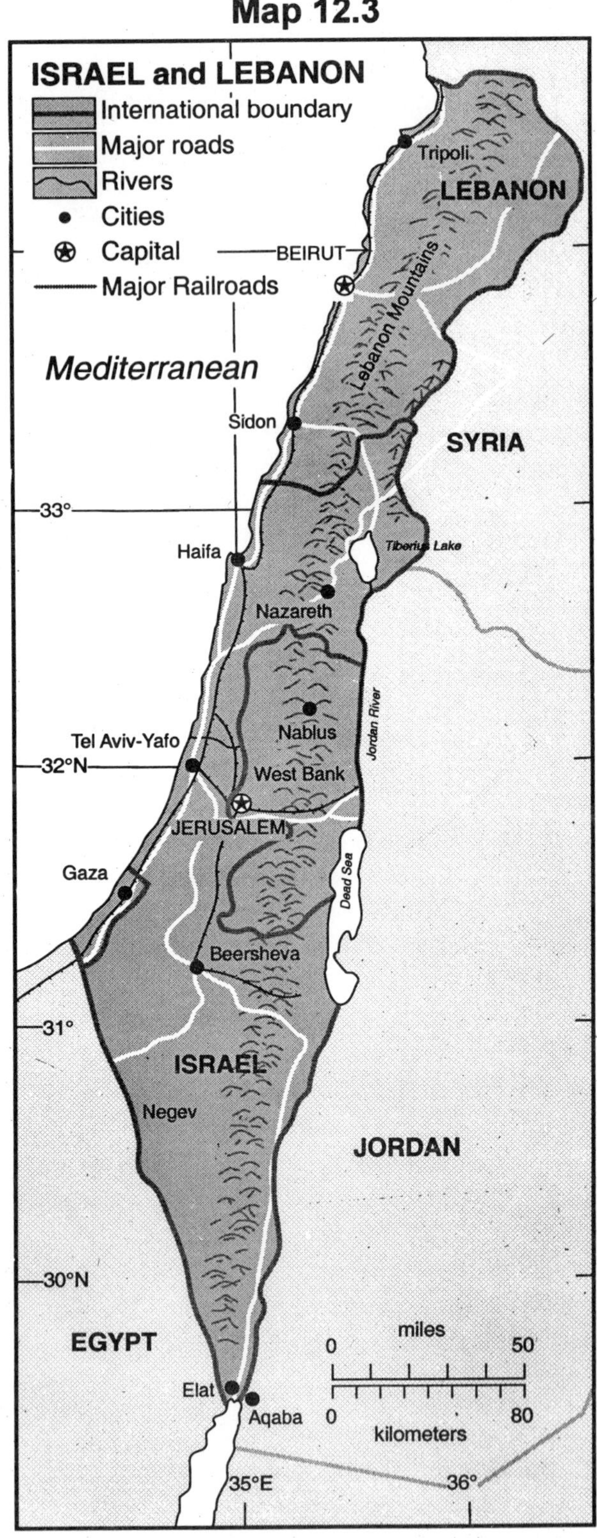

Physical Setting The state of Israel is a small country that is located at the eastern end of the Mediterranean Sea, and has a total land area of nearly 8,000 square miles and a population of close to 6 million. It is bounded to the north by Lebanon, to the northeast by Syria, to the east and southeast by Jordan, to the southwest by Egypt, and to the west by the Mediterranean.

Physically, the country may be divided into four regions: the Mediterranean coastal plain; the northern and central hilly region; the Great Rift Valley in the east; and the Negev Desert in the south. The coastal plain is a narrow band about 110 miles (177 km) long and 15 to 20 miles wide. The northern hilly territory of Galilee contains the highest elevations, reaching to about 4,000 feet (1,220 meters) at Mount Meron. These mountains drop as an escarpment to the east overlooking the Great Rift Valley. The mountains of Galilee are separated from the hills of Samaria and Judea to the south by the Plain of Esdraelon that connects the coastal plain with the Rift Valley. The Great Rift Valley is an elongated fissure in the earth which runs from the northern frontier of the country to the Gulf of Aqaba at the southern end. The Jordan River runs from the north, where it is 500 feet above sea level, to the Sea of Galilee at 689 feet below sea level, and finally into the saltish end at 1,312 feet below sea level. The Rift Valley contains large deposits of rock salt and gypsum. The Negev Desert is a wedge of territory that comes to a point at the port of Elat on the Gulf of Aqaba.

The climate of the country is mainly Mediterranean, but extreme in some parts, and might almost be described as semi-desert in the southern region where precipitation amounts to about one inch (25 millimeters) annually south of the Dead Sea. By contrast, in the Upper Galilee region it increases to over 44 inches. Most of the precipitation falls between October and April. Summers are dry and hot, but the coastal areas exert a moderating influence. The Jordan Valley is hotter and drier than the coast. Temperatures depend largely on elevation and distance from the sea. In the coastal area these range from 90°F (32°C) in August to about 65°F (18°C) in January. In the south, however, at Elat, daytime temperatures reach 70°F (21°C) in January and may soar up to 114°F (46°C) in August.

Most of the original vegetation of evergreen forests has largely disappeared on account of centuries of cultivation and goat herding. The hills are now covered by shrub vegetation. Millions of trees have been planted since 1948 under a reforestation program. North of Beersheba most of the land is either under cultivation or used for hill grazing; where irrigation is available, citrus groves and eucalyptus and conifer plantations flourish.

Historical and Cultural Background

The idea that the Biblical territory of Palestine was destined to be the Holy Land of the Jews, sanctioned in the ancient scriptures of the Jews and promised to them by God, is rooted in the Jewish psyche. It was in Palestine that the last sovereign Jewish state existed until Roman legions destroyed the Jewish temple of Herod in Jerusalem and dislodged the Jews from their Holy Land. Since then, during their long *Diaspora* (forced exile), Jews have lived as alien and generally been persecuted minorities, scattered largely in Europe, Russia, and the Arab countries.

In the late 19th century, after centuries of persecution, a group of Jewish leaders known as Zionists started the movement to secure the creation of a state in the territory of their ancient homeland. Following World War I, Britain gained control over that territory, to be known as the Mandate of Palestine. Britain which had already been in favor of the creation of a Jewish homeland in the region, announced its intention in the form of the Balfour Declaration in 1917. The declaration, as well as the systematic persecution of the Jews in Germany, accelerated the already growing influx of Jews into Palestine from Europe. Because of the Nazi persecution of Jews, international support for a Jewish state also gained momentum. Britain referred the matter of the creation of a Jewish state in Palestine to the United Nations, which called for a partition of the Palestine Mandate into a Jewish state and an Arab state (present day Jordan). Jerusalem was to be administered internationally under U.N. control. Nearly 56 percent of the British Mandate was awarded to the Jewish sector, although only 7 percent of the land was actually owned by Jewish people (who comprised about one-third of the population). The Jews agreed to the partition plan, but the Arabs did not; both communities prepared to fight.

Following the United Nations partition of Palestine, a new state of Israel was established in 1948, although the Arabs refused to recognize it. Immediately war between the Arabs and the Jews began. Arabs demanded the dissolution of the "unwanted" colony. At the root of the present Arab-Israeli conflict is the contention between Israel and the neighboring Arab states regarding the rightful possession of the partitioned territory. The Arab states believe that the land belongs to the Arabs who resided in Palestine before the new state was formed. Nearly a million of the Arabs who had lived in Palestine for centuries were uprooted to make way for Israeli immigration. Most fled to Jordan, which was particularly unable to support them. Others went to Gaza, Syria, Lebanon, and Egypt. Many of these refugees have not established permanent new homes or are still living in refugee camps. Most hoped to return to the "occupied" territories in Israel, their former homes, and were supported in this hope by the governments of the Arab states. In their place, Jews from North Africa, the Arab countries, Europe, and the Orient started pouring into the newly-created Israel.

By 1962, nearly 2 million Jews lived in Israel; now they number close to 6 million. Close to three-fourths of a million non-Jews (mostly Arabs) still live in Israel. The descendants of the displaced Arabs are called Palestinian refugees. Through population growth the displaced Palestinians now number over 2.5 million and live in Arab countries or the territory that Israel occupied since the end of the 1967 Six-Day War, known as the West Bank and the Gaza Strip, sometimes in squalid refugee camps.

Hostility between Israel and the neighboring Arab states have prevailed from the outset and erupted into several armed conflicts in 1948, 1956, 1967, 1973, and 1982. After the 1948 war, the Arab nations refused to absorb the three-fourths of a million homeless Palestinians because to do so would acknowledge the permanent status of the State of Israel.

In 1956, when Egypt threatened to nationalize the Suez Canal and block Israeli shipments through the canal, Israel occupied the Sinai and the Gaza Strip, but returned the territories under international pressure. But when in 1967 the canal was nationalized by Egypt in actual practice, warfare started on various fronts along the Egyptian, Syrian, and Jordanian borders. Israel obtained a decisive victory in a short, conclusive war that resulted in the Israeli occupation of the Sinai, the Gaza Strip, the Golan Heights in Syria, and the entire area west of the Jordan River known as the "West Bank". Nearly 1.5 million Palestinians became refugees in the occupied territories, as the Israelis initiated an extensive program of settling the rapidly increasing number of Jewish immigrants flocking into Israel from Europe, Africa, and the Orient. In 1973, Egypt and Syria attacked the Sinai and the Golan Heights to regain their territories lost during the 1967 war. A ceasefire was quickly arranged, but in the aftermath the Arab nations imposed an oil embargo on Israel that had a devastating effect. Israel pulled back from the Suez Canal.

In 1979, with the support of the United States, Egypt signed a peace treaty with Israel at Camp David, and recognized Israel's rights as a sovereign nation. The Suez Canal was finally opened. The disposition of the West Bank and Golan Heights occupied by Israel since 1967 was left to future negotiations as part of an ongoing peace process. After securing the western border, Israel turned to the Lebanon border, where the Palestinian Liberation Organization (PLO) group, formed by many of the refugees from the West Bank who had taken refuge in Lebanon, was engaged in activities against Israel. In 1982, Israel invaded the PLO in their hideouts and inflicted a crushing defeat on them and their supporters, the Syrians. In the face of mounting casualties on all sides and under pressure from the United States, Israel finally pulled out. PLO and Syrian strength was largely eroded, although it continued to exercise close surveillance over a "security zone" in southern Lebanon.

Realizing the futility of long drawn-out warfare, both Israel and the neighboring Arab states had by the mid-1990s started peace negotiations. The PLO and Israel signed a peace treaty in 1994 in Washington, DC to establish a process that may lead to the solution of some of the knotty problems concerning the West Bank and the Gaza Strip. Soon after, Jordan also officially put an end to hostility and accorded Israel recognition as a sovereign state. In 1998, Syria and Israel started to negotiate the transfer of the Golan Heights that Israel captured during the 1967 war.

The question of the distribution of the waters of the River Jordan, the headstream of which is controlled by Lebanon, Syria, and Jordan, is yet another dispute between Israel and its Arab neighbors. Water, being scarce in this part of the world, is a precious commodity, desired by all the parties. After independence, Israel expressed a desire to use water from the river for irrigation and hydro power development, and share it with Jordan, across whose land the lower river flows. Syria controlled the headstream, but Israel went ahead and made plans to divert the stream after occupying the headstream during the 1967 war, and then effectively denied the use of water downstream in the Jordanian territory of East Ghor. By 1997, the two nations of Israel and Jordan had reached an amicable settlement of the problem.

The status of Jerusalem is yet another thorny issue between Israel and the Arab states. Recognizing the sacred character of the city, the original United Nations partition plan intended to make it an internationally administered territory, separate from both Israel and Jordan and safe for all religions. Even before partition was to go into effect in 1948, fighting between Jews and Arabs led to the surrender of the Old City to Jordan, while the New City remained in Israeli hands, and Israel quickly made it the capital of the new state. After the Arab-Israeli Six-Day War of 1967, Israel annexed the Old City and placed Jerusalem under a unified control. Israel views the reunification as permanent, whereas the Arabs see it as temporary and long to repossess the Old City. Most of their holy places, as well as those of others, exist here.

Jews account for 80.5 percent of Israel's population. Muslims, mostly of the Sunni faith, account for about 14.6 percent and Christians, Druze, and others make up the remaining 4.9 percent. The Jewish population is linguistically less cohesive than the other groups. Jewish immigrants differed in racial origin and brought with them languages and cultures from a variety of sources, although national consciousness is growing, particularly among the young for learning Hebrew, the official language of the country. Among the Jews, there are two main groups: the followers of the Ashkenazic denomination primarily from eastern and central Europe and their descendants, and the followers of the Sephardic and Oriental branch. Minority Ashkenazic Jews of European and American background dominate the high income professions, while the majority Sephardic Jews of Asian-North African origin mostly occupy lower economic and social status. Constant friction between the groups reflects the difference in their cultural values, residential segregation, and educational backgrounds. The Orthodox Jews, who advocate the leading role of religious law over civil Israeli law, are often at odds with other groups who favor a more secular society, a division that often results in violence. Israel is officially a secular state, but religious Jewry of the Orthodox groups forms a significant and politically vocal segment of the population. Disputes between the religious bodies and authorities often permeate national life.

In 1948, the Jewish population in Israel numbered about 650,000. Between 1948 and 1986, nearly a million and a half Jewish immigrants entered the country, half a million of whom subsequently left it. During the two years of 1989 and 1990, more than 200,000 Soviet Jews immigrated to Israel. Of the Jewish community, the largest proportion was born in Israel, followed by those born in Europe, America, Africa, and Asia.

In 1948, only 155,000 Arabs decided to remain in the country and the rest, numbering perhaps three-fourths of a million, left. Growth rates of the Arabs have, in general, been relatively greater than those of the Jews. Muslims form the largest minority group and constitute more than three-fourths of the Arab population. Practically all the Muslims subscribe to the Sunni tradition. Most Muslims live in towns and enjoy considerable freedom in internal social code, although the state subsidizes and supervises their religious institutions. The Bedouin Muslims live primarily in the Negev and the rest in Galilee. The Carcassian Muslims belonging to the Sunni tradition number a few thousand and are the descendants of the immigrants from the Caucasus of the 1870s. Older Carcassians speak Arabic, but the younger generation speaks Hebrew.

Members of the Christian community belong to several branches: Greek Orthodox, Roman Catholics, Uniates (Melchites, Maronites, Chaldean Catholics, Syrian Catholics, and Armenian Catholics), Russian Orthodox, Evangelical Episcopal, and Lutheran Churches. The majority of these groups speak Arabic. Their churches generally are headquartered in Jerusalem. The Druze mostly live in villages in the Gallilee and Mount Carmel area, and are mostly Arabic speakers. They are primarily agriculturalists and proud of their distinctive way of life. The Baha'i faith, like Judaism, has its world headquarters on Mount Carmel in Haifa. They number a few hundred, and are employed in Haifa.

Resources and Economic Development

Israel is an advanced industrial state with high levels of technological development.

Except for a few oil-rich Gulf states (Kuwait, Qatar, and the United Arab Emirates) whose economies have prospered from the revenues of a single commodity, it enjoys the highest standard of living among the nations of Southwest Asia. Perhaps the most remarkable aspect of its development is the speed and efficiency with which the Israelis have transformed their country into a strong and prosperous nation out of extremely limited natural resources. Heavy subsidization has come form Jewish communities in the West, particularly in the United States and Germany, but the Israelis have built their economy out of practically nothing by hard work and careful planning. The technical expertise of the thousands of skilled and professional European and Russian immigrants during the 1960s and 1970s provided an essential core of workers for the country's economic reconstruction.

Mineral resources of the nation (particularly of coal and iron one) are limited but include potash, bromine, and magnesium, the latter two being obtained from the waters of the Dead Sea. Copper ore, phosphates, gypsum, and oil are also found. Oil and natural gas exist in modest amounts in the northern Negev and south of Jerusalem, and near Beersheba. Limited amounts of these have been exploited since the 1950s. Some hydro power resources are potentially available in the Jordan River Valley and the Golan Heights. Electricity is generated primarily from thermal stations and is under government control; the government provides it to agriculture and industry at favorable rates. Israel has built two atomic reactors with U.S. assistance near Tel Aviv and in the Negev with French help.

Industry forms a significant component of the country's economy, and nearly 20 percent of the labor force is engaged in manufacturing. Close to 80 percent of all the exports consist of manufactured goods. Nearly one-third of the country's total imports are in industrial raw materials. To reduce the cost of these imports, Israel has built a sizable merchant marine that grossed over 650,000 tons in 1992. Deficient in most minerals, Israel has vigorously pursued precision industries (diamond cutting and polishing and optical goods, for example) with the help of a labor force that consisted largely of skilled European immigrants. Israel is the world leader in diamond-cutting; cut and polished diamonds form the largest export item (25 percent of all exports) centered primarily in Tel Aviv and Jerusalem. Oil refining is located at Haifa. A small steel plant built at Tel Aviv for the manufacture of heavy machinery and defense weapons has not proved to be an economic proposition but has been kept for strategic purposes. The manufacture of chemical fertilizers from the potash of the Dead Sea and from the phosphates found in the Negev; copper-based manufacture near the port of Elat; and construction of buildings to accommodate a large number of immigrants are other major industries. In addition, a wide array of light industries are located in all the major cities (Tel Aviv, Yafo, Jerusalem, and Haifa) and craft goods are produced on kibbutzim and moshrava. The light industries include the manufacture of precision instruments (electronic and electric goods, watches, optical materials), shoes, ready-made clothes, plastic products of various kinds, and a large number of consumer goods. Foreign investment is vigorously pursued and low rates of taxation and favorable terms for setting up industries are provided to foreign and domestic investors. A close working relationship exists between workers, owners of the industries, Histadrut (a trade union federation), and the government, although high wages of the workers and the high cost of imported raw materials often tend to price many Israeli goods out of the world market. Israeli industry has approximately tripled in extent since 1950, and along with construction and trade account for about 40 percent of the value of the nation's total domestic production.

Although agriculture forms a minor part of Israel's economy and only 3 percent of the labor force is engaged in agricultural activities, it is highly mechanized and despite adverse natural circumstances, has been developed to a degree that compares favorably with the agricultural techniques used in the advanced countries. The area of land under cultivation has been greatly increased since the founding of the state in 1948, and extensive irrigation has been provided to develop farmland. In 1948, Israel produced only one-third of the food it needed; by the early 1990s it produced enough fruits and vegetables, poultry and eggs, and milk and dairy products to meet all domestic needs, although a small amount of the grain required is still imported.

The chief commercial crop and most important agricultural product is citrus fruit, primarily oranges, ninety percent of which are exported. Citrus production, severely affected during the wars, remains valuable for export purposes. Non-citrus fruits (such as melons, bananas, and peaches), and vegetables are also grown and exported to Europe in winter. A variety of other commercial crops that registered significant gains during the last three decades are sugar beets, peanuts, cotton, sisal (used in the manufacture of paper), and flowers.

A major agricultural problem is scarcity of water. Irrigation facilities were greatly expanded during the 1970s and 1980s. Water from the Jordan and Yarqon rivers and from Lake Tiberius (El Ghor area) is piped to farms even in the Negev Desert. Practically all of the country's potential water resources have been tapped. Further expansion of agriculture involves intensifying the yield from land already irrigated or getting more water by cloud seeding, reducing the amount of evaporation, desalinizing sea water, and expanding desert farming in the Negev. Nearly one-fifth of Israel's territory is given to cultivation. Forests cover only 7 percent of the surface; and fully two-thirds of the land is unsuited for permanent cultivation.

A special feature of Israel's agricultural development is the *kibbutz*, an agricultural settlement in which most farm families and agricultural workers live. Many of the early immigrants who entered the agricultural labor force before independence were settled in these communal farms or rural settlements, each one of which was known as a *kibbutz* ("ingathering"). Farming is done collectively, and management is by majority consent. No wages are paid, and no personal property is permitted except for married members who are given separate living quarters. The largely self-sufficient kibbutz usually practices a mixed form of agriculture, growing vegetables, cotton, grains, and producing meat. These kibbutzim are oriented towards agriculture but some have developed small manufacturing enterprises producing such goods as plywood and plastics. The sale of products is through cooperative organizations.

There are also villages where land is privately owned. In this second form of agricultural settlement, personal property is allowed and family members are housed separately. Produce is marketed though cooperatives collectively. These settlements, in part communal or cooperative, are *moshava*. The number of Jewish rural settlements increased greatly during the first few decades after independence; in 1960 these numbered close to 1,000, and in large part consisted of *moshava*. A large proportion of later settlers were of urban origin and preferred to opt for cities, but many were reluctantly placed in rural settlements and trained as successful farmers. As a tiny fraction of the labor force is engaged in agricultural activity, the government has incorporated farm training into regular army service, which applies to all able-bodied men and women.

Israel's most fertile cultivated land is in the plains along the Mediterranean, the Plain of Esraelon, the Jordan Valley south of Lake Tiberius, and the Beersheba area. The largely arid Negev region is now being developed to extend the agricultural land.

Ninety percent of Israel's population lives in cities, and urbanization has rapidly increased since the inception of the nation. The largest increase has been in the two larger conurbations of Tel Aviv-Yafo-Haifa and of the city of Jerusalem, which together contain nearly a quarter of Israel's population. The authorities are attempting to reduce overconcentration of population in these areas by building new towns, primarily to accommodate new immigrants and to serve as centers for regional settlement or to fulfill special economic tasks such as the manufacture of textiles, clothing, or machinery.

The ancient and historic city of **Jerusalem** (population: close to 600,000) is the national capital and has distinct Jewish and Arab sections, the New and Old cities. As a holy city for Christians, Jews, and Muslims, it contains numerous shrines for them such as Mount Zion, the Church of the Holy Sepulcher, Mount of Olives, the Wailing Wall, and El Aksa Mosque, which are concentrated in the Old City. Israel occupied the Old City (previously under Jordan's control) during the 1967 war, and now controls the unified city. Jerusalem has a sizable manufacturing component that includes diamond cutting and polishing, plastic goods, and cement making.

An important city and industrial center is the twin city and port of **Tel Aviv-Yafo** (population: approx. 360,000). Yafo (historically Jaffa) is an ancient city, dating back to the 15th century B.C., but the newer component of Tel Aviv is a modern city dating from 1907. Tel Aviv is the nation's premier cultural and commercial center, along with Jerusalem. The twin cities have an artificial harbor, and next to Haifa perform important port functions. Nearly one-half of Israel's manufacturing plants are located in the Greater Metropolitan Area of Tel Aviv. Main industries are: textiles, clothing, food processing, tobacco, engineering, metals, furniture, printing, and electric and electronic equipment.

Haifa (population: approx. 250,000) to the north has Israel's best natural harbor and is another important manufacturing center. Formerly Haifa was the terminus of pipelines from Iraq, but oil flow was stopped in 1948 during the Arab-Israeli conflict. In 1957 an oil pipeline was constructed from the Israeli port of Elat on the Gulf of Aqaba (an inlet of the Red Sea) as a possible alternative to the Suez Canal, and to feed Haifa's oil refining industry. Several industries such as steel manufacturing, diamond cutting and polishing, ship building, automobile assembling, cement works, and consumer goods production are located here.

Tourism is a growing industry of Israel, and is one of the chief sources of foreign exchange. It grossed nearly 3.0 billion dollars in revenues in 1990. Shipping is a vital factor in both the economy and in communications with other countries. Because of the Arab blocking of Israel's land borders since the inception of the country, shipping has played an important role in the transportation of supplies.

Prospects Can Israel survive as a viable state? It is a tiny nation of limited resources, surrounded by hostile neighbors. Artificially created nearly half a century ago in a region that has for long been known for political instability and turmoil, the country has been valiantly struggling for recognition. Nonetheless, despite heavy odds, it has been able to build a vibrant modern economy that commands admiration. The overall standard of living for its people is now among the highest in the region. As the most powerful military nation in the region it has not only been successful in defending itself, but also in extending its territory. But, the phenomenal economic and military success has come at a great price. Heavy subsidization from foreign Jewish communities and aid from international agencies have pushed the public debt to over 55 billion dollars and the balance of trade has become highly unfavorable.

Since inception, Israeli society, though sharing a common faith, has been anything but homogeneous. The assimilation of North African Jews who immigrated since 1948 and who were for the most part unskilled, illiterate, deeply religious, politically immature, and unfamiliar with western culture and democratic institutions, has caused tremendous strain on the new nation's relatively modest resources, and the problem of their assimilation has proved to be a difficult task.

The nation's biggest problem of viability is political. Recognition of the nation has yet to come from several Arab states and other nations. A permanent peaceful solution of the "Israeli question" depends not necessarily on the development of a strong Israel's economic health, military strength, and democratic institutions, or on internal cohesion which the nation has been able to develop successfully, but on secure relations with its neighbors. To achieve this, the nation is now engaged in talks with the PLO for the disposition of the West Bank and the status of Jerusalem. Although Israel has signed an agreement with the PLO and Jordan, the explosive problem of the administration of Jerusalem and the return of the Golan Heights to Syria yet remains to be tackled. As the strongest and most high-developed state in Southwest Asia, Israel has to make greater concessions to its Arab neighbors if a peaceful solution is to be achieved.

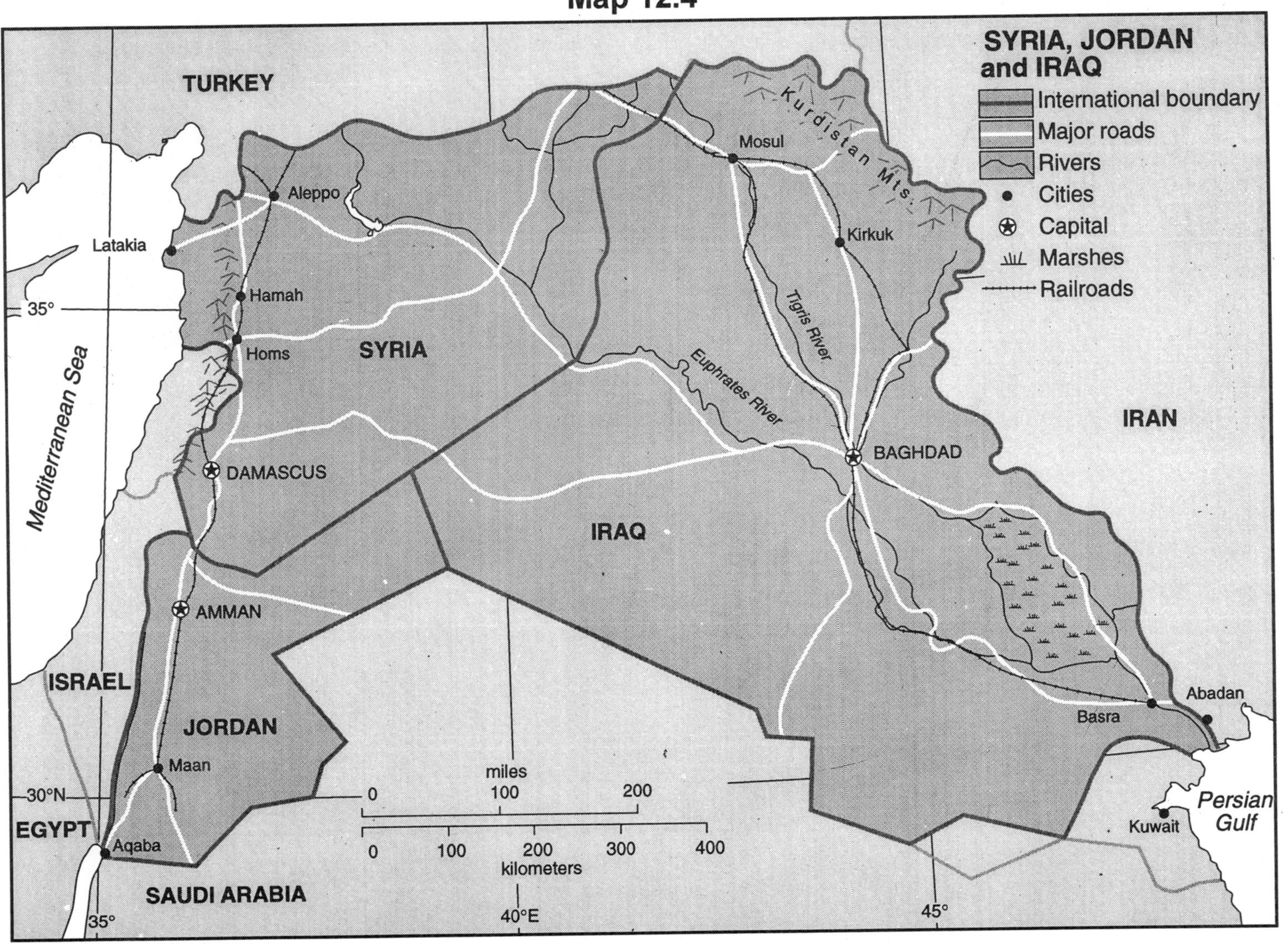
Map 12.4
SYRIA, JORDAN and IRAQ
International boundary
Major roads
Rivers
Cities
Capital
Marshes
Railroads
TURKEY
IRAN
SYRIA
IRAQ
JORDAN
SAUDI ARABIA
ISRAEL
EGYPT
Mediterranean Sea
Persian Gulf
Kurdistan Mts.
Tigris River
Euphrates River
Latakia
Aleppo
Hamah
Homs
DAMASCUS
AMMAN
Maan
Aqaba
Mosul
Kirkuk
BAGHDAD
Basra
Abadan
Kuwait
miles
0 100 200
0 100 200 300 400
kilometers
35°
40°E
45°
30°N
35°

JORDAN

Jordan was established as a constitutional monarchy in 1946 after it gained independence from Britain which controlled the territory as a mandate after World War I. When Israel was created in 1948, the area west of the River Jordan (known commonly as the "West Bank"), was controlled by it, but Israel occupied it after 1967. In 1988, Jordan renounced claims to the West Bank. Thus constituted, Jordan has a territory of 34,342 sq miles (88,946 sq km) and a population of 4.6 million. Jordan has a coastline of 12 miles (19 km) on the Gulf of Aqaba, where the country's only port (Aqaba) is located.

Jordan is a small nation with meager natural resources and is beset with several pressing political problem—all the more so because the country has provided a haven for thousands of Arab refugees who fled from their homes as a result of the Arab-Israeli wars. From the beginning, the country has depended upon foreign aid and has maintained friendly relations with the West despite strong internal and outside pressures. Foreign aid has benefited the country in its development programs. Aid has come largely from the United Kingdom, the U.S., and, since the 1967 war with Israel, from Saudi Arabia and other Arab countries. Since the 1970s, the country has successfully implemented plans for economic and social development with a view to reducing the need for outside aid.

The country is primarily a desert and semi-desert that occupies more than eighty percent of the country's surface, largely in the eastern and southern parts. The desert is rocky and bare and made up of volcanic materials of sandstone and granite. The uplands lie east of the Jordan River with altitudes between 2,000 and 3,000 feet (600 and 900 meters) increasing to the south. In the northern uplands there are several valleys and some perennial streams. The Jordan Valley is part of the great African rift-valley system and lies some 1,312 feet or 430 meters below sea level at the Dead Sea, which is the lowest point on the earth's surface. The Jordan River flowing southward drains the waters of the Sea of Galilee, and the Yarmuk River and streams of the uplands into the Dead Sea.

Climatically, the country belongs to the Mediterranean type in the west to the desert type in the east and south. The proximity of the Mediterranean Sea modifies, to an extent, the otherwise generally arid climate. The prevailing winds are westerly to southwesterly, but the dry, hot, dusty winds (known as *khamsin*) from the Arabian Peninsula, frequently bring uncomfortable weather. Average temperatures range between 46° and 78°F (80° and 26°C) in the capital at Amman, while the port of Aqaba in the far south experiences greater seasonal ranges (between 60° and 90°F). The average rainfall in the uplands east of the Jordan River totals about 15 inches (375 mm) annually, and that drops down to 8 inches in the valley and less than 2 inches in the desert areas of the south.

Cultural Patterns Ninety-nine percent of the population is Arabic, half of which are Palestinians, either from Palestine or the West Bank, who fled their homelands and took refuge in Jordan. Circassians, Armenians, Turks, and Kurds combined make up for the remaining 1 percent. Most native Jordanians belong to the several Bedouin tribes (King Hussein belongs to one of them) who support the King. Palestinian refugee settlers perhaps now outnumber the native Bedouins, and occasionally are a source of the nation's political and economic problems. Most Palestinians are employed and hold Jordanian citizenship; a few thousand live in refugee camps and receive aid from the United Nations.

Ninety-three percent of the population belongs to the Sunnite Muslim tradition, and five percent is Christian. The Christian groups include Greek Catholics, Roman Catholics, Syrian Orthodox, and some Protestants. The Druzes (an Ismaili Shiite sect) number a few hundreds. The Armenians, Druze, Baha'i, Turkmen, and Circassians each number a few hundred and form small non-Arab communities.

Economic Activity The Jordanian economy has changed dramatically during the last three decades. Although agriculture is still an important occupation, only 6.5 percent of the labor force is engaged in it and it accounted for only 8 percent of the country's gross product in 1995. Only 4.5 percent of the land is under cultivation; forests cover less than 1 percent, whereas 86 percent of the country's surface is a wasteland. Pastureland is limited and is overgrazed so as to barely support the nation's livestock. The government has made some attempts to enlarge the pasturage area by digging artesian wells in several parts of the country. Wheat and barley are the main crops of the uplands where there is some rainfall. Irrigated land in the Jordan Valley produces citrus and other fruits (such as grapes) and vegetables (including tomatoes and cucumbers); some of it is even exported to other Arab countries. Livestock is raised by Bedouins in the desert. Sheep and goats are the most important livestock, although other livestock like cattle, camels, and horses are also raised. There is some fishing in the Gulf of Aqaba.

Mineral resources include deposits of phosphates, potash, limestone, marble and salt. Gypsum, copper, uranium, and some shale oil have also been discovered. A little over 10 percent of the labor force is engaged in mining and manufacturing, which collectively accounts for nearly a quarter of the value of domestic production. Most of Jordan's manufacturing consists of extraction of phosphates, petroleum refining, cement production, and the production of a variety of consumer goods and processed foods, that are mostly concentrated around the capital, **Amman** (population: close to a million in 1998). Amman is linked with Syria in the north and Saudi Arabia in the south by a railroad and roadways. The seaport of **Aqaba** in the south at the Gulf of Aqaba and the West Bank and Jerusalem in the west are also connected to Amman by a roadway. The Royal Jordanian Airline links Jordan to the outside world.

The expansion of the Jordanian economy, especially the export of phosphates and the regional transit trade with Iraq during the 1980s and 1990s, has substantially boosted the growth of the port of Aqaba. Food items, including cereals, mineral fuels, machinery, clothing and textiles are the major imports. Phosphate fertilizers, cement, fruits and vegetables, and synthetic fibers, are the main exports. The U.S.A., Iraq, the European countries, Saudi Arabia, Japan, and India are the chief trading partners. Tourism is an important industry that pumps about half a billion dollars into the country's economy; another half a billion comes from Jordanian nationals living abroad. The economy has shown remarkable resilience and growth after losing a third of its domestic income in 1967 when it lost West Bank territory, but revenues increased due to the growth of its trade with Iraq during the Iran-Iraq War when Iraq sought and obtained access to Jordan's port of Aqaba. Again, the economy slumped during the Persian Gulf Crisis (1990-1991) when Iraq agreed to the United Nations' trade sanctions against Iraq; but recovered after 1991 when many Palestinians brought much capital back into the country after they were expelled from Kuwait. Jordan's future depends largely on foreign aid and its policy of good relations with the West is most likely to continue.

LEBANON

With an area of a little less than 4,000 sq miles and a population of a little over 3.5 million, Lebanon is one of the world's smaller nations. Located at the site of some of the oldest human settlements in the world, the Phoenician ports of Tyre, Sidon, and Babylon were major centers of trade and culture in the third millennium B.C. the contemporary nation was created in the 1920s when France, which administered Lebanon as a League of Nations mandate, relinquished control over it. In its present form it became an independent country in 1943. Though an Arab republic, Lebanon is characteristically different from many of its Arab neighbors. It is densely populated, has one of the highest rates of literacy in the area, is cosmopolitan, and is highly urbanized. Although relatively poor in natural resources, the nation has long been able to serve as a commercial and cultural center for the Middle East. Prior to the civil war of the mid-1970s and the Israeli invasion in 1982, it was one of the more prosperous, progressive, and secular nations in the Middle East, and its various ethnic and religious groups lived in comparative political and social harmony.

Physically, this small nation is a land of contrasts. The coastal plain along the Mediterranean Sea is narrow and discontinuous, nowhere more than four miles wide, at places interrupted by sea cliffs, resembling the coast of southern California, that has, with the exception of Beirut, few good harbors. Next to the coast in the east are the snow-capped Lebanon Mountains, varying in width from six to thirty-five miles, rising to elevations of 10,138 feet (3,090 meters). East of the mountains is a rich, alluvial plain, 110 miles long and six to ten miles wide, and the Bekka Valley, the scene of much political unrest in the 1908s and 1990s. The Bekka Valley is drained by the south-flowing Litani River, which is important for the development of hydroelectricity. Next in the parallel sequence and east of the Bekka Valley are the Anti-Lebanon Mountains, dominated by 10,000 foot Mount Hermon, that exhibits strikingly beautiful scenery.

Ethnic and Religious Complexity

Throughout history, the rugged, mountainous terrain of the country has been a haven for diverse religious and ethnic groups. Since the census of 1932, no official data have been taken. At the time of independence, Maronite Roman Catholics were in all probability the largest religious group and had a slight majority. No official census has taken place since then, in part due to a suspicion that the original ethnic balance no longer holds. Since then, there has been an influx of Muslim Palestinian refugees, primarily during the 1950s and 1960s. The official estimate of the religious afffiliations of the population for 1995 was: Christians 37.6 percent (divided by Catholics, Greek Orthodox, Armenians and Protestant groups), Muslims 55.3 percent, and Druzes 7.1 percent. Ethnically 93 percent were listed as Arabs (84 percent as Lebanese, 9 percent as Palestinians, 6 percent as Armenians, and 1 percent as Kurds).

Based on the population structure of the 1920s, the French who controlled Lebanon after World War I, had worked out a political formula under which the claims of different minorities were balanced by assigning top government positions to various groups. The President was always to be a Maronite, the Prime Minister a Sunni Muslim, and the president of the Chamber of Deputies a Shiite. The key ministerial positions were also reserved for particular religious groups. In the late 1950s, the "balancing act" broke down, as the ancient animosities between the different areas and villages surfaced. Meanwhile, a heavy influx of Sunni Palestinians from the West Bank, and Shias, mostly from Iran, complicated the existing complex demographic and political structure of several areas, particularly in the Bekka Valley.

The uneasy situation persisted for some time, as the United States, through military intervention, tried to patch up the seething internal hostilities. A civil war finally broke out, fomented to some degree by the outside pressures. By 1975, numerous factions were pitted against each other, and the city of Beirut became a major battleground, divided into hostile zones controlled by the contending groups. In 1976, Syrian forces entered Lebanon as "peacemakers" and started operating from bases in the Bekka Valley. In 1982, Israel invaded southern Lebanon to destroy the Palestine Liberation Organization in southern Lebanon where they had taken sanctuary and were considered a grave security risk to the northern border of Israel. But Israel withdrew, after inflicting a heavy blow to the P.L.O. and the Syrians. Since then, the U.N. peace-keeping forces have maintained peace in the "security zone" established by Israel in southern Lebanon, that was withdrawn in 2000.

Economy The largest sectors of the economy are manufacturing, trade, and the service industry; three-fourths of domestic income is derived from these sources. Agriculture and mining activities are more restricted, and account for a smaller proportion of income. A relatively larger share of the service industry (nearly a third of the labor force and the same proportion of contribution to the nation's economy) is attributed to international commerce, and to the preeminent position of **Beirut,** the national capital (population: 1.1 million) as a center of international banking and tourism. These statistics hold under normal conditions, but since the civil war of 1975-76, and the later Israeli invasion of 1982, the continuing violence has left the economy shattered. Nearly all sectors of the economy have been severely damaged, and the productive capacity of the nation drastically reduced, especially the sectors of housing, trade, public services, and manufacturing.

The mineral resources of the nation are few, limited to some high-grade iron ore, lignite, building stones, and high quality sand. The only important power plant is that of the Litani River hydroelectric project, which appreciably increased the amount of irrigated land for agriculture.

Although the area for cultivation is limited, there is an abundant supply of water from springs, favoring the cultivation of a variety of crops on the mountain slopes and in the coastal areas. Vegetables, citrus fruits, and bananas are cultivated in the irrigated coastal areas, and olives, grapes, figs, tobacco, and almonds are produced in the foothills. At higher altitudes fruits such as peaches, apricots, plums, and cherries are grown. Sugarbeets, cereals, and vegetables are the principal crops of the Bekka Valley. Poultry is an important source of income to the farmers; goats, sheep and cattle are also raised. However, the country is not self-sufficient in foodstuffs. Wheat, cotton, and sheep are the major agricultural imports. On account of continued violence, many small farmers own livestock and there has been a noticeable decline in the production of most crops. However, in the Bekka Valley there has been a noticeable increase recently in the growth of hemp, the source of hashish which is exported illegally through ports on the coasts.

Many of the industrial plants have, on the other hand, survived the civil war, although Beirut's industrial belt was practically destroyed. Manufacturing activity has revived to a smaller extent since the 1970s on account of difficulties of obtaining labor, working capital, and credit.

Lebanon's greatest economic asset has been its commerce, dating back to the period of the ancient Phoenicians. Beirut's well-developed seaport and airport, and the country's free economic and foreign exchange systems and banking secrecy law (modeled after Switzerland) have all contributed to the traditional supremacy of trade and services. Tourism has been an important industry, spurred by the country's scenery, mild climate, its historic sites, its hotels and restaurants, its seaside and mountain resorts, its cultural events, and its banking facilities.

Lebanon's geographical situation makes it a vital crossroads between East and West. The road network in the country includes international highways, which form a part of the major land routes connecting Europe with the Arab countries and the East. Several ports dot the coast. Oil tankers were built at Tripoli and Sidon where the pipeline terminals are located. Beirut is the chief cargo and passenger port, which has a free port zone. During the 1950s and 1960s the port was greatly expanded and deepened. Beirut also has an international airport and was one of the busiest in the Middle East before the civil war.

Traditional exports of the country have been textiles, jewelry, vegetable products, and pharmaceutical goods directed primarily to the Middle East countries (Saudi Arabia, Kuwait, Jordan, Turkey), Europe, and the United States. Consumer goods, machinery and transport equipment, petroleum products and food come mostly from western Europe. A high trade deficit has been generally offset by "invisible" trade (foreign aid, government loans, tourism). Since 1975 widespread smuggling, covert foreign aid to armed groups and illegal drug production has, however, characterized the traditional pattern of trade.

SYRIA

The name Syria was applied loosely to a large territory between the Taurus Mountains and the Sinai Peninsula by ancient writers such as Strabo, Pliny, and later on by the Arab geographers. Present-day Syria emerged as an independent nation in 1946 when the French relinquished control over the territory (of nearly 71,500 sq miles or 185,180 sq km) when the League of Nations mandate in Southwest Asia expired at the end of the World War I. Since inception, political life in the country has remained unstable, resulting in a large measure from the intense friction between the country's social, religious and political groups. In 1967 it lost a part of its territory (the Golan Heights) to Israel, following a short war between Israel and the neighboring states. The present authoritarian rule, committed to regain the lost territory and to achieve dominance in the Arab world, has spent enormous amounts of money to accumulate a large arms buildup, which has put a severe strain on the national budget, leaving little for development.

Physically, the country has a small coast (110 miles or 180 km) along the Mediterranean Sea between Turkey in the north and Lebanon in the south, and a narrow coastal plain, bordering which to the east is a mountain range from north to south with average heights of 3,000 feet (900 meters) in the north declining to 2,000 feet in the south. To the east of the range is a longitudinal trench that contains the valley of the Orontes River. The highest mountains bordering Lebanon rise to over 9,000 feet (Mt. Hermon). The rest of the country, to the east, is an undulating plain, consisting of mostly a desert topography of rock and gravel. Most of the country's drainage flows underground. The most important water source is the Euphrates River that originates in Turkey and flows southwestward across Syria, changing eastwards into Iraq. A big dam on the Euphrates was built in the 1970s. The Orontes, the other important river, rises in Lebanon and flows northwards into Syria.

The coastal areas and western mountains have a Mediterranean climate, marked by a long, dry summer and winter rainfall, with a total of 30 to 40 inches (762 to 1,016 mm) annual precipitation. Rainfall decreases rapidly eastward; the eastern deserts receive as little as 10 inches a year. Damuscus and Aleppo have average daily maxima of 91° to 99°F (33° to 37°C) in summer and average daily minima of 34° to 40°F (1° to 4°C) in winter. On the coast conditions are milder, and summers are much hotter and the range of temperatures is much larger in the eastern part of the country. Snow occurs on the mountains in winter.

Cultural Characteristics Nearly 90 percent of the country's 16 million population is ethnically Arab. Although the Syrians trace their origin from several sources, the predominantly Arab strain is unmistakable, with a mixture of some Greek, Roman and Turkish elements. The Turks, like the Greeks and Romans before them, influenced primarily the economic and political structures of the area they subjugated, but failed to produce noticeable change in the dominant Arab character of the Syrian people. Next to the Arabs are the Kurds who account for about 9 percent of the population. The Armenians who account for nearly one percent of the population are mostly

Arabized whereas the later immigrants such as the Turkmen who arrived after World War I have retained their cultural and linguistic identity. The Assyrians who have intermarried with the locals and mostly live in the cities have also become Arabized. Arabic is the mother tongue of most of the people. Other languages in order of usage are: Kurdish, spoken in the northeast and northwest; Armenian, spoken in Aleppo and in the larger cities; and Turkish, spoken in the villages east of the Euphrates and along the border with Turkey.

Islam is the predominant religion, professed by nearly 86 percent of the population, three-fourths of which are the Sunnites. The Alamites, an extreme Shiite sect, are the next most important Muslim group. Most of the Alamites live in Damuscus, Aleppo, and a few other cities. Christians form nearly 9 percent of the population, and are divided into several denominations---Greek Orthodox, Greek Catholics, Syrian Orthodox, Armenian Orthodox, Syrian Catholics, Maronites, Protestants, Nestorians, and Chaldeans. Most of the Jewish population moved to Israel in 1948 and only a few thousand of them now remain in the country. The complex ethnic and religious pluralism of the population accounts for much of the country's current problems.

Like the neighboring Islamic nations (Iraq, Jordan, and Saudi Arabia) the population is growing at a high rate of increase, about 2.6 percent a year. Nearly half the population is below 15 years, despite the fact that family planning is becoming popular among the educated. In 1948 some 80,000 Palestinians fled to Syria, and most of the population of the Golan Heights moved to other parts of Syria in 1967 when Israel occupied that region. Increasing numbers of workers and professionals have been leaving the country for other Arab nations, the United States, and western Europe due to the disturbed political conditions.

Resources and Development The discovery of oil in the mid 1950s in the country diminished the traditional importance of agriculture in the Syrian economy; and oil became the country's major resource and its chief export, accounting for forty percent of all revenues derived from all exports. The oil fields are a geologic extension of the Iraqi fields of Mosul and Kirkuk. Natural gas was also discovered but its exploitation has been limited. Syria is also rich in iron ore, chrome, and manganese, but production is small. Phosphates and asphalt are mined in modest quantities. Forests occupy a small proportion of the country's area (about 3 percent) as compared to over 30 percent of the land given to agriculture, Hydroelectric power in small amounts has been generated from the Euphrates at Tabqah Dam since 1978 but oil remains the chief supplier of domestic power.

Providing a quarter of the nation's income and employment to one-fifth of its labor force, agriculture remains an important if not the major segment of the nation's economy. Wheat is by far the most important crop, equal to the combined acreage of all other crops, followed by barley, millets, and cotton. Since the 1960s, with greater attention to other sectors of the economy (particularly the oil industry), the country's production of foodgrains has decreased, in part due to insufficient funding and political instability. The nation that was self-sufficient in its food requirements before the 1950s is now an importer of foodgrains. One-fifth of the country's imports are those of food and beverages. Raw cotton and cotton textiles are the most important exports. Aridity is the major handicap to farming; nearly all cultivation depends on irrigation. The Barada River, the Orontes, and the Euphrates waters are used for irrigation. The western plains and hills near Aleppo, Hama, and Homs rank high in agricultural production accounting for nearly a third of Syria's wheat and barley production. Other crops are sugar beets, vegetables, and fruits. Olives and grapes are grown on the slopes of the mountains facing the Mediterranean. Except for the irrigated tracts, non-intensive agriculture, based on dry-farming methods, is the rule. The Jezira area of the Euphrates in the northern part of the country is one of more promising sections of Syria for agricultural development, although it suffers from lack of adequate transport facilities and shortage of labor. Raising livestock including sheep, cattle, camels and poultry is also a significant economic activity.

Wool, cotton, and nylon textiles are the country's most important manufactures. The factories are located mainly in Aleppo, Damuscus, Homs, and Hamah. Natural silk is produced at Latakia. Chemical and engineering industries are located in Damuscus. Other industries include food-processing, pharmaceuticals, and fertilizers. Syria has an unfavorable balance of trade, offset by revenues from tourism, foreign aid, and earnings of Syrians abroad. Major exports are petroleum, textiles, vegetables and fruits, leather goods, and chemicals. Russia, France, Italy, Germany, Saudi Arabia, Lebanon and Iran are the chief importers. Imports include food, machinery, chemical products and metals, and the major exporters to Syria are the western European countries, the United States, Turkey, China, Japan, and Russia.

Nearly half of Syria's population lives in the cities, reflecting a dramatic shift from three decades ago when the population was predominantly rural. During this period there has been an effort to sedentarize the Bedouins and other nomadic groups. With agriculture pressing into the steppelands, the nomadic herders have been deprived of their grazing lands, and have taken to farming or have been moving to cities in search of work.

The Romans, Greeks, Arabs, Ottomans, and the French have left their distinctive marks on the Syrian towns. While Latakia and Palmyra exhibit Roman and Greek cultural influences, Damuscus and Aleppo are distinctively Islamic in look although other influences also are noticeable to a lesser degree. The continuation of old commercial and cultural links have enabled the cities to maintain their economic and cultural supremacy under the four centuries of Ottoman rule. Palmyra was a fortified garrison and a major intercontinental transit center under the Romans and later on the Ottomans, and now functions as an oasis and a way station along the Iraq-Mediterranean oil pipeline and preserves many of the historical monuments. **Aleppo** (population 1.6 million) is located between the Orontes and Euphrates rivers, and is a trade and light-industry center. It still retains some traits of the Roman rectangular street patterns and some Arab influences. Next to Constantinople among the cities of the Ottoman Empire, it supported a large textile industry by the 18th century. Now, it is an important commercial and manufacturing center with factories producing such items as silk cloth, woolen carpets, tractors, and chemicals. As a major railroad and highway focus, it reflects the locational advantages which the city enjoyed historically as a center of caravan trades.

The national capital and the largest city of **Damuscus** (population 1.6 million) is a rival of Aleppo as the major center of education, culture, and industry. Its important manufactured products include brocades, engraved wood, gold and silver jewelry, and carpets. It is well served by transport facilities. The city is located in a large oasis that enjoys the natural advantages of fertile soil and available water of the Barada River, and the assured advantages of agricultural and regional exchange functions. As a historic town which experienced many fluctuations in its fortunes, it is now the focus of the nation's political, social, and economic planning, besides being a leading commercial and manufacturing center. The Mediterranean port of **Latakia** (population: 306,000) is important as a seaport and tourist center. **Homs** (population: 645,000) located in the middle of a fertile plain east of the Orontes River, is the hub of the country's road and railroad systems. **Hamah** (population: 254,000) to the northeast of Homs is an important commercial and agricultural trading center.

Further Readings

Beaumont, P. et al. *The Middle East*. New York, 1988.

Cressey, G.B. *Crossroads: Land and Life in Southwest Asia*. Chicago, 1960.

Fisher, W.B. *Middle East: A Physical, Social, and Regional Geography*. London, 1978.

13

The Arabian Peninsula

Located in the extreme southwestern corner of Asia, the Arabian Peninsula, commonly known as "Arabia," occupies a little over half of Southwest Asia and contains nearly a fifth of its population. It stretches from Jordan and Iraq in the north to the Arabian Sea in the south and southeast, and from the Gulf of Oman and the Persian Gulf (also called the Arabian Gulf) in the north to the Red Sea to the west and southwest. Four-fifths of the peninsula is occupied by the Kingdom of Saudi Arabia. Several smaller states-Kuwait, Bahrain, Qatar, and the United Arab Emirates-are strung along the Persian Gulf, and Yemen borders the Arabian Sea.

Despite political turbulence in the past, and a variety of national experiences in the present, the states of the Arabian Peninsula share an underlying unity of environment, society, culture, and faith. The physical cohesiveness is reflected in a shared environment of desert, and the fact that most of the peninsula is unsuited for sedentary agriculture is of great importance. Habitable land is scarce, and efficient use of water is critical to the welfare of each state. The geophysical homogeneity has also helped to create a predominant cultural environment among the people: a degree of similarity in language, religion, and political experience. Most of the states share common political systems. Nearly all are or have been monarchies based in a large measure on the principles of religious legitimacy, which are being tempered slowly under the impact of western social and political influences.

Most of the inhabitants are ethnic Arabs, and a large number can trace their ancestry back through many generations living in the same area. Almost all speak Arabic, although a large number of non-Arabic immigrants, mostly from southern and southeastern Asia, have in recent years settled in the Gulf states. They comprise a substantial proportion of the population, ranging from 10 percent in Saudi Arabia to 45 percent in Kuwait. Most of these immigrants are short-time workers, and have adopted Arabic speech. They carry on their activities through the universally acceptable English among the elite and the commercial communities, while such immigrant tongues as Hindi, Urdu, and Farsi may also be heard.

Arabia has been the cradle of the Arab and Islamic cultures whose major contribution to history has been as the homeland of Islam. Since the 7th century, almost all the Arabs have adopted the Islamic faith. The non-Muslim incursions into the Peninsula's population have been a phenomenon of the 1970s and 1980s, and directed mostly at Bahrain, Kuwait, Oman, Qatar, Saudi Arabia, and the United Arab Emirates by the south Asian

Map 13.1

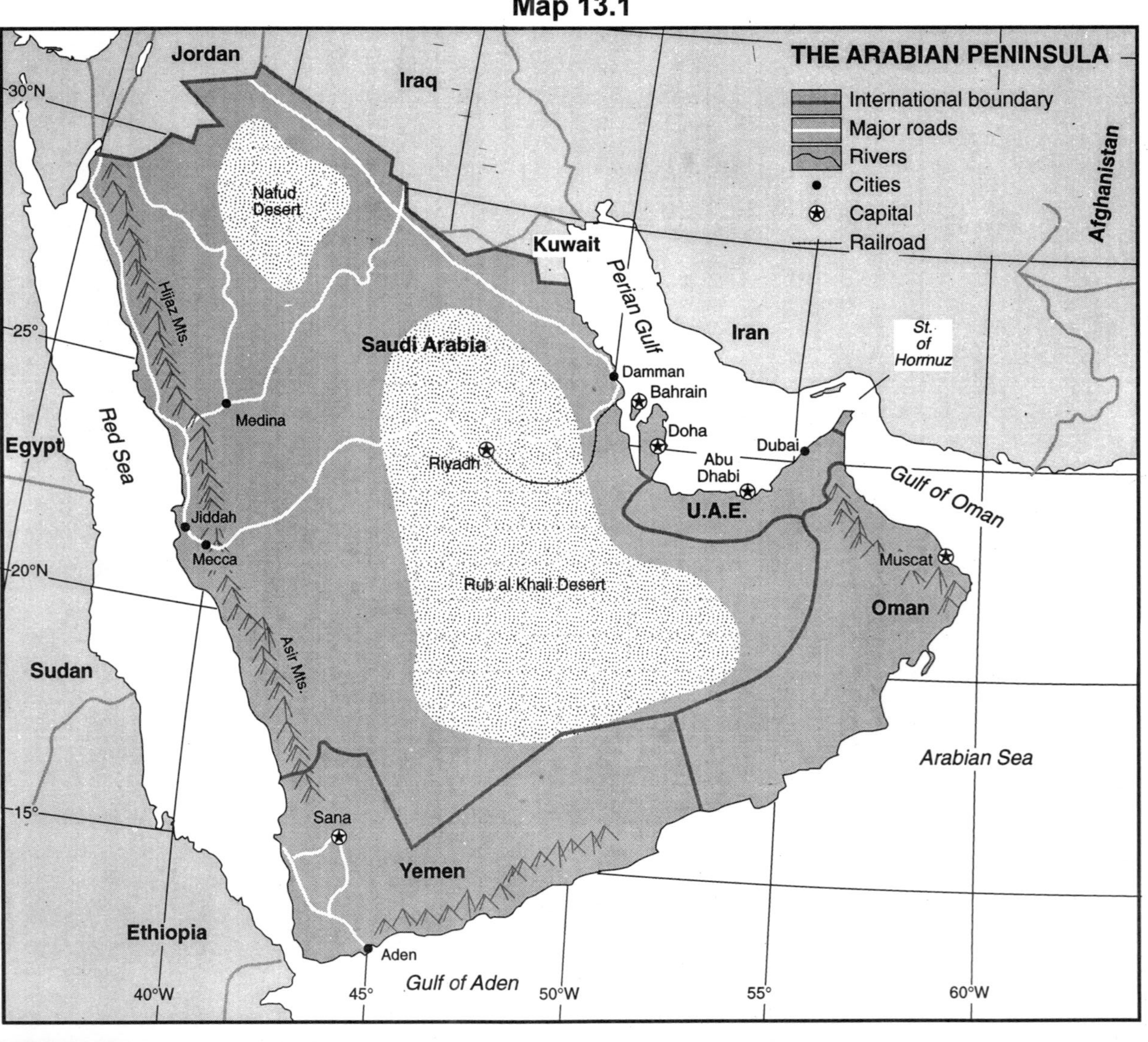
THE ARABIAN PENINSULA
International boundary
Major roads
Rivers
Cities
Capital
Railroad
Jordan
Iraq
Kuwait
Afghanistan
Iran
Saudi Arabia
U.A.E.
Oman
Yemen
Egypt
Sudan
Ethiopia
Perian Gulf
St. of Hormuz
Gulf of Oman
Arabian Sea
Gulf of Aden
Red Sea
Nafud Desert
Rub al Khali Desert
Hijaz Mts.
Asir Mts.
Medina
Jiddah
Mecca
Riyadh
Damman
Bahrain
Doha
Abu Dhabi
Dubai
Muscat
Sana
Aden
30°N
25°
20°N
15°
40°W
45°
50°W
55°
60°W

workers. Thirteen percent of Oman's population in the mid-1980s was listed as professing the Hindu faith. (Other states reported fewer non-Muslims.) From the advent of Islam the religion of Islam and the Arabic language were spread from the Arabian Peninsula throughout much of Southwest Asia and North Africa, and developed close links with them through commercial, religious, social, military, and political interactions. Islam became a powerful, cohesive force; the differences in sects were of local significance. The Ibadhi sect of Islam is important in Oman and Shia is the dominant faith in Bahrain, while small Christian and Hindu groups in some countries reflect other immigrant concentrations. The historic commitment of the peninsula to the Prophet Muhammad, the founder of the faith, transcended the local differences, and has done more to unite than divide the peninsula.

Politically, the peninsula has rarely been united under one government. In the 16th century, the Ottoman Empire conquered most of the coasts; the central interior and the southeastern ports remained independent under the Saudis. Later, in the 19th and early 20th centuries, the British controlled most of the coastal portions but the interior sections remained beyond their reach, and much of Arabia remained a backwater as the waves of modern civilizations passed by. The discovery of petroleum in the early 20th century has, however, led to increased contacts with the West, and catapulted Arabia with dramatic suddenness into the international arena, which has presented the region with both opportunities and problems. Despite the uneven distribution of resources among its states, a somewhat similar economic transformation has now been taking place in all parts of the peninsula as contacts with the West continue to grow. Urbanization, greater access to healthcare and education, secularization, and the programmed settling of many nomads have been changing the character of daily life throughout the region. The Arabian Peninsula continues to share an underlying unity of physical environment and society.

Climate is a great limiting factor. A large part of the peninsula is a tropical desert, with very meager moisture and high temperatures that result in extreme aridity. Summer temperatures often exceed 120°F (48°C) in the shade, while high humidity along the coasts adds to human discomfort. In winter, temperatures moderate a little and occasional frosts may occur in the higher altitudes. Summer is also the main season for dust and wind storms. In general, therefore, winters tend to be somewhat pleasanter, and except for the southern coastal areas winters may occasionally record temperatures below freezing during the nights. The diurnal range of temperatures tends to be large. Precipitation is scanty everywhere, averaging two to five inches (51-127 cms) annually, except for the mountainous regions, notably the highlands of Yemen in the southwest, where probably precipitation between 20 and 40 inches may occur during summer, and the mountains of Asir north of Yemen. When the rain does arrive it may be in the form of a very heavy downpour but may only last for a short time, and the *wadis* fill quickly. The run-off often is impounded and utilized for irrigation. There are no permanent streams on the peninsula. Arabia is characteristically Mediterranean in its precipitation distribution, in that the rainfall comes in winter, except for Yemen which experiences a summer maximum that is monsoonal in character.

Economic Activity Settlement is thus concentrated along the coasts, where contact with the outside world and economic development are the maximum, and in a few scattered oases inland where important historic cities exist. The city of **Taif** (population: 400,000) at about 8,000 feet (2,432 meters) in the mountains near the west coast, has become Saudi Arabia's fourth largest city with close to half a million inhabitants. It is the country's summer capital, and its cool temperatures, green and forested hills, and proximity to the holy places, has made it a rapidly growing tourist attraction. **Jiddah**, the port city on the Red Sea, performs several of the administrative functions of Saudi administration, and has overtaken Riyadh, the administrative capital, in population and now contains more than 1.5 million inhabitants. It is not only a major commercial center, but houses foreign embassies and legations as well. The capital of the nation, **Riyadh**, located in the interior on a large oasis, is growing rapidly (population 1.8 million), is closer to the Persian Gulf coast, and is connected to it by a railroad. The holy city of **Makkah** (population: 650,000) is the birthplace of Prophet Muhammad and is located inland, not too far from the coast. The cities of the Gulf states (Kuwait City, Manamah, Abu Dhabi, Dubai, and Mascat) owe their origins and growth to their port functions, proximity to the oil-refining centers, and administrative roles for the sheikhdoms.

Urbanization is on the increase in the peninsula as the traditional economic activities such as nomadic herding and some forms of conventional agriculture continue to decline. The rural population has been drifting to the cities which have come increasingly to represent opportunity and prosperity. Eighty percent of Saudi Arabia's population is classified as urban, and half the total population of the country now lives in seven large cities. Riyadh, the capital, and Jiddah, the diplomatic and commercial center, together contain more than one-sixth of the Saudi population. A little less than half of the United Arab Emirates' population lives in the city of Abu Dhabi and another quarter in Dubai, while in other Gulf nations, more than half the population is concentrated in the capital.

The Arabian Peninsula has a population of nearly 40 million, close to a half of which is in Saudi Arabia and the rest divided among the other nations of the region. Given the contributions that it has made to the world through Islam of Arabic language, culture, art, and ideas, and the enormous geopolitical and economic significance of its vast oil reserves, the population remains surprisingly small. Although people are predominantly Arab, along the Gulf the unprecedented economic growth based on oil has created far more jobs than the small indigenous populations can fill, despite the city-ward movement of rural populations and the large numbers of foreign minorities who are present. South Asians make up 34 percent of Qatar's population, and Iranians another 16 percent. The native Qataris account for less than 30 percent of Qatar's total population. Only one-fifth of the United Arab Republic's current population is classed in the "citizen" category. Among the other groups represented in the Gulf states are Iranians, Pakistanis, Thais, Filipinos, Sri Lankans, Baluchis, and Zanzi-baris from Africa's east coast.

One result of the influx of non-native peoples to the Arabian peninsula has been the introduction of a cultural diversity that creates stresses in the Arabian and largely traditional societies. The religious differences also create cause for concern. Saudi Arabia, as a "keeper" of the two most sacred places of Islam, Makkah and Medina, has come under pressure from the Shia extremists. Occasional disturbances during the annual pilgrimage (*hajj*) to Makkah underlie such tensions in the region.

SAUDI ARABIA

In area, population, and historical importance, the Kingdom of Saudi Arabia dominates the Arabian peninsula. It covers over 860,000 sq miles (2,240,000 sq km) of territory and contains over 18 million inhabitants. The country has the world's largest petroleum reserves (nearly a quarter of the world's total) and after the Soviet Union and the United States is the third leading oil producer. Oil and natural gas dominate the economic scene, although some other minerals such as high-grade iron ore, gold, copper, zinc, platinum, and lead are also produced in small quantities. The production of natural gas amounts to about 4 percent of the world's total output.

Most of the country is a vast, barren desert and plateau area, an eastern section of which is known as *Rub-al-Khali* (the empty quarter). It is fringed on the west by a mountain range, where nomadism has been the traditional economic activity. Less than two percent of the land is given to permanent cultivation, and the contribution of agriculture to the national economy is about 6.5 percent of the gross domestic product. Under the Five Year Plans that started in 1970, the administration is trying to diversify economic activity and encourage the agricultural sector by affording subsidies and introducing dry-farming technologies to the farmers. About 10 percent of the labor force was engaged in agriculture in 1990, a significantly higher proportion from two decades ago when the Five Year Plans were initiated.

Water is scarce, but where it is present, principally at the several oases, the desert blooms, and irrigated crops in great variety are intensively grown. Except for dates, which may be exported in some years, the production of crops is for local consumption only. The classic plant of the oases is the date palm. Its high-energy fruit, which dries well and tends to resist spoilage admirably, was the ideal "portable" food for the nomads, and trees were cultivated as a virtual monoculture. But with the decline in recent times of the caravan and nomadic herding as a dominant lifestyle, demand for dates has diminished and many groves have been replaced by more profitable crops. In 1991 wheat replaced dates as the leading crop. Other cereals and crops include barley, rice, corn, millets, sorghum, and vegetables. Sheep, goats, and cattle are raised on pasturelands (nearly one-half of the land is classified as pastureland) on which some vegetation can be obtained, but the camel, perfectly adapted to be the beast of burden across the arid landscapes, retains its dominant position as a draft animal of the desert and a good part of its historic importance. The critical role of water in the future planning has led to several interesting ideas, some as farfetched as towing giant icebergs from Antarctica into the Gulf or collecting melting fresh water from the mountains and piping it to desired

Map 13.2

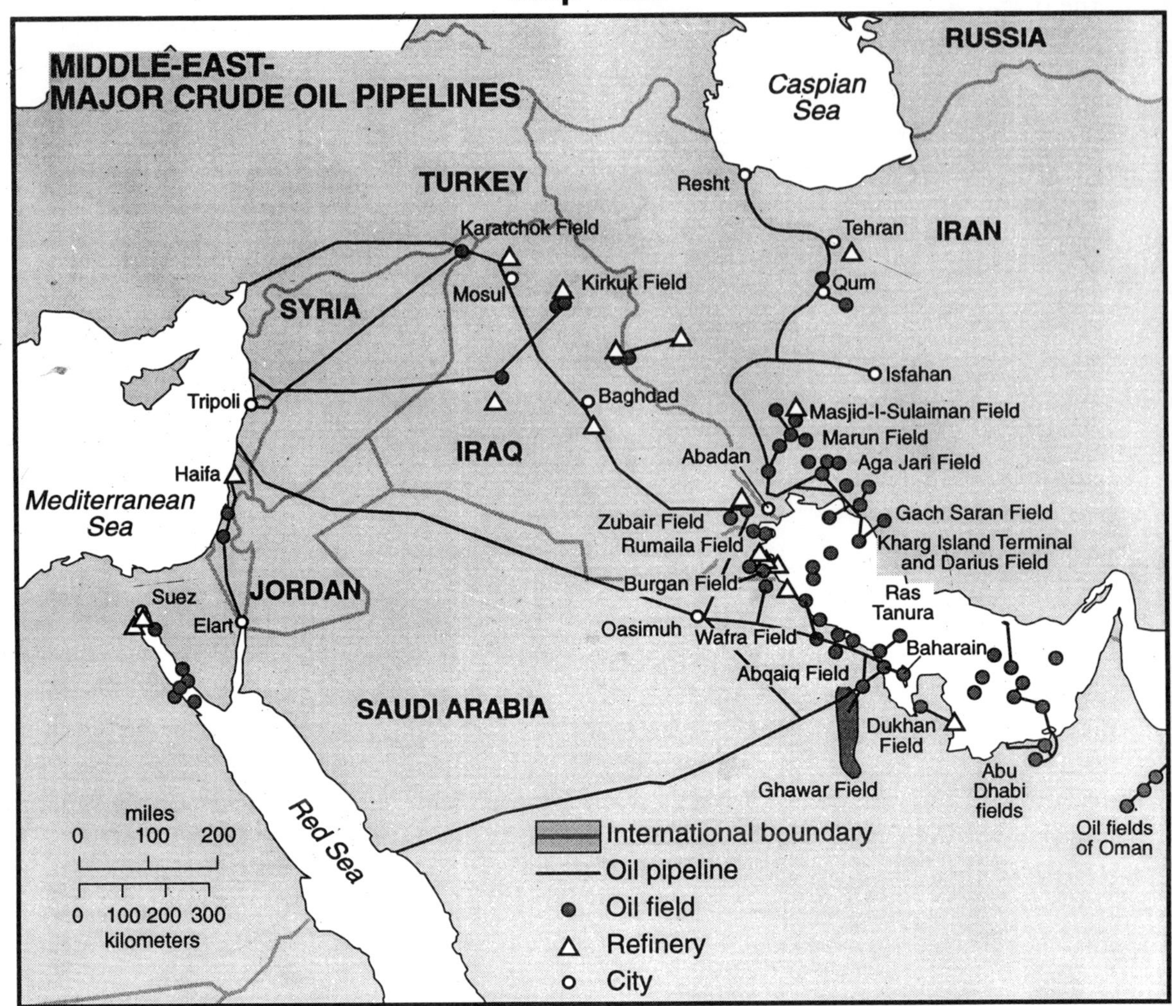

locations for cultivation of crops or for drinking have been seriously considered by the Saudi administration! While some of these ideas remain at the consideration stage, improvements in agricultural productivity have been significant during the last few decades, and a near sufficiency in several food items (wheat, eggs, milk, barley, and millets) has been reached and food is no longer one of the principal import items.

Oil has undoubtedly fueled all of the economic development and infrastructure improvements. Saudi Arabia is, by far, the leading oil exporter in the world. Over ninety percent of the country's export income is derived from oil and this represents 73 percent of the government's total revenues. While the oil lasts (estimates run from 20 to 50 years) the Saudi administration is trying to ensure the country's long-term well-being by recycling oil income into overseas investments and domestic growth. The manufacturing sector has been substantially diversified and widened to include the production of rolled steel, petrochemicals, fertilizers, aluminum goods, truck assembly, refrigerators, plastic materials, metals, cement, furniture, and printing presses. The country has invested heavily in the industrial cities such as Jubail on the Gulf and Yanbu on the Red Sea, where a pipeline delivers oil from the eastern fields. Material progress includes new highways, dams, port facilities, factories, modern universities, and advanced desalinization plants that convert sea water into fresh water for the growing population. **Riyadh** and the chief port **Jiddah** have been modernized and serve as growing centers for manufacturing and commerce. Larger towns are becoming increasingly cosmopolitan in character. **Daharan** on the Gulf coast and the center of the Aramco, is like an American city transplanted into the Arabian desert. This well-planned town contains modern, air conditioned buildings, schools, hospitals, supermarkets, and a variety of recreational facilities.

THE OTHER GULF STATES OF THE ARABIAN PENINSULA

The tiny constitutional monarchy of **Kuwait** (area: 6,880 sq miles and population 1.9 million) is among the most prosperous nations in the world. The per capita income of its inhabitants approximates its gross national product at $21,000 and rivals that of the most developed nations of Europe and the U.S. Wedged between Saudi Arabia and hostile Iraq at the mouth of the Gulf in an oil-rich belt, the location is particularly significant as the Gulf War of 1990 demonstrated. Kuwait can easily control the entry of Iraq (albeit with the help of military powers) into the Gulf, and just as easily Iraq can over-run Kuwait from the land frontier.

The country is largely a sandy desert except for the few oasis-patches. It has virtually no agricultural land, and pastures account for less than 8 percent of the land where sheep, goats, and cattle are raised. Although some vegetables and fruits are produced, and commercial fishing is locally important (pearl fishing has declined in recent years as easy money can be obtained from oil), almost the entire foodgrain requirements are met through imports. Food and live animals account for one-sixth of all imports.

Oil is the nation's major resource. Nearly 10 percent of the world's total reserves exist here, next only to that of Saudi Arabia. Production is a little more than 2 percent of the world's total. Natural gas reserves and production are much smaller, but are still quite substantial. The economic transformation of the nation has been dramatic and spectacular from the sleepy, pearl-diving, small-trading, and fishing economy of the 1950s to a highly developed welfare state. At the turn of the 20th century there were no schools or hospitals and much of the drinking water had to be hauled in by boats from neighboring Iraq. Today, all that has been replaced by modern air-conditioned schools and hospitals equipped with the latest facilities, the world's largest water distillation plant, large boulevards, modern banks, and shopping centers. The entire length of 2,700 miles (4,300 km) of roadways is paved. All this has been possible through the enormous oil wealth that has flowed into the country.

Since 1970, the government has owned all the petroleum and natural gas undertakings and derivative industries, electric-generation companies as well as desalinization plants, whereas trade, manufacturing of building materials, and banks are in the private sector. Apart from various petroleum products, the manufacturing sector includes the production of plastics, cement, ceramic goods, metal pipes, electric cables, and furniture. Petroleum and petroleum products account for over 90 percent of all exports. In the 1970s and 80s, exports have exceeded three times the value of imports in normal years (production and world prices of oil would affect trade balances). Western Europe, the U.S., Pakistan, Singapore, India, China, and Taiwan are the major export destinations. Imports include machinery, manufactured goods, food, and chemicals. The U.S., Japan, Germany, the U.K., India, and Saudi Arabia are the principal suppliers. Capitalizing on oil revenues, and recognizing limitations of domestic development, the administration has made enormous overseas investments that should provide substantial future revenues (Kuwait owns shipyards in France, automobile companies in Germany, commercial properties in London, and office buildings in the U.S.).

Iraq's invasion in 1990 destroyed most of the nation's infrastructure (including 1,300 of the country's oil wells) and depleted the nation's treasury. Following the U.N. assisted peace, a swift reconstruction and recovery program was put in place. **Kuwait City** is the national capital, which together with the suburban metropolitan areas contains a population of over one-half million. It is a modern city, complete with such modern facilities as super markets, automobile dealerships, elegant hotels, banks, schools, and parks. The city has attracted thousands of immigrants from other Arab nations, and foreign companies.

Two tiny, diamond shaped **Neutral Zones** border Kuwait; the western one is jointly controlled by Iraq and Saudi Arabia, and the eastern one is under the joint jurisdiction of Saudi Arabia and Kuwait. Each covers roughly 50 sq miles (130 sq km) of barren desert land. They were created in 1922 when boundaries between Kuwait and Saudi Arabia could not be agreed upon by Iraq, Saudi Arabia, and Kuwait. Initially useless territory, the zones became hot real estate property when in the 1970s and 80s substantial oil reserves were located in this tiny territory. In 1990 the eastern Neutral Zone came under heavy Iraqi bombardment and most of its working oil wells were destroyed; they have since been repaired through help from the U.S.

The tiny archipelagic principality of **Bahrain** in the Gulf contains less than 270 sq miles (695 sq km) of territory and has a population of a little over 650,000, a third of which consists of foreigners working in the oil industry, and in commerce and business enterprises. Most densely populated among the Gulf states, ninety percent of its population lives in the four major cities. **Manamah** (population: 152,000) is the capital. Only three percent of the land is arable, and agriculture occupies a position of minimal importance; only 2 percent of the labor force is engaged in it and accounts for less than 2 percent of the country's gross domestic product. Petroleum dominates the economy, and is the major export. Almost all the foodgrain requirements are imported as well as most of the manufactured and consumer items. But manufacturing, in addition to petroleum extraction and refining, is growing in importance as the administration tries to diversify economic activities in face of dwindling petroleum reserves. Nearly 12 percent of the labor force is engaged in manufacturing; the remaining workers are employed in construction, transport, trade, public utilities and service sectors. Exports consisting almost entirely of petroleum and natural gas products are directed toward Saudi Arabia, the U.S., the United Arab Emirates, Japan, South Korea, India, Canada, and Singapore. Imports include machinery, food items, chemicals, and petroleum products. The U.S., U.K., Japan, Germany, and Saudi Arabia are the major sources of the imports.

Qatar is a small peninsula into the Gulf with a population of 580,000 and an area of less than 4,500 sq miles (11,700 sq km). The principality is especially important for the collecting and storage of oil for its three off-shore oil fields. Topographically, much of the surface is a flatland of sand dunes and less than 1 percent of the land is arable. Underground water is salty and unsuitable for drinking. Distillation and desalinization of seawater provides half of the country's water supply. Oil and natural gas provide 85 percent of the country's revenues, and the per capita income is nearly $16,000 a year, one of the highest in the Gulf region. The government owns all the agricultural land. Vegetables and fruits are grown, enough for domestic needs; most other food crops must be imported. Fishing is an important economic activity.

Currently, oil production is less than 1 percent of the world's total output, and the proven reserves are limited. The prospect of declining reserves makes natural gas reserves, which are more abundant, most likely to afford the long term basis for the Qatari economy. The administration has been trying to diversify the economy and set up a fertilizer plant, a steel mill, and a petrochemical plant.

Eighty percent of the country's labor force consists of non-Qataris while the short-term workers from India, Pakistan, Bangladesh, and Sri Lanka make up a third of the labor force; Iranians, non-Qatari Arabs (Palestinians, Egyptians, Jordanians, and Syrians) and Africans make up the remaining two-thirds of the labor force. A 65-mile (105 km) paved highway goes from the capital city, **Doha** (population: 250,000) to the Saudi border. While there is no railroad, Doha has an international airport linking the country with the rest of the world.

The United Arab Emirates is a federation of seven emirates (formerly independent sheikhdoms): Abu Dhabi, Ajman, Dubai, Fujairah, Ras-al-Khaimah, Sharjah, and Umm-al-Qaiwain that lie across the eastern coast of the Arabian Peninsula. The country covers an area of 30,000 sq miles (77,700 sq km) and has a population of a little over 2.8 million, a substantial proportion of which consists of Pakistanis, Indians, Iranians, Baluchis, Zanzibari Africans, and other groups who have immigrated to work in the oil industry and in the service sector of the economy. The capital city, **Abu Dhabi**, has grown into a large urban conurbation of over three-quarters of a million people to accommodate the immigrant groups and developed into a modern, cosmopolitan center. **Dubai** and **Shariqah** (population: 600,000) and 125,000 respectively) are other larger cities that have grown enormously during the last two decades.

The country is mostly a low-lying desert plain of below 500 feet (150 meters) except for the Al-Hajar Mountains rising to nearly 10,000 feet (3,050 meters) in the east. The U.A.E. is an oil-rich nation, whose petroleum and natural gas reserves are among the world's richest (nearly 10 percent of the world's total). Annual production amounts to nearly 2 percent of the world's output. The oil and natural gas industry accounts for one-third of the gross domestic product and is concentrated primarily near Abu Dhabi. Enormous oil revenues have made the U.A.E. one of the richest nations in the world, with a per capita income close to that of the U.S. ($23,000 annually in the mid-1990s).

Agriculture plays an insignificant role in the economy, and less than 1 percent of the land is given to cultivation, two-fifths of which is under irrigation. Date palms, mangoes, melons, wheat, alfalfa, millets, and vegetables are the main agricultural products. Most of the food grain and meat requirements have to be met through imports which account for one-sixth of all imports. Fishing along the coast is an important economic activity, and supplies all domestic requirements.

Besides industry based on petroleum and natural gas, the manufacturing sector has grown substantially during the last two decades and accounts for nearly 9 percent of the domestic revenue. The construction industry has also expanded enormously, made possible by oil revenues. Extensive construction of new buildings, roads, and industrial plants has taken place during the last two decades. Manufactures include such products as aluminum, plastic and building materials. Traditionally, exports have exceeded imports. In addition to petroleum and natural gas, dates, aluminum, dried fish and pearls are the main exports. Imports include machinery and transport equipment, chemicals, and food grains. Japan, United Kingdom, United States, Germany, the Netherlands, Taiwan, Singapore, and Turkey are the major trading partners.

The Sultanate (Kingdom) of Oman lies almost entirely outside the Persian Gulf, except for a tiny finger of its territory between the Persian Gulf and the Gulf of Oman at the strategic Strait of Hormuz, and thus it can control the route to the oil-rich region of the Persian Gulf. Most of its 118,000 sq mile (306,000 sq km) territory is a desert that merges into the Rub-al-Khali of the Arabian Peninsula and remains undemarcated. A mountain range that parallels the Gulf of Oman coast rises to about 10,000 feet (3,000 meters).

Although less than 1 percent of the land is under cultivation, as a contrast to other states in the region, one-third of the labor force is engaged in agriculture. Dates, mangoes, bananas, melons, alfalfa, and onions are the major agricultural products. Goats, sheep, and cattle are raised in the Dofar region. Fishing is important in the Gulf of Oman and along the Arabian Sea coast. Manufacturing accounts for 6 percent of the labor force, and consists mainly of the making of cement blocks, furniture, plastics, and fiberglass. The country has a tradition of maritime trade between the East and the West, and was a major base of Arab traders with East Africa. One-sixth of the labor force is still engaged in trading activities that account for over 16 percent of the nation's gross domestic product. The old trading patterns of pearls, incense, and silks, are changed now, but trade remains important. Major imports are manufactured goods, machinery, food grains, and chemicals. Oman is the least urbanized state in the Arabian Peninsula; only 13 percent of the population in the mid-1990s was listed as urban. **Muscat** (population: 52,000) located on the Tropic of Cancer is the capital and the nation's major port.

The Republic of Yemen includes the former North and South Yemens and has an area of a little over 200,000 sq miles (515,000 sq km) and a population of 17 million. The island of Socotra, a military check-point in the Arabian Sea, is 600 miles (960 km) due east of Aden and is politically a part of Yemen, and is of considerable strategic significance.

The present political structure dates to

1990 when the former North Yemen and South Yemen were joined to form a republic. The capital is at **Sana** (population: 950,000) in the former North Yemen. North Yemen was an authoritarian old-style Muslim monarchy, remote and unknown until the 1960s when Egypt took control of it. A military junta subsequently seized the government. Yemen has been the battleground of the leftist guerrillas, mainly drawing support from the former South Yemen which provided the Soviet Union with a foothold on the Arabian Peninsula at the strategic Bablel-Mandeb, joining the Red Sea with the Indian Ocean.

Most of the country is a highland and a broad desert plateau which merges with the Rub-al-Khali of Saudi Arabia, that leaves only a very narrow coastal plain nowhere wider than 25 miles (40 km). The mountain ranges are rugged and dissected, ranging in altitude from 1,150-5,000 feet in general, but climbing to over 12,336 feet at the highest point northwest of Sana. Precipitation in the northern desert region adjacent to Saudi Arabia is less than 4 inches annually but is much higher in the southwest highlands where more than 16 to 20 inches is the norm. Thus, in contrast to the generally barren character of most of the Arabian peninsula, the highlands of western Yemen are surprisingly green with vegetation and are fertile. Terraced farming in these areas is important for the cultivation of such crops as cereals (especially sorghum and corn), vegetables, grapes, dates, Mocha coffee, and *khat* (a narcotic stimulant grown on irrigated land). As a sharp contrast to the rest of the Arabian Peninsula, 56 percent of the labor force is engaged in agriculture. Fishing is important along the coast.

Industry and mining account for nearly 10 percent of the labor force but bring in 15 percent of the nation's revenues. The manufacture of cement blocks, textiles, bricks, rubber goods, plasticware, soft drinks, and some dairy products are the chief industries. Yemen is relatively rich in mineral resources; iron ore and salt are mined, and some oil (discovered in the mid-1980s) is produced, giving some relief to the desperately poor nation. Foodgrains, manufactured goods, machinery, and chemicals are the major imports. Food items (particularly fish), coffee, cigarettes, leather, and grapes are the principal exports. Germany, Japan, the U.S.A., Saudi Arabia, France, Italy, Australia, and Thailand are the major trading partners.

The former British colony of **Aden** (population: 562,000) is a major commercial center. It has a deep water natural harbor and is located at the important sea route to the Red Sea and the Indian Ocean, and is a refueling station for ships. During the closing of the Suez Canal (1967-75) it declined but has since regained much of its commercial importance.

Further Readings

Ali, S.R. *Oil, Turmoil, and Islam in the Middle East*. New York, 1986.

Al Farsi, *Saudi Arabia*, London, 1986.

Cressey, G.B. *Asia's Lands and People*. New York, 1960.

Stone, R. (ed.) *O.P.E.C. and the Middle East*. New York, 1977.

14

Iran, Iraq, and Afghanistan

IRAN

By virtue of its size, location, and distinctive cultural orientation, the Islamic Republic of Iran qualifies for a special position among the nations of Southwest Asia. In area, it is the largest after Saudi Arabia, and second only to Turkey in population. As an imperial power since the 16th century, it has been an important factor in the Middle East, and with increased involvement of the western powers in the Persian Gulf, it is likely to play a significant role in the politics of the region.

Iran is unique in many ways, and quite distinct from the Arab nations. The adopted official language of the nation is Farsi or Persian which belongs to the Indo-European family and is distinct form Arabic or Turkish, and employs a modified form of the Arabic script. But more importantly, Iranians adopted an independent stream of Muslim tradition, the Shiite branch, that advocated a strong support for the house of Ali (the Prophet's son-in-law) and rejected the Sunnite majority's injunctions, although such a course was considered heretical in the early history of the Islamic faith. Despite the universal persecution of the Shiites by the mainstream Sunnites, the Shia faith was made the official religion of Iran in the 16th century, that persists to this day. Iran proudly maintains a close link with its ancient heritage that goes back thousands of years.

Historical Background Iran (formerly known as Persia) has a long and rich history. Traces of the world's most ancient settlements have been found in the Caspian littoral and in the interior plateaus; village life probably existed there before 4000 B.C. The Aryans came into Iran about 2500 B.C. and split into two main groups, the Medes and the Persians. The first known period of glory goes back to the 6th century B.C. when Cyrus the Great founded a large well-organized empire that included much of modern Turkey, Iraq, Syria, Jordan, and Afghanistan. The 2,500th anniversary of his empire was celebrated with great pomp in 1971 at Persepolis. After Cyrus, a succession of Greeks, Parthians, and the Sassanids ruled the country. The Arab invaders occupied the capital, Ctesiphon, in 641 and brought Islam to Persia, and it was in Persia that the Shiite sect was developed. The Turks included Persia as one of the Turkish states in the 10th century, and were followed by the Mongols in the 13th century. Persia again reached great heights under Shah Abbas who ruled from 1588 to 1627. Under him the society prospered through commerce and trade, and political ties were developed with several countries of Europe and Asia. He built several public buildings, gardens, mosques, bridges, and roads throughout the country. The capital city of Isfahan under him was adorned with Persia's finest and most graceful buildings and gardens, and became one of the most splendid cities of the world. Later, during the Zand dynasty in the 18th century, Shiraz was made the capital, and several fine buildings were constructed there.

Following the Zand dynasty, the country steadily lost territory to neighboring countries and fell under the increasing pressure of European nations, particularly Czarist Russia. Iran was forced to give up its Caucasian region in the northwest to Russia, and the city of Herat in the east to Afghanistan. A series of campaigns to reclaim the lost territory in the east ended with the intervention of the British and resulted in the recognition of Afghan independence in 1857. The discovery of oil in the early 1900s intensified the British-Russian rivalry for power over the nation that culminated in an Anglo-Russian agreement (revoked later by Iran) dividing Iran into spheres of influence. The designs of Czarist Russia and the British Empire thus clashed in Iran. Its growing importance as a major oil producer in the 20th century further highlighted the significance of its geographical position. Although Iran's importance as a vital link in the east-west land route had by then substantially declined, Russia's desire to move into the warm waters of the Persian Gulf and the Gulf of Oman remained a primary concern, at least until World War I, and Russia continued to interfere in the internal affairs of the country.

In the aftermath of World War I, the administration of the chaotic Iranian nation was overthrown by Reza Khan (later, Reza Shah) Pahlavi who abrogated the previous treaties with the British, reorganized the army, and introduced many reforms, somewhat similar to those carried out in Turkey under Ataturk. The Muslim clergy were stripped of official privileges, and women were encouraged to participate in public affairs. Universal education was introduced, and a standard form of the Persian language was declared as the official medium of communication to help unite the country's several tribal groups.

By the early 1940s, the Reza Pahlevi's administration had grown dictatorial, and Britain again intervened (with the assistance of the U.S.) in the administration, partly to deny German access to the east, and partly to handle the delivery of war supplies to its ally, Russia. In 1943, the Declaration of Tehran which was signed by the U.S., Britain, and the U.S.S.R., guaranteed the territorial integrity of Iran. However, the U.S.S.R., dissatisfied with the refusal of the Iranian government to grant oil concessions, fomented a revolt in the northwest, which led to the establishment of the People's Republic of Azerbaijan, and the Kurdish People's Republic. In 1951, the oil industry was nationalized, and the National Iranian Oil Company was formed. By the 1960s, the U.S. had become actively involved and Iran received substantial economic and military aid from the United States, and undertook a broad program designed to improve the economic and social conditions of the masses. Substantial industrial and commercial development followed during the 1960s and 1970s until the overthrow of the Shah government in 1979.

By the late 1970s, the masses were

Map 14.1

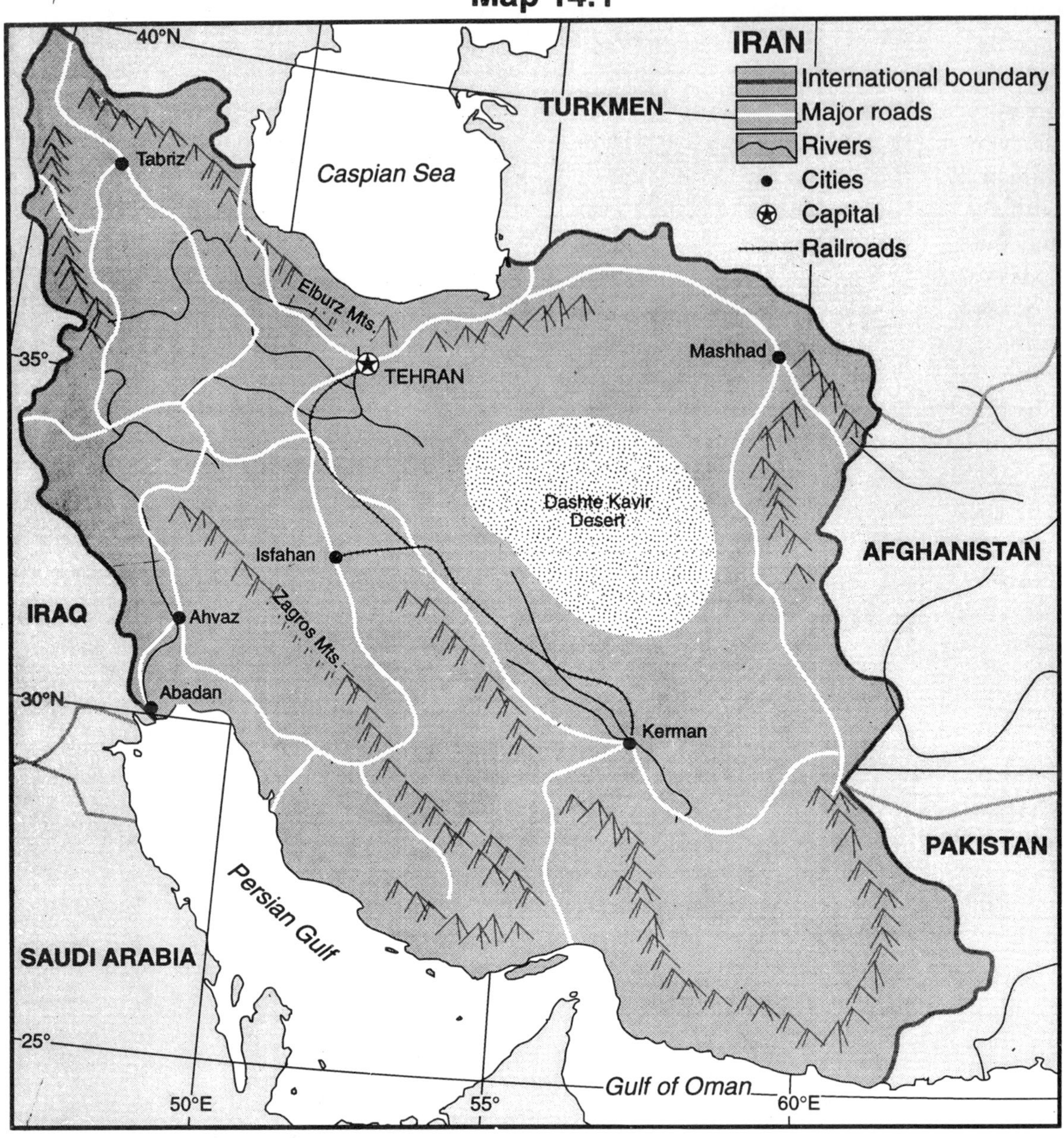

seething with discontent under the harsh rule of the government. Millions had flocked to the cities and oilfields for employment, but failed to find adequate housing and service, and were cut off from family, friends and the traditional lifestyle of their villages. The Shah held absolute power, and paid little attention to widening political participation and democratizing the administration. Many were dissatisfied with the erosion of spiritual values, and the religious leaders (the "Mullahs") whose role in society had been sharply reduced in the emerging Westernized social order, finally demanded an abolition of the monarchy and a return to spiritual traditions. Support for the Shah quickly vanished. The Shah, no longer able to rule effectively and under mounting pressure, left the country in 1979 paving the way for establishment of a new administration. Iran now sought to identify with the fundamentalist tradition of Islam, and quickly revamped the civil and legal codes of the administration to conform to strict scriptural injunctions of the Muslim faith.

Physical Features Most of the country is above 1,500 feet (460 meters) with one-sixth of it above 6,500 feet (1,976 meters). The largest part is the vast dryland of the interior plateaus which are virtually enclosed by high mountains, effectively cutting off sources of moisture. Outside the mountain ring are the coastal plains. In the north the 400-mile (644 km) narrow strip along the Caspian Sea, nowhere more than 70 miles (112 km) wide, falls abruptly from the 10,000 foot (3,040 meters) summits to below 90 feet (27 meters) below sea level; and in the south the land drops away from the 2,000 foot (608 meter) plateau and the rugged mountains nearly three times as high towards the Persian Gulf and the Gulf of Oman leaving a small, elongated coastal plain.

The Zagros Mountains stretch from the border with Armenia in the northwestern corner of the country to the Persian Gulf, and thence eastward to the Strait of Hormuz and further into Baluchistan. These mountains are made up of a series of parallel, rugged ranges that are high and broaden up to 200 miles, and in their northern section lie partly in Iraq. In the north is the Elburz Mountain chain that runs along the southern shore of the Caspian Sea, towards the Kopet Mountains bordering Turkmenistan and eastward to the Afghan border. The highest of the several volcanic peaks, some of which are still active, is the snow-clad Mt. Demavand at an elevation of 18,400 feet (5,612 meters).

Lying between the Elburz and Zagros systems in the northwest are the fault-block ranges and volcanic peaks of Azerbaijan. The volcanic belt extends from the Azerbaijan border to Baluchistan in the southeast. Earthquake activity is frequent and violent in the mountains bordering Azerbaijan.

The interior plateaus cover nearly half of the country and contain some of the world's most desolate areas. The relatively flat surface of the plateaus is broken locally by low hills and mountains that contain a series of basins of interior drainage of varying sizes. At the lowest level, the basins remain poorly drained where the evaporation of accumulated water produces salt wastes known as *kavirs.* The largest of the basins are the plateau's most remarkable features, the Dashte Kavir (Kavir Desert) and the Dashte Lut (Lut Desert). The two are largely uninhabited, and practically impassable during winter when temporary lakes are formed. The drainage of the vast interior plateaus is largely of the inland type, consisting of a few streams that empty into the arid plateaus. Outside the interior plateaus, the drainage pattern is down the outward slopes of the mountains, terminating in the sea. Among the large rivers only the Karien is navigable. It originates in the Zagros Mountains and flows south to the Shatt-al-Arab and empties into the Persian Gulf. The largest inland body of fresh water is Lake Urmia in the northwestern part of the country; other lakes are mostly seasonal, and have a high content of salt.

Climate The climate ranges from subtropical to sub-polar. Altitude, marine influences, seasonal winds, and proximity to mountain ranges or desert play a significant role in the fluctuation of temperature, which varies from a high of 131°F (55°C) in the Khuzistan area (at the head of the Persian Gulf) to a low of -35°F (-1.7°C) in the Azerbaijan region in the northwest. Southern Iran has mild winters, though only a few coastal areas are frost-free. Areas below 5,000 feet are hot in summer, and excessive heat is common in the lowlands of the south, where the daily maximum is mostly above 100°F for one or two months of summer.

Precipitation also varies greatly from less than 2 inches (55 mm) in the southeast near the Baluchistan border to about 80 inches in the Caspian region. The national average is 16 inches. Winter is mostly the rainy season for the country as a whole when over 50 percent of the annual precipitation falls, and is either cyclonic or orographic in origin. Central and eastern Iran receive less than 10 inches. For most of Iran, long hot summers and short, cold-to-mild winters are characteristic.

Cultural Patterns Iran is a multilingual and culturally diverse society in which less than one-half (45.6 percent) of its 61.2 million people speak Farsi (Persian) of the Indo-European family. Farsi speakers are the descendants of the Aryan tribes who probably moved from Central Asia as early as the first or second millennium B.C. Besides the Persian people, other groups such as the Kurds, Luris, Gilakis, Mazandaranis, Baluchis, and Bakhtiaris form substantial minorities. Their languages also belong to the Indo-European family. The Kurdish speakers account for about 9 percent of Iran's population. They are a fiercely independent and nomadic people inhabiting the Kurdistan mountains along the western borders of Iran, and the adjacent region in Iraq and Turkey. The Iranian government has tried to bring them into the mainstream of national life, but has had little success.

The Semi-nomadic Lurs and other groups such as the Bakhtiaris live in the Zagros Mountains. These groups speak a language distinct from, but related to, Farsi. The Baluchis, a small minority (about 2 percent of the population) inhabit the southeastern region of the country that borders on Pakistan. A very small minority of the Armenians numbering half a million have also maintained their Indo-European identity, and are concentrated in major cities, and are engaged in commercial activities.

About one-fifth of Iran's population speaks Turkic languages, the largest of which, Azerbaijani, is spoken by 17 percent of the country's population, and Turkmen, occupying the two border regions in the northwest and northeast corners of the country respectively. A few scattered remnants of Brahui-speakers (related to the Dravidian tongues of Southern India) are found in the Sistan region to the southeast. The Arabs form the largest group, over 1.4 million, in the Semitic family occupying primarily the Persian Gulf islands and the Khuzistan area close to the Iraqi border. The Arab demand for autonomy was among the factors that led to the Iran-Iraq War of the 1980s. The Jews have retained their linguistic and religious identity and are clustered mostly in the large cities. Most of the ethnic groups are concentrated in certain regions, the Persians in the Elburz, much of the Zagros, and the peripheral areas of the central plateaus; the Turkish groups in the northwest; the Arabs in the southwest; and the Baluchis in the southeast. In practically every instance, however, these groups overlap. There are substantial numbers of each group living in the major areas of the others.

Although Iran is predominantly an Islamic nation, it is not free from divisive religious forces. The majority of the population (93.5 percent) professes the Shiite branch of Islam (it is also the state religion), while Kurds, Arabs, and Turkmen belong to the Sunni tradition. Other religious minorities include Christians, Jews, and Parsis (Zoroastrians). The Parsis are largely concentrated in Yadz in central Iran, Kerman to the southeast, and Tehran. However, the preponderance of the Shia sect and the strong official backing accorded it has done much to subdue the ethnic differences. Since the Islamic revolution in 1979 and change in the administration, much of the religious toleration practiced under the monarchy suddenly came to an end. While Christians, Jews, and Zoroastrians are recognized in the constitution of 1979 as official minorities, the revolutionary atmosphere in the country is not particularly conducive to equal treatment of non-Muslims. Most of the Jews, who formed a significant minority in the cities, have left. Under such circumstances, a strong, cohesive national identity covering all ethnic groups has yet to develop in the country.

Resources and Economic Activity

Oil and natural gas are the two most important natural resources of Iran that have been developed. Most other mineral resources such as chromium, copper, gypsum, lead, zinc, manganese, antimony, and uranium remain largely undeveloped. Major coal mines are in Khorasan, Kerman, Mazandaran and Gilan; smaller ones operate near Tehran, in Azerbaijan, and in Isfahan. Cement is produced in the provinces of Tehran and Khuzistan.

Production of oil and natural gas is concentrated in the southwest, but oil has also been found near Qom and Dashte-Kavir, and in the off-shore areas of the Persian Gulf. Iran was the earliest among the nations of Southwest Asia to produce oil on a large scale. Commercial production was started in 1912 originally by the Anglo-Persian Oil Company at Abadan, but was heavily damaged during the Iran-Iraq War in 1990. Pipelines to inland oil refineries also exist at Isfahan, Shiraz, Lavan, Tehran, and Tabriz, where a variety of refined products are manufactured. Oil is produced both for export and for domestic needs. Abadan is accessible to small tankers and long distance hauling for export is made from the terminal of Kharg Island in the Persian Gulf where oil is piped from Abadan. From the Kharg Island terminal oil is shipped by tankers to western Europe, Japan, and other countries. Iran possesses nearly 8 percent of the world's petroleum reserves, and those of natural gas are even higher. In normal times, Iran is the world's fourth largest oil producer (after Russia, the U.S., and Saudi Arabia). Oil forms the nation's most valuable export; along with petroleum products and natural gas it accounts for over 80 percent of the country's total exports. Natural gas is found in the Elburz Mountains and piped to all major cities. The petrochemical industry is concentrated in the south at Bandar-e-Khomeini and at Kharg Island producing ammonia, phosphates, sulfur, sulfuric acid, and other products. Many of the manufacturing plants producing petrochemical products were established as joint ventures with Japanese, American, German, and other European companies.

Although close to one-fifth of Iran's labor force is engaged in agricultural activities (still making it the primary sector of the economy), agriculture accounts for a relatively small part of the nation's gross domestic product. Due to the shortage of water and the poverty of the soil in many areas, less than 11 percent of the agricultural land is under cultivation. Only one-third of the cultivated land is irrigated; the rest is devoted to dry farming. Nearly one-half of Iran is unproductive desert or mountain land, 7 percent is forest area, and 6 percent is classed as grazing land.

The basic problem of livelihood in rural Iran is the shortage of cropland. The farms are small; most are less than 25 acres (about 10 hectares). Farming areas are mostly concentrated on the several patches of oases surrounded by tracts of unexploited deserts or mountain slopes, or in the narrow, coastal plains of the Caspian and the fertile Khuzistan region at the head of the Persian Gulf.

A unique method of providing water to the farm, which probably was developed first in Iran over 2,000 years ago, is the *qanat* system. The *qanats* are essentially horizontal tunnels with a slight gradient that carry water over long distances. To give access to the tunnel, a number of vertical shafts are dug at closely spaced intervals and thus water is tapped for irrigation. A long *qanat* may have several wells. The use of the *qanat* is widespread especially on or near the areas of gentle slopes leading from the rainfed mountains, and is possibly one of the most efficient dry systems of irrigation as the underground channels keep both evaporation and seepage to a minimum.

A wide range of temperature conditions in the various parts of the country makes it possible to cultivate a variety of crops. These include cereals (wheat, barley, rice, corn), fruits (dates, figs, pomegranates, melons, grapes), vegetables, cotton, sugarcane, sugar beets, nuts, olives, spices, tea and tobacco. Agriculture is largely of subsistence type, and wheat is the major crop; the regions of its greatest production are the Mashhad region, the valleys in the northwest Zagros, districts of Lake Urmia, Karman, Isfahan, and the well-water tracts of Khuzistan. Barley is grown in the same areas as those for wheat. Rice is grown in the Caspian region and Gilan where water is adequate and irrigation facilities exist. Fruit growing is of considerable importance, since fruit forms an important part of the everyday diet of most people. Iran is known for its apricots, and many varieties of grapes (Shiraz is reputed to produce some excellent varieties) are grown. Other fruit crops are: figs, peaches, melons, oranges, and citrus fruits confined to the Caspian coast and the foothills of the central Zagros Mountains. In southern Iran dates gain importance, particularly on the coastal plains of the Persian Gulf and the Gulf of Oman.

Cotton is the chief cash crop grown largely in the Azerbaijan area, in Kerman, Fars, and Khuzistan province. Sugarbeets, tea, tobacco, and opium are some other crops. Silk, a traditional product of Iran, is important on the eastern section of the Caspian coast, where mulberry growing is locally widespread.

Most of the agricultural products of

Iran are consumed locally and foodgrains have to be imported (in 1990 grains and derivatives accounted for 7 percent of the country's imports). Fishing is an important activity in the Caspian Sea area where several varieties of fish such as sturgeon (yielding Iran's best known roe caviar), salmon, carp, perch, and mullet, are caught. Persian Sea fishing also yields several varieties that include shrimp and prawns. Of the country's livestock, sheep are by far the most numerous, followed by goats, cattle, horses, water buffalo, and camels. During the rule of Reza Shah Pahlevi, the Iranian government had undertaken steps enticing the pastoralists and semi-nomads to take up sedentary cultivation, which created much social upheaval and human suffering. It also brought substantial economic loss to the pastoralists and the carpet industry which was dependent on local wool. The revolutionary administration since 1979 has tried to initiate several reforms to revive pastoralism.

One-fifth of Iran's labor force is engaged in manufacturing and construction activities. Most factories employ traditional handicraft or cottage style methods except for oil and petrochemical plants. Modern manufacturing dates to the mid-1930s when Reza Shah Pahlevi began a program of industrialization to produce manufactured goods for the domestic market. Earlier Iran did produce a moderate volume of manufactured goods, made entirely by hand or with the use of primitive machinery. During the brief period before World War I, state-owned plants were well located in relation to raw materials and markets, and in general, were fairly successful. But the bulk of manufacturing activity remained in small-scale, traditional, craft-oriented, bazaar-style industries such as carpet weaving and textile manufacturing that are less susceptible to mechanization. Carpet making, a predominantly cottage industry, has traditionally been very important. Iran has its own designs and colors of carpets (Tabriz specializes in pastel tints, and vivid colors are employed by other cities, whereas heavier woolen carpets come from Kerman, Tehran, and Shiraz. Finished products find ready markets in the U.S., Germany, France, U.K., Japan and many other nations all over the world. Although the industry is controlled now in part by large companies, it is still carried on by hand in small shops and homes.

Since 1954, industrialization has registered considerable gains and Iran has been producing a wide range of products such as automobiles, electrical appliances, telecommunications equipment, industrial machinery, paper, rubber products, steel, food products, wood and leather goods, textiles, and pharmaceuticals.

Urbanization and Cities Fifty-seven percent of Iran's population lives in the cities. Urbanization proceeded rapidly under the Pahlevis as the cities became the focus of an emerging class, the professionals, and hundreds of thousands of peasants looking for a better life. Cities have a long history; several have served as the nation's capital, or as regional service centers. Each of the major cities contains an old section and at least one major covered bazaar, teeming with craftsmen selling their wares, and modern, well-planned sections built in the 1940s. The largest of these, **Tehran** has a population of about 7 million and is the national focus. It is located in the north at the base of the Elburz Mountains on a broad alluvial fan at an elevation of 4,000 feet (1,295 meters), and was made the nation's capital in 1788 although it did not develop much until the mid 19th century. Most of its development occurred during and since the reign of Reza Shah Pahlevi (1925-1941). The city and the suburbs now contain a large share of the nation's commercial, industrial, cultural, and political activities. Tehran is an important manufacturing center, accounting for a third of the nation's industrial units specializing primarily in light engineering, chemicals, textiles, food processing, and consumer goods. Its rapid development has made it a crowded, congested metropolis with serious air and water pollution problems.

Isfahan (1.3 million) is more centrally located in a fertile plain in the middle of the Zagros Mountains, commanding routes towards Iraq. It was the capital of Persia under Shah Abbas in the 16th century, and is notable for its fine monumental buildings that reflect Iran's past splendor. **Tabriz** (1 million) lies in the extreme northwestern part close to the borders of Turkey and Azerbaijan, on the ancient east-west routeway between Europe and Asia. It is also an ancient city and known for its elegant monuments and buildings such as the Blue Mosque. Tabriz has played an important role in Iran's political affairs, mainly as the center of Kurdish and Azerbaijani dissident movements against the central government. The city is commercially important and is connected with Tehran by railroad and road. The principal industries of the city deal with the manufacture of carpets, cement, and agricultural machinery. The city lies in the earthquake zone liable to frequent and severe shocks. **Mashhad** (1.9 million) on the historic eastward route to Afghanistan has experienced phenomenal growth as a major religious and political center of the Shiites. It is located in a rich agricultural region, specializes in the wool trade, and is connected with Tehran by a railroad. **Shiraz** (1 million) is located in south-central part in the Zagros Mountains near the historic site of Persepolis, the seat of the ancient empires of Cyrus, Darius, and Xerxes. Persia's two great poets, Saadi and Hafiz, were born here and praised its beautiful gardens and excellent wine. The city is a trading and road center, and is linked to Bushire, its port on the Persian Gulf, and to Tehran. Cement, sugar, fertilizer, and textiles are the major industries. **Abadan** (300,000), the nation's only large port, is located in Khuzistan province in the rich oil-producing region on the Karun River (part of the Shatt-al Arab delta) some 33 miles (52 km) from the Persian Gulf near the Iraqi border. Iran's major oil company is located here, and the city is the terminus of oil pipelines. The city's development and economy are entirely based on petroleum refining and shipping. In 1980 the city was overrun by Iraq and the refinery was largely destroyed; it was partially reconstructed in 1988 following the end of the Iraq-Iran War.

Assessment and Prospects After World War II Iran embarked on a program of economic and social reconstruction, but progress was slow until the 1970s when oil prices escalated and large oil revenues made it possible to undertake unprecedented industrial, economic, and social development. Most of the industrial development was centered around Tehran, and controlled by a few families or foreign companies. Attempts were also made by the Shah regime to bring about land reforms by transferring large amounts of

Crown and other lands purchased by the state from large private estates and religious holdings to landless peasants. There was also some success in raising agricultural productivity through better methods, tools, seeds, and chemical fertilization of crops. Provision of increased irrigation and electric supply in favored areas such as the Karun River basin in Khuzistan province by the construction of storage dams was also undertaken. But the greatest beneficiaries of these undertakings were the giant plantations in Khuzistan. Most farmers remained unaffected by these reforms, which proved ineffective in changing the generally depressed state of the Iranian economy The poverty-stricken families from the countryside started to pour into Tehran and other large cities. Much of the support for the revolution of 1978-79 that ousted the Shah came from these devoutly religious new urbanites.

The political change dramatically altered the social and economic climate as well. The new rulers—the handful of revered and powerful *Ayatollahs*—declared that the country would be governed according to the Islamic fundamentalist canons based in the Shiite code. Thousands of clerics called the *mullahs* would supervise all aspects of Iranian life and perform functions allotted to civil servants in most nations. Although the revolution had capitalized on grievances against the widespread corruption, inflation, and extravagant display of pomp during the Shah rule, the real driving force of the revolution was the furious Shiite outrage against modernization, Westernization, western imperialism, and the exploitation of the underprivileged section of society. Iran turned suddenly into a theocratic state administered by strict Islamic principles as laid out in the Koran. Political and religious dissent was crushed. The economic reforms introduced during the Shah regime were quickly revoked, and ties with western companies were snapped. While the new economic order had yet to take shape, the 1980-88 Iran-Iraq War further upset the economic plans of the revolutionary government as enormous sums were funneled into the war chest.

Except for oil, which provides modern Iran with enormous wealth, exploitation of the other rather limited resources is limited. But oil faces an uncertain future, dependent as it does more on Iran's relations with other nations, particularly the Arab OPEC members, who do not always agree with Iran's policy regarding the fixing of oil production and pricing, as also with its oil customers (Japan and the western nations). Besides oil and petroleum products the country's foreign trade has been relatively unimportant (carpet export is another big exception) and Iran's foreign trade is limited. The nation is not self-sufficient in foodgrains. Most heavy machinery (non-electric machinery, iron, steel and transport equipment) have to be imported. Relations with the U.S. (an important partner before 1979) were completely broken off by the revolutionary government; the U.S. imposed an embargo on the import of Iranian oil. Since the early 1990s Iran adopted a more relaxed social atmosphere and the administration has been seeking better relations with the West.

Some of the nation's problems stem from its natural environment with difficult terrain, vast desert expanse in the interiors, and scarcity of water in most regions. But considerable expansion in hydro development is possible, now exploited to a small degree in the Elburz and the Zagros Mountains, although seasonal precipitation requires large storage facilities.

If Iran is to develop economically, better transport is essential. At present there are a number of paved roads and these form the principal means of transport. However, there is less than 3,700 miles (9,583 km) of railroad track in a large country like Iran, so most of the domestic cargo is carried by trucks and busses. The main railroad track leads from Bandar Shahpur at the head of the Persian Gulf across the Zagros to Tehran and from there across the Elburz to Bandar Shah on the Caspian Sea. From Tehran railways extend east to Mashhad and west to Tabriz.

In the interior plateau region Kerman is connected with Qom and Tehran. Some large cities such as Shiraz and Isfahan are still not connected by railroad. Given the difficulties of terrain, air transport is somewhat better developed; all the leading cities have airports and scheduled domestic flights. The five leading cities have international air connections.

Physical incoherence has exacerbated national divisions in Iran. The presence of diverse ethnic groups consisting of more than one-third of the population are a source of constant irritation to the central administration. Many of the groups (Kurds, Luris, Baluchis, Bachtiaris) are fiercely independent, and are armed nomads. They do a little cultivation, but are essentially herdsmen and wander for water and fodder and for markets for their products (tents, rugs, etc.) The Kurds who form a substantial minority have been asking for a separate homeland. During the Iraq War the central administration of Iran lured them to its side by promising to give them greater autonomy, but in general relations with them have been far from harmonious. Iran's revolutionary government thus faces many intractable economic, social, and political problems which need to be addressed.

IRAQ

Iraq was created in 1932 when the British relinquished control over their former mandates in Southwest Asia. It has an area of 167,975 sq miles (435,652 sq km) excluding the neutral zone jointly controlled by Saudi Arabia, and a population of 22.4 million. It was ruled as a monarchy by the Hashemite family until an army-led revolution overthrew it in 1958, when it became a republic. Occupying the territory that was known as Mesopotamia ("Land Between the Rivers") of classical times, the region's extensive alluvial plains of the Tigris and Euphrates gave rise to the world's earliest civilizations. Iraq is one of the world's leading oil producers. In recent decades it has built one of the most powerful armed forces in the Arab world; its strength was demonstrated during the Iran-Iraq War (1980-88) and in the 1990 invasion of Kuwait. Since the establishment of the present regime, it has been an authoritarian state in which internal dissent has been ruthlessly suppressed, particularly among the Kurdish minority in the north. The administration is keen to play an effective role as a regional power in the Middle East.

Physical Features Iraq contains four distinct physical regions: Lower Mesopotamia or the alluvial plains of the central and southwestern parts of the country; Al Jazirah ("The Island") or Upper Mesopotamia, the arid plateau north of Lower Mesopotamia; the western desert; and the mountainous region of the north and the northeast. Lower Mesopotamia and Upper Mesopotamia cover the territory between the two rivers, the Tigris and the Euphrates, south of the northern mountainous region. Both are part of an old geosyncline, still subsiding in the south, and together they make up the core of modern Iraq and are most important from an agricultural point of view. Lower Mesopotamia is a poorly-drained alluvial plain of low elevations, below 300 feet (100 meters) and thus subject to large-scale flooding. The plains contain extensive marshlands, some of which dry up in the summer to become salty wastelands, and a few lakes. In many parts, salinity is a serious problem. A high water table, poor drainage, and a high rate of evaporation have caused the alkaline soils to accumulate at or near the surface, rendering the ground unfit for cultivation. The uplands of Upper Mesopotamia (Al Jazirah) rise to about 500 feet (152 meters) elevation. The soils, built by the silt of the rivers, are very fertile and are very important to Iraq's agriculture. The combined waters of the Tigris and Euphrates form a delta near Basra and flow as Shatt-al-Arab on the last leg of their journey before draining into the Persian Gulf.

The western desert is a vast, arid region adjoining the deserts of Syria, Saudi Arabia, and Kuwait and covering almost 40 percent of the country. The land rises to more than 1,600 feet (534 meters) sloping, in general, towards the Euphrates. The topography is complex and includes rocky desert, sanddunes, *wadis*, ridges, and depressions. The northern and northeastern mountains occupy nearly a fifth of the country. The relief of the mountain region rises from the Tigris towards the Turkish and Iranian borders in a series of rolling uplands, river basins, and hills until the high mountains of Kurdistan (between the Turkish Taurus and the Iranian Zagros Mountains) are reached. The general alignment of the mountains is northwest to southeast. The summits rise to 10,000 or 11,000 feet (3,040-3,618 meters) in places, and the mountains are deeply dissected by the several tributaries of the Tigris.

Climatically, Iraq may be divided into two provinces: the hot, arid lowlands including the alluvial plains and the deserts; and the mountainous, damper northeast where temperatures are lower. Here, rainfed cultivation is possible, but elsewhere irrigation is essential. Summers are hot with mean daily temperatures of about 95°F (35°C) and dry with 4 to 7 inches (100 to 180 mm) of precipitation. Winters are cool with average daily temperatures between 35 and 60°F or 2 to 15°C for Baghdad. About 90 percent of the precipitation falls from December to April. In the northeastern province summers are shorter and some 5 to 10 degrees cooler than those of the lowlands. Here, winters can be much colder. In Mosul, January temperatures range between 24 and 63°F (–4 to 17°C). In higher elevations, rainfall may exceed 40 inches annually, but in the foothills 12 to 22 inches (30.5-55.4 cm) precipitation is the norm.

Historical Development Iraq is rich in history that dates back to the Sumerian civilization of 3400-3000 B.C. Some of the oldest remains of urban civilization have been unearthed in ancient Mesopotamia, the land between Iraq's two principal rivers, the Tigris and the Euphrates. The northern half is ancient Assyria and the southern half is the flat delta of ancient Babylonia. Both rivers have so built up their beds that successive waves of deposition lie above the surrounding flood plain, with the Euphrates generally higher than the Tigris. The Babylonians (1800 to 1500 B.C.) created a mud civilization. Sun-dried bricks were used for building, whereas the Assyrians (2025-600 B.C.) built buildings of stone, as stone was in greater supply in northern Mesopotamia. Excavations near Ur and Nineveh give us a glimpse of the beginnings of incipient urbanization during these civilizations which were primarily based on the use of river waters for cultivation of crops and for transport. The cities had all the hallmarks of present-day urban centers: residential streets and commercial areas containing restaurants, blacksmith shops, and chapels. Many of the homes were built around a paved central court.

We know little about Iraq's history from these ancient beginnings to its next golden age (750-1260) associated with the Abbaside Caliphates, when the center of Islamic and Arab culture moved from Damascus to Baghdad. During the 8th and 9th centuries, Baghdad had become a great emporium of trade, and a political center of international importance. Agriculture flourished in the fertile, alluvial Mesopotamian basin during the rule of the Abbasids. Old canals dating back thousands of years were de-silted and new ones dug. Great irrigation works were opened up. The staple crops of Abbasid Iraq included barley, wheat, rice, dates, sesame, cotton, flax, fruits, and vegetables, much the same as those of today.

The next important phase in Iraqi history is that of the Ottoman Turks who ruled from 1514 to 1918. Between the Abbasid period and the arrival of the Turks, Iraq was overrun by the Mongols, whom the Iraqis blame for the destruction of irrigation channels and Iraq's great culture. Under the Turks, Iraq came under the pressure of two strong conflicting neighbors, Turkey and Persia. In religion, southern Iraq was Shiite like Persia, but Sunnite in the north and west. This conflict in religion and political orientation still remains today. The great Shia shrines of Kerbala and Najaf are located in Iraq near Baghdad. Turks brought their Turkish culture to Iraq and Turkish remained the language of the elites. During Turkish rule, however, several tribal groups known as the Kurds had become powerful, and they established control over many northern mountainous sections and some of the western desert area.

Cultural Patterns Seventy-seven percent of Iraq's population consists of Arabs, twenty percent are Kurds, and the remaining three percent consist of small minority groups. The Arab conquests of the 7th century brought about the Arabization of central and southern Iraq, whereas the northern and northeastern sections (Iraqi Kurdistan) remained under the control of the Kurds. A mixed population of Kurds and Arabs inhabits the transition area between central Iraq and Iraqi Kurdistan.

The Kurds constitute a separate and distinctive cultural group. They are mostly Sunnite Muslims who speak a Farsi-related language. They are of Indo-European antecedents, different from the other, largely Semitic groups. They are strongly clannish, and are distinctive in costume, music, and dance. Concentrated primarily in the relatively inaccessible mountains of Iraqi

Kurdistan, the Kurds spill over into the adjacent areas of Turkey, Syria, Iran, and parts of Transcaucasia. In order to win an independent state, the Iraqi Kurds rose in open rebellion against the government from 1961 to 1975, aided in part by military support from Iran. But their attempts to gain autonomy failed. In 1970, Iraq agreed to accept the bi-lingual character of the country and granted the Kurds political and cultural autonomy, but it was never implemented. In 1975 the Iraqi administration decided to create a Kurdish Autonomous Region covering an area of about 14,700 sq miles, nearly one-half of the area demanded by the Kurds, but excluding the oil-rich Kirkuk area. Having failed to gain freedom for their territory, the Kurds continued their rebellion. Following the Iran-Iraq War (1980-88) a brutal repression of the Kurds was resumed (Iran had provided support to the Kurds during the War). At least 300,000 Kurds were deported from their homes, effectively depopulating one-third of the Iraqi Kurdistan territory. Thousands of Kurds fled into Turkey and Iran. By 1990 the Kurdish opposition had largely been eliminated. Turkmen, Persians, and Assyrians are some other groups, each forming less than one percent of the population.

Islam is the predominant religion, divided between its two major sects: Shia and Sunni, representing 63 and 34 percent of the population, respectively. About three percent of the population consists of Christians. Yazidis, Jews, and Bahais are other minor groups. The Jewish community traces its origin to the Babylonian Exile (586-516 B.C.). In 1948 the Jews constituted about 2 percent of the population; most have moved to Israel since its establishment in 1948. The Christians are the descendants of an ancient population that escaped conversion to Islam in the 7th century. They belong to several sects---Chaldeans, Nestorians, and the Eastern Orthodox church---and are found in all parts of the country. The current administration, belonging to the Arab Baath Socialist Party, has aspired to leadership in the Arab community and has asserted its loyalty to Arab solidarity, opposition to Israel, and the creation of a self-dependent, secularized, and socialistic nation. Although it professes to be non-religious, the ruling party is largely Sunnite and is often disturbingly unprotective of the interests of the other religious groups, including the Shiites who are concentrated in the southern part of the country.

Population and Settlement Patterns

Iraq has the third largest population (22.4 million) in Southwest Asia (after Iran and Turkey). But the densities are considered low, and the country is considered underpopulated. The nation has a potentially large, productive area of the Tigris-Euphrates system, and access to large oil revenues. The irrigated area could be increased substantially, as the existing water is now most inadequately utilized. The government strongly encourages population growth, which will remain high, as the birthrates at 45 per 1,000 population and fertility rates are among the highest in the world. Forty six percent of the population is below the age of 15, and the population is likely to double in about 20 years (despite the large losses in the Iran-Iraq War of the 1980s). But the workforce (15 to 50 age groups) remains small. Many Iranians, most of whom had settled in the south and east, were deported in the 1970s, and the subsequent

Map 14.2

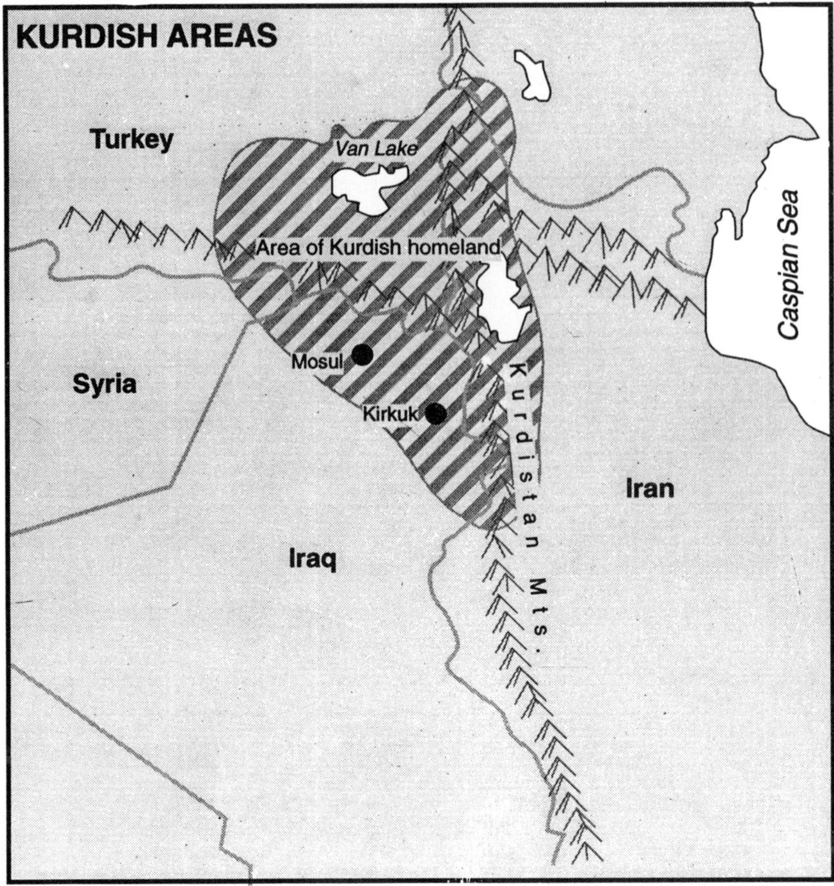

labor shortages resulted in the arrival of large numbers of foreigners, particularly Egyptians.

Most of the population is concentrated in the lowlands, particularly along the fertile Tigris and Euphrates. Almost a third of the population lives within 90 miles of Baghdad, and a substantial proportion of it is in the densely-populated area between Baghdad and Basra along the Tigris and Euphrates. Although these settlement patterns were set thousands of years ago, there has been, during the past few decades, increasing migration to the cities. Ten percent of the labor force is engaged in agricultural activities. Only a small portion of the population is classified as *bedouins* who raise camels, sheep, and goats and are concentrated primarily in the western desert region. Some of the Kurdish tribesmen in the mountains and the foothills are also semi-nomadic graziers, although most of the four million Iraqi Kurds are sedentary farmers.

Baghdad (population: 4.5 million), the capital of the nation, is situated near the center of the nation at the head of the Tigris delta at a point where the Euphrates comes closest to the Tigris. It is a historic city, and was the seat of the Babylonians, Arabs, Mongols, Persians, and Ottomans. The site, however, presents some problems: the Tigris is surrounded by natural levees, adjoining the flood plains, but given to recurrent seasonal flooding. Another major problem is the unceasing influx of people from the countryside. They come to find jobs in the metropolitan area, but continue to add to the ranks of the frustrated unemployed. Baghdad is the nation's dominant commercial, industrial, and cultural center. **Basra** (Diyala) is the second largest city, and the nation's only seaport; it has a population of close to half a million, and a growing metropolis, despite suffering a setback during the Iran-Iraq War of the 1980s and the Gulf War of 1990. As a port, it has excellent arrangements. **Mosul** (population: 670,000) in the north, at the edge of the mountains, lies at the Tigris and is an important oil-refining center.

Resources and Development Oil is Iraq's greatest economic resource. The country possesses one of the world's largest known oil reserves, and has been one of the world's leading exporters. Most of the oil is obtained from the Kurd areas. Prior to the Gulf War of 1990, oil accounted for nearly 98 percent of the country's total exports, a third of which was shipped to the United States. Following the discovery of an oilfield near Kirkuk in 1927, and other fields near Basra and Mosul in the 1940s and 1950s, production went up rapidly during the 1960s and 1970s. The northern fields in the Kurdistan area have been producing more, although reserves are larger in the southern fields near the Kuwait border. New fields in central Iraq have also been found recently near Baghdad, which started production in 1989. Most of the oil is exported directly from tanker terminals in the Gulf, although pipelines carry oil to the Mediterranean terminals as well. Iraq also possesses large reserves of natural gas. Although Iraq has a range of other minerals, including coal, iron ore, lead, copper, and sulfur, exploitation has lagged behind. Emphasis on oil production has meant that development of the nation's waterpower resources has been neglected; although several dams were built on the rivers, the potential remains largely unutilized.

Traditionally, Mesopotamia (the land between the rivers) has been best known for its agricultural productivity, which was recorded by Herodotus as early as the 5th century B.C. The total area given to cultivation at one time or another may have been much greater than today. Agriculture, based largely on irrigation, has been practiced in Iraq for a long time, perhaps dating back to 5000 B.C. Centralized governments built and carefully devised irrigation systems capable of tapping the seasonal flow of the rivers. By the time of the invasion of the Mongols, most of the hydraulic works had fallen into disrepair, and during the Ottoman rule progressive siltation of the irrigation system meant that it remained neglected. The area devoted to cropland drastically declined until the mid-20th century, when efforts were made to arrest the trend. New water diversion and flood control measures have been adopted, but the centuries of silt build-up and salinity problems have greatly reduced agricultural efficiency.

Wheat, barley, rice, dates, vegetables, millets, sugarcane, tobacco, cotton, and fruits are the chief crops grown. Wheat is the major crop overall, grown primarily in northern Mesopotamia; rice, fruits, and dates are grown in the southern parts of the Tigris-Euphrates basin. Barley is widespread as it better tolerates aridity and soil salinity. Grain production is inadequate for domestic requirements, and food items account for a quarter of the country's total imports. Dates are the chief commercial crop, grown west of Shatt-al-Arab. Next to oil, dates are the country's most important export. Cotton is an important commercial crop, grown primarily for export. Nearly one-eighth of the labor force is engaged in agriculture, accounting for one-fifth of the nation's gross domestic product. Despite the government's promotion of agriculture, the cultivated area declined by about one-half during the 1970s, but increased during the 1980s as a number of reclamation projects were completed. Iraqi agriculture faces several problems, including soil salinity, floods, droughts, and siltation that impedes the efficient working of irrigation.

The manufacturing sector developed rapidly since the mid-1970s, when the government policy moved toward heavy industrialization and import substitution. Petrochemical and iron and steel plants have been built at Basra and near Baghdad. Tractors, electrical goods, telephone cables, tires, trucks, aluminum goods, and fertilizers are the main products. Production of advanced military hardware (and chemical weapons) got a boost during the Iran-Iraq War. After 1973, increased oil revenues helped finance new social programs, large construction projects, and creation of state-controlled new businesses, new high-rise buildings, and busy expressways that are largely concentrated in the metropolitan Baghdad area and in the urban axis between Baghdad and Basra. However, punitive military strikes,primarily by the U.S. in retaliation to Iraq's occupation of Kuwait in 1990 caused large scale destruction of Iraq's economy and infrastructure. Economic recovery since then has been stalled by the UN sanctions against Iraq, dimming changes for future development.

AFGHANISTAN

Tucked in the heart of south-central Asia, Afghanistan is a land-locked and remote nation. The nearest coast is that of the Arabian Sea, about 300 miles (483 km) away to the south. Its longest border, of 1,125 miles (1810 km), is with Pakistan to the east and south. To the west, there is a 510-mile border with Iran, and to the northeast a 200-mile border with the part of Jamnu and Kashmir disputed between India and Pakistan, as well as a short 50-mile strip bordering China at the end of the long, narrow Wahan corridor. In the north is a 1,050-mile border with the three states of Turkmenistan, Uzbekistan, and Tajikistan. These boundaries were created in the late 19th century by a treaty between Britain and Russia, the two rivals gaining political domination in the region. In the late 20th century Afghanistan became a pawn in the struggle over political ideology and influence by Russia and suffered greatly from the Soviet invasion (1979) and a prolonged civil war (1979-89).

Physical Features The most prominent feature of this largely highland country is its mountain range, the Hindu Kush, a formidable barrier that roughly divides the country into a relatively flat and sloping area to the north, and the remaining uplands and deserts in the south. The Hindu Kush is one of the mountain chains that radiate from the Pamirs near the China border and extend westwards towards the Iranian border; at its highest, it reaches over 21,000 feet (6,565 meters). To the east, towards Pakistan, several ranges

Map 14.3

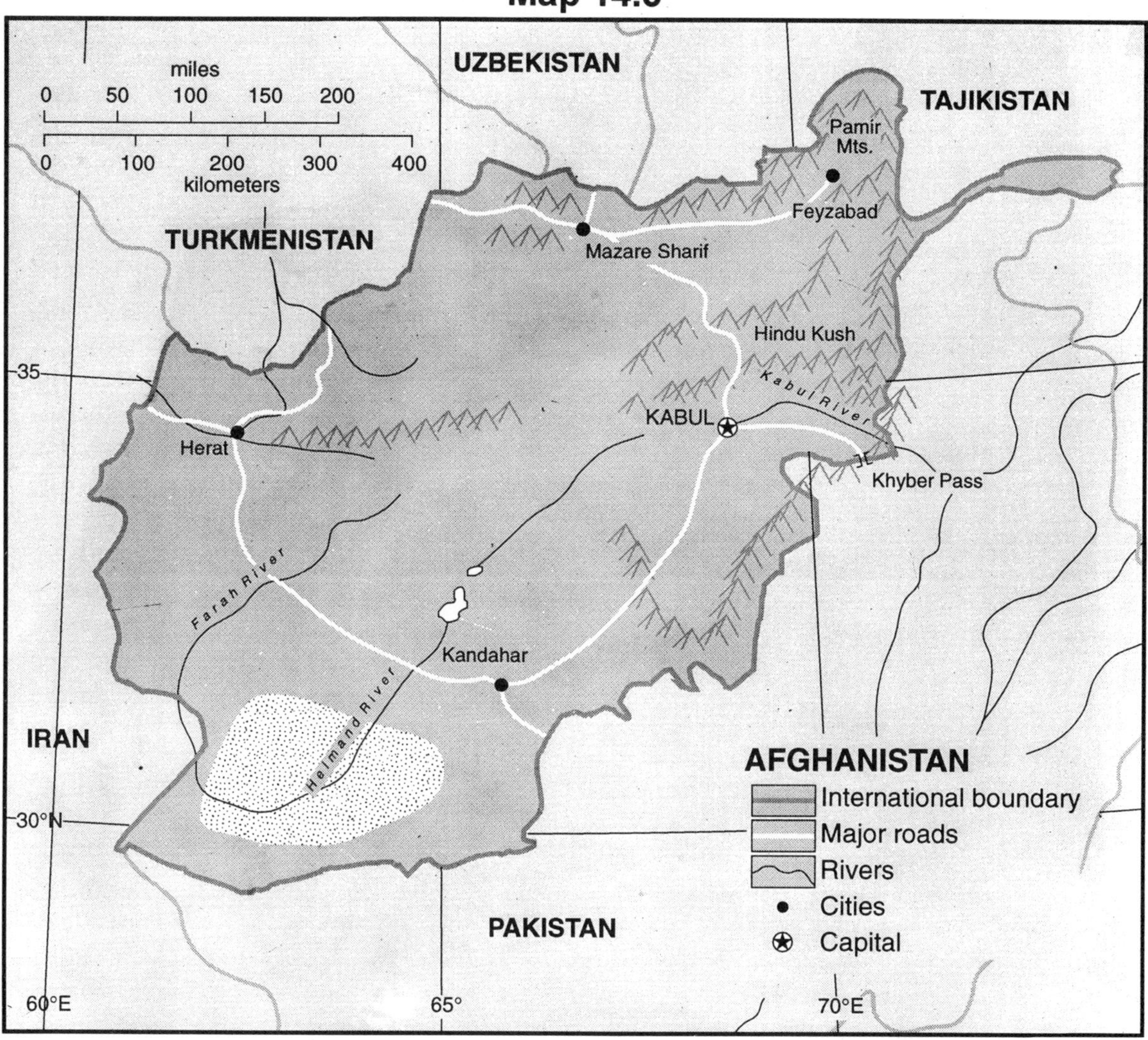

effectively isolate the interior of the country from the outside world. The mountains are the highest in the northeast section of the country where high mountain passes are at between 12,000 and 15,000 feet altitudes; the region in general has deep narrow valleys and lofty mountains. To the south of the ranges is the Khyber Pass, which leads to the Indian subcontinent. Northwest, and north of the Hindu Kush ranges, is a somewhat flatter and more fertile region that borders Tajikistan and slopes gently towards the Amu River. The average elevation of this essentially Central Asian steppeland is about 2,000 feet. South of the Hindu Kush and its branches is a region of high plateaus, sandy deserts, and semi-deserts with average altitudes of about 3,000 feet. The drainage is entirely inland.

In general, Afghanistan experiences very cold winters and hot summers, typical of semi-steppe climates, with regional variations due to altitude and exposure.

While the mountainous regions of the northeast have a subarctic climate with dry, cold winters, areas close to the Pakistani border are influenced by the Indian monsoons. Much hotter summers and a larger range of temperatures are experienced in the southern and western desert areas. Precipitation ranges from only 3 inches in the western and southwestern deserts to over 50 inches in the northern Hindu Kush area. Most of the country's precipitation occurs from December to April, while the summers are mostly hot and dry.

Cultural Patterns Historically, Afghanistan has been a crossroads that has linked China with Rome, and Central Asia with India. Greek culture, Buddhism, Mongols, Persians, all have left their mark on the land. The region has experienced invasions time and again. Greeks, Scythian, Dravidians, Aryans, Arabs, Mongols, Turks, and Persians have all passed through it with India as their ultimate target and have inhabited it at different times to make it an ethnic and linguistic mosaic. Intermixture of the two principal linguistic groups is evident in such people as Hazaras and Chahar Aimaks, who speak Indo-European languages but have pronounced Mongolian physical features and cultural traits, usually associated with Central Asia.

The Pashtuns are the principal group, forming 40 percent of the population of 25.8 million. They occupy mostly the eastern and southern parts of the country. Their traditional core area is around Kabul, the capital, and spills over into the contiguous territory of Pakistan. Some of the Pashtuns (Pathans) are sedentary and others nomadic. The Durrani group among the Pashtuns, who inhabit the region between Kabul and Kandahar, essentially form the traditional nucleus of Afghanistan's social and political elite. The Tajiks, mostly farmers and artisans, account for one-fourth of the country's population and live mostly in the northeastern part. Uzbeks, Turkmen, and Kyrgyz inhabit a region north of the Hindu Kush and collectively account for one-sixth of the population. Uzbeks are the largest Turkic-speaking group in the country representing about 6 percent of the nation's population. In addition, three percent of the population consists of Chahar-Aimaks, one percent of Baluchis, and less than one percent of Nuristanis, who speak Dardic tongues. Brahui speakers are related to the Dravidian tongues of southern India, but are racially dissimilar to the Dravidians.

Nearly 99 percent of the people in Afghanistan are Muslims, of whom three-fourths are members of the Sunnite sect. Shiites account for 15 percent of the population. The Hazaras belong to the Shia sect. Nuristanis are recent converts to Islam (mostly in 1895). Afghanistan is among the least urbanized of the world's nations. Only 20 percent of the population lives in the cities. The Soviet invasion of the country in 1979 and the ensuing civil war and destruction of towns and villages resulted in mass movements of people in two major directions—in the east to Pakistan and in the west to Iran—or to the relative safety of the capital city of Kabul, causing great disruption of the population patterns. It is estimated that nearly 3.5 million people escaped to Pakistan and some 2 million to Iran, and the population of Kabul nearly doubled in size between 1980 and 1990. It is estimated that Kabul now contains over a million people, representing a large proportion of the country's urban population.

Economic Activity Afghanistan is primarily a rural and agricultural or pastoral country. Nearly 60 percent of the labor force is engaged in agricultural pursuits. Agriculture and animal husbandry consist mainly of subsistence farming and pastoral nomadism. Since much of the land is arid or semi-arid, only one-eighth of the land is under cultivation. Meadows and pastures account for about 45 percent of the land surface. About half of the cultivated land is irrigated. Cereals, of which wheat, corn, barley, and rice are the principal crops, occupy most of the cultivated area. Cotton, for the domestic textile industry, and fruits and nuts are also important export items. (Pistachio nuts are produced on the northern slopes of the Hindu Kush.) Opium poppies and *cannabis* are grown for the illegal international drug trade on substantial scale. Animal husbandry yields meat and dairy products for domestic needs, skins (particularly the famous karakul) and wool are important products for export and domestic carpet weaving.

Although several minerals are known to exist, exploitation has been poor, arising partly from the country's unsettled political climate. Coal has been found in the northern slopes of the Hindu Kush, and high-grade iron ore 60 miles (96 km) northwest of Kabul. Copper is also mined near Kabul in some quantities. The most important recent discovery is that of natural gas, near the Turkmenistan border about 75 miles (120 km) west of Mazar-e-Sharif. Pipelines from there deliver natural gas to Uzbekistan and Tajikistan and to thermal and fertilizer plants in Mazar-e-Sharif. Afghanistan's petroleum reserves are poor. Some lead, zinc, silver, and chrome are also mined.

Afghanistan is potentially rich in hydro-electric resources, which remain mostly untapped because the nation's demand for electricity is negligible and projects are unprofitable. These have, therefore, been exploited only in the demand-areas of Kabul-Jalalabad. Manufacturing activity is only limited to items based on agricultural and pastoral raw materials. Cotton textile is by far the most important industry, although some rayon and acetate fibers are also manufactured. A fertilizer plant, based on natural gas, is located at Mazar-e-Sharif. Natural gas, dried fruits and nuts, carpets, rugs, hides, and cotton are the major exports; and machinery, basic manufactures, vehicles, petroleum products, vegetable oils, tea, and textiles are the most important imports. The Soviet Union has traditionally been the leading trading partner. Japan, United Kingdom, Germany, Singapore, India, Pakistan, and China are the other trading partners. The country remains at a much lower stage of economic development than most of its neighbors. Between 1956 and 1979, economic growth was guided by several five-year and seven-year plans, and supported by extensive foreign assistance, primarily from the Soviet Union and the United States. The Soviet invasion of 1979 and the subsequent civil war, however, severely disrupted the country's development plans. Agricultural production declined and food shortages were reported, and industry stagnated.

As a landlocked country, Afghanistan depends on transit facilities from its neighbors for its international trade. It has no railroads and only a few navigable rivers, relying mostly on roads as a mainstay of its transport system. Despite the high cost of road construction in difficult terrain, efforts were made, with foreign assistance in the 1960s, to upgrade the highway system and to integrate the main trading centers of the country with one another and with the railheads or road networks of the neighboring countries. The nation's road network now connects the railheads of Turkmenistan, Uzbekistan, and Pakistan. Paved roads connect the nation's leading cities (Kandahar, Herat, Mazar-e-Sharif) with Kabul and frontier towns of the neighboring countries.

Kabul (population: 700,000) is the nation's administrative capital and a major metropolis, located south of the Hindu Kush at the crossroads of the trade routes between the Indian subcontinent and Central Asia and between the Middle East and the Far East. It is also the country's main center of economic and cultural activity. **Kandahar**, second to Kabul in population (225,500), is located close to the Pakistan border on the Asian Highway in the south-central part of the country. **Herat** (178,000) is the regional center in the western part of the country on a highway leading to Iran. **Mazar-e-Sharif** (131,000) is an important commercial and manufacturing center, on the highway leading to the Central Asian republics of Uzbekistan and Tajikistan. The city has been the traditional entry point for Soviet merchandise.

As a final note, Afghanistan has been, in modern times, a buffer between the two great imperial powers, Britain in India and the Soviet Union. With the partitioning of the Indian subcontinent in 1947, Britain's interest in the nation virtually disappeared. The Kremlin has historically looked southward, but with the Russian withdrawal in the mid-1980s, Pakistan's interest in Afghanistan has been aroused, partly resulting from the influx of millions of Pashtuni refugees from Afghanistan and partly to secure its own unruly and tribal borderlands that lie between the two countries. Presently, the country remains in a state of on-going civil war between the feuding tribals and the extremely conservative central administration. The United Nations' policy towards Afghanistan recognizes the desirability of strengthening its domestic economy and of preserving its national integrity. The main roadblock is the unstable political situation that continues unabated in the country.

Further Readings

Fisher, W.B. *Middle East: A Physical, Social and Regional Geography*. London, 1978.

Khadduri, M.T. *The Gulf War: The Origins and Implications of Iraq-Iran Conflict*. New York, 1988.

U.S. Government Printing. *Afghanistan: A Country Study*. Washington, D.C., 1986.

U.S. Government Printing. *Iraq: A Country Study*. Washington, D.C., 1990.

15

South Asia: Physical, Cultural and Economic Patterns

The label "South Asia" can be conveniently applied to the countries which lie between the two broad realms of Southwest Asia and Southeast Asia. This broad region extends from Afghanistan in the west to Myanmar in the east and covers 1.7 million sq miles (4.4 million sq km) and includes the nations of India, Pakistan, Bangladesh, and Sri Lanka; the Himalayan kingdoms of Nepal and Bhutan; and the tiny country of the Maldives in the Indian Ocean. India, Pakistan, Bangladesh, and Sri Lanka, and the Maldives formed a part of the British Indian Empire prior to attaining independence in 1947-48. Nepal and Bhutan, never formally annexed by Britain, were regarded as within the British sphere of influence and were once British protectorates. The entire territory coincides to a remarkable degree with the geographic concept of the Indian subcontinent, the ancient, mythic *Bharatavarsha* (the official name of India is **Bharat**), a title that is rooted in the territorial lay-out of the subcontinent and the disposition of physical features within it. Hemmed in effectively by the Himalaya Mountains' influences from the north, and circumscribed by the peninsular coasts, the subcontinent developed a unique form of civilization, popularly known as the Indian civilization. The region juts into the Indian Ocean as a peninsula, with the two arms of the Indian Ocean: the Bay of Bengal and the Arabian Sea, flanking its coasts. The Equator passes through the southern part of the Maldives, while the northernmost extremity extends as far as 39° North. The Tropic of Cancer divides the nations of India and Bangladesh into two parts, leaving the peninsular part of India and Sri Lanka in the tropical zone.

In reviewing the long history of South Asia one clearly sees a cultural and geographic unity that transcends the diversity of languages and religions within it. The histories of these nations have been so inextricably intertwined as to include them all within the influence of Indian culture. The Indian Culture represents a third distinctive stream in Asian history---the other two being the civilizations of China and Islam. In the north, the Himalaya Mountains acted as an effective barrier, bolstered in large measure by the elevated and forbidding Tibetan plateau. The two civilizations of China and India matured independently and exclusively.

On the western side, isolation was less restrictive. Pakistan, and the west of it—Afghanistan lay at the crossroads between Islamic penetration from the west and the heartland of the Indian subcontinent. The Afghani mountains, breached by several

Table 15.1 *South Asia: Basic Data, 1999*

Country	*Area (Thousand sq km/sq ml)*	*Population (Millions)*	*GNP ($/Capita)*	*Capital*	*Urban Population percent (est.)*
Bangladesh	147/57	127	360	Dhaka	21.0
Bhutan	47/18.1	0.6	470	Thimpu	7.0
India	3,279/1,266.3	1001	370	New Delhi	26.8
Maldives	0.3/1.1	0.3	1180	Male	27.0
Nepal	147.1/56.8	22.5	220	Kathmandu	15.0
Pakistan	796.1/307.3	152.2	500	Islamabad	33.0
Sri Lanka	65.6/25.3	18.9	800	Colombo	23.0

Source: Encyclopedia Britannica, Book of the Year, 2000.

passes, could not deter a succession of periodic invasions starting with the ancient Aryans to the Mughals during medieval times. In due course, the Mughals, as did the invaders who had come to India prior to them, became Indianized, absorbing many of the native mores, and transmitting their forms of art, music, literature and philosophy to India. Through this interaction a unique and enduring Indo-Islamic synthesis took place. By virtue of her size, population, central location, and cultural preeminence, India has played a pivotal role in the region throughout history. Even as South Asia moved into a period of independence after the dissolution of British rule, India's dominant role remains.

Broadly speaking, the physical characteristics, historical evolution, and cultural, economic and demographic developments of South Asian nations have been so intricately intertwined that it is best to treat them in an integrated chapter. Thus, this chapter gives an overview of the physical, cultural and economic patterns of South Asia as a whole, whereas the distinctive details of the South Asian nations are explored individually in the next chapter.

Natural Environment The general arrangement of South Asia's landforms is simply laid. In the north rises the massive mountain wall of the Himalaya (literally "Abode of Snow") system together with its eastern and western offshoots giving the subcontinent a sort of geographic unity. South of it lies the Indus-Ganga Plains, with an extension into the Brahmaputra Valley in the east, making it one of the largest stretches of lowland in the world. The southern and peninsular part of the subcontinent is occupied mostly by the Deccan (literally "the South") Plateau, flanked by mountains and narrow coastal plains. Within this threefold division exists a considerable diversity of structure, drainage and topography.

The Himalaya forms a part of an extensive and complex mountain system, the hub of which lies in Central Asia in the Pamir Knot. From the Pamirs it swings in an unbroken arc of several parallel ranges for over 1,500 miles (2,400 km), rising abruptly to form the northern, western, and eastern borders of the subcontinent. Its central and the highest range, the Great Himalaya, contains 40 peaks which rise to altitudes between 25,000 and 29,028 feet (7,600 and 8,848 meters). The highest

Map 15.1

SOUTH ASIA-POLITICAL DIVISIONS
International boundary
Itanagar Cities
DECCA Capital
ASSAM Major Political District
AFGHANISTAN
PAKISTAN
NORTHWEST FRONTIER
BALUCHISTAN
PUNJAB
SIND
CHINA
NEPAL
BHUTAN
BANGLADESH
MYANMAR
SIKKIM
JAMMU AND KASHMIR
Occupied by Pakistan
Occupied by China
HIMACHAL PRADESH
PUNJAB
UTTARANCHAL
HARYANA
RAJASTHAN
UTTAR PRADESH
BIHAR
JHARKHAND
WEST BENGAL
MADHYA PRADESH
CHATTISGARH
ORISSA
MAHARASHTRA
ANDHRA PRADESH
KARNATAKA
TAMIL NADU
KERALA
ARUNACHAL PRADESH
ASSAM
MEGHALAYA
TRIPURA
NAGALAND
MANIPUR
MIZORAM
Gilgit
Srinagar
Muzaffarabad
Peshawar
Lahore
Quetta
Hyderabad
Shimla
Chandigarh
Almora
DELHI
Agra
Jaipur
Lucknow
Patna
Ranchi
Kolkatta
Gangtok
Itanagar
Dispur
Shillong
Kohima
ImpHal
Aizwal
Agartala
DHAKA
Chitagong
Gandhinagar
Bhopal
Raipur
Bhubaneshwar
Mumbai
Hyderabad
PANAJI
Goa
Bangalore
Chennai
Pondicherry
Trivandrum
35°
30°N
25°
20°N
15°
70°E
75°
80°E
85°
90°E
miles
0 100 200 300 400 500
0 200 400 600 800
kilometers

elevation in the world, Mount Everest (Sagarmatha) 29,028 feet (8,848 meters) lies on the Nepal-Tibet border. The Himalayan scenery is among the most spectacular and grand in the world. These ranges, together with the high-altitude Tibetan Plateau further north, have exercised profound climatic influence on the subcontinent. The mountains act as an effective barrier blocking the inflow of cold, dry air-masses into north India during winter. During summer they deflect the moisture-laden monsoon winds along the Himalayas' southern flanks, and cause the moist winds to bring precipitation to the North Indian Plain and the Brahmaputra Valley. The Himalaya is also a feeding ground of numerous rivers which are utilized for year-round irrigation in Pakistan and on the North Indian Plain.

The western offshoots of the Himalaya, in general, extend in a north-south direction along the Afghanistan-Pakistan border, extending southward all the way to the Arabian Sea. The eastern mountain arcs lie beyond the gorge of the Brahmaputra, running in a series of mountain ranges along the Myanmar border and continuing into Myanmar. These are not as rugged or high as the main Himalaya but are covered with dense forests. Historically, this disease-ridden and rain-swept land has been a difficult zone of movement.

Just south of the Himalaya, lies the vast Indus-Ganga plain of India, Pakistan and Bangladesh. It includes the lowlands of the catchment area of the two major rivers, the Ganga and the Indus, together with their tributaries. Occupying only one-fifth of the area, but supporting half the population of South Asia, these plains have historically been the core region of its political, economic and cultural activities. The Ganga-Brahmaputra Delta forms the southern projection of the lowlands at the head of the Bay of Bengal. East of the lower Indus section of the plains lies the Thar Desert, an extension of the adjacent plain. The topography of the Indus-Ganga plain is essentially flat, averaging a drop of less than a foot per mile, with a long, nearly straight horizon. The topography should not, however, be considered as absolutely flat. Locally, however, terraces may rise as much as 100-200 feet (30-60 meters) relieving the general monotony. Partition of the subcontinent in 1947 created physical and human problems, both in the Ganga-Brahmaputra Delta and on the Indus Plain. The main problem hinged around the division of the waters between the new countries. In the Punjab, a name which means five rivers and derives from the five tributaries of the Indus, political partitioning created problems by disrupting existing irrigation systems, and transport lines.

The peninsular part of the subcontinent is mostly a raised tableland of old, stable structure, known generally as the Deccan Plateau. The Deccan, in fact, includes a number of basins in addition to several plateaus of subdued relief, with elevations ranging between 1,000 and 3,000 feet (300 and 900 meters) above sea level and is flanked on the west and east coasts by the coastal ranges. In the south, near where the two coastal ranges appear to merge, are two high granitic massifs, the Nilgiri Hills and the Cardamon Hills, rising to elevations of about 8,500 feet (2,600 meters). The western flanking ranges, the western Ghats or the Sahyadri Mountains rise as a bold escarpment of 4,000 to 5,000 feet (1,200 to 1,500 meters) above the western coast, leaving a very narrow coastal plain. Seen from the interior the western Ghats assume a hill-like appearance. The discontinuous ranges flanking the eastern side of the Deccan Plateau are collectively termed the Eastern Ghats and with altitudes of 2,000-2,500 feet (600-750 meters) are neither as high nor rugged as the western Ghats. These are merely discontinuous hills.

Climatic Characteristics A distinguishing characteristic of South Asia's climate is the prevalence of monsoonal (from the Arabic word *mausim* meaning "season"), or seasonal wind systems, the alternating dry cold continental winds in winter and moisture-laden, tropical, oceanic winds in summer. The term monsoon is applicable to most, but not all, parts of the subcontinent. The idealized perception of the monsoon climate best applies to the areas of northern India and Pakistan. In southern India, in Kerala, and in southwestern Sri Lanka, a tropical rain forest climate prevails. Along the southern parts of the eastern coast, the monsoon is not dry but crossing the Bay of Bengal brings rainfall. In Pakistan and the Punjab plains, rainfall is brought by eastward moving cyclonic disturbances originating over the Mediterranean basin during winters. These exceptions apart, the monsoonal winds have almost a universal effect in bringing summer rains and are of great human significance. Most wet season crops depend on rainfall associated with the monsoon winds.

A convenient division of the calendar in areas where monsoons have their greatest impact is: (a) the rainy season of southwest summer monsoons, (b) the dry winter season of northeast monsoons, and (c) the intervening hot season between winter and arrival of the summer monsoons. The hot season in the subcontinent begins in March and lasts until June. From March to May, during the northward movement of the path of the sun's vertical noon rays, the increasing length of day causes a belt of intense solar radiation to move progressively northward over the tropical areas of South Asia, advancing by June into the extra-tropical zone. By June the highest daytime temperatures 104°-113°F (40°-45°C) are recorded in central Deccan and the Indus-Ganga plains. The highest temperatures are, however, recorded in the Thar Desert and on the Indus Plain, and occasionally in the middle Ganga plains.

During summers winds move quickly into peninsular India and Bangladesh toward the low-pressure system of northwest India from the Arabian Sea and Bay of Bengal. The moisture carried by these southwesterly winds from the Arabian Sea and south and southeasterly winds from the Bay of Bengal bring rain to the subcontinent, leaving Tibet dry in the "rainshadow" of the Himalaya. The rainy season sets in by the end of June and lasts until the end of August. This is the "summer monsoon." The amount of summer rainfall varies in the different parts of the subcontinent, depending on their distance from the coast or from the Himalaya, their altitudes, the direction of the monsoon winds and whether the locations lie on the windward or leeward side of mountains. A part of the Bay of Bengal monsoon stream which strikes the Himalaya after crossing the Ganga delta, is uplifted, and sheds copious precipitation. The lofty Himalaya also diverts the moist stream westward into the Ganga plains of north India. This diversion, or channeling effect carries rainfall farther west. As the monsoon stream moves westward in the Ganga plains, it continues to lose moisture, becoming increasingly dry. Most places in the middle Ganga plains receive between 25 and 50 inches (625-1,250 millimeters) of rainfall during the rainy season. In the west of peninsular India most moisture is shed between the sea and the crest of the western Ghats is reached, leaving the Deccan Plateau practically rainless in the "rain-shadow" of the mountains.

By September the systems change, responding to southward movement of the sun toward the Southern Hemisphere. The development of a high-pressure system over the subcontinent sets in motion a stage for this retreat of the monsoon winds. Winds are now northeasterly and are dry and bring rain only to a southern section of the eastern coast, after picking up moisture while crossing the Bay of Bengal. From October until March is the cool season when the entire subcontinent experiences mild and sunny weather. The mean January temperature is about 70°F (21°C) in southern India and 60°F (15.5°C) in Pakistan and north Indian plains. In the northwest occasional light frosts may also occur. In the Himalaya, temperatures decrease with altitude.

Historical evidence indicates that an almost continuous expanse of forest covered much of north India at the time of Alexander's invasion in the 4th century B.C. Even during the 10th to 14th centuries, the Ganga plain contained vast patches of forest. Human modification of interference with the "original" plant cover has resulted in its removal or degradation, creating serious problems of soil erosion and the growth of stunted "jungle" vegetation (thickets, bushes and dwarf trees). At present, forests occupy only about 15 percent of the total land of the subcontinent.

Natural vegetation now exists only in a few areas at high altitudes or coastal swamps, or in the "protected" forest reserves under government control. Of the several forest types identified, three major vegetation zones are: an evergreen broadleaf forest, deciduous broadleaf forest and thorny scrub, with their distribution related to climatic conditions and altitude. Evergreen broadleaf trees are located mostly along the western coastal lowlands, Bengal and Sri Lanka, in areas of annual rainfall of more than 60 inches (1,500 millimeters). The deciduous broadleaf zone is located in areas with annual rainfall of 750 to 2,000 millimeters falling mainly between June and October. The thorny scrub zone lies in semi-arid areas in Pakistan, peninsular India and parts of interior Sri Lanka. The Thar Desert supports some shrubs and grasses.

Soils follow the climatic and vegetation zones. One-third of South Asia is covered with water and wind transported surface materials showing little profile development. About 20 percent is covered with alluvial (river-borne) sands, silts, clays, or gravels, deposited by the Indus, Ganga and other rivers. The Indus-Ganga deposits are among the thickest in the world, attaining thicknesses of several thousands of feet and are very fertile. In the northwestern and north central part of peninsular India dark-colored *regur* soils are found. These are derived from the Deccan Lavas. Rich in nutrient-content they are used widely for growing such crops as cotton and millets. In areas where annual rainfall is more than 60 inches (1,500 millimeters), laterites are found. Laterite may be used as building material and for road surfacing.

Historical Development There is abundant evidence that man has been in continuous occupation of South Asia since the Middle Pleistocene period (several hundred thousand years ago) of geological history, although beginnings of mature urban civilization may be traced to the Indus Valley culture in the third millennium B.C. This civilization persisted for about 1,000 years and covered an area of over half a million sq miles stretching from Baluchistan in Pakistan to the Ganga Basin, northeast of modern Delhi and south to southwest Maharashtra in India. Remarkably uniform in several cultural traits (agricultural products, inscriptions on seals, figurines and artifacts of presumably religious significance), the Indus Valley culture region was an urban and commercial civilization. Internal communication links were developed mainly along the river. Trading connections with the outside world were established through several ports on the Arabian Sea. Cities were supported by a vast agricultural base in the fertile river basins.

Between 2500 and 1700 B.C. the agricultural settlements and the associated urban economic structure of this civilization dominated the Indus Basin. As external commercial contacts developed, the civilization diffused to the Ganga Basin and parts of western India. For reasons not properly understood, this urban civilization declined quite suddenly by 1500 B.C. Geological and ecological factors, such as changes in river courses, salinization and waterlogging of soils, expansion of desert, and climatic changes which made the area physically inhospitable, have been advanced for its nearly complete disappearance. Another theory is that the civilization fell victim to invading Aryans.

Despite its impressive development, the Indus Valley Civilization left no enduring mark on Indian civilization. It is significant to note that contemporary Indian culture stems largely from the confluences of three major streams of tradition: the establishment of Hindu thought starting with the Indo-Aryans but developed, refined and greatly transformed subsequently, the Indo-Islamic synthesis, and the host of alien ideas brought by the British.

The establishment of Hindu tradition starts with the Indo-Aryans. Little is known about them. Linguistic and literary evidence, however, points to their unmistakable links with the Indo-Europeans as well as to the fact that they came from the northwest, the direction of the Iranian Plateau, in successive waves between 2000 and 600 B.C. Early Aryan literature known as the Vedas, as well as the contemporary Iranian *Avesta*, attest to such migrations although precise source-regions and timing are still obscure. Most scholars favor the idea that the migrations originated in Central Asia. They were probably a part of the great and complex folk movements which brought ancient Persians into Iran and Greeks into Greece. The early Aryans were nomadic pastoralists. As they moved into the north Indian plains and finally penetrated into southern India, they gradually cleared the forests and established agrarian communities.

During the 6th century B.C., two sects, Buddhism and Jainism, stood up against the ritualistic and polytheistic forms which had crept into society. Each also struck against the entrenched caste system within society. In this last effort they ultimately failed, for most participants in the system clung to the norms of Hinduism.

In 326 B.C. Alexander of Greece invaded northwestern India through the passes from Afghanistan. He could not advance beyond the Beas River in present-day Punjab. His fleeting stay and sudden departure left few marks on Indian society. Shortly thereafter, the Mauryan kings (321-150 B.C.) consolidated the various tribes in northern and northwestern India into a large unified empire which in times covered all of the subcontinent. Their administration centered on a bureaucracy headed by the emperor. The imperial system was sustained financially by the income derived from land taxes, internal commerce and foreign trade. Cities flourished and were connected by a well-developed system of highways. Artisans formed a major section of the urban population. The state maintained a large standing army.

After a lapse of several centuries, large sections of the country, at least in north and central India, which had disintegrated into several principalities, were united by the Gupta Kings who attempted to revive the imperial idea of a *Chakravartin* (Supreme Ruler) of a pan-Indian empire. In addition to its military glories, the Gupta Age (320-540 A.D.) achieved notable successes in artistic and humanistic spheres and is often acclaimed as the Classical Age of India. Peninsula India over the centuries, developed its own linguistic, philosophic, and cultural traditions. Only rarely did the military arm of the north Indian emperors interfere in the political and administrative affairs of the south.

By the 8th century, Islam was firmly rooted in Arabia and Iran, and started to encroach upon the lower Indus Basin. Until the 10th century Islamic invasions of northwestern parts of the subcontinent were sporadic, but during succeeding centuries the petty, warring Hindu principalities could no longer withstand their onslaughts. Islam finally came to South Asia, but not mainly from Central Asia, rather than from its Arabian homeland. By the 10th century the Turks had begun penetration from the northwest, from present-day Afghanistan and Pakistan, through the several passes in the area.

The establishment of Mughal rule in India in 1526 A.D. opened a new and long chapter of Islamic factor in Indian history. Originally natives of Central Asia, the Mughals established a vast empire encompassing not only the northern part of India, but during its greatest extension, almost all of south India as well. Gradually they intermarried with the natives and were, to a degree, assimilated into society. Under them India became one of the world's richest and most powerful countries, with which Persia, Turkey and several European nations sought friendly relations and commercial links.

The Mughal rulers were patrons of architecture, art and literature. They introduced a new court language, Urdu which arose out of an interaction of Hindi (the existing language) and Persian brought by them. More importantly, the final outcome of the Muslim interaction with the Indian society was the ongoing development of a distinctive Indian civilization in which Muslims became a significant element in Indian national life, although Hindus remained a primary and dominant factor in a plural society.

The Europeans arrived in South Asia in search of profitable trade in spices, gems and textiles. After passing through stages of intrigue and rivalry with the other European companies, the British East India

Company had emerged victorious as the single most important force in the country by 1750. The company successfully interceded in the internal affairs of the weak rulers of Bengal and swiftly gained a stronghold in the lower Ganga plain, as well as controlling the strategic ports of Channai and Mumbai.

Between 1757 and 1803 British supremacy was consolidated in the Indus-Ganga plains and the coastal areas, and was eventually extended to the entire subcontinent. Although most of the country had fallen to British control, within the British-occupied territories lay over 572 states which were internally administered by native princes. The British not only controlled the external affairs of these states, but also their internal security. By the establishment of military stations (known as "cantonments") at key locations and by the institution of grants and favors to the rulers, the Crown was able to control the princely states, but after periods of political disorder, and in the absence of a male heir to the princely throne, it annexed many princely territories into British India.

An elite class of English-educated Indians emerged during British rule. Trained in the British system of jurisprudence, justice, and equality, these young men started agitation for participation of Indians in the administration and formed the Indian National Congress in 1885. Later, their objectives changed from mere legislative participation to complete independence and a long and strenuous struggle to achieve that objective began. The Indian National Congress, which had mobilized a broadly based nationalist support, felt hampered in dealing with the British by itself. In 1940 the Muslim leadership opted for a separate independent "homeland" for the Muslims to be named Pakistan and to be carved out of the subcontinent. Two independent nations were thus created out of the subcontinent in 1947, principally according to the geographical concentrations of Muslims and non-Muslims in the subcontinent. The initial years of independence were full of turmoil for both countries. Even before the partitioning of the subcontinent, Hindu-Muslim riots had erupted in those areas which were likely to be divided between the countries, notably Punjab in the west, and Bengal in the east. Over 12 million persons, Hindus Sikhs, and Muslims left their ancestral homes and migrated to areas of their religious affiliations across the newly created international borders. Over a million perished in the tragic events of the partitioning of the subcontinent. The new borders posed other problems as well. The international boundary of the new countries had little regard for established networks of railroads and irrigation canals, or for the linkages between manufacturing areas and the areas that supplied them. Fortunately, well-trained Indian and Pakistani officers quickly brought civil order to the two new countries.

The preceding overview of the history of Indian civilization points out that a recurrent pattern of territorial instability and development of subregionalism in the subcontinent has been a major theme of Indian historical geography. The cherished dream of establishing a pan-Indian empire was accomplished only a few times, most notably under Ashoka, Aurangzeb, and more recently during the British period.

Cultural Patterns In reviewing the language types of South Asia one is at once struck by the diversity and complexity of the pattern. In 1927 *The Linguistic Survey of India*, listed 179 languages and 544 dialects that are spoken within the subcontinent. It excluded Sri Lanka, where linguistic pluralism is also the rule. Of the six-language families identified on the map for South Asia the two most widely-spread are Indo-European and Dravidian. The Tibeto-Burman family is distributed along the northern and northeastern borders of India, Munda and Mon-Khmer languages are scattered in less accessible and tribal parts of northeastern India, and the Ural-Altaic family covers a part of northern Afghanistan.

In the Indo-European family, Hindi (and the closely-related Urdu, Hindustani, and Rajasthani tongues) claims the largest number of speakers (nearly 430 million) and is spread over a wide territory in North and Central India. Urdu grammatically akin to Hindi but uses Persio-Arabic script and vocabulary, is the official language of Pakistan, though it is the mother tongue of relatively few Pakistanis. Bengali, Marathi, Gujarati, Punjabi, Oriya, Assamese and Kashmiri each claim large number of speakers, ranging from 5 million for Kashmiri to over 150 million for Bengali including over 170 million of Bangladeshis. Each is located in a compact territorial block. In India this enabled state boundaries to be drawn mainly on linguistic grounds, the greatest changes being made in 1956. Bengali's territory is divided between India (in the state of West Bengal) and Bangladesh, a consequence of the partitioning of the subcontinent in 1947. The Dravidian language family is located in mainly southern India. It includes four major languages: Tamil, Telugu, Malayalam and Kannada, each claiming between 20 and 70 million speakers.

Following independence, India gave recognition to 15 (currently the constitution recognizes 18) of the major languages belonging to the two language families, and soon undertook a sweeping reform to reorganize its state boundaries. By 1956, most state boundaries coincided with the language regions. In addition, the Constitution invested Hindi with the status of a national language by making it the sole administrative language of the central government. To allay the fears of the non-Hindi speaking states, the central government decided to keep English, along with Hindi as a medium of transaction of government business and as a "link" between the states for a period of 15 years, but extended several times later. Pakistan after independence declared Urdu as the official language for the entire nation. Though it was not the principal mother tongue of any section of the country, it is widely understood, and is acceptable to the educated elite. Among all South Asian countries, Bangladesh is linguistically the most homogeneous. Even before 1971, when it was a part of Pakistan, few of its people understood Urdu, and fewer still were willing to accept it. Bengali had to be declared a second language. After independence Bengali took its rightful place as the national language of Bangladesh.

In Sri Lanka language has been a powerful divisive force. Tensions have long existed between the Sinhalese-speaking, mainly Buddhist majority, and a Tamil-speaking, largely Hindu minority, the latter concentrated mainly in the northern and eastern part of the country. Many other Tamils emigrated from India in recent times to work on tea plantations and have not yet gained citizenship rights. The

Map 15.2

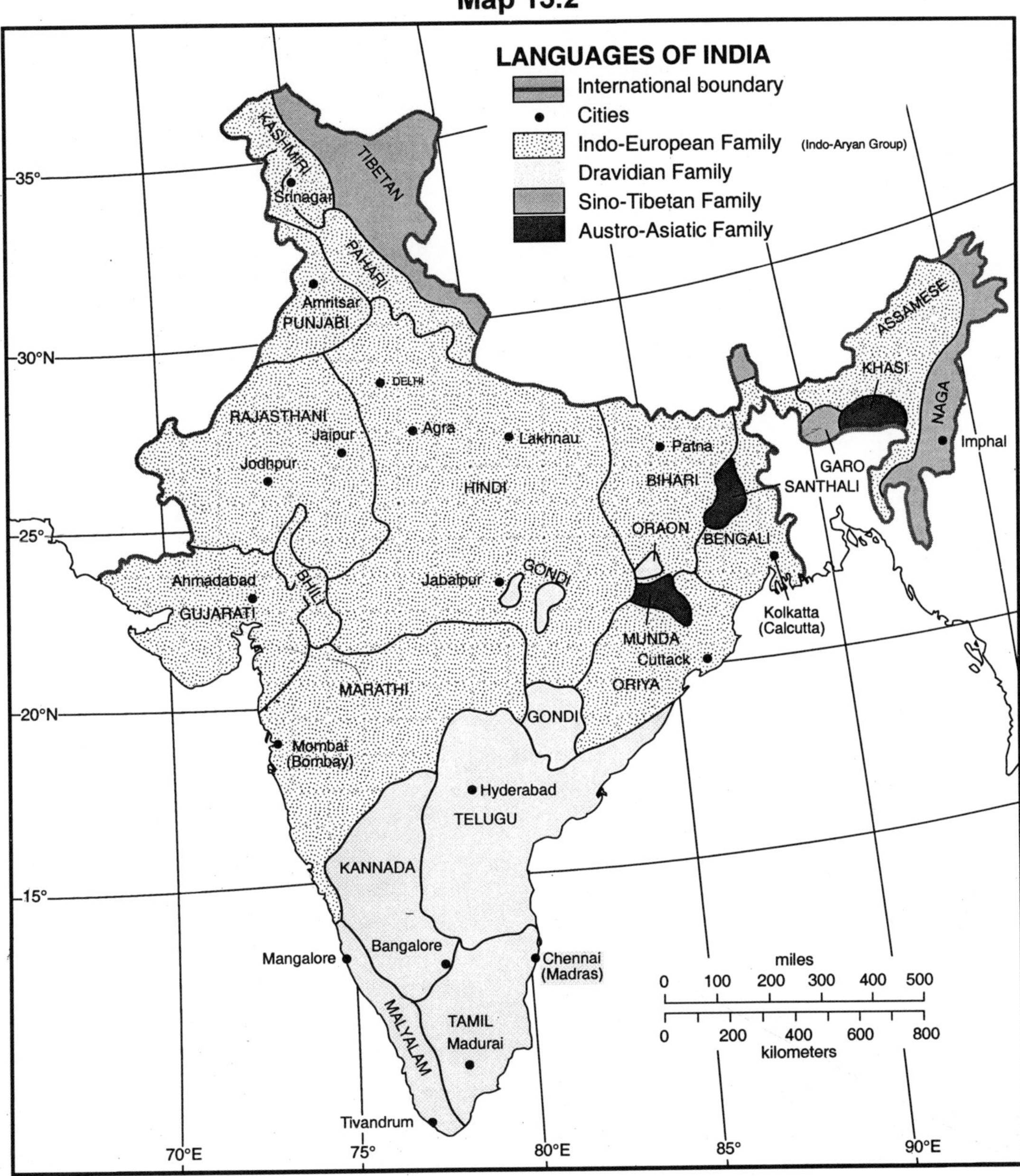
LANGUAGES OF INDIA
International boundary
Cities
Indo-European Family (Indo-Aryan Group)
Dravidian Family
Sino-Tibetan Family
Austro-Asiatic Family
KASHMIRI
TIBETAN
Srinagar
PAHARI
Amritsar
PUNJABI
DELHI
RAJASTHANI
Jaipur
Agra
Lakhnau
Jodhpur
HINDI
Patna
BIHARI
ORAON
BENGALI
SANTHALI
GARO
KHASI
ASSAMESE
NAGA
Imphal
Ahmadabad
GUJARATI
BHILI
Jabalpur
GONDI
MUNDA
Cuttack
ORIYA
Kolkatta
(Calcutta)
MARATHI
GONDI
Mombai
(Bombay)
Hyderabad
TELUGU
KANNADA
Bangalore
Mangalore
Chennai
(Madras)
MALYALAM
TAMIL
Madurai
Tivandrum
miles
0 100 200 300 400 500
0 200 400 600 800
kilometers
35°
30°N
25°
20°N
15°
70°E
75°
80°E
85°
90°E

long-established Tamils have been seeking a political "homeland" within the country and agitating for greater recognition for their language. This has led to violent confrontation with the central administration.

Elsewhere, in Nepal, the Nepali-speaking majority of the Indo-European family, has to deal with large numbers of mountain people speaking various languages of the Tibeto-Burman family. In Bhutan, dialects of Tibetan are predominant. The language of the Maldives Islands is Divehi and is akin to Sinhalese with an infusion of Arabic words.

Religion has been a powerful force molding the destinies of the people in South Asia. In recent times the separation of Pakistan from India stemmed principally from the political loyalties arising out of the religious and socioeconomic differences between Muslim and non-Muslim populations. The current demands for a "homeland" by Sikhs in the northwestern part of India, and for independence for the Tamil-speaking Hindu-dominated areas in Sri Lanka are two of the several instances of loyalties-based largely on religious and linguistic preferences.

In each South Asian country, significant religious minorities exist. Not only are these minorities widespread, but in some areas of India, Bangladesh, Nepal, and Sri Lanka, they constitute local majorities. In India, Muslims form nearly 12 percent of the population giving it, after Indonesia, Pakistan and Bangladesh, the fourth largest Muslim community in the world. Nearly eleven percent of Bangladesh's population professes Hindu faith, almost all the remainder is Muslim. During the war with Pakistan in 1971, an estimated 9-10 million of the Hindu population left the country to take refuge in the neighboring states of India, especially in West Bengal. Nearly 8 percent of the predominantly Hindu kingdom of Nepal are Buddhists, while as much as one-fourth of Bhutan's population may be composed of Hindus of Nepali origin, (the majority of Bhutanese are Buddhists). In Sri Lanka nearly 70 percent of the population is Buddhist, the remaining is divided between Hindus, Christians and Muslims in that order of importance. Pakistan, and the Maldive Islands, are nearly homogeneous in religion, claiming no more than two percent of their populations of non-Muslim faith.

While most South Asian countries face the vexing problems of integrating their minorities into the dominant national stream, no where else is this internal problem more difficult than in India, where the population is divided among several religious groups. An overwhelming majority of the population (80 percent) professes the Hindu faith. With the exception of Jammu and Kashmir, and Punjab, Hindus form majorities in all the remaining states of India. Christians and Sikhs predominate at local levels. Other minorities----Muslims, Jains, Buddhists and Parsis----are dispersed among Hindu populations. Many practices, characteristic of the Hindu society, have been retained among them, as most of whom are descendants of converts from Hinduism.

Of the minority groups, Muslims remained more distinct from the majority group of Hindus, and presented the biggest obstacle to national unity during British rule, culminating in the partitioning of the subcontinent into two countries in 1947. The Hindu-Muslim rift is deep-rooted, going back to the period from the 13th to the 17th centuries, when Muslim rule was established by conquest. Here, the Islamic civilization clashed with Hindu civilization as it had clashed in Spain and the Balkans with Christianity. Islam presented a socio-economic structure based strictly on social freedom and equality, a concept entirely antithetical to the prevailing social rigidity of the Hindu caste system. Lured by the prospect of social freedom, as well the desire to escape the threat of extinction by the sword, many lower caste and some tribal people were attracted to Islam. Among the higher castes, political advantages occasionally proved irresistible, and paved the way to conversions.

Among other religious groups of India, Sikhs warrant particular attention. Nearly 90 percent of the Sikh population of 18 million lives in a critical area bordering Pakistan, in metropolitan Delhi, and adjacent areas of the states of Uttar Pradesh and Rajasthan. An offshoot of Hinduism, the Sikh religion retains many Hindu practices, and many Sikhs intermarry with Hindus. An eclectic, monotheistic faith, and initially a pacifistic religion, Sikhism began as a reform movement in the 15th century against the Hindu practices of caste and rituals. In the 17th century, when Sikhs were persecuted by Muslim rulers, they assumed a militant posture. The partition of the Indian subcontinent cut directly through their territory, and about 1.5 million Sikhs, who found themselves in Pakistan, moved to India mainly to the state of Punjab. This influx of Sikhs into the northwestern part of India pushed them to a majority status in areas bordering Pakistan. Within India, Sikhs have been politically the most active among religious minorities. They successfully agitated for the creation of a Sikh-dominated Punjabi-speaking state in 1966. Highly cohesive, enterprising and proud of their traditions, many Sikhs have recently been vocal and often violent in their demand for Khalistan, an independent "homeland," thus posing a serious problem for India's internal security.

Although the regional aspirations of most major linguistic groups in India had been met by 1970, regionally important cultural or religious groups continued to agitate occasionally for the creation of new states or even for independence. In the northwest, Sikhs demanded and succeeded in the break-up of the original state of Punjab into two parts: a truncated Sikh-dominated Punjab in the north, and a Hindu state of Haryana in the south (both created in 1966). The Sikh problem was, however, defused only temporarily, for a group of extremists continued to agitate for a separate Sikh homeland---a new Punjab state to be called Khalistan (literally "land of the pure") in which they would be totally independent of India. Other Sikhs would settle for enhanced autonomy. Violent confrontations between Sikh extremists entrenched in their holy temple in the city of Amritsar, and government forces led to tragic loss of lives and destruction to the holy shrine in 1984, and to increased Hindu-Sikh polarization. India has since then been able to defuse this explosive problem. In early 1994, much of the tension had subsided and Punjab had become peaceful. Pressures for regional autonomy among several others have also existed since independence. Most of the groups were pacified by creating several states in eastern India bordering Myanmar. It is remarkable that the Indian federation has been able to survive the "dangerous decades" since independence. Other South

Asian nations did not follow the Indian pattern of linguistic states despite their apparent ethnic complexity (only Bangladesh is linguistically homogeneous).

In addition to the linguistic and religious diversity in South Asia, there was a considerable political fragmentation within its territory, and danger of eternal threats since their inception. At the end of British rule in India the subcontinent was subdivided into a fragmented patchwork of large and small political units. In addition to the territories directly ruled by the British, known as provinces, there were 572 dependent native or "princely" states (562 in India and 10 in Pakistan). After attaining independence the two new nations faced with the difficult task of national integration, could ill-afford the continued existence of such states scattered throughout their territories which could pose a serious threat to their political stability and economic development. With the exception of Kashmir state, that became a bone of contention between India and Pakistan in the aftermath of partition, all the princely states had been totally absorbed into one of the two countries by 1950.

The South Asian nations have never been entirely free from the specter of internal and external conflicts after gaining independence from the British in the late 1940s. India's frontiers with Pakistan and China have been the scene of warfare in 1947-48, 1962, 1965 and 1971. A dispute between India and Pakistan over the state of Jammu and Kashmir remains unresolved despite the state's *de facto* partitioning between them. China also claims a large territory in India's northeastern borderlands and in eastern Kashmir, and engaged in a limited war with India over these areas in 1962.

Population Problems South Asia is home to over 1.3 billion people, one in every five persons in the world, although it accounts for only one-thirteenth of the Earth's surface. India alone counts over a billion inhabitants. Pakistan and Bangladesh contain 152 and 127 million each. No other region of the world has such serious population problems as does South Asia; and no other area of comparable size and cultural attainments is as poor. These numbers are certainly a cause for concern, but even more disturbing is the fact that they continue to escalate. Recent declines in fertility in Sri Lanka, Pakistan and India give only scant comfort, since the total numbers being added to the existing population keep increasing. Some demographers project a population of 1.5 billion for the South Asian region in 2000 A.D.

Population pressures began to build in the region during the later decades of British rule. Improvements in public health measures, medical care, agricultural productivity, the establishment of law and order, and reduction in the frequency and severity of famine, pushed the shadow of death further into the background. Mortality declined, and life expectancy rose. Fertility remained high, as the traditional preference for larger families was the rule. This created a widening gap between fertility and mortality, and a consequent rise in the rate of natural increase (difference between birth rates and death rates expressed as a percentage), particularly since 1921.

The year 1921 marks a watershed in the demographic history of the subcontinent. Population began to increase, steadily at first, but later at a faster tempo. Only in the last decade has the rate of increase begun to slow. Presently, population is growing at a rate of 2.2 percent annually

(Sri Lanka is an exception, recording less than two percent growth) as compared to one percent per year in 1921. The progressive increase in the growth rates after 1921 contributed to a massive accumulation of numbers—from 256 million in 1921 to 1.3 billion in mid-1990s---an increase of over four times in 67 years. Looking at the future, we find that South Asia will double its population in about 30 years if the current trends of population growth continue. Even by medium projections it will contain nearly 1.5 billion people by 2012 A.D. India alone will have 1.2 billion inhabitants; both Pakistan and Bangladesh will contain close to 150 million people each.

The demographic structure of South Asian nations is represented by an age-sex pyramid of India which clearly shows that the youthful age groups are very large. Nearly 40 percent of the South Asian population is below 15 years of age (for Nepal and Bangladesh the figures are 42 and 45 percent respectively). Only Sri Lanka has as little as 35 percent of its population in the youthful category. Such age structures are particularly troubling, as the provision of services, schools, food, hospitals, and housing for the young must consume a large part of national expenditures. An examination of population pyramids is particularly useful in assessing future trends as well. As the youthful population moves into the adult age bracket (that is, enters into the working age group) public welfare programs, meager as these are in these countries, will be further strained. Problems of unemployment, food shortages, illiteracy and poor standards of living, characteristics of these nations, will be further aggravated. The present situation is serious enough; but prospects for the future are even more disconcerting, given the rate of increase of population, which for the entire region is over 2 percent a year.

Under given conditions, large populations can be advantageous. A larger and more skilled pool of workers can be productive and useful in meeting the demands of a large consuming market. However, the consumption demand of a vast population in the South Asian countries, working under conditions of scarcity, poor technology, and an incentive-free environment, is low. Furthermore, the ratio of labor to both land and capital is high. This has led to an enormous underutilization and inefficient use of labor. For example, on a typical farm or in a government office, five or more workers may be available to do the work that two could easily perform. In factories, many operations are handled manually because of the abundance and cheapness of available labor. Still, millions remain unemployed.

Given the current rates of fertility and mortality, and the increasing dependency load (essentially that of youthful population below 15 years of age), the situation threatens to deteriorate in the next decade. Results have been uneven, though promising in the more literate nation of Sri Lanka. Progress has also been more noticeable in cities, where children are perceived less as an asset to their parents and where it is relatively easy to contact people regarding birth control.

A crucial aspect of the population problem is the related matter of food sufficiency, a factor on which the survival of millions depends. These nations in their post-independence period have experienced critical foodgrain shortages, because of the ever-increasing numbers of mouths to be fed. India alone imported between 10 and 15 million metric tons of foodgrains annually during the time of its greatest shortages during the 1960s. Most other countries imported smaller amounts. Fortunately, the "Green Revolution" (the introduction of high-yield, fast-growing varieties of rice and other cereals) has led to increased production in parts of India and Pakistan. This has narrowed the gap between population growth and foodgrain requirements. India has now become self-sufficient in its foodgrain requirements. Pakistan and Sri Lanka have also achieved substantial success in their foodgrain productions. Pakistan not only attained self-sufficiency in its foodgrain requirements, it became an exporter of rice in the late 1980s. Sri Lanka has to import only moderate amounts. Recent successes must, however, be considered in the context of rapidly growing populations. The future remains problematic, as South Asia's population keeps growing by 30-32 million every year. Most South Asian nations are cognizant of these problems, but have failed as yet to develop effective policies. The main reason is the general scarcity of two critical resources essential to the development and implementation of such policies: capital and technology.

These nations (with the exception of Bhutan) have adopted population policies directed toward attacking the problem at the source: i.e., in achieving reductions in fertility, rather than waiting until the capital and technological resources have been securely developed. India was the first nation of the region to initiate a family planning program in 1951. Pakistan, Sri Lanka, and Bangladesh followed, each within the framework of its national planning effort. Motivating millions of couples, traditionally hostile to birth control methods, to use contraceptive devices, or to submit to sterilizations or intrauterine device (IUD) insertions, has been a complex and frustrating process. Capital outlays for such programs have progressively increased over the years, often with the aid of international agencies, but are still very low.

Urbanization A little over a quarter of the South Asian population is classified as urban (that is, as living in towns and cities of over 5,000 population), with Pakistan leading at 33 percent urbanized and the Himalayan kingdoms at merely 5 to 7 percent. Those low levels are, however, deceiving considering the total numbers involved. For example, in India 270 million people belong to the category of urban population---a figure substantially greater than the combined populations of West Germany, Italy, and Great Britain, the three largest countries of western Europe. Twelve cities were classified in 1981 as "millionaire" cities in India. Four of the world's largest cities, predicted to emerge as superconurbations (super-large cities) by the year 2000, are in South Asia: the Indian cities of Kolkata, Mumbai, Delhi, and Karachi.

The distribution of cities predictably conforms fairly closely to the overall pattern of population distribution. Thus, most cities are located in the densely populated Indus-Ganga plains, and on the costal lowlands. Most of these were established as centers of commerce or industry or as trading posts by the Europeans (Kolkata, Mumbai, Chennai, Karachi, Dhaka, Colombo). Elsewhere, inland cities often began as administrative centers of the native provinces or the British provinces.

Map 15.3

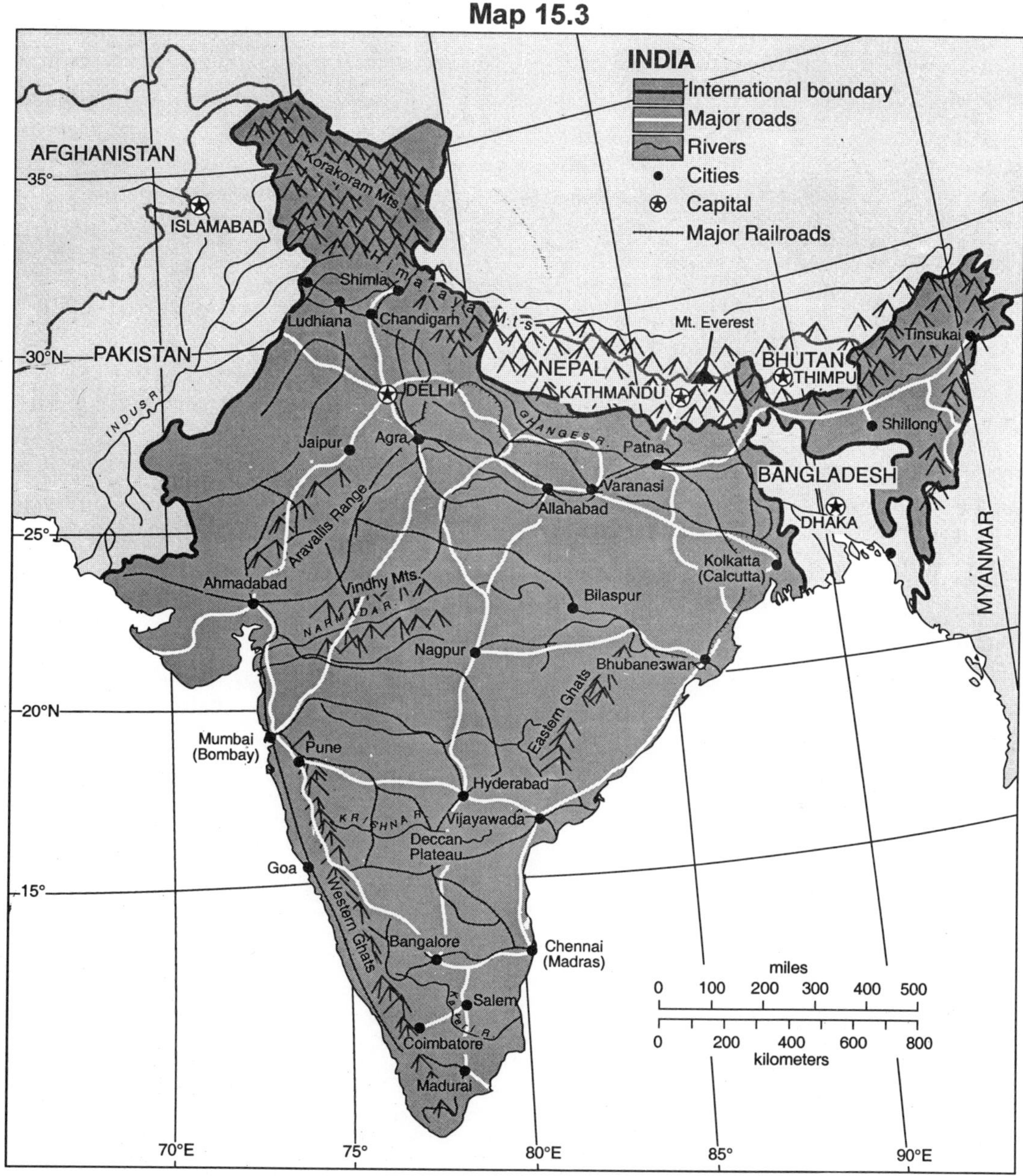
INDIA
International boundary
Major roads
Rivers
Cities
Capital
Major Railroads
AFGHANISTAN
ISLAMABAD
PAKISTAN
Korakoram Mts
Himalaya Mts
Shimla
Ludhiana
Chandigarh
DELHI
NEPAL
KATHMANDU
Mt. Everest
BHUTAN
THIMPU
Tinsukai
Shillong
BANGLADESH
DHAKA
MYANMAR
INDUS R.
GHANGES R.
Jaipur
Agra
Patna
Varanasi
Allahabad
Aravallis Range
Kolkatta (Calcutta)
Ahmadabad
Vindhy Mts.
NARMADA R.
Bilaspur
Nagpur
Bhubaneswar
Eastern Ghats
Mumbai (Bombay)
Pune
Hyderabad
KRISHNA R.
Vijayawada
Deccan Plateau
Goa
Western Ghats
Bangalore
Chennai (Madras)
Salem
Kaveri R.
Coimbatore
Madurai
miles
0 100 200 300 400 500
0 200 400 600 800
kilometers
35°
30°N
25°
20°N
15°
70°E
75°
80°E
85°
90°E

Examples of this category are Jaipur, Poona, Nagpur, Hyderabad, Kabul, Kathmandu, Delhi and Lahore. For the most part, smaller cities and towns are strongholds of cultural orthodoxy; by contrast, larger cities are centers of cultural and economic advancement. Larger cities are the foci of political and social power and are likely to play an increasingly influential role in development of these countries.

The basic element in the design of cities is the sharply differentiated old and new urban components. The old sections generally consist of congested settlements, tortuous and narrow alleys, generally caste and religion-based neighborhoods, a mix of commercial-residential functions in the central business section (*bazar*); and contain a melange of transport modes. In sharp contrast to the chaotic world of the old cities are the modern, and well laid-out recent sections. Built mainly during the 20th century to house British officers, government personnel, and the growing elite community of businessmen, these sections present an aspect of comparative spaciousness, comfort, greenery, and peace. Here, wholesale, retail, and recreation areas tend to be clearly separated from the residential neighborhoods. Built in a distinctive colonial style of architecture, they are commonly separated from the old cities by open spaces. Since the independence of the nations of South Asia there have sprung up, especially around the larger metropolises, vast suburban developments. These new suburban dormitory settlements depend on a very inadequate mass transit system to transport workers to their jobs in the urban cores. A direct consequence of the rapidly growing urban areas is the chronic and ever-increasing problem of over-crowding, housing shortages, and high rents; and the accompanying racketeering and loan sharking. So acute is the shortage of housing that in the larger metropolises, thousands of people must sleep in the streets. A related phenomenon is the mushrooming growth of shantytowns, tin-roof structures, often juxtaposed with the well-developed modern and more open sections of the cities, but lacking in city services. If urbanization has been an instrument of economic, social and political progress, it has also led to serious socioeconomic problems.

Economic Activity: Agriculture

Throughout history, agriculture has played a central role in the economies of South Asian countries. Over two-thirds of the population still depends on it for a living, and it accounts for nearly one-third of the region's exports. Such major problems as food shortages, rural unemployment, and social, economic and political discontent are directly related to the agricultural systems.

In statistical terms, the region occupies a major position in the world in several agricultural commodities, with India logically contributing a large share. India is a major producer of foodgrains in the world; third in wheat production, second in rice and millets; and first in the production of tea, jute (used as a fiber), pluses (beans, peas), and peanuts; and holds an important position in the production of such commercial crops as spices, bananas, tobacco, oil seeds, and cotton. It also has the world's largest cattle population. Pakistan is another important producer of wheat, cotton, and millets. Bangladesh is significant for jute and rice; while Sri Lanka is a major producer of several commercial crops, notably tea, coconuts and rubber.

Map 15.4

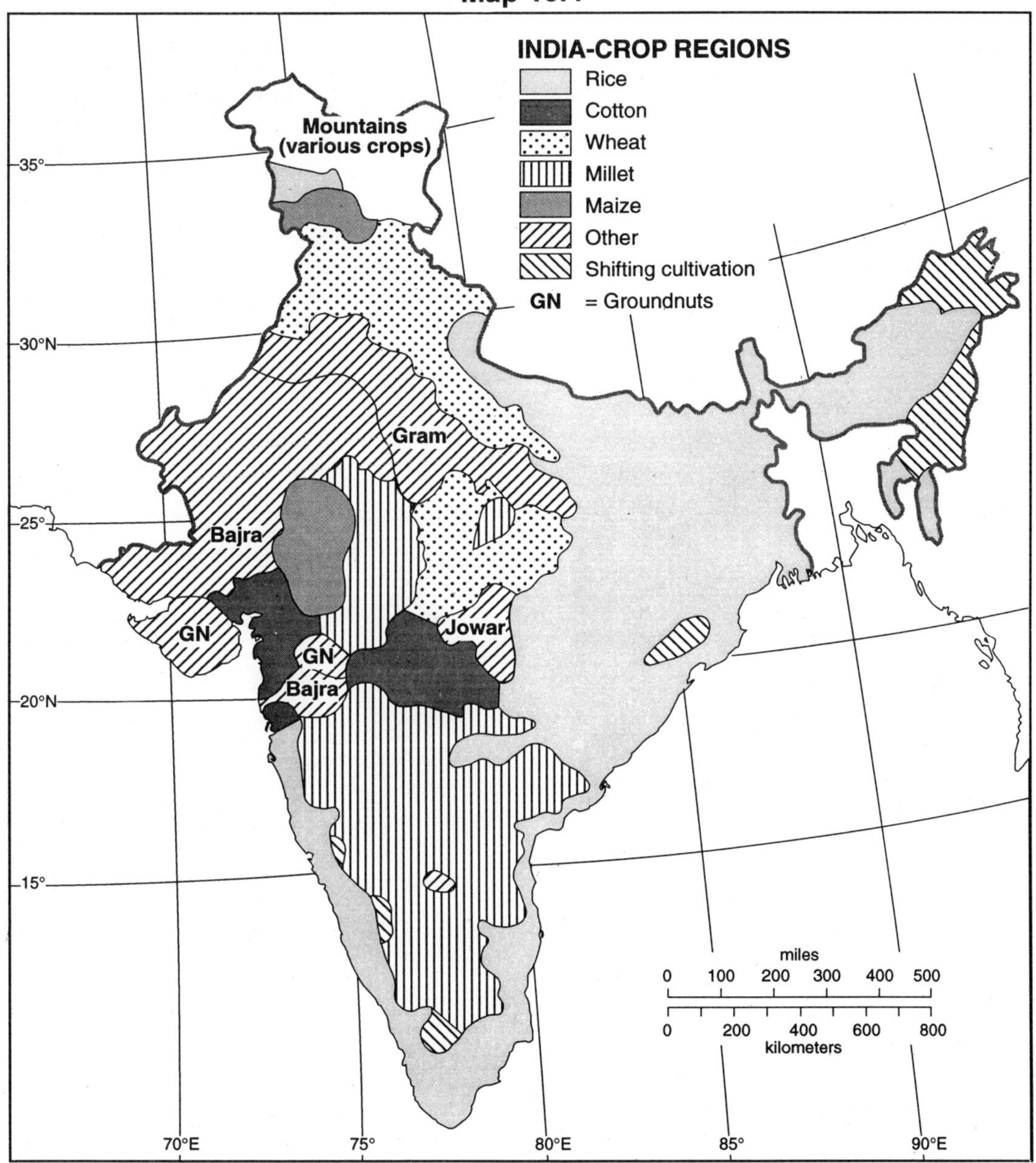
INDIA-CROP REGIONS
Rice
Cotton
Wheat
Millet
Maize
Other
Shifting cultivation
GN = Groundnuts
Mountains (various crops)
Gram
Bajra
GN
GN
Bajra
Jowar
35°
30°N
25°
20°N
15°
70°E
75°
80°E
85°
90°E
miles
0 100 200 300 400 500
0 200 400 600 800
kilometers

In 1966, first India and later Pakistan introduced the newly developed high-yielding varieties of wheat and rice. Dwarf strains of these crops were introduced in areas of favorable environment in parts of the northern plains of India and Pakistan. Initial results were truly dramatic, production registering growth rates of over 20 percent annually between 1966 and 1971. These measures ushered in the so-called "Green Revolution." By the early 1970s more than a third of the wheat planted and half the wheat harvested was from the new varieties. During the next decade three-fourths of the wheat and nearly half the rice produced in India were of high-yielding varieties, and nearly 80 percent of these were grown in Punjab, Haryana, and west Uttar Pradesh in the northern Indian plain and in the state of Tamil Nadu in south India.

Grain crops almost always form the basis of national diets and they support a vast majority of the people. The favored grains have been rice and wheat, with sorghum, millets, barley and maize as secondary food crops. Rice is the leading crop in all countries except in Pakistan, where wheat enjoys by far the most commanding position. Although rice production there substantially increased during the 1980s, the Ganga delta and the coastal lowlands with high temperatures and ample rainfall 40 to 60 inches (1,000-1,500 millimeters) during the growing season, and conditions of clayey loam soils form an ideal environment for rice growth. In areas where rainfall is less than 1,500 millimeters annually, rice is raised with the help of irrigation, as in parts of peninsular India, the middle Ganga plain and Pakistan (where it assumes a secondary position).

In the late 1960s, "the Green Revolution" took hold in northwest India and the adjacent areas of Pakistan, as also in lesser measure in Sri Lanka. Under government patronage, the progressive farmers of this region took to newer, high-yielding varieties of grain crops introduced by the Green Revolution. As a result, production of these crops increased substantially. Bangladesh and Tamil Nadu State in India also benefited from these new strains of grain crops. Improvements in grain production and undoubtedly been also facilitated by the introduction of scientific farming practices and land reforms that had become necessary with the adoption of these strains of crops. However, crop yields in South Asian nations still remain among the lowest in the world.

Sorghum and millets (*jowar*, *bajra*, *ragi*), are coarser grains, and form the poor man's diet; they are extensively grown in the drier parts of the subcontinent and in areas of poor soil. Among other staple crops *pulses* (peas and beans of various kinds) and peanuts are raised extensively everywhere and are the chief sources of protein. Most *pulses* are eaten as *dal* (puree form) complementing bread, rice or curry dishes in the cuisine of these nations.

Among the specialized cash crops, sugarcane, peanuts, cotton, oil seeds, and jute are grown in India, Bangladesh and Pakistan. Tea, coffee, and rubber are raised on plantations in India and Sri Lanka, and tea in Bangladesh. South Asian countries are the major exporters of several of these cash crops. In perspective, it may be pointed out that India has retained its age-old pre-eminence in the world in several crops. It is the world's leading producer of tea, cane sugar, millets, peanuts, and jute and ranks second in cotton and tobacco.

Mineral and Energy Resources

Among South Asian nations, only India's mineral resources are of considerable importance by world standards. Ranking fairly high among the producers of iron ore, coal, mica, and manganese, it also has extensive reserves of chromite, bauxite, and limestone. A major deficiency is petroleum, which must be imported in substantial quantity in view of rising demand, created in large measure by a recent increase in the production of chemical fertilizers. Current production of oil is limited to small fields in Assam, and Gujarat and to a recently discovered offshore fields, named Bombay High, along the western coast, with seemingly sizable potential, is likely to ease the situation. Offshore exploration near both the Ganga and Indus deltas, among other areas is currently being vigorously pursued. Nevertheless, India is far better endowed in minerals than most industrial countries of Europe, and has resources vastly superior to those of Japan.

Elsewhere, there is general deficiency of mineral wealth, with the exception of some natural gas, petroleum, iron ore, and low-grade coal in Pakistan. The search for minerals in Sri Lanka and Bangladesh is understandably urgent; future economic development in South Asian nations will to some degree depend on the discovery and exploration of energy and mineral resources.

Within India, minerals are distributed in three well-defined areas: (a) the northeastern part of the Deccan Plateau, which is especially well-endowed with coal, iron ore, manganese, bauxite, and mica; (b) the south Indian peninsular section, which is important for mica, chromite, and iron ore; and (c) eastern Rajasthan and northern Gujarat, which contains considerable amounts of bauxite, limestone and iron ore.

In general, energy resources (excepting coal and in some measure hydroelectric power in India) remain largely underdeveloped in these nations. Potentially, several areas have good prospects for the development of hydroelectric power: the submontane Himalayan belt in India and Nepal; and the central mountainous region of Sri Lanka. Development is hampered as much by lack of government priorities and funding, as by the fluctuating climate regimes, inaccessibility of physically suitable locations, and difficulties in construction. The generation of electricity from nuclear energy is still in its initial stages in India and Pakistan. The few existing plants serve the industrial centers of Mumbai, Channai, and Bangalore in India. Since the successful explosion of an atomic device in 1974, India has been working on the extension of nuclear energy production for peaceful purposes. Nuclear technology is well advanced in Pakistan, as well, and quite likely it has already capabilities of building nuclear weapons. India's reserves of radioactive thorium ore and monazite sands appear fairly extensive, but the limiting factor in power generation is the prohibitive costs involved in importing or developing a breeder reactor to provide fuel for the nuclear power plants.

Manufacturing Against the background of underdeveloped mineral and energy resources manufacturing remained at a low level during the colonial era in the subcontinent. There were only two large modern plants, all in India, devoted entirely to the production of iron and steel items (nearly one million ton of crude steel and another million ton of pig iron). Industrialization in the post-independence period has expanded more rapidly in India and to a lesser extent in Pakistan than elsewhere. India now claims a rank among the top ten industrial nations of the world in total industrial output, yet only a little over 10 percent of the 375 million persons in its labor force are employed in large scale manufacturing establishments, (employing 50 or more persons and using mechanical power). Even lower percentages characterize the other nations of South Asia. Over half of the persons engaged in processing or manufacturing activity are in the village-based "household industries," the remainder are in urban-based but small-scale, non-factory handicraft industries, producing locally-used materials such as textiles and tools. These were encouraged and subsidized by local governments particularly in India in the first two decades after independence. The focus was then shifted to heavy industry within the framework of national planning.

Unlike India, heavy industry did not greatly expand in Pakistan or Bangladesh largely as a consequence of their limited resource-base. In all of South Asia, however, cottage industries producing handicraft items like superb Kashmiri shawls, and woodwork, Rajasthani metal ware, Benares Saris, Dhaka Muslim cloth and Pakistani rugs, prized at home and abroad for their truly outstanding workmanship, continued to grow; but face an uncertain future as most small-scale enterprises will ultimately have to give way before the inexorable trend toward more efficient, low-cost factory production.

In India and Pakistan small factories using machinery and modernized equipment are now widespread. Their growth occurred largely in the past two decades in the emerging industrial corridors along the major railroads; they now produce such durable items as electric fans, sewing machines, bicycles and tools. These modest operations continue to grow as electricity is introduced into new areas. In view of their employment potential and low capital requirements, state governments have encouraged such growth by financial support. As a result, small-scale production of soaps, paints, fans, smaller consumer goods, sewing machines, and so forth has substantially increased.

Expansion of heavy industry has been more rapid than for small-scale enterprises because of significant federal support. Air and rail transport, and certain types of manufacturing such as military hardware and nuclear energy equipment, are under direct government control. A large portion of the iron and steel, machinery, mining, and chemical industries are also in the public sector.

In South Asia, India is the most advanced country in industry and manufacturing. The first steel mill was constructed at Jamshedpur in a mineral-rich area of Chota Nagpur early in the present century and went into operation in 1912. The nearby rich iron ore, limestone, and manganese, and the Damodar Valley coalfields serve the industry. Government has favored iron and steel investments and protects domestic production against foreign imports. The Soviet Union, United Kingdom, and West Germany and the United States have all assisted India in the development of its iron and steel industry. Today there are 10 major iron and steel mills, 8 of them government-owned and operated, with a total production capacity of over 12 million tons per year. The engineering, metallurgical, chemical, and electronic-electrical industries have also expanded substantially in recent years. Cement and fertilizer production is being given special governmental support. The stress on agricultural productivity has resulted in the construction of many fertilizer plants, though output lags behind the increased demand generated by the Green Revolution. Production of automobiles, aircraft, and electrical goods (both in the private and public sectors) are protected by heavy tariffs and other types of government aid against foreign competition. Increasingly, India is changing from a total importer to a competitive exporter. The range of her major industrial exports has been broadening from such commodities as cotton cloth and bicycles to textile machinery, machine tools and locomotives.

India and Pakistan have developed textile industry, which is widespread but especially concentrated in Mumbai and Ahmedabad in India, and Karachi in Pakistan. Dhaka in Bangladesh is also an important center. The metropolitan region of Kolkata and several cities of Bangladesh are the centers of jute manufacturing. Woolens and silks are also produced on a large scale in both India and Pakistan. In general, the manufactures of Pakistan, Bangladesh, and Sri Lanka are much less diversified than those of India. Production of cement, and fertilizers; and food and leather processing activities are steadily increasing in India and Pakistan. Jute products, rice milling, and petrochemicals, and a steel plant in Chittagoing in Bangladesh add somewhat to its otherwise small industrial base. The traditional dependence of Sri Lanka on plantation crops (tea, rubber, coconuts) is now steadily being shifted to the manufacturing items based on these cash crops. Although the share of industrial items increased from one-sixth to one-fourth of her exports during the 1970s, manufacturing still remains at a low level in Sri Lanka.

During colonial rule Indian industry was highly localized in a few cities: the major ports of Bombay, Calcutta, and Madras (now Mumbai, Kolkata, and Chennai), and the inland cities of Ahmedabad, Bangalore, and Kanpur. The commercial legacy of over-concentration in a few commercial centers is now steadily changing with respect to light consumer-oriented industries. New industrial corridors are emerging along the major railroad lines between the chief metropolitan areas linking the port cities with the inland agricultural hinterlands. Primary plants for industrial goods and industrial raw materials are now widely dispersed. The South Asian governments, in general, have pursued a policy of industrial decentralization, largely to placate regional demands, but also to reduce congestion in the older industrial centers. This is particularly true in India, where steel mills are dispersed through several states: West Bengal, Bihar, Orissa, Madhya Pradesh, Andhra Pradesh, and Karnataka.

Although South Asia's expanding industrialization holds promise for considerably improved living standards at some future date, the present situation of the mass of factory workers is not a happy one. A large proportion of these workers, especially in the larger cities, is composed of male immigrants from the countryside who have flocked to such industries as cotton and jute milling, leaving their wives and children in the village and sending them regular remittances. These factory workers often have to cope with the high cost of urban living, and lead lives of drudgery, deprivation, and loneliness, while their families in the countryside subsist in scarcely less depressing conditions on the meager sums sent by family members working in distant cities.

Further Readings

Cassen, R.H. *India: Population, Economy, Society*. London, 1980.

Muthiah, S. (ed.) *A Social and Economic Atlas of India*. New Delhi, 1987.

Wolpert, S. *India*. Berkeley, 1991.

16

Nations and Regions of South Asia

INDIA

India has played a vital role in shaping the lives and fortunes of all the people inhibiting South Asia. In its physical dimensions and characteristics, it is truly subcontinental. It is a collection of lands and climates, of diverse races and creeds, and of multitudinous contrasts. Next to China, it is the second most populous nation in the world, and at the current rate of increase will contain over one billion people by the next decade, and will surpass that of China by 2025.

The richness and complexity of its physical and cultural make-up can be best understood by an exploration of its major divisions.

The Northern Lowlands is a vast alluvial plain composed mostly of the Ganga River system, and partly of the Indus River system. It stretches for nearly 920 miles (1,480 km) from Pakistan to Bangladesh and covers over 150,000 sq miles (388,500 sq km) of territory. This is the "heartland" of Indian civilization, the home of glorious kingdoms of the past and of great cities. By religion and tradition, this vast plain has a mystical quality about it. Many of India's major cities, rich and famous in history, centers of immense importance in religion and politics, representing various phases of Indian thought and development, line riverbanks of the River Ganga and its tributaries. These include Delhi, Agra, Kanpur, Allahabad, Varanasi, Patna, and Kolkata.

In general, it is a featureless, alluvial lowland without sharply defined internal subdivisions, where climate and soils are not entirely homogeneous. The eastern part of the plain is wettest receiving 40 to 60 inches (1,000 to 1,500 mm) of annual rainfall. Precipitation decreases to the west; near the Pakistan border it is only 20 inches (500 mm) a year. The deficiency of rain in the middle Ganga basin and in areas west of it in Punjab is met mostly by irrigation canals and only partly by wells and tubewells. Considerable extremes of temperature exist with ranges increasing toward the west. Soils are fertile and deep almost everywhere, and generally two crops are raised. Between two-thirds and three-fourths of the land is arable. Almost all grain crops are grown: rice more towards the east and wheat toward the west. Legumes, cotton, and sugarcane are also important. In the west (Punjab and the mid-Ganga basin) winter rains from cyclones help the growth of winter season (*rabi*) crops, principally wheat, sorghum, millets, sorghums and short staple cotton. It is here that the benefits of the Green Revolution have been most noticeable and crop yields have markedly improved.

Toward the east, in Bihar and West Bengal, rainfall is heavy (40-80 inches or

1,000-2,000 mm annually). Here, rice and jute become more important, although many other crops are also cultivated. Crop yields, in general, are low. The farmer has for centuries been working with primitive tools, on tiny holdings, and has remained in debt. It is a land of poverty, disease, and hunger. It is also a very densely populated part of India, averaging 800 persons per sq mile (310 persons per sq km) increasing to over 2,000 per sq mile (770 persons per sq km) toward the Lower Ganga Basin, India's major port.

In the heart of the Indo-Gangetic plains lies the historic city of **Delhi**, which is the national capital of the country and claims a population approaching 10 million inhabitants for its metropolitan area. The city had strategic "gateway" location for invaders from the west, yet close to the Himalayas, the peninsular India, and Rajasthan desert. For centuries it remained the administrative center for large empires. The British moved the capital from Calcutta to Delhi in 1911, and built a planned city of New Delhi as an adjunct to old Delhi. Since partitioning of the subcontinent in 1947, it grew rapidly, spurred primarily by the economic stimulus provided by the enterprising refugee population from areas that are now in Pakistan, and acquired commercial and industrial functions in addition to its administrative role. Important industries located in its metropolitan area are: machine tools, auto parts, plastic items, bicycles, and a variety of consumer goods.

Between Delhi and Kolkata are many other cities with populations ranging between one-half to one million, including the ancient city of **Varanasi** (formerly Banaras), the holiest of Hindu cities, the industrial city of **Kanpur**, and the historical and commercial cities of **Allahabad**, **Lucknow** and **Patna**. West of Delhi in Punjab lies **Amritsar**, the holy city of the Sikhs, near the Pakistani border, the scene of recent confrontation between Indian government and Sikh separatists. **Kolkata** (population: 11 million) and the second largest city and port of India and is located in the Ganga Delta region. It owes its growth and importance to its easy accessibility to the production of a large hinterland of the Ganga Plain, jute cultivation of Bengal, tea plantations of Assam, and heavy mining and industry of the close-by mineral belt. It is India's major manufacturing center for jute textiles, metal and chemical products, and paper.

The mineral resources of the region are slight, and are mostly located in the Damodar Valley, which contains India's major coalfields and some manganese. Oil production in Assam in the Brahmaputra Valley of Assam is now decreasing; it was never sufficient to meet India's needs. However, the nearby Chota Nagpur Plateau, close to Kolkata's industrial belt, is the most mineralized area (coal, iron ore, copper, mica) in the country.

To the southwest of the Northern Lowlands along Pakistan border lies the Great Indian Desert or Thar Desert. An area of scarce and unreliable rainfall, it was, before independence, ruled by several princely houses, and therefore contains several historic cities such as **Ajmer**, **Bikaner**, **Jaisalmer**, and **Jodhpur**. Jaipur, a planned city of 2.5 million inhabitants is the major administrative, cultural, and commercial center.

Central Highlands and Plateaus It is a physically and culturally transitional zone between Indo-Aryan North and Dravidian South in Central India, and is an upland region of highlands and plateaus east of the Aravalli Hills. Linguistic affiliation to the Indo-European tongues is more pronounced. **Bhopal, Gwalior, Nagpur** and **Pune**—all are "millionaire" cities and were formerly seats of powerful princely houses.

The Deccan Peninsula occupies an area of roughly half a million sq miles (1.3 million sq km) and containing nearly 40 percent of India's population. Most of peninsular India lies south of the Tropic of Cancer. It is a tableland of old, hard rocks, rolling hills, and interior basins, flanked by mountains and coastal plains. The coastal plain along the Arabian Sea is narrow, rainy, and tropical. Rising steeply above it are the mountainous western Ghats, forming an escarpment of 3,000-5,000 feet (915-1,525 meters). The central part of the Deccan Plateau gradually dips eastward, declining in elevation from the western Ghats to the coastal plains along the Bay of Bengal. The peninsular rivers Mahanadi, Godavari, Krishna and Kaveri flow eastwards across the peninsula through the Eastern Ghats, and into the Bay of Bengal. Among major rivers only the Tapti and Narmada in the north flow in the Arabian Sea, following structural troughs. The eastern coastal plain is much wider and drier than the western plains particularly near the delta of the Kaveri River where it is over 150 miles (240 km) wide. The northwestern portion of Deccan, covering an area of over 95,000 sq miles (240,000 sq km) is composed of thick black-colored basaltic lava flows known as the Black regur soils. This is a fertile tableland of rich soils with excellent texture for the retention of summer moisture for cultivation. Cotton, sorghum (*jowar*), linseed, and legumes are the principle crops.

Most of Deccan is in the "rainshadow" of the western Ghats (which receive heavy rainfall from the moisture-laden, southwest summer monsoon winds). In this area annual rainfall ranges from 20-35 inches (500-875 millimeters), and falls mostly in summer. Irrigation from tanks and ponds is widespread. Only the southern section of the eastern coast gets winter rainfall from the northeast retreating monsoons which pick up moisture from the Bay of Bengal. Rice is cultivated all along the eastern coast in the river deltas, as well as on the west coast. In southern parts of the western coast (the Malabar) where rainfall is particularly heavy (above 80 inches or 2,000 millimeters annually) commercial crops of coconut, rubber, pepper and cashew nuts, are cultivated in addition to rice. Major minerals are coal, iron ore (with a very rich ore content), mica manganese, copper, chromite and bauxite. India's main iron and steel centers: **Bhilai**, **Raurkela**, and **Durgapur**, make use of these minerals. In general, population densities are high in the coastal lowlands, reaching close to 1,500 persons per sq mile (575 persons per sq km). Elsewhere these are much lower than the Indian average, ranging between 300-500 persons a sq mile (110-180 persons per sq km). The two large ports of Mumbai and Channai, located on west and east coasts respectively, were developed by the British as trading ports. **Bombay** (now renamed Mumbai) is the largest Indian city with a population of nearly 12 million of its metropolitan area, has a natural harbor and is closest to the oil-producing Persian Gulf and the Bombay High offshore oilfield area in the Arabian Sea. It is a major manufacturing center producing textiles, chemicals, pharmaceuticals and consumer goods. **Madras** (now renamed Chennai) is India's fourth largest city (pop. over 5 million), and has a large trading area in Southern Deccan. It is a provincial capital, a port and an industrial and educational center. **Ahmedabad** (population: 3 million) to the north of Mumbai is another major textile manufacturing center. Inland **Bangalore** (4.2 million) is noteworthy for its engineering, automobile, textile, machine tools, aircraft, pharmaceutical, and computer-based industries and is fast emerging as an example of modern industrial India, and rightly called the "Silicon" capital of the country.

Himalayan Highlands To the north of Ganga-Indus plains of northern India lie the Himalaya mountains that extend from northern-northwestern part of Pakistan, to include most of the nations of Nepal and Bhutan, covering nearly 135,000 sq miles (350,000 sq km) of territory and are inhabited by some 20 million people. Pakistan controls about one-fourth of the territory in the northern and western parts of the state of Jammu and Kashmir, and disputes with India the latter's control of the remaining portions of the state. The structure and topography of the Himalaya are quite complex. In general, there are three parallel zones, known as the Outer, Lesser and Great Himalaya, each attaining higher altitudes successively inland from the plains. The Outer ranges are closest to the plains and reach an altitude of 2,500 to 4,000 feet (760 to 1,200 meters), the Lesser Himalaya are in the middle and rise 7,000 to 15,000 feet (2,100 to 4,500 meters) elevation; whereas the Great Himalayan ranges are the loftiest with an average crest of 20,000 feet (6,000 meters) and are the farthest from the plains. Several others attain elevations of over 22,000 feet (6,600 meters). Tucked in the ranges are the alluvial valleys, of which the Vale of Kashmir, at an altitude of about 5,000 to 6,000 feet (1,520 meters), is a good example.

The Himalayas screen off India's plains, and run for 1,500 miles (2,400 km) long and 150-250 miles (240-400 km) wide. They have acted throughout history as a physical and cultural barrier of tremendous proportions between India and China. Although the Chinese invasion of 1962 did dispel notions of the Himalayas' invulnerability to some extent, the climatic and cultural isolation which this great mountain mass imposes on India cannot be ignored. These mountains are so high and their ridges so unbroken that the cold dry winds of central Asia do not reach India nor do the warm, moisture-laden summer winds of South Asia cross over the mountains. Thus, north India's climate is much milder in winter and much wetter in summer than it would be without these mountains. Railroads from the Indian side penetrate only the foothills (exceptions are the railroads to Shimla, constructed by the British for the establishment of a "summer" capital for the country and to Darjeeling, another eastern hill station). The few motor roads that do exist in the Himalayan interior were mostly constructed since independence.

Climate regimes and vegetation in the Himalaya follow elevation and exposure. The vertical sequence is so sharp that within 100 miles/160 km) of the plains, from the lower slopes 2,000-3,000 feet in altitude (600-900 meters) to 12,000 feet (3,600 meters) climates vary from humid subtropical, temperate or humid microthermal to cold tundra. This may be roughly compared to the climatic transitions from north India to the Arctic coast of Siberia. The south facing mountainsides are warmer and rainier. The Tibetan Plateau lying in the rainshadow of the Great Himalaya ranges remains dry.

Most people are settled in the Lesser Himalaya and the river valleys between the mountain ranges. Many live in isolated and remote valleys. Almost all the tillable area is cultivated by terracing. Among the agriculturally rich interior valleys is the Valley of Kashmir, where rice, fruits, vegetables, and flowers are grown. Horticultural advances are also being made in Himachal Pradesh (one of India's mountainous states). **Shimla** is its capital city, and is a railhead; and the main collecting and distributing center for this produce. It is also one of the several summer resort towns in the Himalaya. Most summer resorts, known as hill stations, lie between 5,000 and 8,000 feet (1,520-2,400 meters) altitudes, and were built by the British to escape the fierce heat and humidity of the plains. Other well-known hill stations are **Darjeeling**, **Nainital**, **Mussourie**, and **Srinagar** (the capital of the state of Jammu and Kashmir). Not all hill stations were established during British times. Thus, Srinagar in the Vale of Kashmir was a favorite retreat of the Mughal emperors from the 15th to 18th centuries. Most hill stations inherited the paraphernalia left by the British----the old hotels, boarding schools, bungalows, race courses, polo fields, botanical gardens, and hiking trails. These now serve as the summer haunts of wealthy Indians and foreign diplomats.

Cultural Patterns The Indian subcontinent has acted as a cul-de-sac absorbing successive migratory waves of peoples. Consequently they got intermingled to such an extent that racially distinct divisions are hard to establish. Therefore, language, and religion rather than ethnic origin, are the primary cultural distinctions between peoples.

Although George Grierson (1851-1941), a British authority, identified 723 main languages and dialects in the subcontinent, most of these were spoken by very few people in the late 20th century. The Indian government recognizes 17 regional languages. In 1981 Hindi, and related dialects claimed nearly 45 percent of population as speakers; Telugu, Bengali Marathi (with 7 to 8.5 percent speakers each); Tamil (6.6 percent speakers); Urdu (5.2 percent speakers); Gujarati (nearly 5 percent speakers); Kannada, Malayalam, and Oriya with 3 to 4 percent speakers each; and Assamese, Kashmiri, Punjabi, Sindhi, and Sauskint with less than 2-3 percent speakers each. Sanskrit of the Indo-European family was spoken by less than 3,000 people (in 1981), although it is credited with the status of being the mother of most of the north Indian tongues. It is the language of ancient Indian epics, religious thought and philosophy, and was accorded the status of one of the languages recognized in the Constitution. The Mughal rule in India (16th to 19th centuries) promoted the use of a colloquial tongue, Hindustani, a hybrid, although the official language was Persian. From it two literary languages---Hindi (showing a strong Sanskrit influence and Urdu (with a Persianized vocabulary)----emerged. Hindi, Bengali, Marathi, Gujarati, Oriya, Punjabi and Assamese are spoken in northern and central India, and belong to the Indo-Aryan branch of the Indo-European family of languages.

Telugu, Tamil, Kanuada, and Malayalam are distributed in southern India and belong to the Dravidian family of languages. Of these, Tamil antedates the north Indian Indo-Aryan languages and possesses a rich cultural history. Besides these two language families are two other families— Austro-Asiatic family in the northeast parts of the country inhabited primarily by several tribal groups; and the Sino-Tibetan family along the northern frontier region.

Hindi was declared the official language of the country at the time of independence, and English, not listed among the 18 official languages, was accorded a special status of an additional language (in 1950) initially for a period of 15 years, but later allowed to continue as a second language for the courts, the Parliament and as a lingua fauca and a link language between the federal and state governments. India's language controversy has been discussed in some details in an earlier chapter on South Asia.

India is the birthplace of many religions and has been the home of many others. The strength of the major religions is: Hinduism (professed by 81.3 percent of the population); Islam (by 12 percent) of a predominately Sunni component, Christianity by 2.3 percent divided equally between Catholics forming nearly 1.5 percent and Protestants, Sikhism by 2 percent, Buddhism by 0.8 percent, Jainism by 0.4 percent, and a Zoroastrianism claiming a few thousand adherents. (The political impact of this diversity is discussed in an earlier chapter on South Asia.)

Economic Activity India is a typically agricultural nation. Two-thirds of its laborforce is engaged in agriculture, but which counts for only one-third of the value of the country's domestic product. The proportion of those engaged since the beginning of the 20th century, although the area under cultivation has steadily increased and now occupies more than half of the nation's total area. In some parts of the Ganga Plain the proportion of cultivated to total area often exceeds 90 percent. The average farm size is small (five acres or two hectares) and due to increasing population pressure the size is declining despite several land reforms since independence. The introduction of several government-sponsored large-scale canal irrigation projects since independence (the canal irrigation projects had been previously set up by the British in the mid-19th century, but were greatly expanded after independence during the late 1960s and early 1970s), and the introduction of new hybrid, high-yielding varieties for wheat (and rice) which dramatically increased the production of foodgrains to over 200 million metric tons in recent years, enough to meet the country's current foodgrain requirements.

Rice is the leading grain, in terms of both area under cultivation, and total yield. It is grown largely in areas of ample rainfall in the fertile Ganga Plain, and the coastal lowlands, as well as some irrigated lands. Wheat ranks next in both area, and total yield, and is grown mainly on the fertile soils of northern and northwestern India (Punjab, Uttar Pradesh, Madhya Pradesh) in areas with 15 to 40 inches (375 to 1,000 mm) of average annual precipitation, often with irrigation. Other grain crops (jowar or sorghum), millets, and corn and pulses of several types, grown on relatively infertile soils and grown all over the country. In addition, a variety of vegetables and

Map 16.1

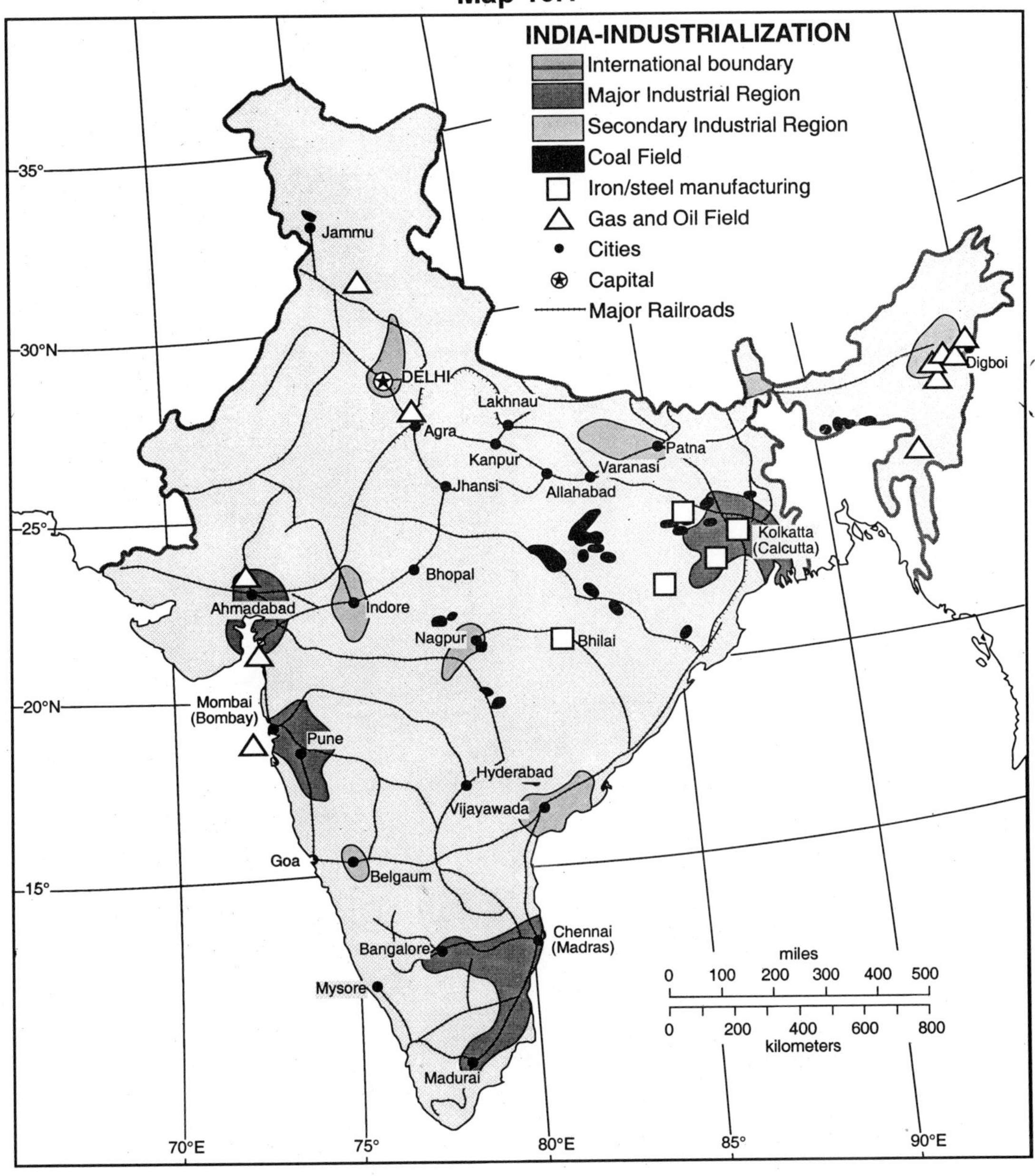
INDIA-INDUSTRIALIZATION
International boundary
Major Industrial Region
Secondary Industrial Region
Coal Field
Iron/steel manufacturing
Gas and Oil Field
Cities
Capital
Major Railroads
35°
30°N
25°
20°N
15°
70°E
75°
80°E
85°
90°E
Jammu
DELHI
Agra
Lakhnau
Kanpur
Jhansi
Allahabad
Varanasi
Patna
Digboi
Kolkatta
(Calcutta)
Bhopal
Indore
Ahmadabad
Nagpur
Bhilai
Mombai
(Bombay)
Pune
Hyderabad
Vijayawada
Goa
Belgaum
Chennai
(Madras)
Bangalore
Mysore
Madurai
miles
0 100 200 300 400 500
0 200 400 600 800
kilometers

fruits are grown in various areas. Crops like sugarcane, cotton, tea, and coffee are grown largely for export on plantations in Assam, West Bengal, Kerala, and Tamil Nadu. Foremost among the commercial, industrial crops is cotton, grown primarily in Maharashtra, Gujarat, and Punjab. After independence the production of jute, formerly grown primarily in areas now in Bangladesh, has increased and grown mainly in West Bengal, Assam and Bihar.

India has the largest bovine population of any country in the world, and livestock raising plays an important part in the agricultural economy. Cattle and buffaloes are used primarily as draft animals, but also as sources of meat (for Muslims, Christians, and Schedule Castes beef eating is not a taboo), and as a source of fertilizer, and as cooking fuel (from dried cow-dung cakes). Cow slaughter is illegal in many states, and many orthodox Hindus are vegetarians. Commercial forestry is not much developed, and restricted primarily to the West Ghats, western Himalayas, and the hilly areas of central India. Fishing is practiced along the coastline, and on the rivers. Marine fisheries, now increasingly becoming mechanized, account for two-thirds of production. Kerala is the leading fishing state. Fish products account for about 3 percent of India's exports.

Mining plays an insignificant role in the economy. Close to 2 percent of the population is engaged in it, accounting for about 2.5 percent of the domestic product. Minerals are relatively important in the country's manufacturing industry, as well as a source of modest export-revenue (polished gems and jewelry formed nearly one-sixth of all exports in 1992-93). Among minerals, petroleum leads in value, followed by coal. The production of petroleum fulfills nearly half of the country's requirements. Almost all of the petroleum is derived from offshore Bombay High, and from Gujarat and Assam. The most important coal fields are in the Damodar Valley in Bihar and West Bengal, which account for about half of the nation's output. Natural gas is of very little importance. Uranium in modest quantities is produced in Bihar. Ironore is an important mineral, mined in Madhya Pradesh, Bihar, Goa, Karnataka, and Orissa. Copper, manganese, mica, bauxite, zinc and lead are other minerals of some importance.

India's manufacturing industry is highly diversified; engages nearly one-tenth of the country's laborforce; and accounts for about one-fifth the nation's domestic product. A large number of people are working in millions of small-scale handicraft enterprises, such as spinning, weaving, pottery making, metalwork, and woodwork. In terms of total output and value added, however, mechanized large-scale industry predominates producing metals, machinery, fertilizers, heavy chemicals, and increasingly engaged in electrical and electronic products.

Given India's large resource base, however, the volume of foreign trade is small. There has been a chronic trade deficit in international trade. Among exports no particular item occupies a dominant position, although gems and jewelry (particularly for the Middle Eastern Market) have usually held the leading position, followed by ready-made garments, leather goods, agricultural products including plantation crops (such as tea), engineering products. Imports include mineral fuels, transport equipment, industrial chemicals and fertilizers. Mineral fuels usually account for one-fourth to one-fifth of all imports. The United States and the former Soviet Union have been the principal trading partners, while Germany and the

United Kingdom have also been important in India's foreign trade.

India has a large railroad track of 39,000 miles (62,950 km). The rail system is the fourth largest in the world after Russia, Japan, and China in terms of distance traveled each year by passengers. The length of the hard-surfaced road has increased nearly tenfold, from 66,000 miles (106,195 km in 1947) to over two million miles (3.5 million) since independence, but still represents half of the country's total road track.

Prospects India celebrated its half century of independence on August 15, 1997. The architects of modern India created a democracy in one of the world's poorest, most populous and socially the most fractured countries. The country contains over a billion people, a third of which fall below the poverty line, and more than half of the adults are illiterate. Its constitution recognizes 18 languages, but in fact there are 35 Indian languages each spoken by more than a million people. The society is divided by distinctions of religion, caste, and language, and prone to occasional violence.

Although the country has made steady material progress since independence and made substantial gains in manufacturing and in increasing agricultural productivity to the extent of eliminating foodgrain imports, population pressures continue to plague its development. It has been persuasively argued by many keen observers that much of the developing world, especially Asia, has recorded much higher rate of development than India has during the last few decades. India's growth is currently below 6 percent a year, but was under 4 percent a year between 1960 and 1990 far below that of other Asian countries such as China, Indonesia, Thailand, Taiwan and South Korea which averaged between 7 and 9 percent growth a year during the same period.

Population pressures, together with the burdensome foreign and economic policies, of course, constrained its economic development before 1990. Since the early 1990s, however, India has liberated itself from the stifling economic environment and chosen to adopt a path of economic liberalization that promises a higher growth, and a more active involvement in world market, calculated to serve its economic and trading interests. The adoption of foreign and economic policy also has begun to open up the rewarding relationship with the West. The country is now self-sufficient in agricultural production and considered to be the world's tenth most industrialized nation. The social, demographic, economic and political pressures will understandably continue to affect its development, but the economic outlook appears modestly promising.

India's large size, strategic location and democratic polity are likely to give it a prominent position in international affairs. Its growing industrial base, military strength and scientific and technical capacity add to its political stature.

PAKISTAN

As a poor offspring of the subcontinent, Pakistan at independence in 1947 inherited a disproportionally small share of its assets, and a larger share of its liabilities; one-fifth of the area and the highways, about one-sixth of the population and railroads,

Map 16.2

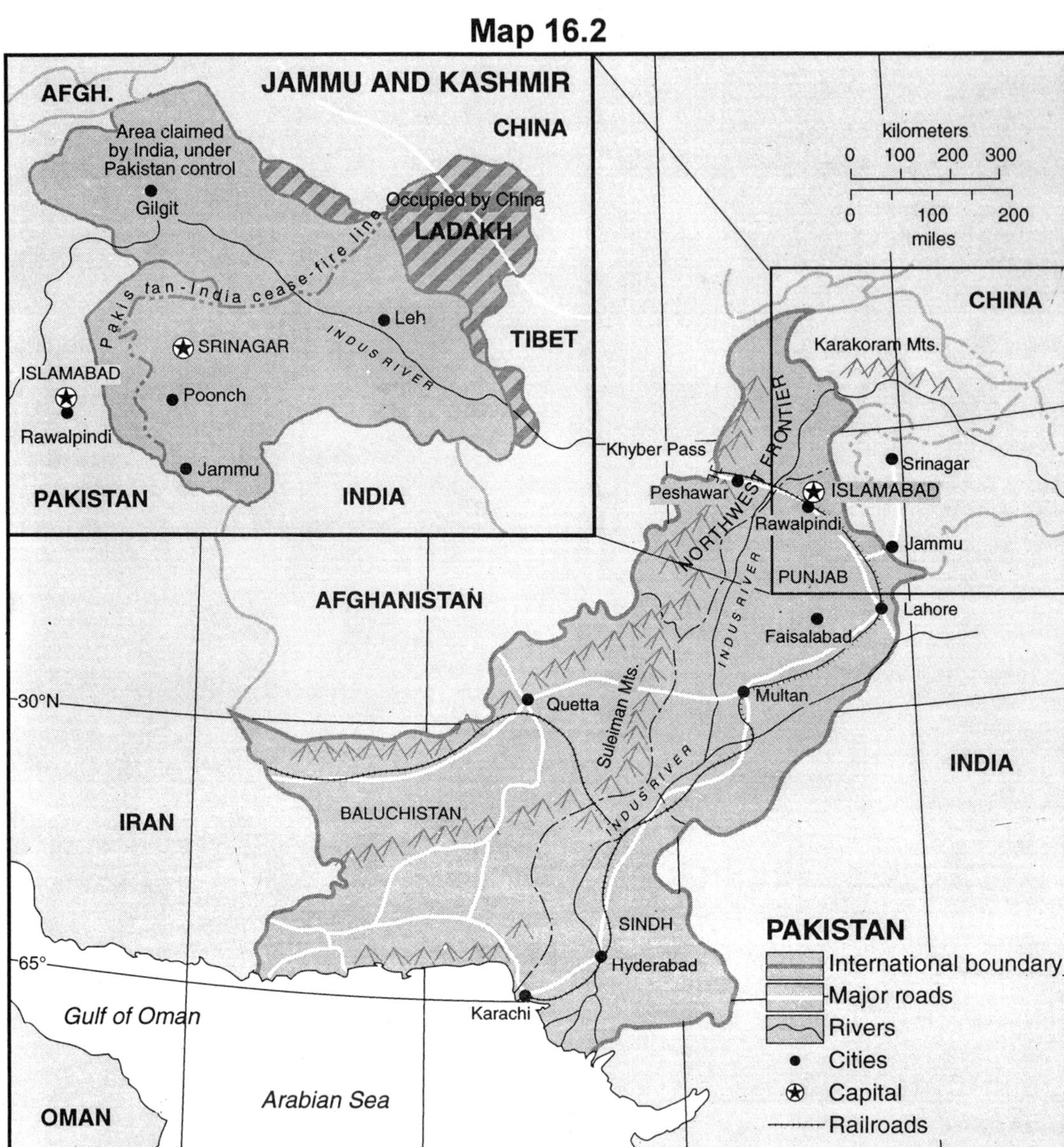
JAMMU AND KASHMIR
AFGH.
CHINA
Area claimed by India, under Pakistan control
Gilgit
Occupied by China
LADAKH
Pakistan-India cease-fire line
Leh
SRINAGAR
INDUS RIVER
TIBET
ISLAMABAD
Poonch
Rawalpindi
Jammu
PAKISTAN
INDIA
kilometers
0 100 200 300
0 100 200
miles
CHINA
Karakoram Mts.
Khyber Pass
NORTHWEST FRONTIER
Srinagar
Peshawar
ISLAMABAD
Rawalpindi
Jammu
PUNJAB
AFGHANISTAN
Lahore
Faisalabad
INDUS RIVER
30°N
Quetta
Suleiman Mts.
Multan
INDIA
IRAN
BALUCHISTAN
SINDH
PAKISTAN
65°
Hyderabad
International boundary
Major roads
Karachi
Gulf of Oman
Rivers
Cities
Capital
Arabian Sea
Railroads
OMAN
60°E
65°
70°E
75°

one-seventh of the cultivated land, a tenth of the army, a small minority of the educated middle class, relatively few of the shopkeepers, tradesmen and professional classes, hardly any industrial workers, very few of the known minerals, a small fraction of the industrial equipment, only a few mills for the cotton crop, and only one major port: Karachi.

The separation of the deficit area of Bangladesh in 1971 did strengthen the food situation of the truncated state and reduced its population problem, but at the same time diminished the revenue base of the export of jute and tea. The two wings of Pakistan were linked by long and costly air-routes across northern India or by the slow sea route of nearly 3,000 miles (4,827 km) around southern India and Sri Lanka. Bangladesh's secession lifted the burdens imposed by this awkward political geography from the shoulders of Pakistan. Pakistan's share of the subcontinent's resources of coal, iron, and other basic materials is insignificant. It has some gas and oil (at Sui fields). Its position in fibers is comparatively strong, and in foodgrains it has normally good surpluses of wheat.

By geography and by history, Pakistan and India are complementary. Political separation based on the concentrations of Muslims and non-Muslims defy several facts of physical and economic geography. The international borders show little regard for the established networks of railroads, irrigation canals and manufacturing. India could, if it wishes, gain control over the water of Pakistan's canals by diverting water to its own territory, thus virtually choking off Pakistan's agricultural base. (The two countries reached an agreement in 1960, the Indus Waters Treaty, regarding the distribution of waters.) This example of Pakistani dependence on India, which controls the headwaters of the tributaries of the Indus River, critical to Pakistan's irrigated agriculture, illustrates the complementary nature of the two countries.

Physical Features The core area of this essentially desert country is traversed by the Indus and its tributaries, forming an alluvial lowland except in the north and west where mountains and plateaus flank its frontiers with Afghanistan and Iran.

The country can be conveniently divided into four main topographic regions: the Northern and western Highlands, the Punjab-Sindh Plains, the Baluchistan Plateau, and the Thar Desert. The Northern and western Highlands cover much of northern and western parts of the country. It is a typical Himalayan region with high, rugged mountains, and several glaciers. The second highest peak in the world, K2, is 28,250 feet (8,611 meters) in elevation and lies in the northernmost part in the Korakoram Range. The highlands in the western part include the Sulaiman Range and other hills containing much lower, elevation, of 4,000 to 10,000 feet (1,219 to 3,049 meters) in elevation but much less dissected than the lofty and rugged northern mountains, and are breached by several historic passes including the more notable Khyber, Gomal and Bolan that are the historic passageways between the Indo-Pakistan subcontinent and Afghanistan.

The Punjab and Sindh plains occupy most of the eastern part of the county, and are traversed by the Indus River and its four tributaries---Jhelum, Chenab, Ravi, and Sutlej. The plains are formed by the alluvium of rivers, and very gradually rise from the Arabian Sea to nearly 1,000 feet

(305 meters) in the north, a distance of nearly 700 miles (1,126 km).

Economically, culturally and politically it is the country's heartland. The non-alluvial portions of the region are mainly in the northwest, where a dissected upland between the Jhelum and Indus rivers averages between 1,500 to 2,000 feet (457 to 610 meters) in the form of a range overlooking the Punjab plains. The Baluchistan Plateau is located in southwestern Pakistan. It is a dry, rocky, and mostly barren plateau, inhabited by Baluchi tribesmen, nearly 4,000 to 7,000 feet (1,219 to 2,134 meters) in elevation, that extends to the Makram coast, and is fringed by low, but rugged hills. The Thar Desert lies in southeastern Pakistan, and is an extension of the adjacent Great Indian Desért. Much of the desert region is a sandy wasteland, parts of which near the Indus River, have been made suitable for farming by the irrigation schemes.

Climatically, Pakistan has dry, hot summers in general, and cool winters, with very cold winters in the northernmost Himalayan regions. Most of the Punjab plains receives on the average 15 inches (375 mm) of rain annually, which progressively decreases to the west from the Indian summer monsoons. Much of Baluchistan and the Thar Desert receive on the average, less than 5 inches (127 mm) of rainfall in a year. Mean temperatures for the summer range from 66°F (18°C) in winter to 86°F (38°C) in summer. Winters are mostly dry, although a slight amount of rainfall is experienced in the plains region, brought by the cyclones emanating from the Mediterranean.

Cultural Patterns Although Islam is a cementing force in Pakistan (its official name is the Islamic Republic of Pakistan and 95 percent of the population is composed of Muslims), several divisive elements---linguistic and cultural---have been a serious threat to the central administration. The Pakistani people are a mixture of many ethnic groups, a result of the occupation of the region by groups passing through on their way to India. The major and politically the most dominant group is that of Punjabi-speakers who predominate the administration and form nearly 48 percent of the population. Pashtu and Sindhi are other major languages. Ethnic plurality is given rise to demands for territorial secession from Pakistan or the creation of autonomous states. Urdu, adopted as the national language, was essentially brought by the immigrants from India at the time of partition and grafted on the new nation, although it claims only 7.6 percent of the population. Sindhi-speakers (nearly 12 percent of the population) do not hold positions in government commensurable with their numerical strength, and have always resented Punjabi dominance. Among the Pashtu speaking tribesmen forming 13 percent of Pakistan's population and inhabiting mostly the Northwest Frontier Province bordering Afghanistan, a large indigenous group, inhabiting areas on both sides of the international border, many would like to establish an autonomous or independent state, to be called Pushtunistan. The Baluchi tribesmen, inhabiting plateaus southwest of the Northwest Frontier Province adjacent to Iran and Afghanistan and covering nearly one-third of the national territory, speak Baluchi language which, like Pushtu, is more related to Persian than to the Indo-European Punjabi speech (one group, the Brahui speakers belong to the Dravidian language family of Southern India). Some of these people have also pressed for the creation of a state that would incorporate parts of Iran and Afghanistan. The Urdu-speaking *Mujahir*, who migrated from India at the time of partition have been involved in bloody ethnic riots in the province of Sindhi-speaking province of Sindh, demanding a greater recognition for them. In the mountainous areas of Kashmir certain tribes speak markedly different languages, related to the Indo-European or Iranian language family. Thus, the entire western flank of Pakistan is a politically sensitive region of divided loyalties. Internally, Pakistan has had several military coups during the four decades of its existence. Predictably, regional demands will likely to grow stronger in the future.

Resources and Economy Agriculture, fishing and forestry are the major sources of livelihood, for the people. Close to one-half of the laborforce is engaged in these activities, but these account for less than one-quarter of the gross domestic production. During the British occupation of the Indian subcontinent in the late 19th and early 20th centuries, the land was transformed by canal irrigation in the Punjab Sindh plains. In addition to wheat, rice, cotton, and sugarcane which became the important crops raised on the "canal colonies," a variety of other crops like tobacco, oilseeds, chickpeas, millets (jowar, bajra), legumes, and vegetables, pulses were introduced.

During the late 1960s (along with India, and Sri Lanka) Pakistan experienced a "green revolution." Introduction of high-yielding varieties of rice and wheat, and extension of the existing irrigation system by the construction of tubewells and canals into the arid lands in the Punjab and Sind increased crop productions substantially. By the end of the 1970s Pakistan had achieved self-sufficiency in food and by 1980's it had become an important exporter of rice, particularly the fine *basmati* variety. Cotton production has also increased significantly since the 1960s.

Pakistan is, in general, poor in mineral and energy resources. Except for petroleum and gas, which provide four-fifths of all energy consumed, few other minerals are extracted. Shortage of energy is a major concern for the country's economic development. Hydroelectricity accounts for only 18 percent of the energy produced, obtained primarily from large dams in the northern mountainous parts of the country, principally at Tarbela and at Mangla Dam on the Jhelum Range. Natural gas reserves are more abundant, located mainly in Baluchistan near Sui and in the Potwar Plateau and in Sindh production is limited. There is one operating nuclear power station near Karachi.

Nearly one-eighth of the labor force is employed in manufacturing, although it accounts for a little less than one-fifth of the country's gross domestic product. Since independence Pakistan made special effort to develop its industries, since most of the manufacturing plants were located in areas that went to India at the time of partitioning of the subcontinent. Initially, most manufacturing activity consisted of the processing of agricultural raw materials, like establishment of cotton textile mills, and production of sugar, paper, tobacco and leather. After the 1960s greater efforts were devoted to the development of heavy industry, and production of intermediate and capital goods including chemicals, fertilizers, and light engineering equipment. In the 1970s and 1980s an integrated iron and steel mill was developed at Pimpri near Karachi. Most of the development was made possible through the use of international aid, particularly by the World Bank. **Karachi**, the largest city (5.3 million), and country's only major seaport, was the first capital of Pakistan. As a result of a huge influx of refugees from India at partition, the establishment of new industries, and expanded regional and international connections, it grew rapidly into a large metropolis, one of Asia's great "boom towns." It has attracted several manufacturing industries such as cotton textiles, chemicals, food products, cement, and fertilizers, because of the advantages for assembly of raw materials and for distribution of finished goods. However, an isolated location away from the nation's cultural and economic heartland prompted the government to move its administrative functions in the early 1970s to a new location at **Islamabad**, near the existing city of Rawalpindi where a planned city was constructed. Islamabad (population now nearing 250,000) is a well-designed, pleasant, and modern city, symbolic of Pakistani determination to build a better future. **Lahore** (3.1 million) an old center of Islamic culture also grew rapidly following independence as a result of increased regional and administrative functions. The city has numerous magnificent buildings, gardens, palaces, and mosques built during medieval times, and retains its importance as a premier headquarters of Islamic culture. Lahore and its metropolitan area gained enormously during the late 1960s and early 1970s and began to industrialize rapidly when Karachi's ethnic problems stemming from the Mujahirs---the Urdu-speaking refugees from India after the subcontinent's partition---who demanded greater recognition---became increasingly strident. In the early 1990s the process was accelerated and Lahore (and Punjab) became Karachi's competitor in industrial output. **Faisalabad** (1.2 million), an important regional commercial center in the heartland of the Punjab plains, owes its origin to the development of the "canal colonies" established in the early 20th century by the British. Since independence it has grown enormously and now contains manufacturing plants such as textile mills, and sugar processing.

Petroleum products, machinery, chemicals and fertilizers are the main imports of the nation, ready-made garments, cotton, rice, leather goods, dried fish, and petroleum products are the chief exports. The U.S., Germany, Japan, the United Kingdom, Hong Kong, and Saudi Arabia are the major trading partners.

Prospects Blessed potentially with a sound economic base that can supply most of the country's needs in consumer goods, Pakistan's natural resources provide materials for such industries as textiles (the largest foreign exchange earner), oil refining, metal processing and fertilizer production. But the nation's current population of 152 million, growing at an annual rate of 2.5 percent, is a serious problem. A large influx of Afghani refugees (nearly 3 million), and an interminable hostility with neighboring India puts an enormous burden on Pakistan's faltering economy.

Politically, Pakistan's democratic framework remains shaky. Since inception there have been several coups, and military rule has often been imposed over the country. Additionally, ethnic problems, particularly in the provinces of Sindh and Baluchistan, have recently assumed dangerous proportions. There is considerable illegal trafficking in narcotics. These tendencies have tended to sap the already fragile democracy.

BANGLADESH

Last among the nations to emerge on the Indian subcontinent after a brief, brutal war in 1971 with Pakistan, of which it had been a part since 1947, Bangladesh (literally Bengal country and formerly known as East Pakistan) is a small, impoverished and overcrowded nation. Before independence it was separated from Pakistan's western region (formerly known as West Pakistan) by some 920 miles (1,554 km) of Indian territory; its political link to West Pakistan was based largely on religious grounds. Dissimilar in most other cultural traits (language, food habits, literary traditions and history), the eastern wing of Pakistan had lived its 24 years of existence in relative neglect, and exploitation. With only one-sixth of the area of Pakistan (54,000 sq miles or 148,000 sq km) but over half of its population (now close to 127 million), the new nation faced seemingly insurmountable problems of exploding population, grinding poverty and lack of overall development. Unable to feed itself, it relied heavily on food, material, and technical aid from many countries, including India, which had helped it to win freedom from Pakistan.

Physical Characteristics Nearly all of the land surface of Bangladesh is below 50 feet (15 meters) above sea level and consists of a flood plain of the Ganga-Brahmaputra rivers and their tributaries. The delta is active and continually silting. The only hilly areas are bordering India where land rises to about 1,000 feet (304 meters), and in the southeastern section near Burma border where the relief is between 500 to 2,000 feet. The climate is everywhere warm and humid. Rainfall is plentiful, varying between 70 inches (1,778 mm) annually in the west and 100 inches (2,540 mm) in the east; the regime controlled by the two monsoon seasons but most precipitation falls during the summer monsoons.

Nature has also dealt harshly with Bangladesh. The nearly flat topography is subjected to annual flooding. To the south, the country lies open to the passage of tropical storms which periodically bring large scale destruction to the densely settled southern sections of the deltaic plains of the three big rivers: Padma (local name for Ganga), Jamuna (local name for Brahmaputra), and Meghna, and their innumerable tributaries. The November 1970 typhoon, which took as many as half a million lives and destroyed 400,000 homes, was among the greatest natural disasters of the 20th century. Similar, if not quite as devastating, events are a common and nearly an annual phenomenon.

Cultural Patterns Ethnically, Bangladesh is one of the most homogeneous nations: 98 percent of the population consists of Bengali speakers. The Bengalis—both Muslims and Hindus—are justly proud of their linguistic culture, for Bengali language possesses a far richer and maturer literature than most other Indo-European languages. Islam is the predominant religion; and Muslims form

Map 16.3

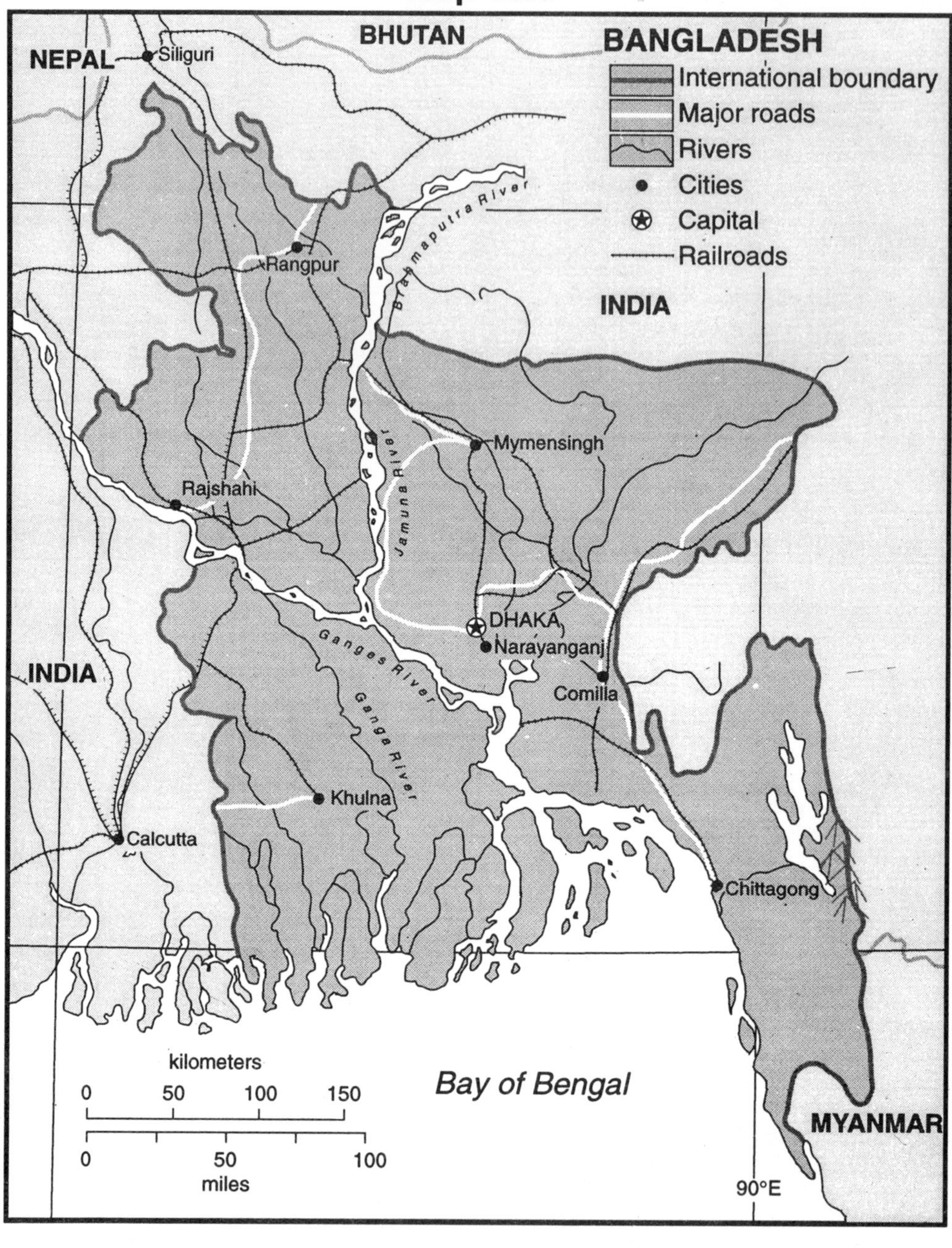
BANGLADESH
International boundary
Major roads
Rivers
Cities
Capital
Railroads
NEPAL
Siliguri
BHUTAN
INDIA
Rangpur
Brahmaputra River
Mymensingh
Rajshahi
Jamuna River
DHAKA
Narayanganj
Ganges River
Ganga River
Comilla
INDIA
Khulna
Calcutta
Chittagong
kilometers
0
50
100
150
0
50
100
miles
Bay of Bengal
MYANMAR
90°E

88 percent of the population. The Hindu minority is substantial, consisting of nearly 11 percent of the total population, and composed primarily of the underprivileged sections, who have, at times, been immigrating to the neighboring India. There was a mass exodus of nearly 10 million persons to India during the 1971 War of Independence with Pakistan consisting almost entirely of the Hindus who felt insecure to remain within the country. Most of them have since returned back, but a small trickle continues to immigrate to the adjoining states of West Bengal, Assam, Meghalaya and Tripura in India. During recent years many have immigrated to cities including the West Bengal in India and India's capital city of New Delhi.

Economy and Resources Agriculture is the dominant sector of economy. Nearly all of the cultivatable area (about 74 percent of its total land) is under crops----rice and to a lesser degree wheat for subsistence; jute and tea for cash. Jute is the mainstay of the economy, producing a substantial part of its export earnings. It is Bangladesh's important export surpassed only recently by ready-made garments which became Bangladesh's leading export in 1988. Tea is another export item. Conditions for rice cultivation are ideal---- ample rainfall, high temperatures, very fertile alluvial soils, and almost level topography. Although two to three crops of rice are raised, low yields and high population densities (now approaching 2,230 persons per sq mile or 6,862 per sq km) have made the country chronically a food deficit area. During the 1970s and 1980s the "Green Revolution" helped boost rice production dramatically through increased acreage in high-yielding varieties of rice. The expansion of wheat as a second grain crop also allowed dietary diversification. Although the country approached self-sufficiency in food grains in the mid 1980s, increasing population remains a serious problem. The Malthusian threat of hunger still looms over a nation already given to recurrent famines, persistent malnutrition, poverty, and typhoons.

The cultivation of jute (produced as a commercial crop) occupies land that could otherwise be used for rice. It s production related to the low-wage hand labor so abundantly available, has suffered lately in competition with the introduction of synthetic fibers into the world market. Lacking in most minerals, Bangladesh's modest resources of peat, natural gas and timber await exploitation as the nation's energies are spent on combating problems of overpopulation and economic underdevelopment.

The capital and port city of **Dhaka** (6.1 million) is centrally located in the delta region. Famous for its fine Muslims and other handicraft industries in the 17th century, it has been a historic/politico-regional center of Bengal. At the time of India's partition, the city contained nearly one-half of its jute manufacturing mills; now it is the premier industrial center of Bangladesh containing besides jute mills, sugar factories, oil pressing mills, and a few glass and cement factories. The jute and cotton textile industry has since undergone considerable expansion. Textiles (ready-made garments, and jute manufactured material) account for nearly two-thirds of country's exports that pass through Dhaka. **Chittagong** (2 million) is the country's other major port. Khulna with a population approaching one million is an important commercial and manufacturing center in the southern region, and Rajshani (over half a million) and Mymensingh are other regional market centers located in the western and eastern parts of the country.

Prospects Bangladesh is a sad paradox. The country possesses rich, fertile land capable of producing several subsistence and commercial crops, but ever increasing population and relative neglect have left the country poor and underdeveloped. An assessment of its future prospects is difficult to make. Some diversification of the economy is possible. Most mineral wealth, modest though it may be, remains untapped. Natural gas reserves might be used in the production of fertilizers. The forest products' industry can be also developed. Diversification of commercial and subsistence crops can relieve pressure on jute, tea, and rice. Fishing as an important resource and a major export, can be further expanded. But population remains a serious problem. Currently, Bangladesh stands close to the bottom of the world's countries in economic development, along with such nations as Chad and Bhutan, which contain only a tiny fraction of its population. The country suffered grievously during its war of independence. Millions streamed across the border into neighboring India. Disruption of transport and of thousands of villages and settlements added to the misery and horror that followed the war from which the country has not since completely recovered.

Since the 1960s the issue of population planning has been taken seriously by the administration. Through its own efforts and the help of international donors (Bangladesh is one of the largest recipient of international aid in population control), there has been a noticeable decline in the average annual increase in population—from 2.6 percent a year during the 1960s and 1970s to 2.3 percent between 1980 and 1995, and by 1999, it had fallen below 1.8 percent a year. During the late 1990's, Bangladesh made major efforts to meet the food needs of its growing population, and the country is now largely self-sufficient in its food requirements.

THE HIMALAYAN KINGDOMS OF NEPAL AND BHUTAN

Landlocked between India and Tibet, the Himalayan kingdoms of Nepal and Bhutan together cover an area of a little over 75,000 sq miles (190,000 sq km). Nepal is three times larger of the two, and shares with India a boundary totaling about 750 miles (1,200 km). Historically, they were maintained within the "inner" ring of the British defense interests as "buffer" states. India's border dispute with China since 1960s brought sharply into focus the strategic nature of their location.

The political pattern of these Himalayan kingdoms emerged recently; for Nepal during the 19th century and for Bhutan even later. Historically, they escaped much of the political turbulence that the Indus and Ganga plains experienced during the Islamic invasions and the consolidation of the British rule. Nepal managed to survive as an independent country during the British annexation of the subcontinent, but kept a close relationship with India by treaties of friendship and protection from external aggression. Bhutan is also protected from external invasion, although it became a fully sovereign nation and a member of the U.N.O. in 1971.

In addition to Nepal and Bhutan, a much smaller Himalayan principality of Sikkim that lay between the two and was a protectorate of India. It was absorbed as a state within India in 1976. Prior to it India exercised control over its externa. Policy as well as its borders and internal administration. Most of the border of these Himalayan states was disputed by China, particularly parts of Sikkim and Bhutan. India formally protested in 1959 against China's claim over Bhutanese territory and maintained its rights to defend its borders. India has been closely helping Bhutan in its defense efforts. Neither India nor China laid territorial claims over the independent kingdom of Nepal, although China disputed a small portion of Nepal's Tibetan border, but relinquished its claims in 1961 following a treaty. Nepal has carefully followed a policy of non-alignment, and has allowed both India and China to construct roads linking its capital city of Kathmandu with its two big neighbors. India considers Nepal to lie within its perimeter of its defense interests and is aware of the dangerous potential of any possible future southward expansion by China made easier by Tibet-Kathmandu road construction in 1968-1970.

The typical Himalayan physiography is well represented in the region, the gently folding foothills in the south to the complex, deeply-folded outer ranges, and the high, crystalline range in the north. The terrain rises abruptly in about 80 miles (130 km) from less than 1,000 feet (305 meters) in the piedmont terai area of the south to 2,600 feet (7,930 meters) in western Nepal to over 2,800-2,900 feet (8,540-8,845 meters) at Kanchenjunga and Sagarmatha (Mt. Everest) in eastern Nepal.

Population densities in general are low, although the central valley of Nepal, where the capital city of **Kathmandu** and other major cities are located, and the terai plains in the south are crowded (over 600 persons per sq mile or 15,500 persons per sq km). The total population for the two countries is estimated to be a little over 23 million of which Nepal accounts for nearly 22.5 million.

Cultural Patterns Ethnic and cultural distributions follow the physiogeophic vertical zonation. The Mongoloid Tibetans generally occupy elevations over 4,000 feet (1,216 meters) in Nepal and only 1,000 feet or less in the valleys of Bhutan. The boundary of Tibetan or Bhutia people rises to nearly 9,000 feet (2,745 meters) in Nepal and descends to 5,000 feet (1,525 meters) in Bhutan. Lamaistic Buddhism is practiced by nearly all the Tibetan speakers at higher altitudes, giving way to tribal or Hindu practices in the valleys and at lower elevations. Nepal is essentially a Hindu kingdom with nearly 90 percent of the population. They are concentrated in the southern and southwestern parts of the country. In 1959 the Bhutanese government banned the immigration of these Nepalese settlers. The discrimination against them is a major internal political problem for Bhutan. The largest ethnic group of Bhutan is that of the Bhutias that make up for over 50 percent of the population and are concentrated in northern, central and western parts of the country. They speak Tibetan dialects, and follow a form of Tibetan Buddhism. Most tribal groups in eastern Bhutan are ethnically related to the hill tribes in adjacent areas of India (Assam and Arunachal Pradesh) and are known as the sharchops. They are Indo-Mongoloid, and practice a loose form of Tibetan Buddhism.

Nearly 60 percent of Nepal's population is that of the Gurkhas, who show a mixture of Tibetan Mongoloid and Caucasian (Indo-Aryan) traits of North India. The Gurkhas speak Gorkhali or Nepali which is the official language of the country, and is a derivation of Sanskirt and akin to Hindi of North India. The Gurkhas provide the ruling family of Nepal, and take pride in their Indo-Aryan and Hindu background. They are principally concentrated in central and northern parts of the country. Several Tibeto-Nepalese tribes—the Tharu, Tamang, Bhutia, and Newar—live in the northwestern and eastern regions, and speak several Tibetan-related dialects and profess Buddhism, adhering to Hindu beliefs, but Bhutan claims about 70 percent of the population professing Lamaistic Buddhism. Both countries are culturally plural. One-third of Bhutanese population is composed of Gurunegs, people of Nepalese origin and professing Hinduism and speaking Gorkhali.

Close to 90 percent of Nepalese and Bhutanese live in small market centers. **Kathmandu** with a population of 535,000 is the capital of Nepal, and is the market center for the central valley, and a major religious and tourist attraction. Smaller urban centers are located in the *terai* along the Nepal-India border, where economic activity has been recently developing. **Thimpu**, the capital of Bhutan, has a population of merely 24,000 and acts as an administrative and a regional market center.

Resources and Development Lacking in substantial resources, landlocked, and hampered by an inadequate transport network, these remote countries of Nepal and Bhutan are among the least developed nations in the world. Their economy is dependent on foreign markets on imports of basic requirements such as fuel, machinery, textiles, most consumer goods, and fertilizers, and their exports of timber and food resources. For geographic and political reasons nearly all of their foreign trade is with India which also acts as their transit conduit to foreign markets.

Economy of these nations is predominantly agrarians and over 80 percent of the labor force is engaged in agriculture. Mineral resources are small, scattered and remain underdeveloped. There are known deposits of coal, iron ore, copper and mica in Nepal; but are virtually untapped. Bhutanese minerals also remain unprospected. The river systems provide immense potential for hydroelectric development. If these resources could well be developed and utilized within the countries and exported to India it could contribute significantly to their economy. Nepal has already started to export a part of its hydroelectric power to the neighboring region of India. In Bhutan, the Chhukha Hydel project financed by India is the principal source of nation's hydroelectric potential most of which is exported to India.

The amount of land available for agriculture is limited by steep slopes, poor soils, and adverse climate. Only a fraction of the total area of Bhutan, and *terai* land and valleys in central Nepal, are utilized for the cultivation of such crops as rice, maize, potatoes, sugarcane, citrus fruits, wheat, and millets. Agricultural productivity is low resulting from the use of inefficient techniques, shortages of fertilizers and improved seeds. Farms are small in size. Terraces are used extensively to raise crops on hillsides. On the whole, Nepal has a small surplus in food grains, due largely to the development of the *terai* land, where greatest potential for increase in agricultural production exists.

About one-third of Nepal's and 70 percent of Bhutan's total area is forested.

Timber represents a valuable resource and a major source of potential revenue despite overcutting and poor forest management in Nepal. Industrial production represents a very small, but growing segment of economic activity. Most industries are small, localized operations based primarily on the processing of agricultural products. In Nepal jute manufacturing, rice milling, and sugar processing are the main industries located chiefly in Kathmandu Valley in Biratnagar, Birgani, and in the *terai* region. Other industries include oil mills, processing of construction materials, paper and food grains. Modern manufacturing plants in Bhutan are of recent origin but growing. These are mostly located in the south, close to the Indian border. Food processing, cement and wood products account for most of the country's industrial base.

Transport facilities in these nations are limited, but are being slowly extended. In Nepal the construction of new roads has been undertaken since the 1970s with aid from India, China, Great Britain and the United States, linking Nepal with India and Tibet. India has been providing aid for the development of a road system in Bhutan. Except for a small stretch along the southern border of Nepal along India, there is no railroad track in Nepal or Bhutan.

SRI LANKA

A small (area: 25,300 sq miles/65,500 sq km), compact, pear-shaped island off the southern tip of India, Sri Lanka has shared historical origins with the subcontinent, but developed distinctive cultural and economic characteristics. Physically, it is a detached portion of the southern India massif with landforms, climate, natural vegetation and soils matching broadly the adjacent areas of India. In general, the island's plateau structure has been eroded into a complex hill-land with broad coastal lowlands and a mountainous central core rising to over 8,000 feet (2,400 meters). Climate is essentially tropical monsoonal type with high temperatures year round and heavy seasonal rainfall, with regimes varying from one coast to another.

Physical and Economic Regionalism

Sri Lanka's lowlands can be conveniently divided into two broad zones. The "west zone" consists of tropical southwestern coastal lowlands which are densely populated, and are devoted to rice-dominated subsistence agriculture, and the production of rubber and coconuts (on plantations). The "dry zone" includes the northern and eastern lowlands with elevation of only 100 to 200 feet (30 to 60 meters); has scrub and grass vegetation, and over large areas is sparsely populated, although it supported larger populations and was intensively irrigated in ancient times. Rice and grain crops dominate the agricultural landscapes of the "dry zone," although in the extreme north around Jaffna, coconuts are also locally important. The remaining section, including the south-central mountainous core and the adjacent hill-lands have been developed for plantations of tea and rubber. This region was linked with the western coast by railroad and road during the British occupation. In total, less than one-third of the country's area is presently productive, although as much as half might be cultivated.

History and geography conspired to make Sri Lanka the "pearl of the Orient" a coveted prize to the Portuguese, Dutch, and British. Its location at the extremity of the Indian peninsula, athwart the trade

Map 16.4

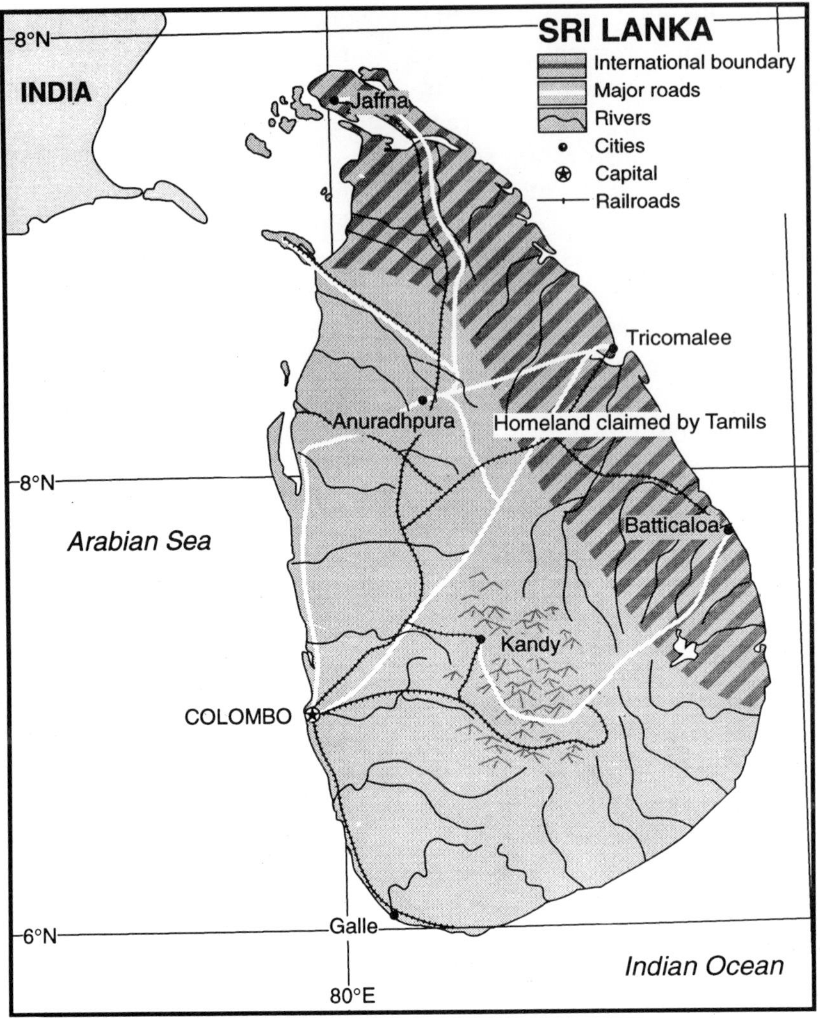

routes of Asia, a wet tropical climate suitable for the growth of commercial crops and lush vegetation, were enticing reasons for Europeans to colonize the area. Britain succeeded in controlling the entire island by 1815. The British superimposed on the native subsistence rice-dominant agriculture an estate or plantation economy based largely first on coffee and later on tea and rubber production for export purposes. The interior hills received the main impact of the new economy. Like Malaysia, "Economic dualism" became the essential character of the country. Today, nearly all of the tea and rubber, and over a half of the coconut production, account for nearly 40 percent of all exports, shipped mainly through **Colombo** (620,000), the country's capital, the largest city, and a leading port which lies on the western coast, on the trading routes to the West. Its major functions are that of a commercial nature; the industry is limited to the production of such items as matches, paper, cement, china, glassware, etc. for local consumption. **Jaffna** in the northern extremity, **Trincomalee** on the east coast, and **Galle** on the south are important port cities and Dehiwala---Mount Lavinia, a religious city, have populations between 100,000 and 200,000.

The present population of 19 million is increasing at a rate of 1.2 percent a year, considerably slower than it had been before the 1980s. Large-scale efforts to resettle the dry zone by restoring ancient irrigation works and adding new ones to increase food production are being made. Between 1945 and 1950 malaria was virtually wiped out in the drive for resettlement, but this in turn led to large population increases. Population growth has abated somewhat since then, but food problem remains a serious one since the foundations of the economy still rest primarily on commercial crops. Sri Lanka spends one-sixth of its import expenditure on grains. In a mineral-deficient country industrialization is not well developed except for some processing of agricultural goods, and the manufacture of such items as textiles, cement, paper, and china, located primarily in Colombo. Recently many foreign firms have invested in electronics and other light industries.

Cultural Patterns Sri Lanka's major political problem stems from the diversity and character of its ethnic distribution. The largest element in its ethnically plural society (accounting for about 82.7 percent of the population) is that of the native Sinhalese who trace their ancestry to the north Indian Aryans. The next biggest group, comprising of 9 percent of the population, is that of Dravidian Tamils who are almost evenly divided between those who have long settled in the Jaffna area at the northern tip of the country and on the east coast and are known as "Ceylon Tamils," and another group of "Indian Tamils", who are relatively recent migrants or descendants of labor imported by the British from southern India to work on tea and rubber plantations in the interior hill country. While the native Sinhalese, who are mainly Buddhists and are widely distributed, the two Tamil groups are mainly Hindus and are relatively concentrated in the Jaffna area, as well as on the eastern coast and in the high-land plantation area. The rest of the population is composed of "Moors" (Muslims) forming 8 percent of the population, and Malays, Eurasians (Burghers) and Europeans, plus a small remnant of primitive tribesmen (the Veddas).

Prospects Since independence in 1948, the Ceylonese Tamils have been demanding a political homeland, *Eelam*, for the preservation of their language and culture within the country. Relations between the Sinhalese and "Ceylon Tamils" have deteriorated in recent years, due as much to country's economic stagnation and as to the political disregard for the "Ceylon Tamils" who are carrying out a bloody conflict with the federal authorities. As a result, there has been much loss of life and economic disruption in the country. The economy is precariously dependent upon the exports of its plantation products which are subject to severe fluctuations in the world market, and internal harmony----both of which have been running a dangerous course for the country. On the other hand, the country is rich in several resources in addition to the plantation crops, such as forest wealth, minerals (particularly mica, gems and limestone), and has a large potential for the development of hydropower. Future development must, however, await political tranquility in the war-torn nation.

MALDIVES

Maldives consists of a chain of over 1,200 small coral islands in the Indian Ocean, 600-1,000 miles (965-1,610 km) west of Sri Lanka. It is the smallest, independent country in Asia, and was a British protectorate until 1965. Most of the population of 254,000 lives on 200 or so of the islands.

Nearly all Maldivians are descendants of Sinhalese people of Sri Lanka, and are Muslims of the Sunni sect. Seventy percent of them are classified as rural. The capital city of **Male** (population: 62,000) contains most of the country's urban population. Growing at a rate of 3.4 percent a year, Maldives is one of the fastest growing countries in the world. The population density of 2,417 persons per sq mile (932 per sq km) is also one of the highest in the world.

The islands are grouped in 12 atolls, enclosed by barrier reefs, most of which are very small, less than 5 sq miles (13 sq km) in area. In topography, the islands are nearly flat, the highest point being at an elevation of 80 feet (24 meters). The climate is typically tropical, hot and humid, with daytime temperatures averaging 80°F (27°C). Rainfall is plentiful, between 100 inches (25,000 mm) and 150 inches (3,810 mm), annually, coming mostly in two rainy seasons.

Fishing and tourism are the two main economic activities. Dried fish, exported largely to Sri Lanka, is the chief source of revenue. One-fifth of the national income is derived from tourism. Coconuts, millets, sweet potatoes are also grown. Coir yarn (made from coconut fiber) is used in the manufacturing of shirts, and sweaters in the island of Gan. Rice, sugar, wheat flour, and manufactured goods are the major imports. India, Japan, Singapore, and Sri Lanka are the country's chief trading partners. With a per capita income of $1,200 a year, Maldives enjoys the highest standard of living in South Asia.

The strategic location of Maldives in the Indian Ocean was used by the British to establish a military base in the island of Gan. During the 1950s a dispute arose between Britain and Maldives over the Gan-territory which was finally settled in 1960 giving Britain rights to establish a military base on the island. In 1985 Maldives became a member of the SAARC (South Asia Association for Regional Cooperation).

Further Readings

Johnson, B.L.C. *Bangladesh*. New York, 1975.

Karan, P.P. *Nepal.* Lexington, 1960.

Majumdar, T. *Nature, Man, and the Indian Economy*. Delhi, 1993.

Rawson, R.R. *The Monsoon Lands of Asia*. London, 1963.

Tirtha, R. *Geography of India*. Jaipur, 2nd Ed., 2000.

17

Southeast Asia: An Introduction

The term Southeast Asia gained currency after the World War II for a vast region situated east of the Indian subcontinent and south of China that includes a sprawling territory of peninsulas, bays and islands, occupying the southeastern corner of the Asiatic continent. Included in it are two dissimilar components: a continental projection on the Asiatic mainland and the insular section that consists of a string of archipelagoes to the south and east of the mainland. Mainland Southeast Asia contains the nations of Cambodia, Laos, Myanmar, Thailand, Vietnam, the small city-state of Singapore (these nations are often collectively known as the Indo-Chinese Peninsula), and Malaysia which is divided between both the mainland and insular sections. The insular section includes eastern Malaysia, the small Sultanate of Brunei, and the archipelagic nations of Indonesia and the Philippines. Parts of Southeast Asia have been variously known as Further India, Indochina, Little China, Extreme Orient, Far East and Greater India, primarily for the mainland section, and the East Indies, Tropical World, and the Island World for the insular section.

Historically, Southeast Asia is a zone of convergence for Indian and Chinese cultures. The two Asian giants, India and China have been indirectly in conflict with each other in this intermediate zone for thousands of years, and many of these lands and their cultures exhibit the cultural impacts of these two large neighbors. Added to these influences was Islamic penetration. Combining commerce with a fierce determination to spread their faith, Muslim traders from Arabia carried Islam as far east as the Philippines from the 13th to the early 16th centuries. Yet Islam was neither indigenous nor dominant as a cultural, economic or political force. The result of these cultural influences, and the physical fragmentation of the vast territory, is a "shatterbelt" of incredible diversity, a crossroad between the Asian East and European West, a zone of mixed cultures. Superimposed on the Indo-Chinese imprints and the Islamic incursions was the commercializing influence of the West. Most of the region's religions, art forms, languages, and economic and political systems had their origins elsewhere on the Asian mainland or in Europe. This was also the region of greatest colonial impact in Asia. With the exception of Thailand all of Southeast Asia was part of one or another of the European empires of the 19th and 20th centuries.

Physical Characteristics Although the physical dimensions are huge----extending longitudinally from 92° to approximately 141° east longitude, a distance of over 3,000 miles (4,800 km) and latitudinally from 11° south in Indonesia to about 28°

Table 17.1 *Southeast Asia: Basic Data, 1999*

Country	*Area (Thousand sq km/sq mi)*	*Population (Millions)*	*GNP ($/Capita)*	*Capital*	*Urban Population Percent (est.)*
Brunei	5.7/2.2	0.3	25,160	Bandar Seri Begawan	58.0
Cambodia	181.9/70.2	10.9	300	Phnom Penh	21.0
Indonesia	1937.1/747.9	206.1	1,080	Jakarta	30.9
Laos	236.8/91.4	5.4	400	Vientiane	23.6
Malaysia	330.4/127.5	22.7	4,370	Kuala Lumpur	57.0
Myanmar	676.5/261.2	48.0	2,610	Yangon	27.0
Philippines	300.0/115.8	74.7	1,160	Manila	58.0
Singapore	0.6/0.2	3.0	32,810	Singapore	100.0
Thailand	513.1/198.1	61.8	2,740	Bangkok	34.0
Vietnam	331/127.8	77.3	310	Hanoi	20.5

Source: Encyclopedia Britannica, Book of the Year, 2000.

north in northern Myanmar, a distance of about 2,700 miles (4,320 km), Southeast Asia is approximately the same size as South Asia and only about one-third the size of East Asia. Its gross area, that of lands and seas put together, roughly equals that of the United States. In total area it comprises 3 percent of the world's territory and in population it has a little over 8 percent of the world's total. Southeast Asia's elongated peninsular projections and the fragmented and extended coastlines of its archipelagoes accentuate its physical dimensions. Thus, Indonesia alone extends over a greater longitude than the contiguous states of the United States. The Philippines, on the other hand, have a north-south extent of 1,152 miles (1,843 km). As a result of the fragmented pattern of lands and seas, of peninsulas and archipelagoes, combined with the prevailing cultural complexity, political unity for several of its nations is a major concern.

Physically, Southeast Asia is one of the most complex regions. Broadly speaking, much of the region consists of complicated groups of mountains, islands and seas, the surface features of which have developed from a considerable amount of faulting, folding, uplifting and volcano activity, as three of the earth's major crustal units---the Eurasian, Indo-Australian, and Pacific plates---converged on one another. The interaction of the plates that gave the present-day landscapes produced strikingly dissimilar physical features in the mainland and archipelagic portions of the region.

On the mainland, the basic pattern is that of a series of mostly north-south-oriented mountains separated by a number of major river valleys and interrupted at places by several old plateaus that have been folded, faulted and dissected over a period of time but now form generally low platforms or massifs. The mountain ranges fan out southward from the southeastern corner of the Tibetan Plateau resembling the spokes of a wheel where the interstices are deep tentacles carved by rivers. Much of the region's surface has been produced by the gradual collision of the Indian subcontinent with the Eurasian Plates over the past 50 million years. Comparatively speaking, mainland South Asia is geologically a more stable area with little or no volcanic activity, except in the northwest and north, and experiences little seismic activity.

On the mainland, prominent mountain ranges that are oriented north-south are: (1) Patkai-Naga-Chin Hills, and the Arakan Mountains through the western margin of Myanmar; (2) a straight north-south axis ranges from eastern Myanmar through northwestern Thailand to south of the Isthmus of Kra on the Malay Peninsula with altitudes of 8000 feet (2,440 meters) on the Chinese border in the north to below 4,000 feet on the Isthmus of Kra; and (3) the easternmost major mountain on the mainland is the Annamite Cordillera in Laos and Vietnam. Among the plateaus of old strata that intervene the series of mountains mentioned above are: (a) the Shan Plateau of eastern Myanmar, some 250 miles (400 km) from north to south, and 75 miles east to west and an average elevation of about 3,000 feet; and (b) Korat Plateau in eastern Thailand and west-central Laos which is essentially a low platform only a few hundred feet above the flood plains on the surrounding rivers.

Between the alternating highlands and the plateaus are the river valleys and the lowlands, covered generally with alluvial sediments of the rivers. From the west to the east the major river basins are: (a) Irrawaddy and Salween in Myanmar, (b) the Chao Phraya in Thailand, (c) Mekong in Laos, and Cambodia and (d) Red River in Vietnam. The three largest river Systems ----Irrawaddy, Salween, and Mekong---- originate in the Plateau of Tibet and make substantial deltas except the Salween which is primarily a highland river in eastern Myanmar. The Mekong is the longest river basin among the river systems of mainland Southeast Asia, and has an extensive lowland that encompasses most of Cambodia and southern Vietnam. Almost in the middle of Cambodia is a large freshwater lake, the Tonle Sap. The river has a large delta that occupies the tip of the entire southern Vietnam. The Irrawaddy of Myanmar and the Red River of North Vietnam also make substantial deltas. Outside the river deltas, the coastal lowlands are generally narrow strips between the mountains and the seas.

The Malay Peninsula pushes the mainland nearly 900 miles into insular Southeast Asia from the Gulf of Thailand to Singapore. The peninsula is primarily a region of north-south trending mountains of altitudes between 5,000 to 7,000 feet, and narrow coastal plains.

The insular Southeast Asia consists of chains of islands----the Malay and Philippine archipelagoes----that were formed at the boundaries of the three crustal plates that meet there, and thus given to crustal instability, earthquakes and volcanic activity. Characteristically, the various islands (Java, Sumatra, Bali) have highland cores, short rivers and narrow coastal plains, and were formed where the Eurasian plate overrides the Indo-Australian plate.

Climatically, all of Southeast Asia falls within the warm, humid tropics; the climates can broadly be labeled as hot tropical of the insular section and monsoonal of the mainland, although there is a high degree of climatic diversity within Southeast Asia.

Temperatures generally remain fairly constant near sea level throughout the year, varying somewhat with increasing latitude and altitude. With the exception of northern Vietnam and Laos, annual temperatures average close to 80°F (27°C). Higher altitudes act to decrease average temperatures.

Precipitation, in general for most of Southeast Asia, is higher than 60 inches (1,500 mins) annually, and several parts receive much more than that. The incidence of rainfall is affected by two seasonal wind systems: the northeast dry monsoon and the southwest wet monsoons. The northeast monsoon is a winter phenomenon (generally November to March) in the Northern Hemisphere. The southwest air stream passes over the warmer sea from May to September and gradually warms up and picks up moisture. Precipitation is particularly heavy where the air stream is forced to rise over the mountains. The eastern coast of peninsular Malaysia, the Philippines, and parts of Indonesia receive the highest moisture during this period. Most of mainland receives its rainfall in this period. Rainfall, in general, is more pronounced over much of southern Malay Peninsula and insular Southeast Asia and there are no particular dry seasons. It is particularly marked in much of the equatorial region and the east coast of the Philippines.

In addition to the monsoonal pattern of rainfall, there are other factors that produce variations over relatively short distances combined with monsoonal affects such as relief, land and sea breezes, convectional rainfall and cyclonic disturbances. Cyclones generally bring moderate rain, but many of the cyclonic disturbances mature into tropical storms----called cyclones in the Indian Ocean and typhoons in the Pacific---that bring heavy downpours and destruction to areas over which they pass. The Philippines are particularly prone to such storms.

The natural vegetation types of Southeast Asia can also be grouped into two broad categories: the tropical evergreen forests of "equatorial" lowlands and the tropical-deciduous or "monsoon" forests in areas of seasonal drought. The equatorial evergreen vegetation consists of a multi-layered vegetation of a variety of trees with proliferation of vegetation species. The deciduous or "monsoon" forests found in eastern Indonesia and those parts of the mainland where annual rainfall does not exceed 60 inches, and there is a marked seasonal rhythm of precipitation. Like the equatorial types, there is a wide variety of species, of which teak is one of the more valued commercial tree found in parts of Myanmar, Thailand, Laos, and Indonesia. Additionally, mangroves occur along the coast where salt is deposited. Upland forests are represented by such trees as maples and oaks.

Cultural Diversity Southeast Asia is a plural world. The plural pattern is particularly striking on the mainland where each nation contains diverse groups based on language, religion, and culture, although there may be a dominant group in its central lowlands. Each state has areas of hill and lowland cultures. This interweaving of diverse cultures within the political framework of present-day states has resulted, in general, from European colonial rule, which established a single administrative control over the diverse ethnic groups included within each colonial empire.

Historically, Southeast Asia has been a giant funnel into which peoples from the north and west have migrated. These immigrations into and through the region left their mark in a variety of ethnic, linguistic and cultural differences, that has resulted in making the present-day nations complicated, plural societies. The initial peoples arrived from the Asian continental interior. Successive movements displaced the initial settlers and created a complex ethnic pattern. The contacts with the region were not entirely through land, however, but also through sea routes. Chinese, Indians, Europeans, and the United States, came over the sea routes.

Centuries before and after the start of the Christian era, both Indian and Chinese cultural agents became active in the region after crossing the Indian Ocean and the South China Sea. After establishing themselves in Southeast Asia, they never collided head-on, or were engaged in struggle for territorial gain, but both exerted cultural pressures as early as two thousand years ago that continues to the present time. In due course, the Chinese greatly influenced architectural forms, social structures, political organizations, local governments, agricultural economies, and handicraft technologies of the region. At times, they assumed lordship over much of the territory and occasionally reaching on to Indonesia as well.

Indian influence was even more direct. The early Indian contacts were in trading goods, technologies, crop plants, religious concepts, and the formation of the religious states. The impact of Indian culture was pervasive and deeper, particularly in religion, architecture, and folklore than that of China. Indian commerce had started as early as the 6th century B.C., and was followed by the establishment of trading posts and communities; Indian religion, philosophy and colonization took root centuries later. In contrast, the Chinese contacts (starting from the 3rd century B.C.) were initially insignificant. The real push did not come until after the 17th century, when the Chinese were driven by demographic pressures at home.

To the cultural influences of Hinduism and Buddhism brought by the Indians and the secular impact of the Chinese was added another cultural influence provided by Islam. Brought by the Arabs in the late 13th century it was mainly focused in Indonesia and Malaysia, where the previously Indianized landscapes yielded quickly to an overlay of Islamic religion. The Islamic superimposition was in the form of a thin veneer over the existing dominant cultural mores. Its diffusion was also uneven, for excepting Malaysia, it never completely displaced Buddhism. In the Philippines its advance was checked by the Spanish. The stronghold of Islamic penetration was the coastal areas. Elsewhere, it coexisted with the indigenous Hindu, Buddhist and animistic cultures.

Later, in the 18th century when the European established colonial rule over these diverse cultures, their domination never changed the Hindu or Buddhist cultures, but was responsible for shaping these landscapes by such activities as the building of plantations, roads, hospitals, and railroads; and the introduction of European languages, legal systems, educational institutions and systems of government. The European impact has been most pronounced in cities, although it is increasingly making inroads in rural areas. In remote upland areas, tribal, animistic and indigenous cultures continued largely unaffected by the religions, languages, and cultures of the lowlands.

Southeast Asia's population now includes a wide variety of ethnic groups and cultures. On the mainland the Khmer peoples of Cambodia remain as the ancestors of the earlier groups. Although the remnants of the Mon group are distributed in parts of Myanmar and Thailand, the ethnic mixture resulted from the overlay of Tibeto-Burman, and Tai, Lao and Shan peoples. The Vietnamese population most likely originated in the Red River area (in the present-day northern Vietnam) and may be a mixture of the Tai, and Malays. In addition to these major groups are the various minorities of tribal peoples such as the Karens, Chins, Shans, and Nagas in Myanmar, who have affinities with other Asiatic peoples. Insular Southeast Asia contains a mixture of Proto-Malayo and Malayc-Polynesian groups. Indian, Chinese and Arabic influences have also affected the ethnic pattern in the insular regions of Southeast Asia.

The Burmans account for more than two-thirds of the ethnic stock of Myanmar; while ethnic Thais and Vietnamese form four-fifths of the respective populations of their nations. Indonesia is clearly dominated by Javanese and Sundanese ethnic groups while Malaysia claims two-thirds of the population to Malays and less than a third to the Chinese, and less than one-tenth to the Indian groups. The Philippines are divided between a number of groups, among them the Tagalogs, Cebuano, Ilocano and Bicol are more prominent.

Nearly 15 million Chinese who inhabit Southeast Asia have been traditionally a source of special problems for almost all the nations except Thailand and Singapore. In Thailand one out of every five persons is said to have Chinese blood; in Singapore the Chinese are in absolute majority. Some of the Chinese are the descendants of those who migrated to the region centuries ago, although most have arrived within the last century seeking economic opportunity. Relatively unassimilated, except for Thailand, they form an alien element in every nation, but exercise significant economic and cultural dominance much beyond their numbers. Many of the Chinese have been middlemen in commerce and business, except for Singapore where they retain key political position as well as economic power. Resentful of their overwhelming numerical superiority and economic power at local levels, non-Chinese nationals in several Southeast Asian nations have often expressed bitterness towards the Chinese.

Anti-Chinese animosity has often taken the form of both official and unofficial persecution. This arises from the fact that many of the Chinese periodically remit large amounts of money to China, their ancestral homeland, and occasionally view sympathetically the policies of the present Communist regime in China. As a result of this, Chinese frequently have been restricted by legal restraints or have been the victims of local violence where they are a small minority as in Indonesia or the Philippines.

The language patterns in Southeast Asia are more complex since not only does each ethnic group have its own tongue, but also local or regional dialects that may be mutually incomprehensible. Broadly speaking, these belong to four major language families: the Sino-Tibetan, Tai, Austroasiatic, and Austronesian. The linguistic diversity is especially noticeable in fragmented areas such as the Philippines and Indonesia, and in highland and remote areas on the mainland. Thus, linguistic complexity has been an inhibiting factor in national unity and development in Southeast Asia. But dominant languages do exist in most of these nations. Burmese and Thai are spoken by an overwhelmingly large number in Myanmar and Thailand respectively. Khmer is the principal language in Cambodia, as is Vietnamese in Vietnam. Pilipino and English are the official languages in the Philippines and spoken by a large majority of the population, but Tagalog and Visayan are also important. Malay and Bahasa-Indonesia are the official languages of Malaysia and Indonesia respectively; these are related to one another and are mutually intelligible. Bahasa-Indonesia is a good example of a truly national language, used all over Indonesia and unlike in Myanmar, has been a unifying element in the country. Throughout Indonesia and Malaysia bazaar Malay has been used as a commercial *lingua franca*, while European languages, especially English, Dutch and French, have been in use on a wide scale.

The Indian and Chinese immigrants in modern times have introduced into the region numerous other languages. The most significant are a number of dialects spoken by the Chinese communities such as Cantonese, Hokkien, and Hakka in Singapore where they constitute the majority population and in most of the larger urban areas of the region. Tamil and Hindi speakers form significant minorities in Malaysia, where Indian immigrants constitute 7.4 percent of the population. Indian communities are scattered throughout the Southeast Asia, and are especially conspicuous in Singapore and Myanmar.

Religious differences have added further variety to the cultural scene. Islam, Buddhism and Christianity are the major faiths, but Buddhism, particularly its more orthodox form, Theravada Buddhism, is practiced in most of the mainland; only in northern Vietnam the more liberal Mahayana Buddhism is common. Islam is predominant in the southern half of Malay Peninsula, East Malaysia, southern Philippines, and Indonesia. In total count it is the religion of two-fifths of the population of Southeast Asia, accounting for 250 million people (21 million in Indonesia alone). The diffusion of Islam began in early 14th century through contacts with Muslim traders in Sumatra. Perhaps more than any other religion, it has been a strong unifying element among its adherents. Christianity arrived in the region with the Europeans. The Spanish and Portuguese brought Roman Catholicism in the 16th century to the Philippines, and somewhat later by the French to southern Vietnam. But Protestantism is also locally important in Sumatra, and among a growing number of people in Singapore. Hinduism, once much more widespread, now is practiced by many people among Indian communities. In addition, various forms of animism are practiced among tribal groups in more remote areas, particularly in central Kalimantan (Borneo), northern Laos, and north Myanmar.

Political Geography: A Historical Perspective The basic structures of Southeast Asian societies have been shaped by the civilizing influences of India and China even though the predominant religion of Indonesia and Malaysia is Islam. But the present political and economic geography of this vast area is a product of the European colonial rule of the 19th and 20th centuries. The complex problems of international boundaries and the existence of ethnic minorities within the states have resulted largely from the way Southeast Asia was divided by the Europeans during the era of colonization. As in Africa, the international boundaries that resulted from the European occupation lumped together ethnically diverse peoples.

The major colonial powers in Southeast Asia were Spain which occupied the Philippines (the U.S. occupation was a brief spell of 50 years); the Netherlands, which controlled the vast archipelago of Indonesia; France, which took possession of the eastern flank of the mainland known as Indochina; and Britain, which took over the Malaya Peninsula along with the northern part of the island of Borneo, and Myanmar. Among the other European powers, Portugal had a minor role, for it controlled a small portion of the island of Timor.

The first Europeans came in search of trade, particularly in spices which were much in demand in Europe. Commerce changed later into conquest. The **Portuguese** were the first to arrive on the scene in the 16th century. The Malay communities scattered along the shores in Malacca were taken over in 1511, but Portuguese rule did not survive the pressures of other European rivals, and in the end the only area that remained under control was a part of the island of Timor.

Spain was next to gain control. In 1565 it began its occupation of the Philippine Islands, not by way of the Indian Ocean but across the Pacific. In addition to the establishment of flourishing trade relations between the Philippines, Spain, Mexico, and China, Spain systematically converted most Filipinos to Roman Catholicism. As a result, the Philippines is the only predominantly Christian country in Southeast Asia. The United States gained control of the Philippines in 1898 and administered it until 1948.

Next in line was the **Netherlands** which gained control of the lucrative spice and pepper trade in the region, particularly focused on Java, and made large investments in sugar and coffee plantations there. Plantations thus introduced an alien form of economic organization to the existing trade in indigenous commodities. Subsequent European intervention in Southeast Asia followed the Dutch example. Plantations were built, port facilities developed, mines opened, and native rule replaced by foreign control. In short, Southeast Asian countries became an extension of European economic interest.

Eventually the Dutch-British rivalry for commerce, and occupation in the region ended in an agreement in 1824, when the island of Singapore went to **Britain** and the Dutch took possession of Indonesia. Singapore was developed by Britain as Asia's largest trading center. After the opening of the Suez Canal in 1869, British interest in the region intensified, culminating in the eventual occupation of the Malay Peninsula. Following the Dutch practice, the British established plantations of oil palm, rubber, and coconut as the Dutch had done with sugar and coffee in Indonesia. In addition, the British enlarged tin mines. The Malays were, in general, averse to these commercial enterprises involving organized labor, and the British recruited laborers from India and China. By 1939 the proportion of Chinese had risen to 40 percent of the population, and that of Indians to nearly 15 percent of the population of Malaya (the present-day West Malaysia). Britain also assumed control of Burma (Myanmar), and administered it as if it were a part of India until 1935. After that it became a separate country within the British Empire.

France's involvement in Indochina had started as early as the 17th century, but remained minimal until 1933. Gradually, its administrative grip was tightened through the use of the French navy, which controlled the coastal areas. The French Empire included Vietnam, Cambodia and eventually the landlocked territory of Laos which had been under the control of Thailand. Though it lost the territory of Laos, Thailand managed to remain independent as a buffer state. Like the other colonial powers, the French ignored native agriculture and industry and concentrated their efforts on the development of plantations of tea, rubber, coffee and sugar cane, lucrative commodities in European trade. The French also imported workers from China to develop commercial plantations of rice. The Mekong delta was cleared for the rice plantations. The Chinese immigrants, as elsewhere in Southeast Asia, not only worked hard on plantations and mines but also gained control over retail trade and banking. However, the Chinese in Vietnam did not arouse antagonism among the native peoples as did the Indians in Burma and the Chinese in Malaya. The Chinese now form the leading minorities in several of these countries.

During the course of the Second World War, Japan's conquest of the countries of Southeast Asia complicated the colonial picture. Japan emphasized the slogan "Asia for the Asians," and stressed its religious and cultural ties with these nations. However, its insensitive control led to regional disenchantment with Japan which began to loosen its control. After the Japanese defeat in World War II, the colonial administration crumbled and the nations of Southeast Asia emerged as independent countries, but the international boundaries had already been determined during the colonial times. This led to serious problems relating to ethnic splintering among several states. The French had still held onto Indochina during the early 1950s.

Meanwhile, the nationalist movements were crystallizing, in these nations. But even as the states of Indonesia, Burma and the Philippines emerged as independent nations, the French were still fighting to retain their foothold in Indochina. Finally, recognizing the futility of a protracted war, they departed in 1954, giving rise to the new states of North and South Vietnam, Cambodia and Laos. When the nationalist forces of North Vietnam were still struggling for a reunited Vietnam under the communist rule during the 1960s and 1970s, the United States was helping South Vietnam to maintain an independent status. A long, brutal Vietnam War eventually was brought to close, when like the French, the United States had to negotiate for peace that led to the emergence of a united Vietnam in 1976. British holdings in Malaya evolved into Malaysia in 1963, with Singapore separating from it as an independent state in 1965. The coastal areas of western Borneo, remained an integral part of Malaysia as East Malaysia, whereas the tiny state of Brunei remained a separate entity, initially as a British protectorate, but gained independence in 1984. Throughout all these changes only Thailand had retained its independence as a buffer state between the competing interests of Britain and France in Indochina.

Economic Patterns: An Overview

Agriculture has for a long time been a major economic activity in Southeast Asia;

only during the last few decades agricultural employment has been showing a declining trend. With the exception of Brunei, Malaysia, Philippines and Singapore a majority of the work force in Southeast Asia is engaged in agriculture. In Cambodia and Laos more than two-thirds of the labor force is engaged in agricultural activities. Only recently the economies of Southeast Asian nations have begun reorganizing toward industrial growth and registering declines in the proportion of the gross domestic product derived from agriculture; this trend is especially noteworthy in Malaysia, Thailand and Indonesia.

Agricultural output has made considerable gains since 1970; and while there are wide variations in the growth within the region, greatest gains have been reported in Malaysia and Thailand, and little or no increase in Cambodia, Vietnam, and Laos. In Cambodia internal strife, and crop failures have seriously affected agricultural output during the last two decades. In the Philippines landlessness among the peasantry has seriously retarded the growth of agricultural production.

Wet rice cultivation is the most dominant agricultural activity. Two crops a year are usually planted everywhere if physical conditions are suitable. In drier areas, instead of a second planting of rice, such food crops as corn (maize), cassava, and pulses are frequently planted. As rice cultivation requires a dependable supply of water, Thailand and the Philippines rely on a rain-fed system, whereas Indonesia mostly utilizes irrigation systems. Since the introduction of high-yielding varieties (HYV) of rice in the region, the role of water-control, along with the use of chemical fertilizers and pesticides, has been a critical element in the diffusion of the so-called Green Revolution. The results have been mixed because the new, hybrid strains of the high yielding varieties have benefited the larger farmers, who can afford these strains and the additional cost of chemical fertilizers and pesticides more than the poorer farmers.

In addition to subsistence crops, a large variety of cash crops are also grown for the local and export markets, both on large plantations and by individual, small-holders. Tree crops such as rubber, coconuts, and oil palms are also important commercially although land devoted to them is restricted to areas closer to the equatorial regions of Malaysia, Indonesia and southern Thailand. Coconuts and sugar are important in the Philippines. Other major commercial crops are cacao, coffee and spices which are mostly exported, while crops such as sweet potatoes, peanuts and tobacco are consumed domestically. The cultivation of opium poppies has lately become fairly substantial for illegal export in parts of Myanmar, and Thailand in the so-called "Golden Triangle" area.

Since the colonial period rubber and oil palms had commanded a considerable importance in agriculture arising primarily from the worldwide demand, and on account of a nearly continuous harvest period that provides year-round employment. Foreign corporations once dominated production, but as the nations gained independence since World War I, much of the production was nationalized. Although government ownership now continues to predominate, private ownership has been on the increase.

Fishing is a significant source of diet and livelihood, although it contributes only a small proportion of the gross domestic production. The production of fish catch has lately increased with newer technology. The maritime nations of Thailand, Indonesia and Malaysia have global fishing industries. Shrimp catches are in considerable demand in the world market. Agriculture is on the increase in the region; such varieties as shrimp and carp are being raised in ponds.

Industrialization in Southeast Asia is a relatively recent phenomenon. Most of the development in manufacturing has occurred only since the early 1960s. In all these countries, except Brunei, industry's share in GDP is small but has substantially increased during the last three-four decades. The greatest growth has been registered by Singapore, Thailand and the Philippines, although the substantial gains in manufacturing were recorded by Indonesia, Thailand and Malaysia as well during the 1980s.

With the notable exception of Singapore agricultural processing is the most important manufacturing activity in practically all Southeast Asian nations. In Singapore electrical, electronic and transport equipment and a variety of products are the main products. Textiles and clothing are major products of Thailand, the Philippines and Myanmar. Chemical industry has lately become important in Thailand and Indonesia. Generally speaking, industrial units are small in terms of number of workers employed and goods produced. Light, labor-intensive goods, such as electrical and electronic products, are becoming increasingly important.

Southeast Asia has modest reserves of oil and natural gas (4 to 5 percent of the world), most of the fields are located in Indonesia, Malaysia, and Brunei. Mineral exploitation of other minerals is limited in terms of world production, although nickel, copper, and chromite are mined in some quantities. Tin is the most important mineral in the region in terms of value; Thailand, Malaysia, and Indonesia account for more than half of the world's production. In Malaysia, most mines are getting depleted, and the remaining ones are less economical to mine. Fluctuating world prices have occasionally depressed the production of tin.

Given Southeast Asia's strategic location at the main trade routes between the West and the Far East. Trade and commerce are of special significance to all nations in this region, although the region's entire trade is about one-third that of the United States. Exports of such nations as Cambodia, Laos, Vietnam, and Myanmar are very small, and of Thailand, the Philippines, and Indonesia are moderate, whereas Singapore, Malaysia and Brunei have a large proportion of trade in exports. Noteworthy is the dominant role of oil in the export trade of Indonesia, Malaysia and Brunei. Indonesia has, however, been able to diversify its exports to include plywood, rattan, coffee, rubber, and textiles during the last decade or so. Though Malaya's exports include a variety of commodities such as palm oil, tropical hardwoods, and tin, it earns most of its export revenue from petroleum products. Thailand's export base is less diversified; as food and manufactured goods account for nearly all of its total trade. Brunei relies almost entirely on its oil exports. Singapore has utilized its unique geographic location and highly educated labor-force to attract multinational corporations which have been responsible for its highly developed manufacturing and service sectors. Interregional trade among Southeast Asian nations is small particularly that of Philippines, amounting to one-fifth of the region's total trade. Philippines' trade reflects its long-term orientation toward the United States. The trade of the region, in general, is with the nations outside the region, dominated by Japan, Europe, and the United States, but is becoming increasingly oriented to Taiwan, Hong Kong and Korea.

Further Readings

Dwyer, D. (ed.) *Southeast Asian Development: Geographical Perspectives.* New York, 1990.

Fryer, D.W. *Emerging Southeast Asia.* New York, 1979.

Ulack, R. *Atlas of Southeast Asia.* New York, 1988.

18

Mainland Southeast Asia

Mainland Southeast Asia includes the nations of Myanmar, Thailand, Malaysia (western part), Cambodia, Laos, and Vietnam. East Malaysia and the Sultanate of Brunei located on the island of Borneo, for convenience, have been included in this chapter.

MYANMAR

In physical dimensions, Myanmar (formerly Burma) is the largest of the nations of mainland Southeast Asia. The country has been a "hermit" nation since independence from the British in 1948. Deeply suspicious of the West, and marked by internal instability and economic stagnation, it has virtually shut itself off from the outside world----a condition stemming from its peculiar geography, and pre-independence history of subordination of the colonial past. Lacking ethnic coherence, the Burmans, who form over two-thirds of the country's population, are concentrated in a comparatively smaller, central lowland "core" area bounded within a framework of peripheral ethnically diverse area of mountains. In the north, these mountains border China, to the west India and Bangladesh, to the east Laos, and Thailand.

Myanmar extends from latitude 10°N to 28°N, and thus is the northernmost country in Southeast Asia, with a considerable territory located beyond the tropics. The southern elongated and narrow extension, the "tail" of the country runs south of the "central core" along the Malay Peninsula----making its total length from north to south about 1,275 miles (2,050 km).

Physical Framework In general, the topography slopes from north to south, from an elevation of 19,296 feet (5,881 meters) in the extreme north, to sea level at the Irrawaddy and Sittang river deltas. The country can be broadly divided into five regions based on relief features: The northern mountains; the western ranges; the eastern plateau; the central lowlands, and the delta of River Irrawaddy; and the southern coastal plain. The orientation of these divisions is chiefly in a north-south direction. (1) The northern mountains border China and are rugged and steep-sided, rising to elevations of 9,000 feet (2,743 meters or so). (2) The western mountains adjoining India contain the ranges of Patkai, Naga and Chin and form a formidable barrier to movement to the Indian subcontinent. These mountains rise to elevations roughly equal to that of the northern mountains and continue as Arakan Yoma close to the Bay of Bengal

Map 18.1

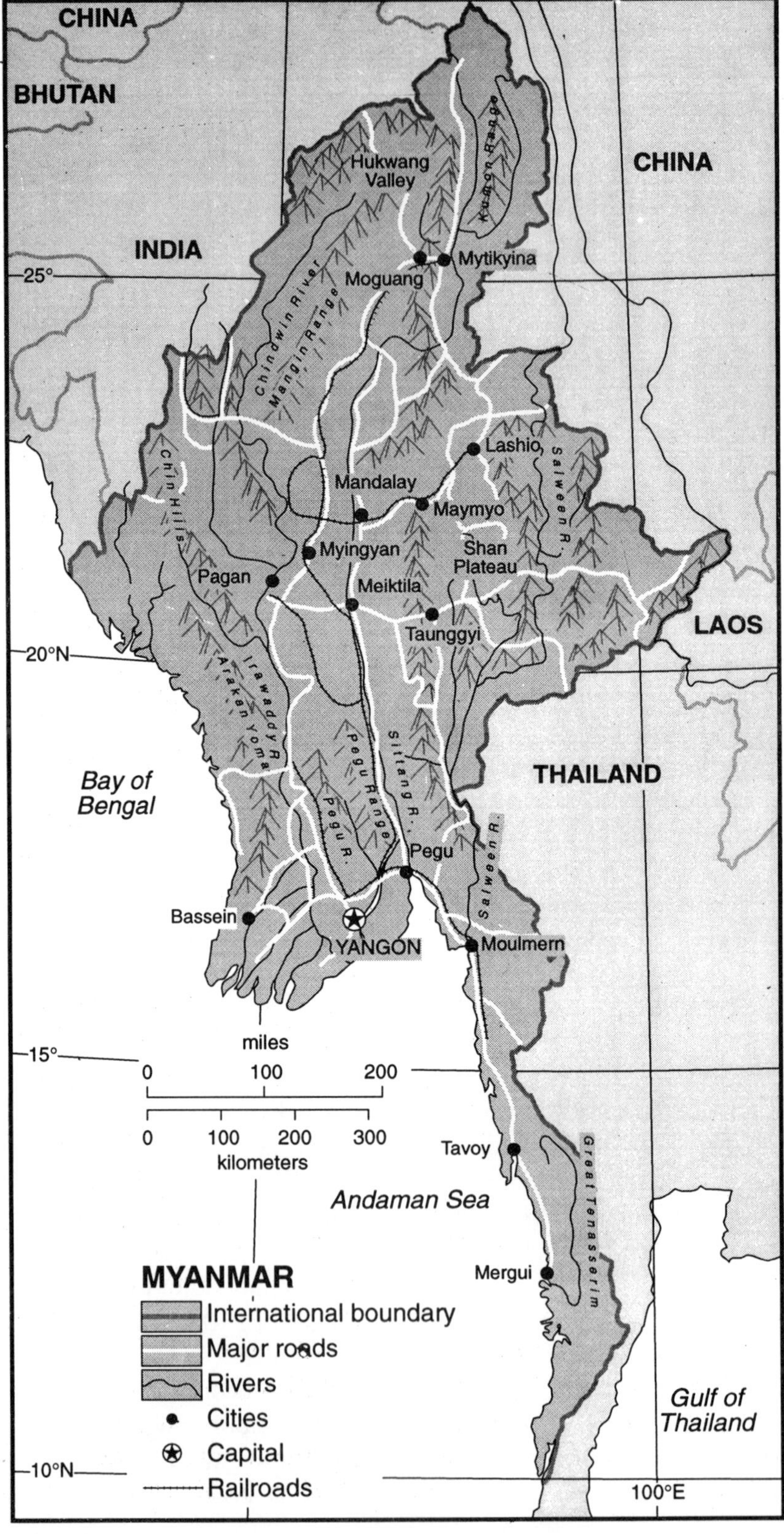
CHINA
BHUTAN
CHINA
INDIA
Hukwang Valley
Kumon Range
Mytikyina
Moguang
Chindwin River
Mangin Range
25°
Lashio
Salween R.
Chin Hills
Mandalay
Maymyo
Myingyan
Shan Plateau
Pagan
Meiktila
Taunggyi
LAOS
20°N
Irawaddy R.
Arakan Yoma
Pegu Range
Sittang R.
Pegu R.
THAILAND
Bay of Bengal
Pegu
Salween R.
Bassein
YANGON
Moulmern
miles
15°
0
100
200
0
100
200
300
kilometers
Tavoy
Great Tenasserim
Andaman Sea
MYANMAR
International boundary
Major roads
Rivers
Cities
Capital
Railroads
Mergui
Gulf of Thailand
10°N
100°E

leaving a narrow coastal plain. To the east of Arakan Yoma are the central lowlands drained by Irrawaddy and Chindwin rivers. The Irrawaddy reaches the Andaman Sea through the many channels of its delta (of some 10,000 sq miles or 26,000 sq km). Both the Irrawaddy and the Chindwin are navigable for a large part of their lengths and make an extensive lowland. The Irrawaddy has been the country's great national highway, and a major commercial artery. (4) The eastern plateau consists of uplands or plateaus with elevations of about 3,000 feet (925 meters), drained by the Salween. (5) The southern coastal plain is narrow and elongated, fringing on the west by the steep Tenasserim ranges that form Myanmar's southeastern border.

The country's climate in large measure follows the pattern of the Indian subcontinent. It is dominated by the southwest summer-season monsoon winds, as it lies mostly in the tropics, and is adjacent to the Indian Ocean. Rainfall is heavy in a narrow, coastal band, particularly in Arakan and the Tenasserim, occurring primarily between May and October, and averaging 200 inches (5,000 mm) annually. As the winds cross the Arakan, rainfall decreases to between 25 and 40 inches (6501,000 mm) in the central lowlands, the socalled "Dry Zone" of the country. By mid-November the winds reverse, their direction to northeast, and are dry. The hot season begins in February and lasts until the rains start. During the hot season the temperatures rise to 100°F (38°C) or more in the Dry Zone, while the coastal areas are about 10°F (6°C) cooler.

Cultural Patterns The country includes a wide diversity of ethnic groups. The geographical heart is the central lowlands and the original home of the Burmans, focusing on Mandalay in the Irrawaddy River basin. Burmans account for about 70 percent of the country's population of over 48 million. The Irrawaddy-Sittang delta region was once inhabited by the Mon people who may have entered the country from the east, and were largely absorbed by the Burmans. They now number about 2.5 percent of the population. Most of them have been integrated into Burman culture and no longer speak their original languages. They are believed to have moved into Burma much earlier than the Burmans, and settled along the coast and came into contact with India, adopted Theravada Buddhism, and Indian law into local practices. The mountainous peripheries of the Irrawaddy lowlands form the homelands of non-Burman peoples, amounting to almost 30 percent of the total population. In the western hills along India's borders are the various groups broadly known by the name of Chin, comprising 2.2 percent of the population. Another important group is that of Nagas, many of whom fall on the Indian side and have been agitating for the establishment of a separate homeland for the Nagas. The upper Irrawaddy valley and the northern hills are occupied by groups such as the Kachins that straddle the frontier with China. The Karens, constituting 6.2 percent of the country's population, have settled in significant numbers in the central lowlands, although ethnically and linguistically they are Tibeto-Burmans in origin. They are the most sophisticated of the non-Burman groups and were pushed by the Burmans into the Irrawaddy delta and the adjacent territory. Additionally, during the British colonial rule, there were sizable communities of South Asians and Chinese, but many of these people left at the time of World War II, and in 1963, when commerce and industry were nationalized.

The official language of Myanmar is Burmese which belongs to the Sino-Tibetan family. The Shan language spoken mostly in the eastern plateau region belongs to the Tai family. The language spoken by the Mons is related to the Mon-Khmer languages. During the colonial period, English was the official language, but Burmese continued as the primary language. Both Burmese and English were made compulsory subjects in schools and colleges. Burmese, Chinese and Hindi (a north-Indian language) were used in commerce. Although English ceased to be the official language after 1963, it was retained as an important language used by the intelligentsia. The country had one of the highest rates of literacy in Southeast Asia (83 percent) during the British rule, particularly among the females (nearly 90 percent) but literary levels have declined somewhat on account of economic and political stagnation that followed the post-colonial period.

The Shans form the largest non-Berman community comprising 8.5 percent of the population, are concentrated primarily in the lower portions of the plateau and hill-territory in the east adjoining China and Laos. The Shan region comprising approximately one-fourth of Myanmar's territory, has traditionally been ruled by several hereditary chieftains (Sawbwas), who enjoyed semiautonomous status. The Shans have little ethnic-linguistic affinity with the Burmans, but most often share the Buddhist faith with them. They have, like other minorities (the Chins, Kachins. Nagas and Karens) considered the possibility of secession but have not posed a serious threat to the central administration.

Except for a small number of Christians and Muslims, almost all of the ethnic Burmans are Buddhists. According to official sources, nearly 90 percent of Myanmar's population in 1983 professed Theravada Buddhism. The Karens are divided between Buddhism and Christianity. The other hill peoples are mostly animists, although most of the Kachins and Chins profess Christianity, intermingling it with a degree of animistic practices. In all, Christians and Muslims form nearly 5 and 4 percent respectively of the country's population.

Economic Activity Today Myanmar's economy is one of the least developed in Southeast Asia. Prior to World War II, Burma was a thriving British colony—a prime exporter of rice, timber, and minerals. During most of the British rule, the prosperity of the Burmese, together with that of the immigrant laborers and merchants from India and China, seemed assured. And its physical geography appeared to be unusually favorable for the development of economic well-being and national cohesion. World War II severely disrupted its economy and left most of the economic facilities inoperative. By the 1960s, economy nearly collapsed. All large industrial enterprises, the banking system, insurance, foreign trade, and domestic wholesale and retail trade were nationalized by the administration. Ever since, strict regulations enacted hostile to foreign investment and skepticism of the foreign labor, internal political strife and suppression of various ethnic and tribal groups, have largely resulted in Myanmar's economic decline.

Basically, Myanmar is an agricultural country, and agriculture accounts for about 70 percent of the labor force, and 62 percent of the value of gross domestic product. Rice cultivation predominates in the delta region, but a wide variety of crops are grown in the Dry Zone, and the peripheral hill and plateau regions where forestry and shifting agriculture are also practiced. About one-half of all cultivated land is devoted to rice cultivation. The total production amounts to about 21 million metric tons annually. Rice production of Irrawaddy delta provides much of the country's export earnings. During the colonial period, the indigenous labor-force was considered insufficient to support the largely rice-based economy; Indian and Chinese immigrant-labor was recruited during the early part of the 20th century, which by 1942 made up about 13 percent of the country's population. But by the late 1940s and early 1950s most of the Indian community left the country.

In addition to rice, several crops including corn, wheat, millets, cotton, peanuts, sesame, legumes, tea, tobacco, sugarcane and some jute are grown. Most of Myanmar's irrigated land is in the Dry Zone, and most of it is planted in rice, and jute.

Nearly two-thirds of Myanmar's cultivated area is in the hill and plateau lands that lie peripheral to the Dry Zone and the Irrawaddy delta, and is home of many of the country's non-Burman ethnic group. They usually practice shifting cultivation (called *taungya* in Burmese), although sedentary agriculture has recently been on the increase with the advance of agricultural technology and central planning.

Although fish is the second most important element in the Burmese diet after rice, marine fisheries are not well developed. The country has a valuable and rich resource of timber, and forestry is important as a source of foreign exchange. Its hill and plateau areas are estimated to possess a bulk of the world's exploitable teak supplies. The forests are owned and regulated by the state. Forest products account for nearly a quarter of the nation's export earnings, whereas the agricultural products, primarily consisting of rice, make up for another 30 percent or so.

Myanmar's mineral endowment is considered to be one of the richest of mainland Southeast Asia. Minerals and gems account for about 10 percent of export by value (a decline from 40 percent in 1940 due primarily to depletion in petroleum production during the last fifty years). Silver, lead, zinc and copper are mined in northern Shan Plateau; and tin, tungsten in Tenasserim area. Rubies and sapphires are also extracted in Shan Plateau. Oil and natural gas are produced in the Irrawaddy for domestic purposes; their production has substantially decreased as most of the refineries located north of the Irrawaddy delta, were destroyed during the World War II by the Japanese. Coal is found in the upper Chindwin valley. Hydroelectric resources of the nation are fairly extensive; the administration has built several hydroelectric plants in northern Arakan state, and near Mandalay, but there remains a good deal of unutilized potentials.

Industrialization before independence was very meager. Yangon, Myingyan (in the Dry Zone) and Arakan states were selected by the government to be the new areas for industrial development. Textile factories were located in these centers. Yangon has also steel processing and pharmaceutical plants. The administration encourages the cottage industries by providing subsidies. Manufacturing remains underdeveloped and accounts for less than one-tenth of the GNP. The principal manufactures include refined petroleum products, cement, sugar, peanut oil, fertilizers, textiles, paper, food (mainly rice mills), and lumber.

The Irrawaddy River, navigable year round up to Myitkyina in the north, is the major transport artery of the nation, and essential to rice trade which is largely dependent on water transport. The Chindwin is navigable for about 500 miles from its confluence with the Irrawaddy below Mandalay. The various streams of the Irrawaddy delta are navigable. The country has a railroad track of about 2,000 miles----the major one connects Yangon, the chief port with Myitkyina in the north----and runs parallel to the Irrawaddy.

Myanmar is a land of villages. Only a quarter of its population was listed as urban in the mid-1990s. With the exception of a few large cities such as **Yangon** (previously Rangoon), **Mandalay and Moulmein** the towns essentially are extended villages. **Yangon** (population 2.5 million) is the capital, a transportation and commercial hub, the chief port and the largest city of the nation. Its rise from a small fishing village until the 19th century to a large metropolis has resulted primarily from its former position as the world's largest rice and teak port. Sited nearly 20 miles (32 km) up stream at Yangon River, it has the advantage of controlling the rail and road approaches to the Irrawaddy valley. The port is deep enough for ocean-going vessels. Well-located on the delta to the northern and southern parts of the country by roads, rail lines, and waterways, it has dominated its external trade and passenger traffic. Major industries located in and around the city include rice mills, textiles, paper, and fertilizers.

Prospects Today one of the least developed of the Southeast Asian nations, Myanmar has slipped enormously in world economy since independence. Mineral and oil exports have sharply declined and rice and teak exports trended downward during the last three decades. Due to decades of experimentation in centralized planning, and isolationist policies, the nation has suffered heavily. The ordinary Burman today is little, if at all, better off than its pre-independence counterpart.

Potentially, Myanmar has several advantages over many developing nations. It is self-sufficient in its food requirements, and possesses abundant forest and mineral resources. The basic problems confronting the country are: recovery of pre-independence export trade in rice, teak and oil; nurturing of mining and manufacturing; political integration of non-Burman minorities and the resuscitation and rejuvenation of the disorganized economy in general. Solution of these problems depends largely on government policy regarding the management of economy, particularly in matters of prioritization, the foreign investment in trade and industry and distribution of land ownership. In 1988, large-scale discontent over administration's socialistic policies based on political centralization and military rule led to a massive revolt and serious disruption of the administration and the demand for the creation of a more liberal, and democratic government. A recent partial liberalization of the economy and a more open-door policy toward the ASEAN (Association of South East Asian Nations) and Japan have brightened some prospects for the economy, but the military government still clings tenaciously to the insulating, restrictive, and outmoded policies that have brought the nation to economic stagnation and political instability.

THAILAND

Next to Myanmar, Thailand is the second largest country on the Southeast Asian mainland. Its territory of 198,115 sq miles (over 513,117 sq km) shelters a population of over 62 million. Geography and history have conspired to make the country a unique nation. There has been a major Thai state in the present territory of Thailand for the last six hundred years, and the country is one of the very few in Asia to have escaped European colonialism. Lying off the major historic sea lanes, it was spared the influences that shaped the maritime world to the south and east----notably the Muslim religion and the European rule. Thailand has thus acted as a buffer between the conflicting interests of France and England for control of the region, and partly because of this competition succeeded in preserving its independence during the European colonial partition of Southeast Asia. The monarchy became and remains a potent symbol of the country's historical continuity and national identity. For over a century, the country has practiced a neutral stance in world affairs, and its post-1950 dependence on the West is a sharp break with tradition.

The non-colonial development of the economy is illustrated by the fact that emphasis was placed almost entirely on non-estate agriculture except for the rubber plantations in the southern peninsula. The great rubber, tea, coffee, coconut, palm, and other plantation estates of Indonesia, the Philippines and Indonesia are conspicuously absent in the nation as has been the European and American capital. It is only in the exploitation in tin extraction in its peninsula region and in the growing manufacturing sector that the American and European involvement has been significant. Thailand, for a long time, remained neglected by the West, partly because the colonial powers were engaged elsewhere, and due in part to Thailand's location off the historic routes of maritime trade.

Physical Characteristics Thailand's physical configuration is simple: a south-facing river basin enclosed on the west, north, and the southeast by mountains, and a long, slender peninsular finger in the south. The northern and western mountains are the southward continuation of the complex mountain system of the Himalayas from eastern Tibet curving to the south that, in part, form the boundary between southern Myanmar and Thailand. These mountains are a series of north-south ranges, rise to nearly 8,000 feet (2,440 meters), and trend southward into Malaysia. To the north are the hills and dissected plateau region of Myanmar that contains caves from which remains of prehistoric humans have been excavated. The Khorat Plateau in the northeast, covers a third of the country that gently tilts toward the east, and lies in the drainage of the Mekong. The Plateau is enclosed on

Map 18.2

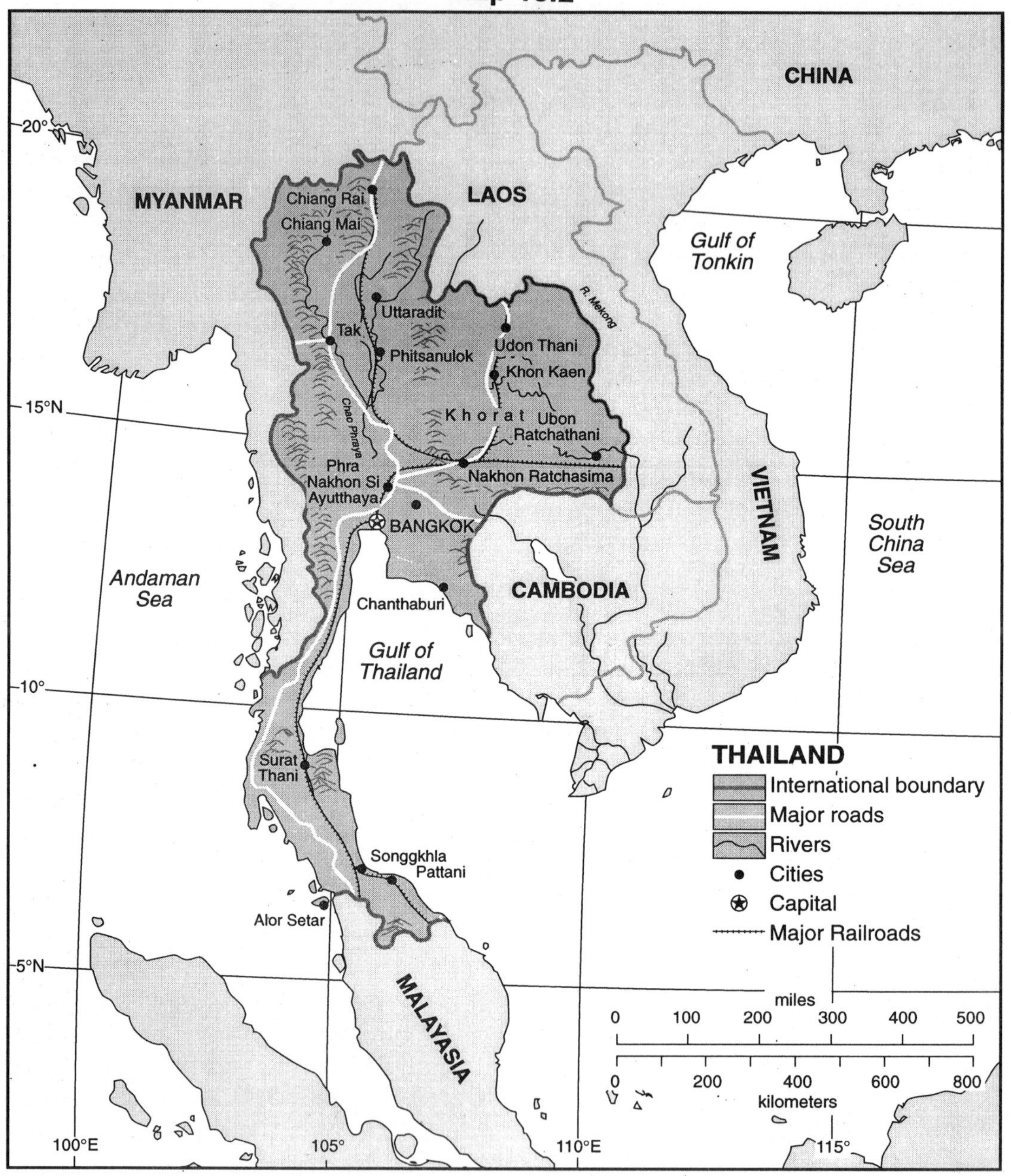

CHINA
MYANMAR
LAOS
Gulf of Tonkin
Chiang Rai
Chiang Mai
Uttaradit
Tak
Phitsanulok
Udon Thani
Khon Kaen
R. Mekong
Khorat
Ubon Ratchathani
Chao Phraya
Phra Nakhon Si Ayutthaya
Nakhon Ratchasima
BANGKOK
VIETNAM
South China Sea
Andaman Sea
Chanthaburi
CAMBODIA
Gulf of Thailand
Surat Thani
Songgkhla
Pattani
Alor Setar
MALAYASIA
THAILAND
International boundary
Major roads
Rivers
Cities
Capital
Major Railroads
miles
0 100 200 300 400 500
0 200 400 600 800
kilometers
20°
15°N
10°
5°N
100°E
105°
110°E
115°

the west and south by low, linear hills. Surface elevations on the Khorat range from 650 feet (198 meters) in the northwest to 300 feet in the southeast. Lying between the northern and western mountain ranges and the Khorat Plateau is a sizable Chao Phraya River basin, which is the cultural and economic heartland of the country, known also as the Central Lowlands. This region consists of rolling plains in the north and a low-lying flood plain and delta of the Chao Phraya formed by the large deposits of alluvium brought by the tributaries of the rivers. The alluvial deposits of the river valleys are the most fertile in Thailand, as these are replenished year after year with river sediments swollen with annual monsoon rains. The topography of the peninsular arm is rolling to mountainous, with little flat land. Higher mountains rise to about 5,000 feet (1,524 meters) on the west. off the rugged and indented coast, lie several small islands, including the Phuketl Island, which is rich in tin.

The climate of Thailand may be described as tropical monsoonal. The major influences are the country's location in the tropics, monsoon zone and the topographic features affect the distribution of rainfall. In early May the southwest monsoons flow from the Indian Ocean, and bring large amounts of rainfall, which reaches a maximum in September. The wind system is reversed between November and February, when a northeast monsoon brings cool, dry air. Occasionally, typhoons may come across the China Sea and bring some rain but fades out across Thailand. The amount of rainfall varies from 40 inches to 120 (1,016 to 3,048 mm) in the various parts of the country. In the southern peninsular region a dry season seldom occurs and receives as much as 160 inches of precipitation annually, whereas Bangkok gets 55 inches (1,397 mm) and Khorat, sheltered by hills on all sides even less than 30 inches (762 min) and almost the whole of the peninsular region receives over 80 inches distributed throughout the year. Temperatures are, in general, moderate to high, averaging between 77° and 84°F (25° and 29°C). The season of highest temperatures is in late March, April and early May. In central, peninsular and southeastern Thailand, maximum temperatures seldom reach 100°F (37.7°C), while minimum temperatures are lower than 65°F (18.3°C). In northern Thailand, temperature range tends to be much larger. Soils of the river valleys are fertile, and the most fertile land is in the flood plains of the lower Chao Phraya basin because it receives large amounts of the rich, alluvial deposits of soil every year. Relatively flat areas elsewhere and parts of the coast also have fertile soils. Elsewhere, soils tend to be poor, highly leached laterites of the humid tropics.

Cultural Patterns Among Southeast Asian countries, Thailand is the most identity-conscious nation. Relatively homogeneous, the country does not possess the multiplicity of languages found in Indonesia and the Philippines, nor contains a complex ethnic mix as in Malaysia. Eighty-five percent of the population speaks Thai, which is a member of a large cluster of languages spoken in all bordering countries as well as southern China and northern Vietnam. Like the people of Myanmar, Cambodia and Laos, the Thais are Buddhists of the Theravada school. In 1991 ninety-five percent of the population was listed as adhering to Buddhism. The minorities include Muslims (who account for four percent of the population), Hindus, Sikhs, and a few Christians, which are concentrated chiefly around Bangkok. The national government plays down regional loyalties, and the Thai language is taught in schools throughout the country.

Non-Thais number nearly 12 million or 20 percent of the population. The largest ethnic minority, comprising over 8 million or 12 percent of the total population are Chinese, who have been assimilated to a far greater degree than in either Malaysia or Indonesia. There are no barriers to intermarriage, and most have embraced Thailand's Buddhism.

The next largest minority is that of Malays, who profess the Muslim faith, and are largely concentrated in the southern peninsular neck of the country close to the Malaysian border. In the northern and northwestern part of the country along the Myanmar border are several hill people----the tribal groups, chief of which are Karens, and Shans (numbering over one million each). Most are shifting cultivators. Also included among Thailand's minorities are Vietnamese, who moved and settled in the northeastern part of the country in the 1940s and 1950s to escape Indo-China war with the French, and Khmers (Cambodians) who fled their homeland as refugees after the 1979 Vietnamese invasion of Cambodia. Such border areas inhibited by the minority groups in the north, northwest, northeast and the southern peninsula are imperfectly integrated into the Thai state, and are economically backward as well. The Thais dominate the lowlands, and there are layers of non-Thai people in the mountainous borderlands.

Economic Activity Traditionally, agriculture has been the dominant sector of Thailand's economy. Although through government encouragement to small industry, its contributions to economic growth have declined consistently since 1950. The proportion of the agricultural labor force has declined from 88 percent in the 1950s to less than 50 percent. Agriculture's contribution to the national economy relative to manufacturing has also declined from more than 50 percent in the 1950s to less than 11 percent in the 1999. Despite this shift to manufacturing, agricultural production has continued to expand, and Thai farmers continue to produce enough rice for the country's needs as well as a surplus for export.

Today, Thailand is the world's fifth largest producer of rice and its largest exporter (exporting one-third to a quarter of rice exports of the world). Agriculture is overwhelmingly associated with rice cultivation, and close to ninety percent of the country's cultivable area is given to it, nearly one-half of which lies in the Chao Phraya basin where the flood waters of the river provide irrigation and silt-laden fertile soils to the fields. During the 1960s movement toward crop diversification became popular and the farmers began growing such other export crops as maize, sugarcane, pineapples, tobacco, coconuts, and *kenaf* (a substitute for jute) on a larger scale than before. These crops have since been slowly acquiring greater prominence. In addition, large quantities of vegetables and fruits are also grown. Cattle breeding is important in the Central plains, and pigs and poultry are widely raised. Fishing is also of considerable importance, and constitutes a growing export commodity. Rubber production---introduced into the country during the 19th century---is important in the southern, peninsular section. Thailand ranks third in the world in natural rubber production. It produces nearly one-sixth of the world's production of hardwoods---particularly teak. Its major forest products, are now exported in small quantities, following a government ban on logging imposed in 1989.

Mining constitutes a small segment of nation's economy, with only 0.2 of laborforce engaged in it and contributing less than 2 percent to the domestic gross product. Tin, mined mostly in the peninsula, has long been a valuable mineral resource, and the country has become one of the world's largest tin producers, producing on the average about one-tenth of the world's total output. Coal, zinc, gypsum, tungsten, and limestone are some other minerals produced.

The manufacturing sector has dramatically grown during the last four decades, representing primarily the large investments made by private firms; the larger ones have been financed by foreign and Thai capital. Japan, South Korea, Taiwan, and Singapore have been the major sources of investment for industry, that is particularly oriented to producing consumer goods such as clothing, canned goods, and electrical products. Japanese capital is increasingly invested in the manufacture of textiles and machinery. At the same time, growth of the traditional, factory-type industry including that of rice milling, sugar and timber, the manufacture of tobacco, jute sacks and cement as the production of textiles (especially based on silk), clothing, furniture, and footwear owned primarily by domestic investors has also registered substantial gains. Factory industry is heavily concentrated in the Bangkok area.

Thailand's imports include electrical machinery, minerals and fuels, iron and steel, vehicles, plastics, and organic chemicals---items necessary for its growing industrialization and domestic needs. Its major exports in the mid-1990s in the category of manufactured items were electric machinery, textiles and apparel, and nuclear reactors that collectively accounted for nearly forty percent of all exports, whereas the traditional exports of rice, tin, rubber, and teak made up for nearly 22 percent of the nation's export earnings.

Physical and Economic Regionalism

Physically, and economically, Thailand is composed of several distinctive natural units, although the key area is the central lowland, the plain of the Chao Phraya, which accounts for about one-fifth of the country's territory and two-fifths of its population. This is the area of most compact Thai settlement and most important agriculture. Population densities are highest of any region: over 600 persons per sq mile (230 persons per sq km). It was formerly forested but now consists of unbroken paddy (rice) fields. Soils are extremely fertile, composed of rich alluvium brought by the river. Despite receiving a relatively low total rainfall of a little over 50 inches (1,250 millimeters) a year, it is the country's agricultural heartland and the rice basket. Cassava, maize and other crops are also grown here. For most of the nation's history, the capital has been located here and the people of the central lowland have been the dominant group in the country.

Most of Thailand's commercial, industrial, and service industries are located in the central lowland, focused largely on **Bangkok,** the capital. The most important theme of the nation's modern history has been the steady concentration of political authority and economic power in a centralized government and at a single place: **Bangkok,** the capital (population 5.6 million), which has come to concentrate all facets of Thai life to a remarkable degree unsurpassed elsewhere. In the process, the city grew to be a classic example of a "primate city," collecting nearly 10 percent of the national population; its metropolitan area is nearly 30 times larger than the next biggest city---**Nakhon Ratchasima** 250 miles to the northeast in the Khorat Plateau. Containing more than 300 Buddhist temples, the royal place, and other cultural attractions, it is a tourist Mecca. Most of the country's trade passes through its port, and the manufacturing sector is growing rapidly. **Chiang Mai** (population: 1.6 million) located in the north, is another tourist center outside the capital.

The vast northeastern region, separated from Laos by the Mekong River, is the plateau area of Khorat. Not blessed with the fertile soils and adequate precipitation of the central plains, it is the poorest area of Thailand, and contains about eight million people who are officially designated as living in poverty. Like the northern region, this area had a history of semi-autonomy until the late 19th century. The people speak a language similar to the Lao, and have often displayed discontent with the central Thai administration, which has recently been trying to bring them into the national fold. The long peninsular tail to the south which joins central Thailand with Malaysia is less fertile, but is the country's major rubber-growing and tin producing region.

Prospects Thailand has recorded some of Southeast Asia's most impressive economic gains (averaging between 6 and 7 percent a year) during the last three decades. The fastest expansion has been seen in the manufacturing, service and trading sectors. Domestic markets have expanded and production of such commodities as cement, soft drinks and textiles has continued to grow. American military expenditure during the 1960s and 1970s and Japanese investments further bolstered the economy.

Between 1950 and 1970 a rapidly growing population particularly in the Central lowlands and around Bangkok had caused great concern, and the administration which had previously supported population growth reversed its policies. Since 1970s the family planning programs of the government helped to substantially reduce the population growth rates, which now stand at 0.9 percent a year at nearly one-third of those prevailing during the 1960s and 1970s. The country is now a model for other developing nations seeking to reduce their rates of population increase. However, a third of Thailand's population belongs to the youthful age group (between 20 and 40) that creates high demands on the nation's education, housing, health and employment systems, but the government is trying to utilize its highly literate human resource (with a literacy rate of over 90 percent) for economic development.

MALAYSIA

Malaysia is a nation of two parts: West Malaysia, the peninsular part on the mainland bounded on the north by Thailand; and East Malaysia on the western coast of the

Map 18.3

10°N
5°
0°
5°
THAILAND
South China Sea
Banda Aceh
Alor Setar
Kedah
Kota Baharu
Medan
Pematangsiantar
KUALA LUMPUR
MALAYSIA
Johor Baharu
SINGAPORE
SUMATRA
Sumatera
Palembang
Telukbetung
BANGKA
BELITUNG
INDONESIA
Sunda Strait
BRUNEI
Kota Kinabalu
Sandakan
Bandar-Seri-Begawan
Miri
Bintulu
Iran Mts.
Kuching
E. MALAYSIA
Pontianak
KALIMANTAN
INDONESIA
Samarinda
Balikpapan
Banjarmasin
Java Sea
Ujong Pandang
GREATER SUNDA ISLES
JAVA (see inset)
BALI
LAMBOK
FLORES
FLORES
Celebes Sea
HALMAHERA
MALUKU
TIMOR
LESSER SUNDA ISLES
IRIAN JAYA
PAPUA NEW GUINEA
Arafura Sea
AUSTRALIA
Indian Ocean
MALAYSIA AND INDONESIA
International boundary
Major roads
Rivers
Cities
Capital
Railroads
100°E
105°
110°E
115°
120°E
125°
130°E
140°E
JAKARTA
Serang
Bogcr
Bandung
Cirebon
Semarang
Surakarta
Yogyakarta
Surabaya
Malang
Banywangi
miles
0
200
400
600
0
400
800
1000
kilometers

island of Borneo. Between these two parts lie 400 miles (645 km) of South China Sea that gives the country an oddly dislocated appearance on the map. With an area of 127,584 sq miles (330,442 sq km), it is a little larger than Great Britain which formerly ruled it. Malaysia represents the political union of the territories under British rule that included Malaya on the peninsular part on the mainland Southeast Asia, the island of Singapore, and the states of Sarawak and Sabah on the island of Borneo. Malaya became independent in 1957, and in 1963 it joined with Sarawak and Sabah, and Singapore in creating the new nation of Malaysia. Singapore withdrew from the federation in 1965 and became an independent republic. The Sarawak and Sabah unit was officially named East Malaysia which covers about 60 percent of Malaysia's land area, although less than 15 percent of Malaysia's 23 million people live there. East Malaysia shares a land boundary with Indonesia's Kalimantan province.

Physical Environment West Malaysia has a mountainous core, with hills aligned in a north-south direction; the most prominent range rises to elevations of 7,200 feet (2,190 meters). The peripheral U-shaped coastal plains and lowlands, 10 to 50 miles (16 to 18 km) wide along the west coast and narrower and discontinuous along the east coast, surround the highland core. The coastline is over 1,200 miles (1,920 km) in length, and is mostly lined with mangrove swamps. East Malaysia also has a mountainous interior, with the usual swampy coastal plain of 10 to 40 miles wide. Sabah's coastline is rugged and indented, while that of Sarawak's comparatively more regular. The mountainous eastern boundary forms the divide between East Malaysia and Indonesia. Both East Malaysia and West Malaysia are covered with tropical evergreen rain forest, although that of East Malaysia is far more extensive than that of West Malaysia; indeed, these forests make up the largest tract of untouched rainforests in Southeast Asia. In West Malaysia, the unbroken forest is confined to the central mountain region. At lower elevations, stretches of jungle are interspersed with villages surrounded by rice fields, with plantations of rubber estates and settlements.

West Malaysia is located between 2° and 6° north latitude and most of the land is less than 100 miles (160 km) from the sea. It has an equatorial climate characterized by uniformly high temperatures averaging 78°F to 82°F (25° to 28°C), and by over 80 inches (2,000 millimeters) of annual rainfall. All of Malaysia is frost-free. Climatic conditions in East Malaysia are quite similar to those prevailing in West Malaysia, but the range of seasonal temperatures and the amount of annual precipitation are greater in East Malaysia. The soils of both East and West Malaysia have been exposed for a long period of time to intense tropical weathering and have been leached out and are poorer except for small areas along the coast. Only Sabah has some areas of fertile soil near the rivers.

Cultural and Historical Factors

Malaysia, like its neighboring nations, has a rich and diverse ethnic mix, but is unique in that its predominant Malay group has been in an uncomfortable minority until recently. In addition to its various indigenous groups, it also accommodates Chinese and Indian populations large enough to make the Malays a slim majority in their native land. In combination with other native inhabitants---know, like them, as *Bumiputras*, "sons of the soil"---the Malays make up a modest majority of 58 percent. Chinese account for about 25.4 percent, and Indians close to 7.2 percent of the population. This uneasy mix of races causes occasional bitter conflicts, and often dictates all political, social and economic calculations in the country.

Until the middle of the 19th century the country remained a rural, feudal Malay society. Malays were landowners, and peasant agriculturists. The picture changed dramatically when tin became a major item in international commerce and the tin mines of Malaysia proved of great value. The Malays, who themselves did not engage in industry, encouraged Chinese and Europeans to operate tin mines and to import large numbers of Chinese laborers from Southern China to work in the mines. Initially, living in large unassimilated communities, these organized several secret societies, and fought among themselves and the mine owners, often ignoring local laws. These societies became entrenched in the towns, posing a threat to the native Malays. The Malaysian ethnic composition was further compounded when, at the turn of the century, Tamil laborers in vast numbers from India were imported by the British rulers and estate holders to work on rubber plantations. In 1920 Malayan rubber production accounted for 53 percent of the world's total production. The importation of Chinese and Indian labor was originally welcomed by the Malays as it did not compete with the Malay economic structure but produced new wealth from tin mines and rubber plantations. However, political problems arose when the combined Chinese and Indian populations exceeded that of the Malays and when the Chinese and Indians were not allowed landowning rights, they slowly branched out into several commercial and business enterprises. They are now found in large numbers in cities throughout the nation and some parts of the countryside as well. They are especially concentrated on the densely populated western flank of the Malayan peninsula. Indeed, the ethnic mix in the small towns in these parts makes them microcosms of the nation.

In addition to the Malays, Chinese, Indians, and the aborigines, small numbers of Europeans, Americans, Thailand Arabs are residents in Malaysia---a reflection of the crossroads location of the Malay Peninsula and a meeting place for peoples from other parts of Asia. The aborigines (Orang Asli) contain several groups, some of which speak Malay and related languages but others speak Mon-Khmer languages. They are primarily adherents of traditional religions, although many have been converted to Islam. The Malays speak Malay, an Austronesian language (officially called Bahasa Malaysia) which is the national language, and is related to the languages spoken in Indonesia. They are overwhelmingly Muslim. Adherence to Islam generally distinguishes a Malay from a non-Malay; the number of Malays who are not Muslim is negligible. The Chinese are less homogeneous than the Malays in language and religion. They speak several dialects (Cantonese, Hakkai, and Hainanese) reflecting the region of their origins in China. In general, they do not follow a dominant religion, and most of them, while subscribing to Confucian faith, are either Buddhist or Taoist. A small minority is that of Christians.

The peoples from South Asia---Indians, Pakistanis and Sri Lankans, commonly grouped as Indians---can linguistically be divided into speakers of Dravidian languages (Tamil, Telugu, Malayalam) and speakers of Indo-European languages (Hindi, Urdu, Punjabi, Bengali, and Sinhalese). Most of the Indians and Sri Lankans are Hindus, while the Pakistanis are Muslims. The Sikhs from the Punjab portion of India adhere to their own religion, Sikhism.

The Ethnic-linguistic composition of East Malaya is even more complex. The main ethnic groups in Sarawak are: Chinese, Malays and the several aboriginal tribesmen including the Dayaks, all adding racial, linguistic and ethnic heterogeneity to the cultural scene. Sabah, even more than Sarawak has a kaleidoscopic mixture of peoples. The largest groups consist of the aborigines, while a small proportion includes Malays, Indonesians, Filipinos, Europeans and South Asians. A great number of aborigines are animists, and a small number of them have been converted to Islam or Christianity. Most of them are shifting cultivators. Although they are divided into subtribes, their languages are mutually intelligible.

The Chinese have been a special irritant in Malaysian political and economic structure. The Malays and Chinese differ culturally and economically. Most Malays speak Malay, the official national language, which is taught in the schools along with English, while Chinese mostly adhere to their own dialects, while frequently using English. Many Chinese have resisted learning Malay, even as a third language. Malays are Muslims, while the Chinese are Buddhists, Confucionists or Taoists. As elsewhere, dislike of Chinese is fueled by the fact that they control most commerce. Although each nationality practices its own tradition with respect to foods, festivals, music, art and architecture, Malays own most of the land and control most governmental posts. Numerically less significant than Chinese, are the Indians who are less assertive, and are rarely involved in the occasional racial conflicts. When the British controlled the country, they awarded "special privileges" to the Malays in terms of landowning rights, access to jobs in the civil service, recognition of Malay as the national language, and designation of Islam as the state religion. After independence in 1957, racial antagonisms surfaced during the early 1960s leading to serious race riots in 1969. Despite the Malaysian administration's efforts to pursue policies conducive to the development of national unity during the late 1970s and early 1980s, a true Malaysian nationalism has thus far failed to emerge.

Resources and Development Malaysia is rich in mineral resources. Its major metallic ores are tin, bauxite (aluminum), copper, and iron. The most valuable one is that of tin; minor ones include manganese, mercury, antimony, and gold. Tin mining is concentrated largely in the alluvial deposits along the western slopes of the Main Range in West Malaysia, with smaller deposits on the east coast. While a good part of the country's development effort was spent on the production of tin until the 1980's, Malaysia's most valuable mineral resources now are its reserves of petroleum and natural gas, although modest by world standard, are likely to grow. The major fields are all offshore, along the east coast of West Malaysia and Sarawak. Malaysia, in addition, contains considerable reserves of coal, peat, wood, and has a large potential for the development of hydro-electricity.

Since 1970, agriculture, fishing and forestry, the traditional basis of the economy, have been declining in importance. Their contribution to the national product has gone down from one-third in 1970 to less than one-fifth, although a large part of the workforce is still engaged in these activities.

Rice, the primary food crop, is grown on small farms. The total production is nearly two million tons. The production has declined a little during the last decade, primarily due to loss of farm labor to growing urban manufacturing sectors. As a result, the country which has been until the late 1980s, self-sufficient in rice production, has to make up the shortfall with imports chiefly from Thailand. Food items (rice being the major one) accounted for over 5 percent of Malaysia's imports in the mid-1990s.

Traditionally, the most important cash crops have been palm oil and rubber, but their proportion to Malaysia's exports declined considerably during the 1980s and 1990s, resulting primarily from the unpredictable fluctuations in the price of these commodities in the world market. This has also resulted in a decline in the number of plantations. Malaysia is still the world's largest producer of natural rubber, and in normal years produces 35-40 percent of the world's output. Most of the plantations are in the coastal and piedmont zone of West Malaysia and primarily in Chinese and European hands, although the European ownership has considerably declined since independence. Palm oil has become a more important commodity than rubber in terms of value and contributes significantly to the exports in the category of vegetable oils. Cocoa, pepper, and coconuts are other valuable crops raised.

Malaysia's export structure has changed dramatically since 1970, from the one dominated by rubber and tin to one in which manufactured items account for most of the export earnings. Electrical and electronic products, transport equipment, and machinery and other manufactured goods have replaced the traditional commodity goods as major exports. Its chief trading partners are Japan, Singapore and the United States. The newly industrialized Asian nations such as South Korea and Taiwan account for a growing share of its foreign trade. The transportation system has improved substantially since independence; although more attention has been received by the Peninsular Malaysia than East Malaysia. The nation has enjoyed a favorable balance of trade consistently for the past several decades.

In all Malaysia has a railroad track of only 2,222 km, but the total length of the paved roads is more extensive (nearly 70,000 km). The road network in East Malaysia is much smaller and river transport is of greater importance, especially in Sarawak. All the best ports such as **George Town** and **Port-Kelang** are along the west coast of West Malaysia. A new port of **Kuantan** has been developed on the eastern coast of the Peninsula. The **Kuching** in Sarawak, and **Kota Kinabalu** in Sabah serve East Malaysia. Since independence air transport has grown rapidly in West Malaya. **Kuala Lumpur** (1.1 million) in West Malaysia is the nation's capital and the chief inland city, served by Port Kelang on the strait of Malacca. Apart from its administrative functions, it is the country's most important commercial center of tin-mining and rubber production industry. Founded by the Chinese tin-mining in 1957, it became the capital of the Federated Malay states under the British in 1896, and was named the national capital when Malaysia gained freedom in 1957. Among the notable buildings that adorn the city is the modern Parliament building in Moorish style.

The move to industrialize has left only a fifth of all Malaysians still on the land, the remaining have taken to professional, commercial, and other jobs. The nation, however, feels hampered in its drive to industrialize by its small population base. It is encouraging procreation and a large family size in order to create a bigger domestic market and a larger laborforce. Administration's exhortations to women to raise more children, however, make the Chinese and Indians uneasy, as most Malays remain largely rural and tend to rear larger families. Any future increase in population is likely to push the proportion of Malays well above the 50 percent mark, making them an absolute majority.

Prospects Malaysia enjoys the highest standard of living in Southeast Asia with the exception of the such smaller countries as Brunei and Singapore. During the last decade its economy has recorded impressive growth rates averaging 8-9 percent annually. Since independence the administration has pursued calculated policy of economic diversification. An increasingly large slice of the government's developmental effort has been devoted to heavy industrial projects, among them pig-iron production, shipyards, cement plants and engineering complexes. Malaysia is now making electronic components for export. Today, manufacturers bring in over a half of Malaysia's export earnings, while petroleum accounts for 8 percent of exports. The main development goal of the administration enunciated in its New Economic Policy (NEP) has been the manufacture of goods for export, with a lesser emphasis on import substitution. Industrial estates within free trade zones have been established in the less-developed parts of the country.

CAMBODIA

Cambodia is a small country with an estimated population of nearly 11 million, but is culturally the richest and ethnically the most homogeneous nation in Southeast Asia. Nearly 90 percent of the inhabitants are Khmer; the remainder are divided between Chams, Vietnamese and Chinese. The nation boasts an impressive historic civilization focused on the city of Angkor, which contained a complex of magnificent temples built on the precepts of Hindu cosmology. The Khmer Empire, compact in shape, has been considerably reduced in size since its heyday; its modern boundaries were defined by the French which relinquished control over the region in 1954.

Presentday Cambodia illustrates how brutal the collision between politics and geography can be. The country is easily penetrated from the Vietnamese and Thai borders, and the age-old hostilities between Vietnam and Cambodia surfaced when the Vietnam War (1954-1955) spilled over the country as Vietcong soldiers sought shelter in the forests of Cambodia's

Map 18.4

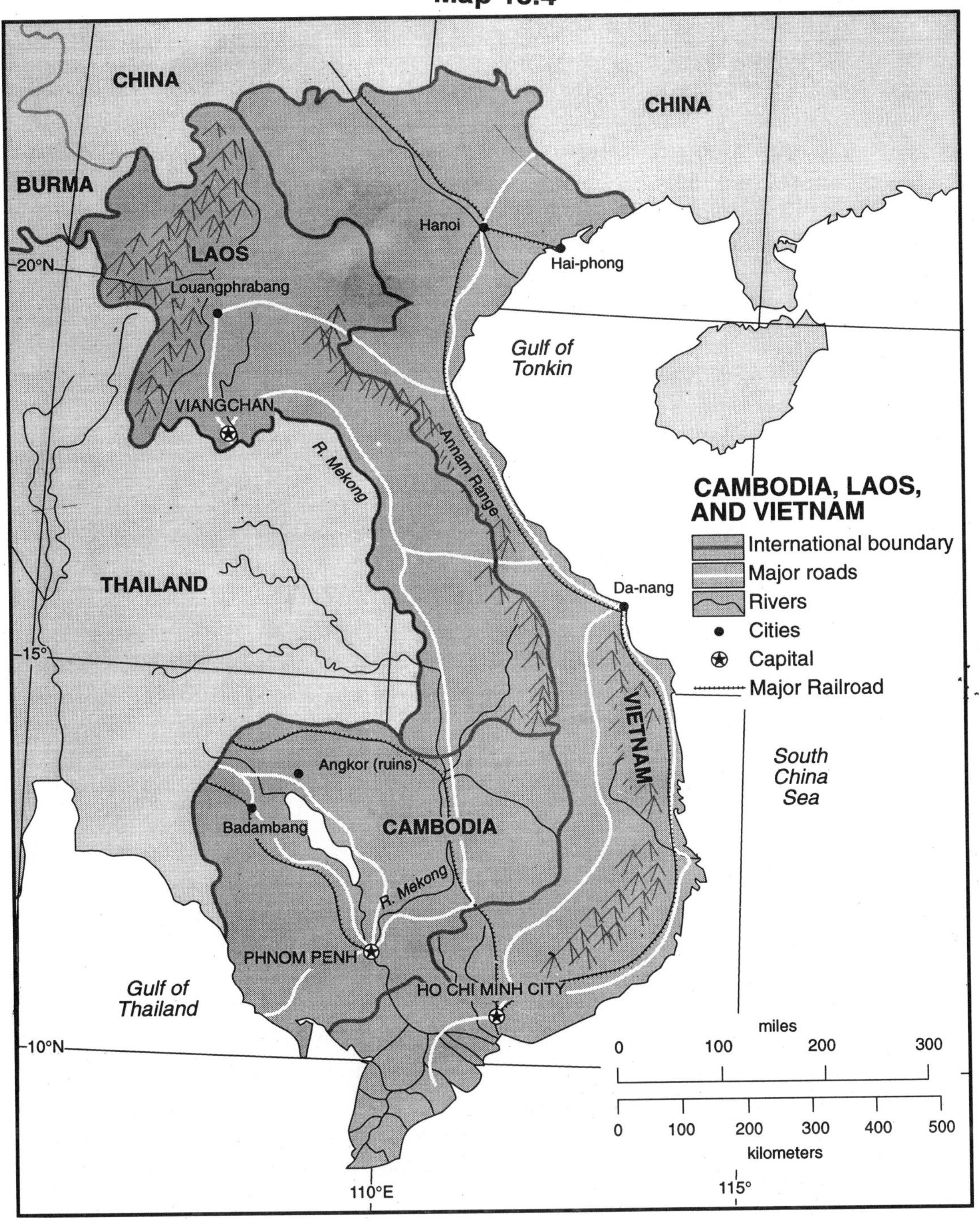
CHINA
CHINA
BURMA
LAOS
Hanoi
Hai-phong
20°N
Louangphrabang
Gulf of Tonkin
VIANGCHAN
R. Mekong
Annam Range
CAMBODIA, LAOS, AND VIETNAM
International boundary
Major roads
Rivers
Cities
Capital
Major Railroad
THAILAND
Da-nang
15°
VIETNAM
South China Sea
Angkor (ruins)
Badambang
CAMBODIA
R. Mekong
PHNOM PENH
Gulf of Thailand
HO CHI MINH CITY
10°N
miles
0
100
200
300
0
100
200
300
400
500
kilometers
110°E
115°

Mekong delta area across from Vietnamese territory. After the overthrow of the king by the Khmer Rouge, the communist revolutionaries embarked upon a course of terror in their own country in the name of national reconstruction. With nearly the entire population of the cities, including that of the capital city of Phnom Penh banished to the countryside, religion outlawed, and the distressed, unemployed, and sick done to death (over 2 million), the country lay in ruins by the early 1970s. In the aftermath of Vietnamese victory in its own country, Vietnam moved into the country to drive out the Khmer Rouge terrorists who had staged a tragic holocaust in a new wave of terror, pushing a stream of Cambodian refugees into Thailand. The Vietnamese-backed government did not, however, gain international recognition and was in constant conflict with the Khmer Rouge and other factions within the country. In October 1991 an interim government was set up through the peace initiatives of the United Nations, which has since been replaced by a newly organized permanent government. Peace in Cambodia, however, remains shaky and the nation's future depends as much on international (essentially of the U.S.) economic aid as on the solidification of its fragile democracy.

Physical and Cultural Characteristics

Physically, the country is essentially an alluvium-covered basin of the Mekong River. The centerpiece of the topography is the Tonle Sap Lake that serves as a huge reservoir for the Mekong. During the rainy season (mid-May to early October) the lake's surface expands from the dry season's minimum of 1,200 sq miles (3,108 sq km), to a rainy-season maximum of more than 3,000 sq miles. As a result of this annual phenomenon and the influx of fresh water, the lake is one of the world's richest sources of fresh water fish. On the north, along the border with Thailand, the Cambodian plain encounters an escarpment, the Dongrak Mountains, making a natural boundary. In the southwest two upstanding blocks (Cardamon and Elephant Mountains) for another highland portion covering the area between the Tonle Sap and the Gulf of Thailand.

The climate is typically tropical monsoonal. From mid-May to early October, the southwest winds bring heavy rains and high humidity; and from early November to mid-March, the drier and lighter winds of the northeast bring infrequent rains and lower humidity. Between these periods, the weather is transitional. Maximum temperatures range from 82°F (28°C) in January to about 95°F (35°C) in April. Annual rainfall varies between 200 inches (5,000 mms) on the southwestern highlands and 50-55 inches in the central lowlands.

Cambodia has always been overwhelmingly a land of villages. Nearly 80 percent of the population was estimated to be rural in 1995. The urban areas primarily date back to the early 20th century during the French colonial period, serving essentially as commercial and administrative centers to the surrounding areas. **Phnom Penh** containing currently nearly a tenth of the country's population (920,000 in 1995) is the only metropolis. Before the outbreak of war in 1970 it had about half a million people. By 1975 the population had swelled to over two million by the influx of refugees from the wartorn rural areas. Since the official depopulation in 1979 it has grown back to its present size.

Between 1975 and 1978, hundreds of thousands of urban people were forcibly moved into rural areas to work in rice fields and maintain irrigation channels. With the exception of Phnom Penh and Batdambang, both located in the central lowlands of the Mekong, few towns have subsequently regained their pre-1970 levels.

The Khmers is the most dominant group accounting for a vast majority (89 percent) of the population. They are concentrated in the lowlands of the Mekong and the Tonle Sap, and speak Mon-Khmer language. The Khmers are a product of centuries of racial and cultural blending since 200 B.C. that in all probability moved from the Khorat Plateau, and was subsequently Indianized by the successive wares of the Indian culture, and later on in the 8th century exposed to the Indo-Malayan influence. Among the minorities the Chinese, now form about 3 percent of the population, and are the most important. They controlled most of the country's commerce and small businesses before 1970. The Vietnamese, number 5 percent, and have a lower social status. The Chams accounting for a little over 2 percent, are followers of Islam, and have been traditionally discriminated against. There are a few tribal groups occupying the forested northeastern part of the country. Most ethnic Khmers practice Theravada form of Buddhism which was the state religion until 1975. The tribal people are mostly animists. The Vietnamese and Chinese are more eclectic and follow generally Mahayana Buddhism. Some of the Vietnamese are members of the Roman Catholic Church.

Resources and Economy The known mineral resources of Cambodia are scarce, consisting mainly of limestone, phosphate deposits, a little iron and some coal. Energy resources are primarily dependent on oil which accounts for 4-5 percent of all imports. Prospecting for oil and natural gas has been started in the offshore areas, close to the sites earlier exploited by Vietnam.

Traditionally, agriculture has dominated the economy. Almost 70 percent of the labor-force is engaged in the agricultural sector (down from the estimated 90 percent of the 1930s), which makes up nearly one-half of the nation's revenue; trade accounts for another one-sixth. Rice is the principal crop. Cambodia produces nearly 3.5 million tons annually and has a modest surplus for export. Although the cultivation encompasses most of the total cultivated land of the country, Cambodia's major rice growing region is surrounding the Mekong, and the Tonle Sap. Under the administration of Khmer Rouge during the 1970s great effort was made to build irrigation systems in the country by forcible resettling of the population in the various regions. Although the results were notable in agricultural production, the irrigation channels were poorly designed and finally collapsed and abandoned after 1979. Other crops include corn, rubber, sugar cane, soybean, oranges, bananas, mangoes, pineapples, papaya and tropical fruits. Fishery and livestock are also important. In the eastern part of the country small rubber plantations developed by the French, and the Chinese, have now fallen on disuse. Forestry, however, remains a significant industry and brings substantial earnings. Sawn and felled timber accounts for over one-half of all the exports.

Industrial development of the county is still in infancy after the massive destruction of the economy during the 1970s. Since independence in 1953, Cambodia had embarked on a program of modernizing its economy. The road system was improved, the railroad connections to Thailand and Vietnam were strengthened, and the nation's only port at **Kampong Saom** (Sinhanoukville) was opened in 1956. But the equipment installed in the 1950s and 1960s became obsolete during the 1970s as the country was plunged into years of internal strife and warfare, which left the economy in ruins. Since then, the country received a modest amount of foreign aid. Some progress was made in the 1980s; and plants were established or reopened to produce soft drinks, paper, cigarettes, construction materials, cement, and cotton textiles. Timber processing and rice milling important before 1975 were also revived in the 1980s. Roads and waterways that constitute the main transportation routes were improved. A sizable maritime trade is also now growing between Thailand and Cambodia, and the port of Kampong Saom underwent substantial renovation. The port is of strategic importance to the nation, and considerable industrial development has taken place in the area. Singapore, Vietnam, Japan, Hong Kong, Australia and Thailand are the main trading partners. Cigarettes, petroleum products, construction materials, and electronic goods were the chief imports in 1994.

In review, Cambodia remains an economically poor and undeveloped nation, with a shattered economy and fragile political institution. There is ample arable land for expansion of agricultural economy if the political conditions were to stabilize. The country does not possess much of mineral resources and is not in a position to utilize its few minerals in domestic industry. Little has also been done with the forest resource much of which was neglected during and since the 1970s. The economic future of the country continues to be closely tied to the trends of internal political events, and the solidification of its democratic institutions.

LAOS

Remote and least developed of Southeast Asian nations is the small, landlocked country of Laos that constitutes the northwestern portion of the Indochinese Peninsula. Making a long boundary with Thailand along the Mekong River. Laos shares borders with five neighbors, including its very sensitive frontier with Vietnam. The access to the outside world is only through Thailand's port of Bangkok. With no railroad, a small network of paved roads, virtually untapped mineral resources, very little modern industry, and an urban population of 23 percent, it is the poorest nation in the region, and is politically the most vulnerable.

Positioned as a "corridor zone" between China in the north and Thailand in the south, the location of Laos has been an important factor in the evolution of an extremely complex ethnic pattern, for its territory has throughout history been exposed to waves of migrating people. Two-thirds of the total population of nearly five and a half million is ethnic Lao, known locally as Lao-Lum, a lowland-welling people, close kin to the inhabitants of adjacent Northeast Thailand. These are the people of Mekong Valley lowlands who are the majorities both in **Vientiane,** the capital, and in **Luang Prabang,** the pre-colonial capital, and who have also traditionally dominated Lao government and society. The remaining one-third of the population falls into three groups. An estimated 16 percent are Lao-Theung, 8 percent are Lao-Tai groups and others are tribal groups who live in the hills and cultivate dry rice, as opposed to the irrigated paddy-rice culture of the lowland Lao-Lum. Most of the groups have a loose affiliation with the Mon-Khmer people making up some thirty or forty different groups including Hmongs who came from southern China during the late 18th century. Ethnic heterogeneity has inhibited the formation of any tradition of cultural or political unity for all of Laotian territory. There is friction between the Lao-Thai of the lowlands and the highland peoples between Upper Laos and Lower Laos. The French created the state of Laos as the last act in rounding out of their colonial empire, but allowed it to languish economically. Since the collapse of the French colonial structure, the last three decades have been one long series of shifts and movements of conflicts and agreements between a small number of individuals, each with outside support (military and economic aid by the West) and encroachments (by Vietnam and China) with little relevance to the interests of the Laotian communities. As a result, the economy, already precarious, has further declined, and political maturity has failed to emerge.

Although most Lao people speak languages related to Thai speech, there are several groups of Tibeto-Burman dialects, who originally came from southern China, live in the north and northwest. Chinese and Vietnamese primarily live in the urban areas. French was the adopted language of the elite and people living in the cities, but by 1970s English had begun to displace it. A majority of the people (58 percent) profess Theravada Buddhism. Most of the remaining people are animists (officially listed as tribal religionists) or spirit worshipers, particularly in the isolated upland areas. Many peoples in the highlands in the northern part of the country, especially those have migrated from China, mix Confucian ideas with Buddhism and animism. The Hmongs are spirit and ancestor worshipers. Only a tiny fraction of the population is Christian or Muslim.

Physical and Economic Base The physical landscape consists of inhospitable, forest-covered mountains in large part which in the north rise to a maximum altitude of 9,245 feet (2,818 meters) above sea level and constitute a major impediment to travel. The principal range has a northwest-southwest alignment and is a part of the Annam Cordillera, but there are several other ranges. In the northern part also lies the rolling grassland which is essentially a plateau but given the name of the Plain of Jars (the name derived from the large prehistoric stone jars discovered there). The only significant lowlands are along the eastern bank of the Mekong River, which flows generally southeast and south along and through western Laos, making a part of the country's boundary with Myanmar. The Mekong forms the country's major transport route, and more than half of the country's population lives along these lowlands and those of Mekong's tributaries.

Laos has substantial mineral resources, including coal, iron ore, copper, lead, gold, tin, gypsum, and precious stones, which have not been well-tapped. Tin has been mined since colonial times, and gypsum is also important, but mining in general contributes less than 0.4 percent in value to the gross domestic product. Although the country's hydroelectric power potential is relatively large, but remains largely underdeveloped. Electricity produced from a dam near Viangchan and supplied to Thailand, is the country's most valuable export after timber. The main resources of Laos are timber, minerals, and hydroelectric potential, which have been exploited poorly because of the difficulty of access and political turbulence arising largely from the Civil War since independence in 1953 at the end of the French occupation. The Communist rule brought a measure of stability, but lacked expertise to plan and execute economic development programs. A major obstacle to the economic and social development has been the nation's poor transportation system.

The country is predominantly rural and agricultural. Three quarters of the population is engaged in agricultural activities, mostly in rice cultivation. It is self-sufficient in rice production, a reflection of private landownership and market incentives provided to the farmers by the administration. A good part of rice production comes from upland dry variety raised by hill peoples using shifting-cultivation methods (i.e., fields are cleared and cultivated for a few years before being abandoned and allowed to revert to forest). Considering this as a major cause of deforestation, the administration has attempted to resettle the hill peoples on plains where they can adopt sedentary practices and take to cultivating a variety of cash crops, such as sugarcane, tobacco, and coffee. Coffee grown on plantations during the French rule, remains an important source of export revenues (nearly 8 percent of all exports). The traditional cultivation of the poppy has become important as a source of cash income to the hill people such as the Hmongs, during the last two decades, and has become Laos' important export. Its cultivation is illegal, but government's effort to curb production has had limited success in view of crops profitability and the inaccessibility of areas that can be put to cultivation of other crops. Wood products and textiles are the other major exports.

Laos contains extensive stands of forests, particularly of the valuable teak wood, cardamom, and stick lac (used for making shellac). Logging of these is mostly government controlled, but regulations to enforce it have been difficult because of poor communications. Fishing is carried on by people in the lowlands, and livestock raising has grown in value since 1980. The industrial sector of Laos is small, and includes rice milling, saw-milling and the manufacture of building materials and some light consumer items. Handicrafts have been traditionally important. **Viangchan** (population: close to 200,000) located on the Mekong is the capital, and the commercial and industrial focus of the nation. It is accessible to the outside world primarily through Thailand's railroad terminus across the Mekong.

Politically, Laos has been a pawn between China, Vietnam, Thailand and the West. Frequent international deals have, only in theory, guaranteed it an independent political status. The state is neutral and its political machinery and economic viability is supported by the World Bank (largely American subsidies). Factional rivalries, and civil unrest, bolstered by outside forces (Vietnam and China), and the residual effects of the Vietnam War have left deep scars on its already fragile political and economic structures.

VIETNAM

Vietnam emerged in its present unified form in 1976, after the French occupation of the territory for nearly 100 years as three French states: Tonkin in the north; an elongated Annam in the middle; and Cochin China in the south, and on the ruins of the two states of North and South Vietnam that existed between 1954 and 1976. The country extends nearly 1,000 miles (1,600 km) along the eastern rim of mainland Southeast Asia, from China's southern borders to the Mekong delta. The two former states of Tonkin and Cochin China are crowded delta lands. With its present population of over 77 million, Vietnam is the most populated country of mainland Southeast Asia, and despite the huge losses of reproductive age groups during the Vietnam War (1954-1976), now records a high rate of population increase.

Despite the unification of North and South Vietnam, creating one of the most elongated states in the world, the age-old divisions between north and south continue to persist. Hanoi in the Tonkin area in the north and Ho Chi Minh City, formerly Saigon in the Mekong delta in the south, continue to dominate the two areas as regional foci of economic, social and political power. In the aftermath of the dissolution of the French rule and the creation of two states: communist-led North Vietnam and non-Communists South Vietnam, the tragic developments of the Vietnam War during the 1960s and 1970s plunged the region into political instability and economic ruin from which it is struggling to recover. An armed push to oust the government in Saigon by communist guerrillas, the Vietcong, helped by North Vietnam led to a brutal war into which the United States was drawn. Fearing an ever-encircling extension of communism, United States forces supported the Saigon regime of South Vietnam with more than half a million soldiers at one time. What had begun as a scattered rebellion by the Vietcong ended in 1975 in the defeat of South Vietnam and its ally, the U.S. despite the heavy U.S. military and economic aid provided to South Vietnam, including an extensive bombing of North Vietnam and its capital, Hanoi.

Following unification, Vietnam introduced important economic and political changes, but recovery proved difficult. First, the economy had been crippled by wholesale destruction of the already poor infrastructure, resistance on the part of the peasantry to collectivization, minimal external economic aid except from the Soviet Union, and the departure of a large proportion of the middle-class entrepreneurs, civil servants, technicians and professional persons. Secondly, Vietnam became embroiled in the affairs of its neighbors, Cambodia, and Laos. Although Vietnam with its overwhelming military, technical and numerical superiority could retain control over these neighbors, this sapped its energy so vitally needed for its own reconstruction. Thirdly, Vietnam's hostility and open conflict with China along its northern border created a border of tension, weakening its military and economic base. As a by-product of these relations, Chinese (who form substantial minorities in all major cities) have been fleeing into China in a steady stream of refugees. The Soviet Union, and its Eastern block satellites, had helped to keep the Vietnamese economy from collapsing by the provision of economic aid to the extent of 2 billion dollars per year (in the mid-1980s). With the final disintegration of its former ally, the Soviet Union in 1991, and the strained relations with China, Vietnam is now looking to the West for foreign aid for economic reconstruction.

A consequence of the emergence of the new regime, in terms of human suffering as tragic as the political upheaval during the Vietnam War, has been a large-scale, and continuing exodus from Vietnam, of those who opposed the rise of the new order. Fearing for their lives, they had to flee the country as refugees to foreign lands. No refugee movements are easy; but none perhaps have been as tragic as the refugee movements from Indochina since 1975.

Physical Features Physically, Vietnam is largely a mountainous nation but can be divided into four distinct regions: One, the Annam Cordillera of mountains extending from north to south through west-central part of the country, and along the Laotian and Cambodian borders in the west and covering most of the country; second, the Song Hoi or Red River delta in the north; third, the Mekong delta in the south, and fourth, the narrow, coastal plain in the east along the South China Sea. The Annam Cordillera is extremely rugged, densely forested and impenetrable. It is a southward extension of the Yungui Plateau of southern China, and covers nearly two-thirds of the country. It rises to more than 6,000 feet (1,830 meters) in altitude, and widens into a plateau-like surface in the southern half. The fertile Red River delta, roughly triangular in shape, and is bounded in the north by rugged mountains that extend to the Chinese border. The river and its several tributaries are in part out for local freight transportation and irrigation, but are subject to frequent flooding. The Mekong delta in the south is nearly four times larger than the Red River delta and given to floods that bring sediments which increase the fertile land each year on the flood plains. The low-lying narrow coastal plains connect two river deltas.

The climate is tropical monsoonal with hot winters in the south and cool winters in the north. From April to the end of October, the summer monsoon sweeps over the southern half of the country, bringing heavy rains, averaging about 80 inches (2,000 mm) a year. In the north, rainfall averages 60 inches or so.

Cultural Patterns Vietnam has one of the most complex ethnic-linguistic patterns in Southeast Asia, and the population is divided into several disparate groups, although 87 percent are classified as Vietnamese. Ethnically, the Chinese, Phos, Tais, Khmers, Hmongs, and Nungs are relatively insignificant in numbers, each claiming between 2-4 percent of the population. The Vietnamese underwent considerable sinicization during a millennium of Chinese rule. They belong to the Mon-Khmer linguistic group of the Austro-Asiatic language family but exhibit strong Chinese influences. Indian influence is noticeable among the Khmer and Cham groups. The Chams, whose language belongs to the Austronesian family, formed an Indianized Kingdom of Champa from the 2nd century to the late 16th century; only small numbers now remain in the south-central coastal plain and in the Mekong delta. The Khmers are scattered throughout the Mekong delta. Other minority groups are the tribal people primarily inhabiting the highlands surrounding the Red River delta and parts of the Annam Cordillera. Several of the small communities that live in the Central highlands are still largely oriented around the family, and were collectively known by the French as Montagnards ("Highlanders"). Many of them link themselves to the Cham, Malay, and Indonesian peoples, and some to the Khmers, and thus are distinctly non-Vietnamese. The Tais speak Tai languages. Most of these tribal minorities have experienced little Chinese or Indian influence, but were exposed to the western (French from the late 19th century and American during the early 1970s) influence. The relations between Vietnam and China became strained during the late 1970s and early 1980s, and a large number of Chinese left the country. There were over 2 million Chinese at the end of the Indo-China war in 1975; now they number a little over a million. Vietnamese is the official language; French, Chinese, English and Khmer are also spoken.

Two-thirds of Vietnam's population is listed as adhering to Buddhism, and 8 percent to Christianity. Confucianism, Taoism and Mahayana Buddhism (clearly Chinese influence rather than that of the Hinayana of the Southeast Asian neighbors) entered and permeated into Vietnamese society over many centuries. In due course, these were intertwined, modified, and Vietnamized to become a Vietnamese folk religion that coexists with animistic beliefs held by the tribal people. Roman Catholicism was introduced in the 16th century and spread rapidly following the French occupation in the mid-19th century. Most of the Catholics were concentrated in the north before the country's partition in 1954. With the conquest of South Vietnam by North Vietnam, all of the foreign Christian clergy were expelled. Although the country's current constitution has guaranteed freedom of religion, in practice government controls have not been relaxed significantly.

Resources and Development Agriculture is by far the most important sector of economy, employing about three-fourths of the work force, and producing a third of the value of the nation's domestic product. Agriculture is also the main source of raw materials for the processing industries and an important contributor to exports. After years of shortages, the country was exporting rice by the late 1980s. It is the most important crop, grown on nearly four-fifths of the cropped land, principally in Red River and Mekong River deltas, the country's primary areas of permanent cultivation. The central coast land is given to destructive typhoons, and the central highlands area are regions of low productivity. The nation is now producing over 29 million tons of rice; other major food crops are cassava (production over 2 million tons annually), corn (1 million tons), soybeans, peanuts and sweet potatoes. Agriculture is labor-intensive and much plowing is carried on by water buffalo.

During the French rule, considerable effort was made to develop commercial plantation agriculture. There are many plantations of banana and fruit trees, coconut, and sugar cane, most in the Mekong River delta and the southern region; and the plantations of coffee, tea and rubber in the central highlands. The production of rubber was particularly disrupted by the war during the 1970s and production of most plantations also declined. Other cash crops include tobacco, and jute. Fishing is important in the Mekong basin, and accounted for 10 percent of all export earnings in 1999 (rice contributed nearly 12 percent). Forestry is a major industry. Furniture, pulp, and paper are the important products, and together with agricultural products are a source of considerable export earnings. Deforestation and forest devastation during the war, and soil degradation, however, threaten the industry, but there is increasing domestic demand for forest products.

Vietnam's mineral resources are substantial, particularly in the north. There are large reserves of coal, petroleum, phosphates, chromite, tin, manganese, antimony, bauxite, gold, iron ore, lead, tungsten, and zinc. Mining has been important, particularly of coal and oil which are major export items. Other minerals are produced primarily for domestic industry but also exported in some quantities. A number of offshore oil deposits have been discovered in the South China Sea, mainly off the southern coast; the production has steadily increased. In 1994, crude petroleum accounted for a quarter of all export earnings. By world standards, the production is yet small, but has added to Vietnam's strategic significance.

Before the French occupation, Vietnam possessed an insignificant commercial and manufacturing sector, although during the colonial period, significant changes to the economy occurred. The French developed export crops such as rubber, tea and rice. Beyond this, their economic interests lay in creating a market for French manufactured goods. Vietnam, therefore, did not take to industrialization on a substantial scale. Since independence the North (North Vietnam) took to socialist ideology, and the basic industries were nationized, and farms collectivized. The South (South Vietnam) with its limited mineral resources, concentrated on light industry and export crops.

After the war in 1975, the unified Vietnam embarked on the road to instituting centrally-planned economy and emphasized the development of heavy industry. A concentrated effort was made to rapidly transform the private, capitalist industry in the South into a state-run activity. The production of coal and the production of off-shore oil and gas reserves, and the development of transport and communication were to receive special attention. In order to curb population pressure in the Red River delta, a substantial number of people were moved to the less-developed sections of the nation, particularly the central highlands and Cambodian borderlands----the goal was to move 10 million people over two decades. But from the outset, the plans ran into problems resulting primarily from mismanagement, corruption, bureaucratic snags and shortage of capital. By 1978 the economy was in a crisis and the state policy was reversed in favor of greater emphasis on the production of agricultural and consumer goods. Food and industrial output has since been steadily on the rise, though progress is spotty. Foreign economic assistance remains inadequate.

Food processing (especially rice milling, processing of seafood, coffee, and tea) is the largest industrial activity. Textiles are of growing importance; silk production revived after a period of decline in the 1970s. Cement and chemical industries (particularly fertilizer production) are other important industries. Rubber is processed primarily for export, but is also used in the manufacture of a wide range of consumer items for domestic needs. Steel manufacturing is a major heavy industry. The production of bicycles, machine tools, electrical equipment, paper and leather goods has significantly increased since 1980; much of their growth may be attributed to the development of hydroelectricity.

Vietnam's population has grown rapidly since reunification in 1975; it has been rising at a rate of 1.5 to 2 percent annually, aggravating the food shortage situation. By the early 1990s, however, the food production situation had considerably improved, and Vietnam became a food-grain exporter again after a lapse of three decades.

Vietnam experienced trade deficits during the war, and deficits continued since at a much smaller scale. Economic reforms of liberalization instituted during the late 1980s and early 1990s became more meaningful with the breakup of the Soviet Union and collapse of the communist governments in Eastern Europe, as the trade with these countries sharply declined. Japan, Singapore, Thailand, and Hong Kong became the major trading partners. As exports of crude petroleum, light textiles, and agricultural products to the new trading partners grew, deficits declined. Mineral fuels, motor vehicles, machinery and food stuffs are the principal imports.

Transportation between the north and south has been difficult, with traffic limited to the narrow coastal corridor. **Hanoi,** the capital, (population over 2 million) and the traditional administrative center and transportation hub of northern Vietnam, is connected by railroad to China, and through the narrow coastal corridor linked to Vietnam's largest city of Ho Chi Minh City, (its former name of Saigon changed after the unification of the country). Hanoi is located on the Red River, and is an important shipping point for agricultural and industrial products of the north. Majority of its industries deal in the production of machine tools, locomotives, plywood, textiles, and chemicals. The city received heavy bombing by the U.S. during the war (1965-68 and 1972), and was destroyed but regained functions in ruined structures.

Ho Chi Minh City (population: 4.3 million) lies in the delta area of the Mekong. It is the country's largest port, and the commercial and industrial center, besides being a focus of the country's highways, railroads, and of the Mekong's waterways. Its major industries are textile mills, paper mills, shipyards, machine assembling plants, rubber and food processing. The city was modernized during the French rule, and became the capital of the state of South Vietnam in 1954.

Prospects Vietnam contains several mineral and agricultural resources, the development of which suffered during a period of turmoil and warfare during the 1960s and 1970s, and since then from mismanagement of the economy. The actual potential for economic growth has been hindered in effect by population growth, environmental degradation and rigid economic policies. Paradoxically, the nation's important economic resource is the literate and energetic population. The physical resources of long coastline, and areas of scenic beauty are well suited to the development of tourism; but lack of infrastructure has inhibited full utilization of these. Since the mid-1990s there have been some hopeful signs for the future development. The nation has been steadily opening up to the west, and there is a move to a partial privatization of the economy and a relaxation of the rigid government controls. Foreign investment is being lured. In 1994 the United States lifted an embargo on trade, and investment. Coca-Cola opened its first bottling plant. Mobil has been exploring for oil, and several American, European and Japanese companies have started their establishments.

Further Readings

Fryer, D.W. *Emerging South-East Asia.* New York, 1979.

McGee, T.G. *The Southeast Asian City.* New York, 1967.

Rigg, J. *Southeast Asia: A Region in Transition.* London, 1991.

19

Island Nations of Southeast Asia

INDONESIA

Indonesia is a "primate" nation in Southeast Asia; it leads its next rival, Vietnam, in population by a ratio of 2.6 to 1. With a population of 206 million and an area close to three-fourths of a million sq miles (1.9 million sq km), it is the world's largest Muslim country and fifth most populous nation. Its strategic location between the Indian Ocean and the Pacific Ocean and Asian mainland and Australia, an essential political stability during its 50 years of independence; and an impressive range of resources, that includes natural rubber, copra, tin, sugar tea, coffee, kapok, pepper, spices and the largest known oil and gas reserves in non-Communist Asia, has given it enormous prestige. In keeping with its size and importance, it is active in such regional and international organizations as the Association of Southeast Asian Nations (ASEAN), the Organization of Petroleum Export Countries (OPEC), and the United Nations.

Physical Characteristics Physically, this archipelagic nation is made up of a galaxy of over 13,600 islands which are strung across more than 3,000 miles (5,000 km) of tropical seas, roughly between 10° north and 10° south latitudes, a spread exceeding the distance from California to Florida and from England to western Iran; these contain great physical, cultural and economic variety. Focusing on the map of Indonesia several major island-groups can be identified. The easternmost Irian Jaya (West Irian) shares its territory with the independent state of Papua New Guinea, and the other four are collectively known as the Greater Sunda Islands. These include: the islands of Java (the most populous and important) in the south-center, Sumatra in the west, and Kalimantan territory (the Indonesian portion of Borneo). Eastward of Java is a group of smaller islands known as the Lesser Sunda Islands, which include Bali and, near the eastern end, Timor. Yet another island chain is that of the Maluku and Halmehara groups which lie between Sulawesi and New Guinea.

In general, relief in the vicinity of the Java Sea (in northern Java) is low, contains several coral reefs, and is not volcanic. The southern side of the chain of mountains from Sumatra through Java contain a string of active volcanoes. Much of eastern Sumatra is a low-lying, swamp and difficult of access. The Barisan Mountain in Sumatra paralleling the southern coast and the mountains in Java rise to over 12,000 feet (3,658 meters) in elevation. The islands of Sulawesi and Maluku appear squeezed between the forces of the two stable masses with Sulawesi's elongated, mountainous peninsula running generally in a north-south trend. Sulawesi is fringed by coral reefs and bordered by deep-sea trough on the south. Halmehara island is the largest of the Maluku group, and is mostly volcanic.

In all there are over 200 volcanoes in Indonesia, 60 are active; the largest concentration is on Java. Volcanoes play an important role in soil enrichment; and there is a strong relationship between agricultural development, density of population and the location of volcanoes. Perhaps the best-known volcano is Krakatau that lies between Sumatra and Java in the Sunda Strait and erupted in 1883 causing large scale destruction. The mountains of the country are, in general, made up of younger formations, and are given to crustal instability, and earthquakes. Because of its insularity, there are no large rivers comparable to those on the Asian mainland. The seas, surrounding the country are a dominant feature, because these have an important effect on climate, transportation, and the development of Indonesian culture. They also serve as channels of communication and as barriers.

The climate of Indonesia is controlled by its island character and its position of astride the equator, giving it high, and uniform temperatures. The location between the landmass of Asia and Australia influences the monsoonal rainfall pattern. The climate of most of the country can be labeled as tropical monsoon. The temperatures are uniformly high and have a small range. Temperatures are a function of elevation rather than latitude. The annual average monthly range at Jakarta, for example, is only a matter of 2 or 3 degrees.

Most of Indonesia receives heavy rainfall, although it varies from place to place, depending on the elevation and exposure to rain-bearing air masses. The greatest amount falls from December to March. From central Java eastward toward Australia, however, the dry season (June to October) is progressively more pronounced. The highest amount occurs in the mountains of Sumatra, Kalimantan, Sulawesi, and Irian Jaya where precipitation is more than 120 inches (3,048 mm) annually. Eastern Java, Bali, southern Sulawesi and Timor receive between 60 and 80 inches, while islands closest to Australia get only 40 to 60 inches. Sumatra and Kalimantan, which are located close to the equator, and far from Australia, have no dry season.

Much of Indonesia is still covered with tropical rainforest. Irian Jaya and eastern Kalimantan are mostly forest-covered, while the densely populated islands of Java and Bali have a much smaller part of the land under forest. In general, the vegetation is similar to that of the Philippines, Malaysia, and Papua New Guinea. The most important vegetation type is mixed lowland-and-hill tropical rainforest, characterized by a large number of species. Above 5,000 feet (1,524 meters) elevations this forest gives way to temperate upland forest of oak and magnolia trees. Another typical feature of the vegetation is the mangrove forest which grows along the shallow seas on eastern Sumatra, southern Kalimantan and southeastern Irian Jaya. Indonesia has some of the world's largest tracts of exploitable tropical forests, and timber industry is well-developed. Rapid exploitation has, however, caused considerable damage, and government has now several reforestation programmes. In areas where the dry season is pronounced, as in Eastern Java, Savanna grasses are found.

Indonesia is a good example of relationship that exists between climate, vegetation, and bed rocks in the formation of soils, which are generally lateritic (containing iron oxides and aluminum) produced under constantly high temperatures and heavy precipitation but given to rapid erosion, weathering, and leaching, resulting usually in poor soils. Minerals leached from the soils are replaced by alluvial deposition as in some areas of Kalimantan, but the most valuable soil in Indonesia is the volcanic ash, transported by wind, generally deposited as a layer of inorganic materials over wide areas. These fertile soils are derived from the ejecta of the volcanoes that cover parts of Sumatra, Java and western Sulawesi.

Historical Background and Cultural Patterns Very little of Indonesia's history prior to 7th century is known. There is strong evidence, however, that such cultures as Srivijaya existed before that time. The discovery of the remains of **Homo-erectus,** the so-called "Java man," attests to the very early settlement of Central Java. Remains of aboriginal people in southern Sumatra have also been unearthed. The present population of Indonesia is believed to be descended from several waves of migrants from mainland Asia especially from southwestern China. Early in the Christian era, Indonesia came under the influence Indian civilization through a gradual influx of Indian traders, and Hindu and Buddhist missionaries. The early recorded history shows profound religious influences from India. The early Indonesian states that centered on Java and Sumatra evolved through many forms of Hinduism and Buddhism. During the 9th century both Hinduism and Buddhism were practiced as coast religions. The spectacular Buddhists temples of Borabudor date from this period, Sumatra was the seat of the Buddhist kingdom of Srivijaya. The blending of Hinduism and Buddhism continued until the 14th century, when Islam was introduced along the coasts by Muslim traders from Arabia. By the end of 16th century Islam had replaced Buddhism and Hinduism as the dominant religion, and the territory broke into several Islamic states. Throughout these political and religions changes, at the court level, the common people adopted part of each new religion as an added layer over their basic traditional and animistic beliefs.

Early in the 16th century the Portuguese, in pursuit of rich spice trade, began establishing trading posts after taking over the strategic commercial center of Malacca on the Malay Peninsula. The Dutch followed in 1596, and the English in 1600. By 1610, the Portuguese had been ousted, and were allowed to retain only the eastern part of Timor. After a series of Anglo-Dutch conflicts the Dutch finally emerged as the dominant power in Indonesia. From their arrival and control, the Dutch administration was designed to exploit the islands. To this end they introduced very strict forced-crop and forced-labor practices which they called the "Culture System ". The system generated the production mostly of commercial crops such as rubber, sugar, coffee, indigo, tea, and tobacco, all geared for the European markets, and all raised on large Dutch-controlled plantations. The harsh "Culture System" did have its benign results. A concerted effort to develop a transport system to reach the coast for the movement of commodities to Europe, the building of hospitals, and the provision of education helped to rid plantations, particularly in Java (which was the major focus of this commercial and political activity) of diseases and hunger. The Culture System was finally discarded in 1870 to pacify native opposition to Dutch exploitation and brutality on the plantations. The Dutch rule also contributed to the nationalist cause by providing a common focus for resentment, and a desire for Indonesian independence. The Indonesian movement for independence began early in the 20th century. The Dutch resisted the nationalists bitterly, and did not give in even after the Indonesian proclamation of independence in 1945 by the nationalists. Independence finally arrived in 1949 under United Nations' pressure except in Irian Jaya (in New Guinea) which was awarded independence in 1969. In 1976 Indonesia forcibly took over the island of Timor.

Before the Dutch occupation, what is now Indonesia had never been a single native state. When the Dutch took control, they found in the country great diversity in race, religion, settlement density, agriculture, economic prosperity, and culture. Java and the "Inner Islands" were the most developed sections. Other islands and regions differed in culture, language and development. Although the majority of Indonesians are Javanese and Muslims, but there are Hindu Balianese, Chinese Buddhists, and aborigines in the interiors of several large islands and other groups, and though the Dutch had provided a unified rule under their colonial occupation, economic and cultural contrasts remained between the various ethnic groups.

Since independence, the government has tried to forge Indonesian nationalism and a sense of unity amid the country's cultural and economic complexity. In order to provide an egalitarian means of communication it declared "market Malay," which had been the *lingua franca* of traders and sailors throughout the archipelago for centuries, as the official language and the medium of instruction in all secondary schools. Now called Bahasa Indonesia, it is Indonesia's national language. This has provided a common bond between the speakers of 250 different languages in a physically fragmented nation. The common language helps to create a powerful sense of nationhood, which is the envy of other large and heterogeneous countries such as India, although the physical spread and fragmentation continue to pose administrative and political problems for the nation.

It might appear that these scattered and numerous islands and people have little in common beyond the experience of Hindu-Buddhist tradition overlain by Islam and Dutch rule. The Dutch did certainly determine the administrative parameters of Indonesia, but the country is far from an artificial creation. The islands had a long history of contact with each other, particularly among the coastal people, and a high degree of common custom and ritual had developed. Despite cultural complexity and ethnic diversity, nearly 88 percent of the people follow Islam, making it the largest Muslim country in the world today. But Islam is tempered by Hindu-Buddhist mystic beliefs, and an eclectic mystical Indonesian culture. Islam is strictly followed in western Sumatra, western Java, and southwestern Kalimantan. Away from these strongholds the people consider themselves as Muslims, but do not practice the orthodox Muslim injunctions of fasting and prayers. Close to one-tenth of Indonesia's population follows

Christian faiths, primarily of Protestant denominations, and are scattered throughout the country. The Hindu population, accounting for about 2 percent of the total, lives mainly on the island of Bali, and partly on Lombok. Most Chinese practice Buddhism and Confucianism.

Resources and Economic Development

Indonesia contains a large variety of mineral deposits including petroleum, tin, manganese, copper, nickel, bauxite and coal, but most of these have not been fully prospected. Prior to independence the Dutch administration paid little attention to mining industries as an instrument of economic development. In 1925 minerals contributed less than 15 percent to the export earnings, which increased to over 34 percent by 1954. By 1992 the share of minerals in export earnings remained at the 1954 figure, as other sectors of economy had grown steadily. Mining now provides one-eighth of the gross domestic product, although it employs less than one percent of the work-force.

Petroleum production and refining is Indonesia's largest economic enterprise. In normal years it produces about 2.5 to 3 percent of the world's output which may not be remarkable by world standards, but makes Indonesia the leading producer of oil and gas in Southeast Asia. During the 1980s and 1990s, the revenues from oil and gas outstripped the revenues from all sources, and brought in 28 percent of the nation's export earnings.

Major oil fields are located in the northern and eastern parts of Sumatra, eastern coast of Kalimantan, and off-shore sites in the Java Sea. Refining is in the hands of foreign companies, although ownership remains with the government of Indonesia. Indonesia is also the world's second largest producer of tin; the output is nearly 12 to 13 percent of the world's total production. Production is dominated by state companies. Tin is mined in a chain of islands off the east coast of Sumatra (Bangka, Billiton and Singkep). Nickel is produced on Sulawesi, in Irian Jaya, and in the Maluku, and bauxite in Kalimantan. Coal is mined in southern and western Sumatra.

Agriculture accounts for one-sixth of nation's domestic product earnings, although it employs a little less than a half of country's labor-force. During the Dutch occupation it was common to classify Indonesian agricultural economy into two categories: one of small holders producing mostly subsistence crops, and the other of the estate or plantations which were primarily responsible for the production of cash crops for foreign trade. The introduction of the small holder into the cash crop production during the early 20th century blurred this distinction. Since independence the official land policy placed restrictions on larger estates, thus increasing the number of small holders. The growth of cooperative movement, especially in Java, further erased the distinctions between the two production systems. The estates have not been completely eliminated, however, in order to sustain the export that at maximum capacity for earning valuable foreign exchange in combination with those of exportable minerals.

However, it may be useful to classify the agricultural economy into three categories primarily on the basis of technological systems into: shifting cultivation, sedentary fields, and plantation or estate agriculture that characterize the Indonesian socioeconomic society. In terms of these categories, distinctions between domestic and export sectors become less arbitrary. Such a classification clearly reveals the differences between some of the Inner islands such as Java and Outer islands. Java has long been the focus of attention, particularly during the Dutch period, and saw the establishment of plantations and Culture System. It is the heartland of Indonesia, the most populous island with nearly 135 million inhabitants or nearly two-thirds of the nation's population, making it one of the world's most densely populated territories.

Shifting cultivation called *ladang, is* the most widespread cropping system in the mountainous forests of the Outer Islands (Sulawesi, Maluku, Irian Jaya and in Kalimantan). It supports millions of people. The sedentary field gardening (peasant) system is widely employed in Java, Madura, parts of Sumatra, and some of the Lesser Sunde Islands. Among sedentary farmers wet-field rice cropping *(sawah) is* predominant. The crop is grown on small holdings primarily for subsistence purposes. Wet-field rice cultivation has evolved into a complex system of terrace-building, field leveling, water control, transportation of seedlings into wet fields, puddling of soils by wading men and animals, and hand cultivation. The major areas of rice production are Java, Bali, Lombok, and parts of Sumatra. In recent years there has been a discernible shift from rice to other subsistence crops, such as cassava, but rice remains the most important crop of peasant agriculture. The administration has, through the implementation of five-year plans, to raise the production by the introduction of high-yielding varieties (HYV) and fertilizers and by the provision of credit to farmers. In the drier sections of eastern Java crops such as corn, cassava, sweet potatoes, peanuts, and soybeans dominate the small farms; although cash crops such as tobacco, and coffee are grown on plantations. Rice becomes less important in Maluku, and Irian Jaya where taro, yams, and sweet potatoes become more important.

Cash crops raised on estates and plantations, are more important in Sumatra and Outer Islands, with Sumatra accounting for more than half of the total area under estates or plantations. Other important crops raised on estates are rubber, palm oil, coconuts, coffee, tea, tobacco, kapok and cloves, although crops such as rice, corn and cassava are also raised to some extent. Oil palms, coconuts and tobacco are lowland crops, while tea and coffee are restricted to the upland. Indonesia is the second largest producer of natural rubber in the world, contributing nearly a quarter of the international production. Output in recent years has been declining as greater emphasis is placed on growing foodgrains by the administration. Since independence most of the larger estates held formerly by the Dutch, have been redistributed, divided into smaller holdings and given to the Indonesian small farmers, who have failed to modernize the farms in contrast to the neighboring Malaysians. Rubber and rubber products account for some 3-4 percent to the nation's export earnings. Oil palms have been increasing in importance, and sugar production has been given impetus during recent years. Indonesia produces nearly 3 percent of the world's cane sugar, which is grown largely on estates. The fishing industry plays a minor role in the economy. The administration has been recently trying to develop fisheries and fish products with the financial backing of Japan with the result that fish production has been steadily rising.

Manufacturing industries employ only 13 percent of the labor, but account for one-fifth of the value of domestic product. As in most of Southeast Asia, manufacturing in Indonesia has not played a significant role in the national economy, although since independence it has recorded substantial gains. Between 1960 and 1992 its share to the economy in terms of domestic product has more than doubled. The biggest industries such as oil refining are state-owned, as well as those engaged in processing of agricultural and mineral products. The oil refining industry is strictly controlled by the government agency Pertamina. A large part of the small and medium-scale privately owned enterprises consist of cottage and home industries that produce consumer items. Much of such enterprises are engaged in the production of such consumer items as furniture, textiles, household equipment, and printed matter, and are largely in the hands of the Chinese community. The main center of such private industry is western Java, and the capital city of Jakarta. The major industry, based on imported raw material, is the textile industry which is partly state-owned or in the hands of foreign companies, and partly owned by local entrepreneurs. Batik production—an Indonesian system of hand-printing textiles—is concentrated in central Java. Other industries include agricultural processing, light manufacturing, cement and chemical manufacture, rubber industries, and paper manufacturing, and are mostly concentrated on Java and eastern Sumatra.

Because Indonesia is an acthepalegic nation, sea transport plays an important role in the movement of raw materials and agricultural produce from their source to market. Many of the Outer Islands are located far from the commercial activity and is therefore focused on Inner Islands such as Java. The Outer Islands have not been adequately served by transport network, a factor that has critically hampered economic development and political integration of the nation. Although road transport is dominant on the islands, the only islands with adequate land transportation network are Java, Madura and Sumatra. Much of the remaining paved mileage is on Bali and Sumatra. Western and central Kalimantan, and Sulawesi have few good roads. Railroad transport is scarce in the country, and covers primarily Java and Madura. Indonesia exports raw material in return for manufactured goods. Oil, timber products, rubber and garments are the major exports. Predictably other exports include agricultural products such as coffee, tea, tobacco and pepper. Machinery, transport equipment and chemicals are the main imports, in addition to such raw materials as cotton, and foodstuffs. Indonesia's most important trading partners are Japan, the United States, and Singapore.

Population Patterns In 1960, Indonesia's population was 97 million; it has grown to more than twice that size during the last four decades. At the current rate of growth it is likely to add another 40 million by 2025. The country is one of the world's most populated nations, ranking fourth after China, India and the United States. The density of population within the nation, however, varies greatly from

Map 19.1

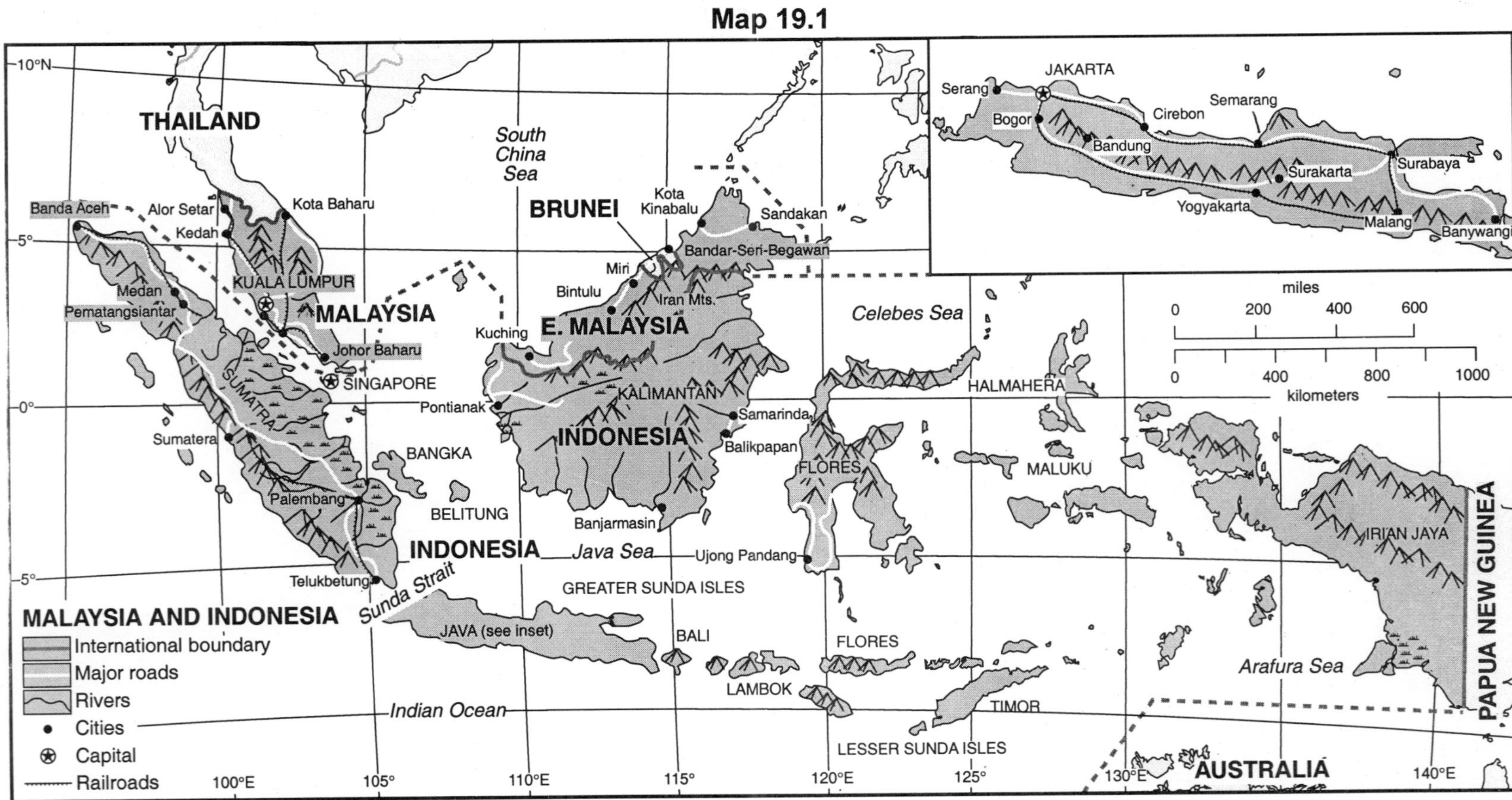
MALAYSIA AND INDONESIA
International boundary
Major roads
Rivers
Cities
Capital
Railroads
THAILAND
MALAYSIA
KUALA LUMPUR
SINGAPORE
BRUNEI
E. MALAYSIA
INDONESIA
KALIMANTAN
SUMATRA
JAVA (see inset)
South China Sea
Celebes Sea
Java Sea
Indian Ocean
Arafura Sea
Sunda Strait
Banda Aceh
Medan
Pematangsiantar
Alor Setar
Kedah
Kota Baharu
Johor Baharu
Sumatera
Palembang
Telukbetung
BANGKA
BELITUNG
Pontianak
Kuching
Bintulu
Miri
Kota Kinabalu
Bandar-Seri-Begawan
Sandakan
Iran Mts.
Samarinda
Balikpapan
Banjarmasin
Ujong Pandang
GREATER SUNDA ISLES
BALI
LAMBOK
FLORES
TIMOR
LESSER SUNDA ISLES
HALMAHERA
MALUKU
IRIAN JAYA
PAPUA NEW GUINEA
AUSTRALIA
miles
0
200
400
600
kilometers
0
400
800
1000
JAKARTA
Serang
Bogor
Bandung
Cirebon
Semarang
Yogyakarta
Surakarta
Surabaya
Malang
Banywangi
10°N
5°
0°
5°
100°E
105°
110°E
115°
120°E
125°
130°E
140°E

region to region; the islands of Java, Bali and Madura account for nearly two-thirds of the population. Java has been the focus of economic development and concentration, and has an average density of about 2,000 persons to a sq mile; as a contrast the nation's overall density of 280 persons to a sq mile. On the other hand, Kalimantan and the island of Sumatra, the largely underdeveloped "Outer Islands" of Sulawesi and Maluku, as well as the territory of Irian Jaya, sustain very small number of people, with very low population densities.

For decades the Indonesian administration, as did the Dutch government before it, has been concerned about the problem of overpopulation on the island of Java, Bali and Madura, and tried several methods of moving population to Sumatra and other islands. In 1970 the administration introduced a nationwide birth control campaign and a relocation plan to resettle villagers from Java and Bali, the most densely populated islands, to the sparsely populated islands of Sumatra, Kalimantan, Sulawesi, and Irian Jaya. During the 1979-84 five-year plan, the government relocated over 2 million people. This "transmigration" of people has not been particularly popular with the relocated families which were moved into an unfamiliar, wilder and new environment where soils are infertile, and the area covered with dense overgrowth.

In spite of intensive cultivation, and techniques of irrigation, and terracing in Java, Bali, and Madura, agriculture production has, in general, lagged behind population growth. However, foodgrain imports sharply declined during the 1980s as population growth was checked. It now stands at 1.5 percent a year, sharply reduced from the figure of over 2.7 percent in 1970. Meanwhile impressive gains were recorded in food production. In 1984, for the first time in 40 years Indonesia became a rice exporter. Now it is nearly self-sufficient in its food requirements and has a small surplus for export.

Nearly a third of Indonesia's population is defined as urban. **Jakarta,** the capital (9.3 million) retains the administrative functions established during Dutch occupation, but has acquired considerable commercial and economic functions during the last 50 years. The island of Java contains three of the other "millionaire" cities: **Surabaja, Bandung, Semarang.** Sumatra's "millionaire" city is **Medan.**

Prospects Indonesia is often described as a "storehouse of natural resources," and has made some notable economic gains in recent years, but still remains poor; the per capita annual income is merely $1,110 despite its relatively large oil holdings. Under the colonial occupation certain resources were exported for specialized world markets such as rubber, and tin, or agricultural products of the tropics, resulting in dependence upon events in distant places. Thus, changes in the demand for these commodities in Europe and the United States caused violent price fluctuations of the agricultural commodities, as well as affected their production and the national economy. The estate crops (coffee, tea, palm oil, sugar, tobacco, etc.) were grown on Dutch and foreign-owned large plantations, and the mineral resources (tin, bauxite) also were largely in the hands of foreign companies.

Since independence most of the agricultural estates have been redistributed among native farmers, and the share of foreign ownership in minerals has also lessened. The Indonesian government policy has followed nationalistic course, although it has not moved toward outright nationalization. There is a vast underlying potential that still remains unexploited, but awaits the introduction of a clear national policy aimed at the exploitation of resources. Indonesia's vast agricultural population lives on a simple basis of self-sufficient food production, carrying on handicraft industries and commercial activity.

Physically fragmented, the large nation is also beset with many problems associated with the centrifugal regionalism. Various groups in the Outer Islands have resented the dominance of the Javanese, and the stability of the administration has often been shaken by the existence of several political groups split along regional, religious, political and economic lines. One militant Muslim group Darul Islam which wants to restructure the state and society according to Islamic injunctions has been active in western and central Java and is often at odds with the central authority. The government has also been carrying on a long struggle against an independence movement in East Timor, a former Portuguese possession that Indonesia occupied after the collapse of the Portuguese Empire in the 1970s. East Timor gained independence after brutal civil war in 2000. In 1965, an attempted communist coup d'etat attempt to take over the administration was put down that resulted in the massacre of probably 300,000 communists and communist supporters by the Indonesian army and by Islamic and nationalistic groups. Nor is Indonesia's relationship with its easternmost island of New Guinea (called Papua-New Guinea) as yet stable. The people of the Indonesian part of New Guinea (Irian Jaya) are ethnically Papuans, and are non-Indonesian. They got a new foreign master (Indonesia) when the old one (the Dutch) relinquished control over their territory. Thus, there has been resistance to Indonesia's administration in Irian Jaya. In sum, as Indonesia moves into the 21st century, it faces enormous economic, social and political challenges which half a century of its independent status has still to overcome.

THE PHILIPPINES

The republic of the Philippines is an archipelago, containing some 7,100 islands and islets that extend across 1150 miles (1,850 km) between the Philippine Sea to the east, and South China Sea to the west and north. The Northernmost island is only 65 miles (104 km) from Taiwan, and the southernmost barely 30 miles (48 km) east of the tip of East Malaysia. Ninety-five percent of the 75 million Filipinos, however, live on only ten of the country's islands. Luzon, the northernmost and largest island, accounts for one-third of the island-chain country's land area, and forms a sort of northern anchor. Mindanao, second in size, forms the southern anchor. Between these two lie the seven major but much smaller islands (Bohol, Cebu, Leyte, Mestabe, Negros, Panay and Samar). Comparable in area to Italy or Great Britain, 11,500 sq miles (29,800 sq km), the country's aggregate coastline is as long as that of the U.S.A. Similar to Indonesia in its archipelagic character and in topography, the islands are mostly mountainous, with few discontinuous, narrow coastal plains, and the climate is tropical monsoonal, but relatively uniform.

Map 19.2

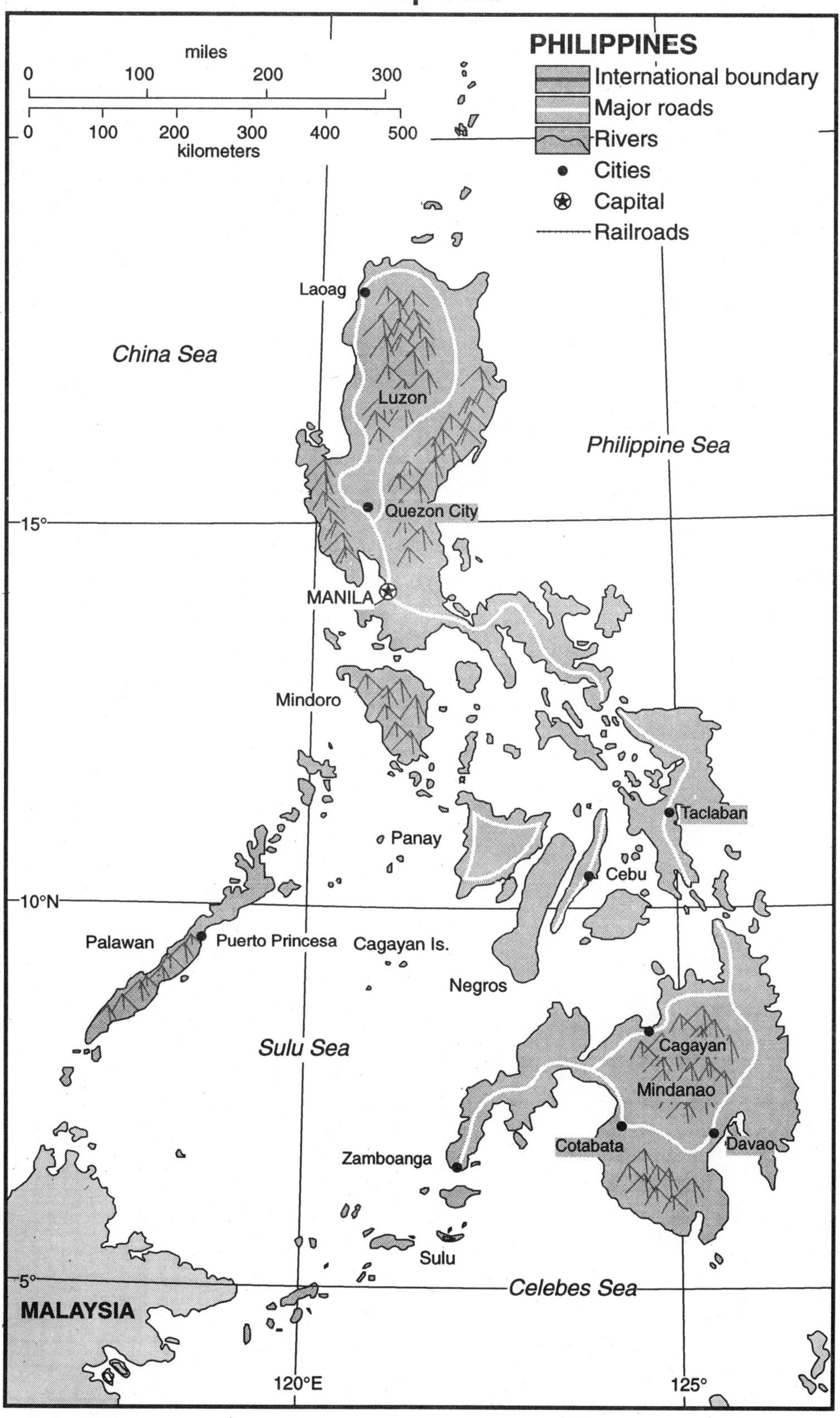
PHILIPPINES
International boundary
Major roads
Rivers
Cities
Capital
Railroads
miles
0
100
200
300
0
100
200
300
400
500
kilometers
China Sea
Philippine Sea
Laoag
Luzon
Quezon City
MANILA
15°
Mindoro
Panay
Taclaban
Cebu
10°N
Palawan
Puerto Princesa
Cagayan Is.
Negros
Sulu Sea
Cagayan
Mindanao
Cotabata
Davao
Zamboanga
Sulu
5°
Celebes Sea
MALAYSIA
120°E
125°

Despite these obvious similarities, the Philippines is unique among Southeast Asian neighbors: distinctive in its religion, history, and national cohesion. Notable among the differences are an emphasis on education and the prevalence of English, a legacy of the American era; literacy is over 94 percent rivaling that of Singapore and higher than that of most Asian countries (with the obvious exception of Japan). Because it was under Spanish control for over three centuries, and the United States for five decades, it developed pronounced cultural affinities with the West. It is, for example, the fourth most populous country in the world where English is an official language, and the only predominantly Roman Catholic nation in Asia. The Filipinos, however, are sensitive to Asian sentiments.

Historical and Cultural Development

The Philippines is predominantly Christian. Converted to Catholicism by Spanish missionaries 400 years ago and by American missionaries in this century, today 83 percent of the Filipinos are Catholics and 5.5 percent subscribe to the Protestant faiths. The remaining 6 percent are divided between adherents of the Islamic faith (2.5 million) known as Moros and concentrated almost entirely in western Mindanao, Palawan and Sulu islands in the southern parts of the country; and animist groups. There are some Buddhists (largely of Chinese extraction). Racially, the Filipinos are a blend of Malay stock with some Chinese and Spanish elements. The addition of Arab, Japanese and American elements into the melting pot has created a distinctive Filipino culture. The infusion of Chinese into Philippine society exemplifies this "melting pot" character of the Filipino culture. The "pure" Chinese minority numbers less than one percent of the population, for most have intermarried, producing a Chinese-Mestizo population which adds up to over 10 percent of the total population, thus the number of racially identifiable Chinese remains obscure. Many of the leaders in the Philippine administration and business have Chinese blood, although they have been otherwise completely assimilated into the Filipinos. The "pure" Chinese are active in retailing trade, and have been the object of much distrust and occasional persecution by the Filipinos, as in Indonesia or Malaysia. A small number of emigrants from the Indian subcontinent has added to the racial mixture. There are small numbers of resident U.S. nationals and Spaniards. The aboriginals constitute a very small percentage of the population.

Despite the 70 or so native languages and dialects that are spoken in the country, Tagalog is the most widely used by almost a third of the population, and is the basis of the official national language. English is the second official language, and taught in the schools as a subsidiary language, and remains the chief *lingua franca,* although an English-Tagalog hybrid "Taglish " is being increasingly used. The Tagalog group of languages is concentrated chiefly in Manila, southern and central Luzon and in the island of Mindoro. Pampango is also used in some parts of Luzon. Several other languages are native to the various islands (Cebu, Bicol, Ilacono, Leyte, Samar).

The Filipino ethnic mix is a product of several historical influences. Although some remote mountain ranges are still inhabited by descendants of earliest human beings to settle in the islands---the negritos, most

Filipinos are descended from later waves of migrants, that of the proto Malays, also from the Asian mainland, and a group of Malay settlers, probably from Indonesia. By 300 A.D., the last Malay groups were firmly established in the coastal areas. The Chinese came later for trade or plunder, but stayed on, initially concentrating in Luzon and later pushing southward to other islands. In the late 15th century Islam arrived through seafarers from northern Borneo to the southern islands of Mindanao and Palawan, where it flourishes to this day. Before Islam could take a firm hold in the northern islands, the Spanish invasion had started in the 16th century, and for 333 years Spain established rule over the country and left a lasting imprint on its religion and society. Most inhabitants were converted to Catholicism. Almost every respectable town in the islands is dominated by a baroque church built during the Spanish period. After 1898, when the islands were ceded to the United States following the Spanish-American War, the United States established its administration, which lasted until the attainment of freedom in 1946. Fifty years of American occupation again produced several changes. Apart from the adoption of English as a lingua franca in the islands, Americans introduced long over-due land reforms which the Filipinos carried on during the post-independence period. These reforms were designed to eliminate the worst aspects of Spanish colonial policies which had created a divided society of landed aristocrats and landless peasants.

Physical Basis The typical pattern for most of the thousands of islands that constitute the national territory of the Philippines is a mountainous core, the narrow and interrupted coastal plains, the generally northward trending of the drainage, and the characteristic irregular configuration. All the major islands possess great physical variety: they are composed of volcanic, coral, and several other rock formations. In northwest Luzon, the highest point at 9,693 feet (2,954 meters) is reached in the prominent range of Cardillera Central which consists of two or more parallel ranges. To the north of and between the ranges is the fertile Cagayan Valley. In the southwestern part of northern Luzon are the rugged mountains, rimmed by foothills and rolling uplands, the seaward margins of which contain narrow coastal lowlands. The pattern is somewhat repeated in some of the Visayan group of islands such as Negros, and Panay. The southernmost large island of Mindanao has huge, rugged mountains and volcanic peaks ascending to nearly 9,000 feet (2,743 meters) and elevated plateaus and a few marshy lowlands such as the Agusan and Cotabato. Everywhere, volcanoes are a conspicuous feature of the landscape covering nearly 40 percent of the surface area of the islands. Fifty of the volcanoes are still active. The most distinctive volcanic areas are in southern Luzon and on the island of Negros, and Mindanao. The remaining 60 percent of the surface consists of the rolling plateaus, the river valleys and the coastal plains.

The climate of the islands may broadly be called tropical humid monsoonal. The temperatures remain in general, fairly constant, averaging between 75° and 85°F (23.8°C-29.4°C) during the year, with few marked seasonal variations. The temperature differences between the north and south are tempered by the seas that engulf the islands. Altitude rather than latitude is the important temperature control. The major distinguishing feature of climate is the precipitation, which almost everywhere is above 60 inches (1,524 mm) annually, and varies seasonally in the different parts of the country depending largely on the direction of the monsoon winds, which blow from the southwest from May to October and from the northeast from November to February. The westernmost areas facing the South China Sea are characterized by a rainy season in summer and fall when the southwest monsoon is blowing, and a marked dry season in winter and spring. The easternmost areas have no dry season, but experience a maximum rainfall in winter when the northeasterly monsoon winds blow strongly across the islands.

An important climatic phenomenon is the typhoon, which occurs some 20 times a year, most commonly between August and November, and is responsible for one-quarter to one third of the entire precipitation amounts experienced. Typhoons are most frequent and distinctive north of latitude of 11°N, and hit the eastern coast of the central islands and Luzon the hardest. Mindanao is generally free from typhoons.

Abundant rainfall and high temperatures have encouraged the growth of forest vegetation, common to most of Southeast Asia, which covers two-thirds of the land surface. Mangrove and nipa palms are common along the coast, and tropical rainforests cover the lower coastal plains and foothills. Mixed tropical evergreen and deciduous forests cover the higher slopes. In comparatively drier areas and areas of shifting agriculture, forests have been largely replaced by bamboo and tall grasses called *cogon*, and the grass-covered area as the *cogonales* exist which cover between 15 to 20 percent of the land surface. In general, Philippine soils are not fertile except in the alluvial lowlands, but even these, have been leached of much of their fertility. On the slopes of the mountains soils are generally thin and eroded. However, along the coast of some central and southern islands soils formed over uplifted limestone reefs, as in Cebu, are mostly fertile.

Resources and Development A little over 36 percent of the Filipino work force is engaged in agricultural pursuits, generating only 20 percent of the nation's domestic production. In contrast, manufacturing and construction activities account for a bout 14 percent of the work force and 30 percent of the national product. Although the nonagricultural sectors of the economy have expanded at a much greater rate since independence, agriculture remains the major activity of the Filipinos.

Three crops dominate the Philippine agriculture: rice, corn, and coconuts. Rice is the principal staple crop, grown in all the parts of the country, but especially in central and north-central Luzon, southern Mindanao, western Negros and eastern and central Panay. Since the 1970s following the introduction of higher yielding strains of the plant, the construction of roads, and irrigation canals, the use of chemical fertilizers and insecticides, rice production has gone up substantially, resulting in some years having some surplus for export. Corn was introduced by the Spanish from Mexico and is another important food crop, and accounts for nearly a third of the area given to food-crops; for many peoples it is a more important cereal than rice, particularly in Cebu and central Visayas islands.

Industrial crops, which include coconuts, sugar, abaca, and tobacco, are grown primarily for their export earnings. The Philippines is one of the world's largest producers of coconuts, and a major exporter of coconut products. It also produces more than one-third of the world's copra. Sugar is one of the major exports and valuable earner of foreign exchange. Sugarcane is grown extensively in central and north-central Luzon, western Negros, and on Panay. Much of the export goes to the United States. Abaca or Manila hemp is the source of plant fiber, and also an important export. It is grown extensively in eastern Mindanao, southeastern Luzon, and on Leyte and Samar. Tobacco is an important commercial crop, though its value has declined during the last few decades due to the decrease in the consumption of cigars in the U.S.A., the country's major importer. The importance of forestry and forest products, traditionally of great value as export items, has significantly declined in recent years. Forest products include high quality timber and veneer, dye woods, rattan, and rubber. Commercial fishing is carried on universally, primarily off Palawan, Panay, Mindanao and Negros, and canned tuna is a major export.

During the Spanish occupation, the allocation of good farmland to loyal citizens of church and state, created a faithful class of plantation holders and landless laborers for the landlords. Most of the country's wealth was concentrated in a handful of landed aristocrats, while the masses remained poor. Meanwhile, crops such as tobacco and maize were introduced from the Americas. Large landed estates producing such crops as sugarcane and rubber were added to the traditional agriculture. In the later 1960s the Philippines successfully experimented with the new high-yielding varieties (HYV) of foodgrains and became a pioneer in the so-called "Green Revolution" and self-sufficient in its foodgrain production.

During the 1940s and 1950s, economic diversification had started. The manufacture of textiles, shoes, and cement began appearing. Mining also grew in importance. The major minerals included copper, bauxite, gold, manganese, zinc, lead, iron ore, and chromite. Some petroleum is also extracted. Several of these are produced for export, and thus are vulnerable to price fluctuations in world market. Although much growth has taken place since independence, industrial production has not been expanding for want of capital investment which must come from the local Chinese, the government, or foreign countries (mostly the U.S.A. and Japan). Major manufactured goods are processed foods, beverages, textiles, petroleum products, wearing apparel, and chemicals.

The land reforms introduced during the United States occupation did help the peasantry but economic hostility between the landed aristocracy, which had acquired political power, and the poor masses surfaced during 1970s and 1980s and led to several uprisings.

Economic discontent was further exacerbated by a high rate of population growth, as population pressures tended to aggravate the hardships of the poor. The census of 1903 counted 7.1 million Filipinos; by 1960 it had increased to 30 million. Today, it stands at 75 million and is growing at a high rate of 2.0 percent a year, and is likely to double in 30 years. Although the

Philippines does not as yet face a major population problem, the threshold level to overpopulation is about to be reached, particularly in the crowded, central plains of Luzon and the Visayan group of islands.

As in other Southeast Asian nations, most Filipino political and economic power is concentrated in the capital, **Manila.** Founded by the Spanish in 1571 on a sheltered harbor, it retains a distinct Spanish appearance. It is truly a primate city in the country, the hub of Filipino politics, and the nerve center of the country's economy, religion, and communications. With a population of over 5.5 million for its metropolitan area that includes Pasay City, Caloogam City, and Quezon City, it is six times larger than the next city Davao, on the island of Mindanao. **Manila** is the home of two-fifths of the nation's industries and international commerce. Most of its industrial establishments are small scale and produce such items as processed food, textiles, and consumer goods. Manila is internationally known for the manufacture of cigars, that utilize the locally cultivated abaca plant. Petroleum products, chemicals and beverages are some other manufactured items. However, like most large cities in developing countries, it contains many slums.

Prospects With the exception of Malaysia and Brunei and Singapore, the Philippines is the most urbanized nation in Southeast Asia (58 percent of its population lives in cities). The country possesses many inherent assets: a considerable endowment of natural resources; a highly literate population; fairly established democratic institutions; a rising entrepreneurial class, strategic archipelagic location; and a close commercial and political liaison with the United States. It is now self-sufficient in its foodgrain requirements. It has already shown substantial economic growth, averaging between 5 and 6 percent a year during the last two decades. But occasional political turmoil often obscures the country's potential for development. The future depends largely on the restoration of political and economic stability by closing the gap between the rich and poor (particularly by addressing rural landlessness and urban poverty), by reducing birthrates, and by strengthening further its democratic political institutions.

BRUNEI

The oil-rich Sultanate of Brunei is a tiny state on the north coast of the island of Borneo, comprising of two enclaves in East Malaysia. With a territory of 2,226 sq miles (5,765 sq km) and a population of about 300,000, it is one of the smaller nations of Southeast Asia but one of the richest in the world. (Its per capita income is estimated to be over $25,000 annually in the late 1990s). Brunei traces its history and royal line to the 14th century when the rulers were converted to Islam. By the 16th century it controlled all of Borneo and parts of Sulu islands (of the Philippines). Most of its territory when the British increased pressure on its territory in the 18th century, resulting in Brunei becoming a small protectorate of Britain containing only a small port town, and the adjoining agricultural hinterland. The discovery of oil in 1929 and natural gas in 1965 pushed the small, sleepy protectorate state into an era of economic prosperity. In 1984 it gained complete independence from Great Britain, and shortly afterwards became a member of the Association of the Southeast Asian Nations (ASEAN) and a member of the United Nations.

Lying between 4° and 5° north of equator, Brunei's land surface is mostly mountainous in the southeast rising to 607 feet (1,850 meters), and is largely covered by dense tropical rainforest. Climate is that of hot, humid equatorial type with temperatures ranging from 76°F (24°C) to 86°F (30°C). Precipitation is heavy---between 100 inches (2,500 mm) annually on the coast to 200 inches in the interior. Rainfall is heaviest during the northeast monsoon (December to March) season.

The population, two-thirds of which is composed of Malays, 16 percent Chinese, 11 percent Indians, and 6 percent of indigenous groups, has been growing rapidly at an average rate of nearly 3 percent a year for the last 20 years. However, it has registered a small downturn in growth rates from the 4 percent a year of the 1980s to 2.1 percent in 1995. Short of manpower, Brunei has depended heavily on foreign workers for its oil industry. Temporary immigrants account for nearly one-third of its entire workforce recruited primarily from South and Southeast Asian nations. Brunei Malays dominate the bureaucracy while the Chinese are largely engaged in commerce. Despite longstanding ties between the Chinese and royal family, tensions between the two communities have always existed, and only 5 percent of the Chinese have acquired Bruneian citizenship. Islam is the state religion professed by two-thirds of the population. Thirteen percent of the population---primarily of the Chinese---is Buddhist, and ten percent subscribe to Christianity. The remainder belong to several indigenous groups living inland and tend to be animists, although many individuals have adopted Islam during the last few decades. Some of the indigenous peoples are Christians.

Oil and liquified natural gas (LNG) dominate the economy, accounting for 98 percent of export earnings. The major export destinations are Japan, South Korea, Thailand, Singapore, and the Philippines, with Japan consuming over 50 percent of all exports. Oil is produced from onshore and offshore wells; but the vast reserves are estimated to last until 2010 if no new oil-bearing areas are discovered. The LNG now equals oil as a major revenue source and the LNG refinery at Lumut is among the world's largest. Enormous wealth from oil revenues has transformed this remote, little-known state into a prosperous welfare state. Education and medical care are free, and there is no income tax. Wage levels are comparatively high for Southeast Asia. Dependence on a single commodity has, however, made the country subject to vagaries of market fluctuations in oil.

Agriculture accounts for only 3 percent of the total value of gross domestic product, and consists of fishing on the coastal areas, the production of fruits, vegetables and rubber on the better-drained portions of the inland area. Agriculture and fishing, the traditional economic activities, and the once-healthy rubber industry, have, however, declined during the past two decades. The area under rice cultivation has also decreased. The timber has been developed for consumption in the home market. Efforts to diversify the economy have failed to attract interest in revitalizing fishing and rubber plantations. Most food is imported. Other major imports are machinery used for the petroleum industry, manufactured goods, chemicals, and consumer items. Major import sources are Singapore, Malaysia, United States, and Japan.

The small size of the labor-force, in general, and people's preference for civil service, has so far slowed the country's industrialization and economic diversification. As part of a long-term economic development plan, the government has modernized the ports, created the world's largest gas liquefaction plant, and built an international airport at **Bandar Seri Begawan,** the capital city (population: 46,000). Around it and its adjacent port city of Muana; and in Seria and Kuala Belait area near the oil fields, nearly two-thirds of the country's population is concentrated. A sign of the development boom is visible in modern apartment houses, shopping centers, and hotels in the capital city, where a splendid 1800-room palace costing half a billion dollars was built by the present ruler. The cities, in general, however, retain the air of a typical Malay town containing *Kampongs,* the Malay water villages with their traditional wooden houses standing on stilts. Some interior portions of the country still remain an area of subsistence agriculture, and rural isolation, virtually untouched by the modernization that characterizes the coastal fringe.

THE CITY-STATE OF SINGAPORE

Located at the southern tip of the Malay Peninsula, the city-state of Singapore is Southeast Asia's smallest nation. It became independent in 1965 after breaking away from the Federation of Malaysia and Singapore. Selected in 1918 as a trading post by Stamford Raffles, halfway along the principal trade route between India and China, it commanded a focal position at the southern extremity of the Malay Peninsula, dominating entrance to the Indian Ocean and China Sea, and became the main entrepot of Southeast Asia. The city lies on a small island, a mere 240 sq miles (600 sq km) in territory, but is blessed with a fine, natural deep-water harbor. Virtually without any natural resources besides the harbor, its meteoric rise to become a major financial, industrial and communications center, and a staging port in the transport of commodities, especially for oil from Indonesia and rubber and tin from Malaysia, is explained largely by its highly literate and energetic population approaching 3 million and the pragmatic policies of its administration.

It is the world's second busiest port after Rotterdam, and Southeast Asia's largest port and an oil-refining center. As a duty-free port it attracts annually more tourists than its permanent resident population. Thanks to its vast commercial success, it can boast of the highest standard of living in Asia after Brunei and Japan (the GNP per capita in 1997 was $33,000 distributed evenly over its population). Singapore has no slums, almost no beggars, minimal unemployment, and a life expectancy of 79 for women and 75 for men. By all normal standards it is, thus, a developed and modern nation.

Following independence, Singapore's government realized that the commercial sector alone could not sustain the island's economic growth in the face of rapid population increase and rising competition from the neighboring countries of Malaysia and Indonesia, the nation quickly embarked upon a crash programme of industrialization during the 1960s and 1970s. The country was singularly successful in attracting foreign investment in export-oriented industries because of its sound and aggressive planning, central location, political stability, an excellent service infrastructure and the availability of skilled laborforce. Its economic growth also profited considerably from the pragmatic, unorthodox, neo-socialistic policy of its administration, which encourages private enterprise and meritocracy; and from tax concessions given to investors, cheap and reliable labor, efficient public services, ethnic harmony, and political stability. There are nearly 1,600 factories manufacturing all types of consumer items, and labor-intensive plants specializing in electronic equipment, industrial hardware, textiles, chemicals, food processing and ship-building. During the 1970s, 1980s and 1990s, its growth rate averaged over 9 percent annually, a rate that was among the highest in the world. Following the international crisis of the mid 1970s, it shifted emphasis to high technology production, such as electronic components, computer hardware, medical instruments, and metal engineering products which form the basis of its exports now. Very little industry is in government hands, yet the state offers more welfare benefits and services than any other nation in Asia. Its chief imports are petroleum, machinery, motor vehicles, and telecommunication apparatus. Although the country imports food items, that category accounts for less than 10 percent of all imports. Singapore's chief trading partners are the United States, Japan, Malaysia, Saudi Arabia and Hong Kong.

Despite ethnic diversity, a happy partnership of races has contributed to Singapore's success story. The population is 77.0 percent Chinese, 14 percent Malay, and 7 percent Indian, with small minorities of Eurasians and other groups. Linguistically, Chinese, Malay, Tamil and English are the official languages; English is widely used as a *lingua franca.* Culturally, the nation is highly eclectic---the society is a religious mosaic composed of Budhists-Taoists, Muslims, Christians, and Hindus that make up for 54, 16, 13, and 4 percent respectively of the population in addition to its 14 percent of the people declaring themselves as nonreligious. In this mosaic each strand is distinct but interwoven into a cultural tapestry. Unlike Malaysia, ethnic and race riots have rarely marred the prevailing political harmony (except for brief riots in 1964).

A remarkable transformation of Singapore's landscape has occurred since the World War II. On the heels of the Japanese destruction of the city and their withdrawal from Southeast Asia in 1945, an ambitious plan of housing projects consisting of high-rise apartment complexes to shelter all sections of society was undertaken. Each complex contained shopping, educational and medical facilities, conveniently placed to serve the neighborhood. A group of several complexes had a town center with a wide variety of services. Today, most of its population lives in this low-cost, state-subsidized housing under strictly enforced neighborhood laws. Only a few sections of the old city still retain the congestion of the past (some preserved as tourist attractions) but most citizens now live in modern apartments with mass transit connecting all parts of the city.

Singapore's planners expect and are making provision for the population to increase to no more than 4 million in future decades. The government has followed a strong----according to some critics intrusive----policy of family planning. Between 1948 and 1958 its population had doubled and by 1965 the annual growth rate was more than 3 percent (usually symbolic of underdeveloped countries). The government took action by enacting strong family planning legislation, including publicizing birth control methods, making abortion readily available, instituting male and female sterilization programs, and economically penalizing large families. In these efforts the administration's campaign proved singularly effective and by the early 1980s, annual birthrates had dropped to one-half the levels of 1965. Today, its rate of population growth is close to one percent a year, almost one-third of 1965s level. However, by the late '90s the administration had become apprehensive of a potential shortfall in the labor force in the future, and began re-examining its population policy in order to make it more liberal. The city-state of Singapore has become in many ways a cultural, political and economic model to the developing nations.

Further Readings

Burley, T.M. *The Philippines: An Economic and Social Geography*, London, 1972.

Dobby, E.H.G. *South-East Asia.* London, 1973.

20

East Asia: China and Taiwan

China (officially the People's Republic of China) occupies most of the territory of East Asia. It is the largest of all the nations in Asia, and the most populous one of the world. Stretching for about 3,100 miles (5,000 km) from east to west, and 3,400 miles from north to south, it covers an area of nearly 3.7 million sq miles (9.6 million sq km), and is almost as large as all the European countries put together. Among the major nations of the world, only Russia and Canada are larger in area. With a land frontier of over 12,400 miles, and a coastline of nearly 8,700 miles, it makes borders with 14 countries including the two other large nations—Russia to the north, and India to the south. It faces Japan and South Korea, across the Yellow Sea, and the Philippines beyond the China Sea.

The Chinese probably justifiably claim that their civilization is the oldest, continuous living civilization that is unique in its longevity and resilience. There is strong historical evidence that the "essential characters" of the Chinese culture would have been evolving at least as early as 2700 B.C. For centuries, and probably for millennia, China has been the most populous country in the world, and has supported one of the most enduring civilizations.

Surrounded by deserts and mountains to the north, west and east, and hemmed in by the sea to the east, Chinese civilization developed in relative isolation. Outside influences were absorbed or modified to conform to the Chinese ethos. Although this made possible the flowering and refinement of a distinctive culture, it resulted in making China ill-prepared to cope with the rest of the world, particularly since the mid-19th century, when the country was confronted by technologically superior foreign nations.

The early name of the country—*Chung Kuo*, "the Middle Kingdom"—indicates the inward orientation of the Chinese and their acceptance of the role as a civilizing agent for the less cultured people. Until the modern times the Chinese were only dimly aware of the existence of higher civilizations elsewhere, an attitude that was undoubtedly contributed in a large measure by the existence of physical barriers, and the intervening distance. The cultural influences from India, Central Asia, during the 7th and 14th centuries, found no lasting place in Chinese life. The traditional Chinese remained essentially unchanged, and China continued to move largely in a cultural orbit of its own. Only during the modern times external challenge presented by the political ideas emanating from the West became a catalyst for a revolution that began in the 20th century against the old regime which resulted in the establishment of a Communist government in 1949. This

Table 20.1 *East Asia: Basic Data, 1999*

Country	*Area (Thousand sq km/sq mi)*	*Population (Millions)*	*GNP ($/capita)*	*Capital*	*Urban Population Percent (est.)*
China	9,572.9/3,696.1	1251.2	860	Beijing	34
Japan	377.8/145.9	126.7	38,160	Tokyo	77.6
Korea, North	122.7/47.4	21.4	970	Pyongyang	61.3
Korea, South	99.3/38.3	46.5	10,550	Seoul	85
Mongolia	1,566.5/604.8	2.4	360	Ulaanbaatar	62
Taiwan	36.1/13.9	22.0	13,900	Taibeh (Teipeh)	74.7

Source: Encyclopedia Britannica, Book of the Year, 2000.

event reshaped world's political geography, and China since came to rank among the most influential countries of the world. Since the 1970s when China decided to increase its interaction with the international economy, its role in world trade has grown and its importance in international affairs substantially enhanced.

Like the Indians the Chinese had attained a high degree of sophistication and maturity in fine arts, music, drama, literature, and philosophy centuries before the Europeans. As early as the 1st century A.D. they had invented such items as paper, compass, iron-ore, steel products and gun powder, the products that shaped the course of human history.

Relief Features Nearly one-third of China's land surface is in the form of mountain chains which have exercised profound influence on the political, economic, and cultural development of the nation. World's highest mountain peak—Mt. Everest which rises to 29,028 feet (8,848 meters) in elevation, lies at the border between China and Nepal. As a contrast, the lowest part of Turpan Depression in the Uigur Autonomous Region of Xinjiang is 505 feet below sea level. The coast of the country contrasts greatly between south and north. To the south of the Hangzhou Bay the coast is rocky and indented with several harbors and offshore islands; to the north, it is mostly flat and sandy except along the Shandong and Liadong peninsulas.

Broadly speaking, the surface features of China are complex, and highly varied that include rugged mountains, high plateaus, gently-sloping river plains, deserts, and areas below sea-levels, all set in an irregular checkerboard pattern. Within these fairly sizable regional units, however, the country can be easily divided into several regions containing a broad measure of coherence in physical features. At the broadest scale the relief is high in the west and low in the east with the result that the major rivers flow generally eastward.

The surface features of the country may be viewed at three levels: first level represented by the Plateau of Tibet with average heights of well over 13,000 feet (4,000 meters) above sea level; the second level contains areas lying north of the Kunlun Mountains descending to elevations between 6,000 and 3,000 feet containing several basins, and plateaus including the Mongolian Plateau, the Tarim Basin, the Loess Plateau region and the Yungui-Guizhou highlands; the third level lie east of the previous two levels, and with few exceptions, is made up of hills and plains lying below 1,500 feet. The broad zones contained within the three levels may be labeled as the *southwest*, the *northwest* and the *eastern* zones. The southwest is cold, elevated, mountainous region containing intermontane plateaus and inland lakes; the northwestern zone is arid, and eroded by winds; whereas eastern China is a region shaped by the rivers which, in most part, have deposited alluvial plains.

The three broad zones, the southwest, the northwest, and the eastern can be further subdivided into several regional units on the basis of geologic structure, climatic conditions and geomorphologic development. A short review of these features of these is given below.

The southwest zone includes the high plateau and the Qaidam Basin in the west and the Yungui-Guizhou highlands in the east. Most of the *Tibetan Plateau* lies at elevations above 13,000-15,000 feet, with the flanking border areas at even higher altitudes; some peaks rising to heights of 23,000-26,000 feet, thus justifying the title of being the "Roof of the World." The interior slopes in the vast plateau of Tibet are generally gentle, and the exterior ones very steep. Much of the interior is without external drainage. In the eastern and southern borderlands, however, several great rivers of Asia---Jiang Chang, Huanghe, Mekong, Salween, Brahmaputra and Indus----find their sources. Within southern Tibet is a series of mountains and basins----its southern rim consists of the lofty Himalaya that extends for 1,500 miles, and contains close to 50 summits exceeding 25,000 feet in elevation. Within the southern part of the plateau is a series of mountains and basins. In the northeast corner, lies the dry, desertic, *Qaidam Basin* at elevations between 9,000 and 10,000 feet, within which gravely sands and salt wastes predominate. The *Yungui-Guizhou* region lies to the east of the Tibetan Plateau. It is a land of great canyons and rugged mountains, with a general northwest-to-southeast orientation. Northern Yungui and the western sections of Guizhou are deeply dissected, whereas the rest of Yungui is more plateau-like with areas of rolling uplands. The general elevations are 9,000 to 12,000 feet. Scattered throughout the region are small lake basins, separated by mountains.

The northwest zone has at least three distinguishable regions. The *Tarim Basin* of Xinjiang lies to the north of the Tibetan Plateau, hemmed in between the lofty mountains. The Kunlun in the south and the Tien Shan in the north, and the Pamirs on the west. The basin occupies one-fifth of a million square miles of desert land at an average altitude of about 3,000 feet. In the middle of the basin is the Takla Makan Desert, one of the world's most barren lands. Two important rivers: Tarim, and Hotien (Kotan) traverse the desert that usually carries water during the flooding period only. North of the Tarim Basin lies another large depression, the *Junggar Basin*. Enclosed between the Tien Shan Mountain in the south and the Altai Mountain in the northeast, the latter separating it from the Mongolian People's Republic, the Junggar Basin is a generally flat surface sloping gently to the southwest with elevations between 1,000 and 1,500 feet and in the lowest part elevations drop to 620 feet. A large part of the Junggar Basin is a desert covered by *barchans* (crescent-shaped sand dunes that move). The *Tien Shan* region is a large and complex system of mountain

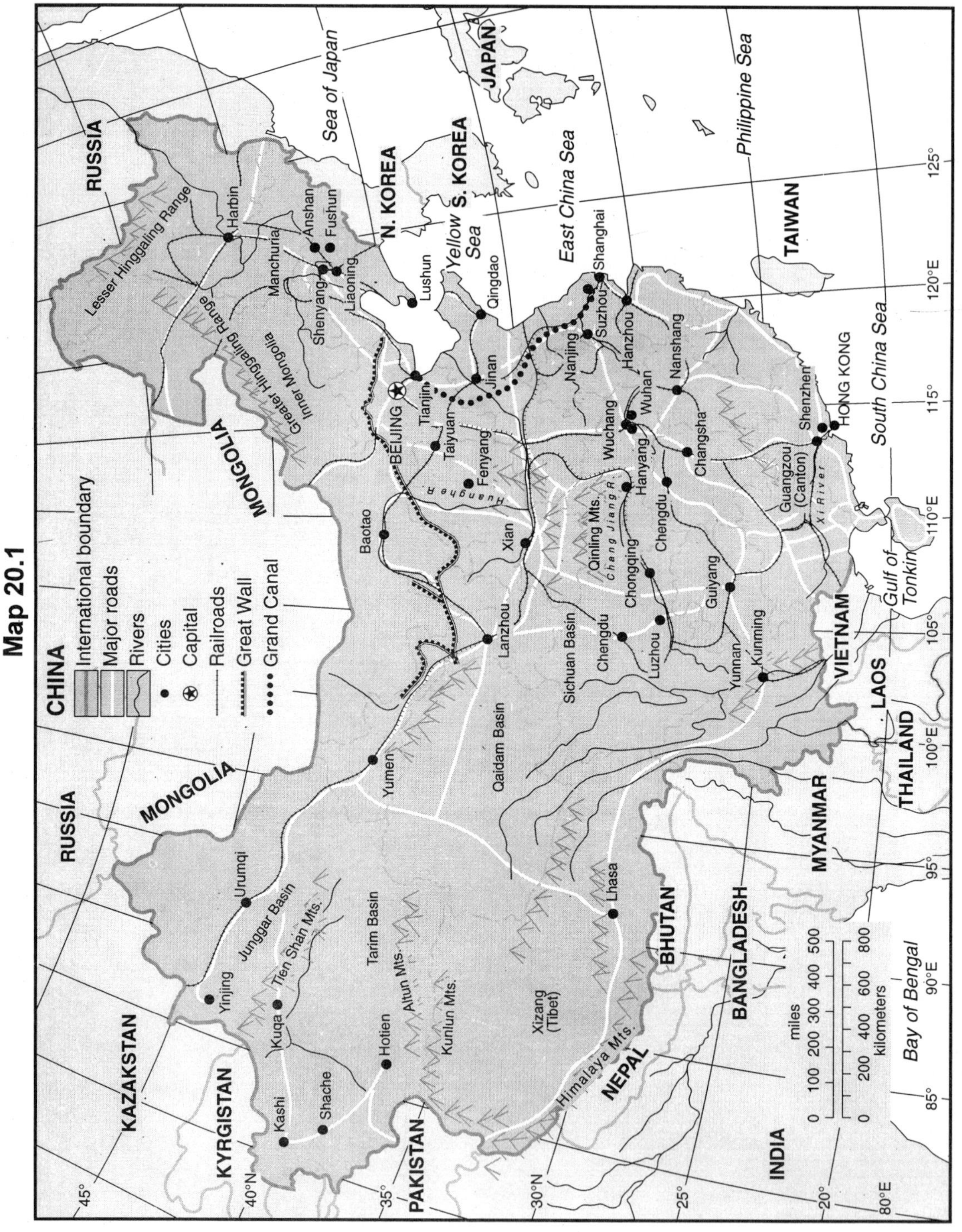
CHINA
International boundary
Major roads
Rivers
Cities
Capital
Railroads
Great Wall
Grand Canal
RUSSIA
MONGOLIA
KAZAKSTAN
KYRGISTAN
PAKISTAN
INDIA
NEPAL
BHUTAN
BANGLADESH
MYANMAR
THAILAND
LAOS
VIETNAM
TAIWAN
N. KOREA
S. KOREA
JAPAN
Sea of Japan
Yellow Sea
East China Sea
Philippine Sea
South China Sea
Gulf of Tonkin
Bay of Bengal
Lesser Hinggaling Range
Greater Hinggaling Range
Inner Mongolia
Manchuria
Harbin
Anshan
Fushun
Shenyang
Liaoning
Lushun
Qingdao
BEIJING
Tianjin
Taiyuan
Jinan
Fenyang
Baotao
Huanghe R.
Xian
Lanzhou
Nanjing
Suzhou
Shanghai
Hanzhou
Nanshang
Wuhan
Wuchang
Hanyang
Changsha
Qinling Mts.
Chang Jiang R.
Chengdu
Chongging
Guiyang
Sichuan Basin
Luzhou
Yunnan
Kunming
Guangzou (Canton)
Xi River
Shenzhen
HONG KONG
Qaidam Basin
Yumen
Urumqi
Junggar Basin
Tien Shan Mts.
Tarim Basin
Altun Mts.
Kunlun Mts.
Yinjing
Kuqa
Hotien
Kashi
Shache
Xizang (Tibet)
Lhasa
Himalaya Mts.
miles
0 100 200 300 400 500
0 200 400 600 800
kilometers
45°
40°N
35°
30°N
25°
20°
80°E
85°
90°E
95°
100°E
105°
110°E
115°
120°E
125°

Map 20.1

ranges and depressions. The altitude of the main chains of the eastern Tien Shan is between 13,000 and 15,000 feet with several peaks rising to over 16,000 feet. The western section of the Tien Shan receives more precipitation, has several glaciers and rivers, and contains alpine meadow slopes that form the best grazing grounds in China. East of the Tien Shan region is the sprawling *Gobi Desert* which spills well into southern portions of the People's Republic of Mongolia and parts of the Inner Mongolian Province of China.

The rest of China, from the Russian border in the north to Vietnam in the south, includes several regions of varying sizes and topographical conditions representing the third altitudinal level, and distinct from the regions of the northwest and the southeast, may be called *Eastern China*. The northernmost region in Eastern China is the *Manchurian Plain* flanked on the west and north by the Hinggalings----the Greater and the Lesser mountain ranges. The Manchurian Plain is divided by the river systems----the Songhua in the north and the Liao in the south. Most of the area has an erosional rather than a depositional surface, but has a deep rich black soil important for agriculture. Southeast of the Manchurian Plain is a series of ranges along the North Korean border, consisting in part of rugged peaks rising to 9,000 feet (2,736 meters), but mostly between 1,500 and 3,000 feet. Almost one and one half times as extensive as the Manchurian Plain, is the *North China Plain* which lies mostly at heights below 150 feet with a nearly flat, monotonous topography. The plain is the largest alluvial plain in eastern China and covers an area of over a quarter of a million sq miles, and was formed by enormous deposits brought down by the Huanghe and its tributaries from the *Loess Plateau* in the west.

Situated west of the North China Plain and east of the deserts of northwest region of China is the *Loess Plateau* region of dry, barren hills and mountains of 4,000 to 5,000 altitudes on the average. To the north is the Great Wall of China, while the southern limit is the Qinling Mountains in the Shanxi Province. Most of the region is covered with loess (windblown glacial deposits) to thicknesses of 150 to 650 feet. Several gorges traverse the region as the loess can be easily eroded by water. The *Shandong Peninsula* that juts into the Yellow Sea is a hilly area of old, crystalline rocks with relief averaging 1,500 feet in the east and a little higher on the west. The coast is largely rocky mountains. In Eastern China, the most prominent mountain chain is that of the high and rugged *Qinling Mountains* that form a traditional divide between the "North" and "South". The mountains are not only a topographic but a climatic barrier and a cultural divide. The altitude of the ranges varies from 3,000 to 10,000 feet, in general; though the western part rises to over 13,500 feet tapering toward the east. The mountains have several parallel ridges, separated by deep canyons, and steep-sided valleys. The Qinling system is high enough to protect the Sichuan Basin from the Mongolian cold weather, and prevents the entry of much moisture into north and northwest China and thus forms a climatic divide between dry, cool north China and moist, warm south China. The Qinling has also remained a major topographic obstacle to communications. South of the Qinling is the *Sichuan Basin* surrounded and thus protected by mountains on all sides. The basin is much warmer in winter than other more southerly areas. Most of it is hilly. The eastern portion contains series of ridges and valleys. East of the Sichuan Basin is the important river in plains of *Middle and Lower Chang Jiang*, the flood plains *of* which is constricted by hills in the north and the south. The flood plain traversed through several lakes (Donting and Poyang) in the middle and lower courses before it makes a delta. Several thousand miles of irrigation canals and waterways crisscross the middle course of the river plains.

Close to China's southeastern coast is the region containing *Southeast Mountains*. The mountains trend northeast to southwest, and at their higher reaches attain the elevations of 5,000 to 6,500 feet. The short, fast-flowing rivers have cut deep valleys. The coast is highly indented, with numerous islands in contrast to the comparatively smoother coastline to the north of Shanghai. To the west lies the *Nam Mountains* region which contains several mountains with a northeast-southwest trend line parallel to the southeast mountains region. The mountains form a watershed between the *Chang Jiang* flood plain to the north and the Pearl River to the south. The highest peaks on the water divide reach the altitude of 6,500 feet or over. Here the only plain of significance is that of the Xi River. The coastline is rugged and irregular; there are several bays and offshore islands, including that of Hong Kong. Xi River is the major river that rises in the highlands of eastern Yungui and Guizhou plateaus.

Climatic Characteristics The climatic patterns of such a large country as China are varied and result from a number of factors; chief among these are: (a) a vast but compact size, (b) a west-Pacific Ocean location, (c) an extensive latitudinal spread, (d) the distribution of relief features within it, and (e) wind systems produced by all of these elements.

Most of the country lies inland, with great distances from its eastern coastal fringe. No large seas indent its vast, compact landmass to moderate the climate of extensive inland locations such as exists in Europe. During the winters due to the rapid and prolonged heat loss through radiation an unusually intense and persistent high-pressure system is developed in the regions of Mongolia and east Siberia. Cold, dry air masses from the north and northwest flow to grip much of the country during long winters, particularly its northern and northwestern parts. The conditions are reversed in summer, when the direct rays of the sun cross Equator and move further north to their limit in the summer solstice, heat absorption exceeds radiation. The rising air brings low pressures in interior western China. The in-rushing warm, maritime, moist air masses from the south and southeast take the place of the retreating cold air masses of the dry interior, bringing moisture with them to most of eastern China. This winter and summer reversal of wind directions are the so-called monsoon winds.

The monsoon effect in China differs from the Indian monsoons. The winter dry season here is much longer in duration than the summer rainy season, whereas in India winter dry season is shorter than the summer rainy period. The gradual onset of summer wind-system is in sharp contrast to the sudden "outburst" of the summer monsoon in India. In China, on the other hand, the winter air drift is markedly more pronounced than the summer one, because of the establishment of intensive winter high-pressure system. The great extremes of winter and summer weather conditions make the weather conditions of China more "continental".

China's west-Pacific location affects the wind direction of the seasonal winds. The waters of the southwest Pacific and the seas bordering China, supply most of the moisture that China receives in summer brought by southerly and southeasterly winds from the warmer, tropical areas. In winter, the winds originate from the Siberian or Mongolian high-pressure and are thus land-based, dry, and northerly.

The effect of latitude on climate is also noticeable. Temperature (isothermal) differences are reflected in the north-south location. In general, 16°C isotherm of average annual temperature roughly divides the eastern China into two parts: the northern one is the colder section and southern one the warmer China. South of this isotherm, which runs roughly along an east-west axis, average annual temperature is high, with mostly mild winters. The lower and middle Jiang Chang Basin have warmer temperature regimes, and the average temperatures increase gradually southwards toward the tropical, coastal areas.

Oragraphical (induced by the presence of mountains) effect on climate are also significant. Four major mountain blocks are climatically significant to China. One is the Himalaya and the associated high Tibetan Plateau region, which block the inflow of moisture into interior northwestern China, from the south. Second mountain block is that borders eastern section of the Tibetan Plateau which creates a similar barrier to Pacific moisture moving westward. The third is the Qinling system which runs west to east to reach the lower Chang Jiang plain. It is high enough to protect Sichuan Basin from the cold Mongolian air masses. It also prevents the entry of moisture into north and northwest China, and forms a climatic divide between dry, cool northern China and moist, warm southern China. The fourth mountain block is the southeast hill-and-mountain complex. During the summer when air masses flow from the south and southeast, those mountains receive much of the oragraphic rainfall.

Rainfall in China generally decreases from the southeast to the northwest. Along the southern part of east coast a few places receive over 80 inches (200 cms) precipitation annually. The Chang Jiang Valley gets about 40 inches; to its south, most areas get between 60 and 80 inches; and to its north 30 inches or less. In the lower reaches of the Huanghe only 20 inches of precipitation are received annually.

By the time the southeast monsoons reach the interior and the distant interiors of the country such as western Inner Mongolia, Junggar Basin, Tarim Basin, Xinjiang and Xizang Plateau, it has lost much of its moisture; precipitation keeps on decreasing, and some remote western sections receive even less than four inches annually. These regions are not only far from the sea but the high mountains prevent the moisture-laden southerly monsoons during summer months from reaching these regions.

Among the various types of air movements that are responsible for precipitation incidences involved in China's climates, four are significant: monsoon-type movements, convectional movements arising from local heating, the large cyclonic storm circulation, and the smaller but the more violent typhoons or hurricanes. The mechanism and incidence of the monsoons for Asia as a whole has been discussed earlier.

Convectional rainfall is associated with the oragraphic lifting of the air on the windward side of the mountains where the rising air masses are moist. The air masses condense as it meets the cold air above. Besides this, the formation of convection precipitation also occurs in the zone of contact between polar Siberian air and maritime tropical air where the frontal (between warm and cold air mass) uplifting takes place. The latter generally occurs in connection with the extratropical disturbances or "cyclonic storms" generated around low pressure center. Such cyclonic storms are frequent in the central Jiang Chang basin. The seas east and west of the Philippines (South China Sea) are important source regions for the typhoons. When the Mongolian-Siberian high-pressure area weakens or disappears as the summer advances from May onward, an increasing number of typhoons strike the China coast, particularly during the late summer, the highest frequency is in August. Typhoons bring heavy rains, and account for much of the precipitation along the southern coast, and often bring destruction on the area. Fortunately very few penetrate far inland; those which enter southeast China usually recurve the coast between Shanghai and Shandong peninsula.

Soils and Vegetation For a major agricultural country like China soils are an important resource and of great significance. Because vegetation and climate factors are important in their formation, and major soil patterns are associated with the climatic and vegetation conditions, although the original character of the soils and vegetation cover has changed in most of the cultivated regions of China due to the activities of man.

All of China south of and including the Qinling Mountains was once in forest; most of it has been cut down, and its slopes cover coarse grasses. The more inaccessible mountainous areas still support spruces and firs. In the densely settled eastern part of the country, it is difficult to identify natural vegetation. Over the millennia, most of the original stands have been cut, burnt, or cleared.

In the dryer, northern and western regions of China the soil cover is thin, and poor in the mountains, on the Tibetan Plateau; and the prairie, steppe and desert soils in the Xinjiang, Inner Mongolia, and parts of northeast Manchuria. These soils of Tibetan Plateau are the result of sparse vegetation, and cold, dry climate. The prairie and grassland soils of the north and northwest have resulted from the accumulation of humus and calcium carbonates in the soils. The soils range from the rich, black chernozem in the northeast; to thin, poor steppe, and gray-brown near desert soils.

China's most fertile soils are found in the river plains of Manchuria, the basins of the rivers Huanghe, Jiang, and Xi. Good soils are also found in much Sichuan Basin, and the loess region.

In eastern humid China, three main vegetational zones can be identified from south to north: (1) a zone of tropical forest and grassland, (2) a zone of subtropical evergreen forests extending north to the Chang Jiang Basin and west to the Qinghai-

Tibet Plateau, and (3) a zone of deciduous forest extending from the Chang Jiang Basin to the Amur Basin in the northern parts of Manchuria.

Forests now cover nearly 12 percent of China's land surface, and the available forest resources are thus quite modest. The Communist government has placed great emphasis on reforestation in recent years, and a large number of trees have been planted throughout the country. Presently, the major commercial forests are found in the northeast where large stands of birch, fir, conifers, and ash are found on the slopes of the mountains of Hinggan.

The dry, western part of the country, in general, resembles Central Asia, a vast area of deserts and steppeland. Vegetation here is sparse, consisting of grasses and shrubs.

Historical Development China's first documented imperial dynasty, the Shang, was centered in the central Huanghe Basin. The Shang rule produced an early urban society, and is credited with the creation of magnificent bronze vessels, the development of horse-drawn war chariots, and establishment of a system of writing.

Since the Shang dynasty, China's history has been characterized by half a dozen great dynasties, most of which were accompanied by periods of stability and progress. Between each has been intervals of chaos and transition.

Over the millennia of Chinese history a unique and fairly uniform culture spread first over most of middle Huanghe basin and later to other parts of China; and its history can be viewed as a continuous development with certain repetitive tendencies. The western margins of the Huanghe basin became the heartland of the Chinese culture, from there it moved to the lower Chang Jiang basin and Guangzhou, and eventually to the southwest. The area of control of the central authority would wax and wane depending on the pressures exerted by the nomadic Mongols and Manchus at different intervals from the north and northwest. While the emperor, as the "Son of Heaven" was the source of all central authority, during periods of weak administration the local chieftains broke loose of the central administration in the "Chinese heartland" (Middle Huanghe basin).

The Zhou people of western China overthrew the Shang and established their rule on North China that lasted for nearly eight centuries (c. 1027-256 B.C.). During the Zhou rule, Confucius, Lao Zi, and Mencius, the greatest of the Chinese philosophers lived. Chinese literature that later became the basis of traditional Chinese education was composed during this period, and the use of iron was first introduced during the Zhou regime.

In 221 B.C. the Qin dynasty took control over the administration and established a centralized imperial system. The construction of the Great Wall of China had been started by the Zhous was extended, and feeder, shorter walls constructed to keep the invaders out. The Han dynasty gained control in 202 B.C. and ruled for nearly four hundred years. The Han period in Chinese history is generally considered as China's imperial age, notable for long, peaceful and prosperous rule, extension of the empire and for great artistic and cultural achievements. Confucianism became the philosophic basis of government. Buddhism which had earlier arrived from India and Daoism (Taoism) also prospered. Indian advances in medicine, mathematics, astronomy, and architecture were accepted. During the Hans, China became one of the strongest empires in the world. The Hans introduced several changes in the systems of landownership as the fendal order broke down and private property rights recognized. The silk trade with the West became the first, large external link of commerce. Han period in Chinese history to this day symbolizes the spirit and glory of China, and most Chinese call themselves as the "People of Han ".

By the early 3rd century, the Han Empire had fallen, and disintegrated. A series of short-lived non-Chinese dynasties ruled, and the short-lived Sui dynasty reunified the country. The Tang dynasty that replaced the Sui in 618 ruled the nation for nearly three hundred years. The Tang period was an age of prosperity and great cultural accomplishments. Containing over a million inhabitants, the city of Changan (Xian) became the largest city in the world and attracted traders, diplomats, poets and scholars from throughout Asia and the Mediterranean region. Art, particularly figure painting, flourished. Buddhism exercised an enormous cultural influence and a distinct Chinese school of Buddhism was evolved. The civil service, based on an examination system on the Chinese classics and a renaissance of Confucianism were other important developments of this brilliant era.

After the Hans, dynasty that ruled for about three hundred years between the 10th and 13th centuries were responsible for two major changes of great importance for the empire: the firm establishment of the existing civil service examination, and the development of neo-Confucianism which combined the moral standards of traditional Confucianism with elements of Buddhism and Daoism. Gunpowder began to be employed for military expeditions. On the material side, early varieties of rice were introduced that made it possible to grow two or three crops a year in the south, an area that had become a part of a large empire.

During the 1200s Mongol warriors had moved into China from the north. Their leader, Kublai Khan, established the Yuan dynasty, under whose rule the most celebrated of the Italian travelers, Marco Polo, traveled widely in China and described in his writings in glowing terms the high level of civilization that prevailed in the country. Improved roads and canals were the dynasty's main contributions to China. After the overthrow of the Mongol rule, the Ming dynasty was established in the later part of the 14th century which lasted until the middle of the 17th century. The Ming rule was marked by stability, peace, and prosperity. Literature and art flourished again.

By the early 15th century, however, European trade and infiltration had begun. The Portuguese established a settlement at Macau in 1557 on the coast, despite the Chinese suspicion and hostility of the foreigners. Meanwhile, the Manchus from the north invaded China and established the Qing dynasty and ruled the country until the early 20th century. Like the Mongols earlier, the Mings had adopted many traits of Chinese culture. They propagated strong neo-Confucianism and retained the political system that the Ming rulers had set up. The Qings extended the empire into Mongolia, Tibet, and Central Asia. During their rule commerce, agriculture and handicrafts prospered and China's population expanded rapidly---from 150 million in 1700 to over 400 million by 1850.

The Qings initially opposed to foreign trade, relaxed ultimately, and in 1834 opened Canton (now Guangzhou) to limited overseas trade. Great Britain dissatisfied with trade arrangements, provoked the Opium War and obtained commercial concessions. Other foreign powers interested in reaping the benefits of commercial deals made similar demands, and eventually succeeded in obtaining several areas of influence along the coast for trading purposes. Meanwhile, Chinese resentment of foreign pressure continued to grow. Several uprisings against the foreigners and a disastrous war with Japan (1894-1895) forcing the Chinese to recognize Japan's control over Korea, and Taiwan, and later the seceding of Hong Kong in 1897 to the British contributed to the weakening of the Manchu rule.

Since the defeat of China at the hands of Japan a movement to set up a republic had been growing, and in 1905 a republican revolutionary organization under the leadership of Sun Yat-Sun was formed. The organization was responsible for establishing the Kuomintang (Nationalist Party).

The end of the imperial rule had ushered in a protracted period of internal turmoil when the central government eventually began crumbling. Russia had forced substantial territorial concessions in the north; British, French, Russia, Germany, and others competed with one another for control of ports, trade, railways, and natural resources. While the rulers continued to hold office in Beijing, the real power was in the hands of warlords (local military leaders). By 1922, the republic had failed and civil war had started. Chinese intellectuals began searching for reasons to explain the inability of the country to set up a national government and sought inspiration in either western society or Marxist philosophy. Marxism was particularly attractive to many as it attributed China's several problems to the degradation of European imperialism. Despite the spread of Communist ideas among the intellectuals and youth, however, a young non-Marxist by the name of Chiang Kai-shek was able to at least nominally reunify several regions of China although order was never fully restored and the warlords continued their hold over independent bases throughout the country, diverting scarce resources and the support so crucially needed to develop a strong, national state.

The Nationalist government was a one-party dictatorship that never gained full control of China. By 1931, the Communists had established a rival government in southern and central China. In 1934 the Chiang Kai-shek government forced the Communists to leave their bases and began their long march. While Chiang was fighting the Communists, the Japanese had occupied Manchuria. After World War II against Japan in 1945, the Communists held an area in northern China with a population of about 100 million, and claimed to have an army of 900,000 soldiers. By mid-1946 a full-scale fighting between the Nationalists and Communists had begun, which eventually resulted in the flight of Chiang Kai-shek to the island of Taiwan, and the Communists established the People's Republic of China over the Mainland in 1949.

From 1949 to 1952, the new Communist government firmly established its control over the entire country and promoted the recovery of the nation's economy. It seized farmland from warlords and redistributed it among the peasants. In 1953, the administration began its First Five-Year Plan for economic reconstruction, and brought all industries under its control, and farmers were persuaded or forced to combine their landholdings into agricultural cooperatives. During the Second Five-Year Plan (1958-1963), this so-called "Great Leap Forward" the country experienced economic depression, food shortages, and a decline in industrial output due to mismanagement, lack of incentives and tight state controls.

During the 1960s when the Soviets began steadily adopting the policy of peaceful coexistence with the West, China accused the USSR of betraying the aims of Communism, and the USSR stopped its technical assistance to China. China broke off its relations with the Soviets, and only after the dissolution of the USSR in 1991, China began to establish relations with the former Soviet republics.

In 1966 Mao Zedong supported the radicals in the Communist which led to what he termed the "Cultural Revolution." The radicals spearheaded by semi-military organizations, called the "Red Guards" who took control of universities, major cities and several provincial administrations triggering violence as competing radical groups struggled for power. Their activities had a severe and disastrous effect on the economy. The Red Guards were not, however, entirely successful, and in 1967 the government had to call out the army to restore order.

In the early 1970s the emphasis of China's foreign policy also changed from revolutionary to diplomatic, and new contacts were developed and efforts were made to improve relations with many governments. China continued to strengthen its influence with other underdeveloped nations. In the early 1980s, the Communist party began economic reforms that led to less government control over business activity and price. By the 1990s, numerous privately owned and operated businesses had been established. Many experts believe that policy of privatization and liberalization has contributed greatly to the present economic progress in China.

Cultural Patterns China is essentially a vast culture area inhabited by the Hans that make up for nearly 92 percent of its population. Only a fraction of its population---roughly 8 percent---is composed of other ethnic and linguistic groups such as the Chuangs, Manchus, Mongols, Tibetans, Uighurs, and Koreans. Officially the government recognizes 55 such minority groups or minority nationalities. Some of which these between 5-15 million people and are distributed over 60 percent of the land, although they are primarily concentrated in specific regions. The Han, the largest group, outnumbers the minority nationalities in every province except Xizang (Tibet) and Xinjiang. Though minorities represent only a fraction of the total population, they number over 100 million people. They enjoy limited political autonomy and self-government, and most of them live in regions set aside in empty pasturelands of the western provinces. The government takes great credit for its treatment of the minorities, particularly in the promotion of their culture, and the introduction of a written language for them, where none existed before, and in raising their level of literacy.

Almost all the inhabitants of China are of Mongoloid stock, and the basic classification of the population is not so much ethnic as linguistic. Overall, four major language families are distributed in China: Sino-Tibetan, Altaic, Indo-European, and Austro-Asiatic. Numerically, and in the extent of its distribution is the Han Chinese of the Sino-Tibetan family which is spoken in hundreds of dialects in various ranges of mutual intelligibility such as the Cantonese, Wu, Min, and Mandarin. Mandarin or Putunghua (meaning "ordinary language") is the northern dialect, and the official language of the government bureaucracy. Prominent among the dialects of the southeast coastal region from Shanghai to Canton are Wu, and Min.

In addition to the ethnic Han people, the Manchu, and the Hui people also speak Mandarin. The Hui are the descendants of those Chinese who adopted Islam when it penetrated into China in the 7th century. They have adopted the Han ways and are concentrated in the Hui Autonomous Region of Ningsia, into which they are also moving from their other scattered communities in Xinjiang, Qinghai, Hoben, Yungui in order to preserve and solidify their religious and cultural traditions. The 9 million Manchus profess to be the descendants of the Manchu invaders of the 17th century. They inhabit mainly the northern and northeastern parts of the country, and have been accorded an autonomous district. The old Manchu language is now almost extinct now. The Chuangs form the largest minority group, and inhabit mostly the Chuang Autonomous Region of Kwangsi, but are also scattered in Yungui and other areas. They are animists and are primarily rice cultivators by profession. The Tais are settled in southern Yungui, and are related to the Shan people of Myanmar. Tibetans are distributed over the entire Qinghai-Tibetan plateau. Outside Tibet the Chinese have created five Tibetan autonomous provinces in Sichuan, Yunghai and adjacent to the Tibetan Plateau where Tibetan-speaking groups predominate. Since the formation of Communist regime, many Tibetans have taken to farming and stock raising, although most of them retain their tribal ways. Inhabiting some parts of the southwest are several ethnic groups that are spatially intermixed, but maintain their own cultural traits and live in relative isolation from one another. In many of these areas the Hans are active in the towns and in the fertile river valleys, while the minority groups eke out living on the primitive farms of agriculture or on grazing their livestock on hillsides or mountains. In general, most of Sino-Tibetan language family is concentrated in the south and southwest regions of the country.

The Altai family is represented by its three major branches----Turkic, Mongol, and Manchu-Tungus in northwest, and northern China. The Turkic branch is the most numerous of the family within which the Uighurs, who are Muslims, form the largest minority, and are distributed over oases in the Tarim Basin and in the Junggar Basin of Xinjiang. They mainly depend on irrigation agriculture for their livelihood. Other Turkic minorities include the Kazakhs, and the Kyrgyz, who are pastoral nomads. The Kazakhs live in northwestern and northeastern Xinjiang, and the Kyrgyz are high-mountain pastoralists concentrated in the western part of Xinjiang. The Mongols, by tradition, are nomads, and inhabit Inner Mongolian Autonomous Region. Most Mongolian tribes retain tribal customs, and are pastoralists. The Mongolian language consists of several dialects, but religion is a unifying force, for most Mongolians profess Tibetan Buddhism.

A few linguistic minorities in China belong to neither the Sino-Tibetan nor the Altai language families. The Tajiks of westernmost Xinjiang are related to the people of Tajikistan, and belong to the Indo-European family. Along the China-Myanmar border are the Kawa people that belong to the Austro-Asiatic family.

Religion in China The religious affiliations of the peoples of present-day China are difficult to determine, as the state does not recognize any religion and is professedly atheistic. However, 52 percent of the Chinese were officially listed as non-religious and 12 percent as atheists in 1980. But these statistics can be misleading as a large number of Chinese observe religious practices at home. A fifth of the population was listed officially as Chinese folk-religionists, 8.5 percent as Buddhists and 1.4 percent as Muslim.

The philosophic and moral codes of the Chinese are derived basically from the writings of two ancient sages, Confucius and Lao Zi, who have spelled out ethical precepts and views of life dating back to more than 2,000 years. Confucius, was born, it is said, in 551 B.C. His disciples put together a collection, called the *Analects,* of what was considered to be master's ideas, which formed the basis for what was considered Confucianism. The basic philosophies presented by Confucianism is rooted in the ideas that people should be content with their situation in life. It maintained the importance of tradition, of loyalty to authority, and etiquette in the regulation of daily life. The state is considered merely an extension of the family. Confucianism as an ethical system is favored by China's current rulers as it teaches obedience to higher authority. In this sense, Confucianism has become more political and practical than ethical or metaphysical. It provides specific guidance how to live and how to rule the state. In a sense, Confucianism appears to be convention-bound and repressive, as compared to the other major current of the Chinese culture, namely Daoism which allows for a more passive and relaxed role for the individual and the state.

The founder of Daoism, Lao Zi, was probably a contemporary of Confucius or lived a little later. He emphasized the love for and worship of nature. His philosophy espouses the simple life of a hermit, and encourages the individual to attain harmony with nature and dismisses worldly strife. Daoism in reality, is complementary to Confucianism, appeals as it does to the human desire for quiet contemplation of beauty, whereas Confucianism stresses the need for loyalty to the needs of family and society. Both are ethical systems rather than formal religions. Neither system speaks of a hereafter----the need for which was provided by the religious philosophy of Buddhism that arrived from India, the land of its birth, to China in the 1st century A.D.

Buddhism, a third stream in Chinese culture contradicted the Confucian ideals of moderation and self-control, although it was highly spiritual and mystical, but recommended quiet contemplation as advocated by Daosim. It was adopted by millions of Chinese over the centuries. And the proverbial Chinese hospitality to different systems of belief became the hallmark of Chinese state. Despite the fact that Confucianism was the state religion during much of Chinese history, many of the emperors were Buddhists. Even under Buddhist rulers, a thorough knowledge of the Confucius classics came to be considered the essential qualification for government employment. A large part of the Chinese bureaucracy, consisting of civil servants, was recruited through examinations testing the applicants' knowledge of the Confucian canon.

Population and Public Policy In the late 1990s China contained over 1.2 billion people, and was growing at a rate of a about one percent annually; adding 13-14 million persons every year to the population. Historical records indicate that, as early as 800 B.C., China already had a population of nearly 14 million, and by the turn of the Christian era, when the first registration of population was taken, it contained 60 million inhabitants. The early official census presumably was imprecise as this was recorded primarily for the levy of poll taxes, and evasion of reporting was probably quite large. By 971 by all estimates, China had over 143 million, and by 1851 it reached 432 million. After 1949 under the Communist regime, sanitation, medical care, and public hygiene considerably improved, and death rate declined much faster than birth rates, resulting in an enormous piling up of people. By the early 1980s population had reached one billion mark. Although official statistics should be used with caution because of the incomplete coverage, it can be safely maintained that over 20 percent of the human race now lives within the borders of China. Both official and private estimates indicate that China has, for long, probably for millennia, been the world's most populous country.

China's enormous population is not spread evenly over its vast surface. With a population density now approaching 350 persons per sq mile (135 per sq km) it is not, on the whole, a very densely populated country. Several smaller industrial nations, and some large ones such as India have much denser concentrations of people and a much higher overall population density. Large parts of the country are thinly populated. Due to a number of complex conditions, the population is quite unevenly distributed. Population density varies strikingly, with the greatest contrast occurring between the eastern half of the country and the lands of the west and the northwest. About 90 percent of its population is concentrated east of an imaginary line that follows the eastern edge of the Great Hinggan mountains south to the Great Wall to the vicinity of Lanzhou, then extended southward along the eastern edge of the Tibetan plateau Exceptionally high population densities are found in four major areas where the natural environment offers most suitable circumstances for the development of intensive agriculture: (a) the middle and lower Chang Jiang River Basin and delta; (b) the lower Huanghe Basin; (c) the Xi River delta, and (d) the Sichuan Basin. In addition to the relatively abundant precipitation and level topography these areas contain fertile, alluvial soils on which intensive agriculture is centered.

In contrast, the rugged, remote, and drier areas of the vast western frontier lands of Xizang, Inner Mongolia, Xinjiang, and Qinghai are sparsely populated or in most part virtually uninhabited. In the 1950s government became increasingly aware of the importance of the frontier regions and embarked on a drive for settling the former military personnel and the young intellectuals in those areas. Following the construction of railroads and highways linking these wastelands with the well-settled eastern sections of the country, the population has been growing rapidly in these areas. Settlement in these lands has also led to the exploitation of considerable mineral (particularly oil) and other resources that lay locked in these backward regions. Prior to the establishment of the Communist regime, the occupation of Manchuria had already developed into the principal industrial region of China. It was left to the Communist administration to sinicize the "wild" west and develop the regions economically.

China's population growing at the current rate of increase will probably reach 1.5 billion by the year 2017, poses enormous responsibility on the state. To feed such a huge population is, of course, the basic need, but for the population that has begun on the course of modernization and where the consumption levels have been steadily rising, the demand for much larger reserves of food supply will continue to grow. During the 1990s China became an important importer of foodgrains in the world market. Although in 1995 the harvest of 465 tons of grain was the largest in history, and China became a foodgrain exporter; the foodgrain situation still remains unpromising in the long run. In normal years China imports the staple cereals—wheat, maize, and rice—roughly 10-11 million tons annually, and spends nearly 5 percent of its import expenditure on the food items. It is quite unlikely that it will be able to become self-sufficient in its foodgrain requirements for its expanding population in the long haul.

Initially, the Communist regime officially refused to recognize the population problem. But faced with the problem of raising adequate food supply, the authorities sponsored a drive for birth control in 1955. Beginning in 1962, the use of contraceptives and advocacy of late marriages became an important part of the official planning program. The outbreak of the Cultural Revolution during the early 1960s disrupted the family planning efforts. In 1970, however, a much stricter program was initiated. The administration made late marriage and family limitation obligatory. Strict rules and a set of punishments for those who did not follow those were forcibly imposed on the public. In 1979 a policy aimed at one child per family was enforced.

The family planning program during the 1980s and 1990s have been, by and large, quite successful, and the state efforts have been much more effective. China's population growth rate, currently at less than 1 percent a year, down from twice this figure during the 1950s and 1960s, is now unusually low for a developing country, although the enormous size of its population, still results in a large annual net population growth.

Resources and Economic Development: Agriculture The administration in China rightly views agriculture as the cornerstone of national economy, although the stated official purpose of economic planning has been the industrial development. The emphasis on agriculture has been regarded necessary, not only to provide food for the expanding population but also to supply raw materials and capital for industrial development, and exports. This is quite understandable for a nation where over a billion mouths must be fed daily. Today, agriculture accounts for one-third of the gross national product, and nearly 60 percent of the labor force is engaged in agricultural pursuits. The problem of adequate food supply has been a serious one, and 2-3 percent of the nation's total imports in normal years are related to foodgrains and live animals.

In commensurate with its size, and the fact that more than one-fifth of the world's population lives within its borders, and the traditional pivotal position of agriculture in society, China occupies a significant position in the world in several agricultural commodities. It is a major producer of foodgrains in the world: first in rice and wheat production, second in millets, corn, and tea; and an important producer of such commercial crops as cotton, jute, peanuts, tobacco, and oilseeds.

The Chinese have been living close to nature ever since Chinese civilization emerged thousands of years ago. The peasants have enjoyed the traditional respect of : social and economic system based on the .ilization of land. Close to three-fourths of the population still lives in farm villages and small market towns, despite the adoption of urban-based strategy by the Communist government during the early 1950's, and the recent drive for agricultural transformation and modernization. Although the economy has moved increasingly toward the industrial and commercial enterprises, agriculture continues to be overwhelmingly important in terms of its contribution to the nation's economy, and as a mechanism for absorbing surplus labor for the nation's enormous population.

Traditionally, Chinese agriculture has been known for its intensity in the use of manpower and intensity in the full utilization of the land. Virtually every bit of available arable alluvial land is utilized in some form of crop production, in terracing of slopelands, in race for a harvest on river beaches or lake beds before the annual flood inundates the lands. This is particularly important as permanent cultivation occupies only one-tenth of the territory. Only a quarter to one-third of China is humid enough, warm enough, and possessed of good enough soils to be considered of value to agriculture. A large part of the country's surface is covered by high mountains, deserts, and forests, and the potential for expanding cultivation has been severely restricted by the relatively modest supply of land suitable for agriculture.

The intensity of cultivation is obviously greatest in certain areas where conditions for it are most suitable. Xizang, most of Manchuria, large sections of Xinjiang clearly are not particularly conducive to cultivation. Close to 95 percent of China's farmlands are found south of the Great Wall and east of the Tibetan Plateau. The addition of the central valley region of Manchuria would add a few more percentage points to the agriculturally suitable lands. The principal regions where the percentage of land cultivation is particularly high are: the Huanghe, Huai basin, the Chang Jiang plains, the Sichuan lowlands, the central lowlands of Manchuria, the central lake region south of the Chang Jiang plains, and the lower Xi valley and its delta. Understandably these areas associated with the great plains rank highest in cultivation. Xinjiang has low density of cultivation because of lack of rainfall and scarce water for irrigation. The abovementioned areas of high density of cultivated land are also the areas that show great pressure of population upon the land. It is clear that extension in cultivated area is constrained by several factors. China is an old country, and all possible traditional methods to increase the output have been applied, hope for future expansion can only be derived through the application of modern techniques and mechanization.

Communizing the Agricultural Systems Immediately taking control of the country in 1949, the Chinese administration embarked on a series of agricultural reforms aimed at revolutionizing China's traditional systems. The administration was of the view that "productive potentials" of agriculture for a long time had been stifled by the traditional pattern of agriculture based upon individual ownership of land. During the mid-1950s the government, therefore, confiscated all land from the feudal landlords and redistributed it among peasants, prescribed cropping patterns, and set the level of prices on all major crops. The transformation of the traditional system was viewed as necessary for raising the agricultural output critically needed not only to provide food for an increasing population but also to supply industrial raw materials and capital for industrialization, and agricultural exports to underwrite the cost of industrial equipment purchased in other countries.

A series of agricultural reforms was started in 1951, and during 1950s the pattern of land ownership was radically changed. The government measures proceeded through four well-defined stages: the initial stage of collectivization; formal communization; the modified communization; and finally the "household responsibility" system. Between 1950 and 1958, during the stage of initial collectivization, farm land was transferred from the landlords and individual owners to the state, and family farms were combined into collective farms. By the end of 1956, 92 percent of all farming households were reported on collective farms. These measures marked the completion of a process in which the traditional Chinese farmland of tiny fields with numerous boundary strips disappeared. Introduced along with the collectivization program was the introduction of large-scale machinery and the development of general mechanization. By 1955 there were only 100 "mechanized farms", and the pace of mechanization was slow due primarily to the lack of manufactured machinery within the country. The Chinese have been manufacturing relatively simple, but efficient small machinery capable of performing such functions as plowing, threshing, planting, harrowing, and seeding. Machines were then introduced widely, but they did not completely replace human labor because the machinery introduced was too simple to operate on large farms.

The formal commune system was introduced in the late 1950s within the framework of "Great Leap Forward" (1958-1960). Soon after, however, the government realized that China was not registering successes as measured against the country's needs. Nearly three-fourths of a million collective farms were combined into one-third of their number into rural communes which were given the responsibilities in agriculture and rural industrial development (including backyard iron smelting), commerce, education, and local administration. The Communist party controlled all the administrative, and policy matters of the communes because it had fixed unrealistic production goals (for example, to double China's food production in a single year between 1959-1960), in actual practice the goals were never realized. Due to bureaucratic stranglehold and lack of private incentives, production actually declined, leaving the nation physically and emotionally drained.

The next stage of agricultural transformation (1960-1975) can best be termed as a "modified commune system." During these years the size of the communes was reduced and their numbers increased. The size of the production teams was to consist of 20 to 30 families (nearly half the size prevailing during the 1950s), who would in the future be the most important local operating, management and decision-making units in agricultural matters, a departure from the previous, strict and formal commune system in which the farm units did not possess such authority. Some economic incentive and reward system was also introduced on the basis of performance of the production teams. A measure of privatization that included small-scale selling of surplus crops to other farm units on a profit-sharing basis was also introduced. By the mid-1970s China's development, particularly in the agricultural area, had not come out of the doldrums.

After the death of Mao Zedong, the prime inspiration behind the original agricultural policies, different ideas regarding economic development gained control among the leaders that had emerged on the scene. Agriculture was viewed by them to be in crisis that needed strong measures of curative medicine in the form of free enterprise to stimulate production through profit motive. They advocated a system, designated as "market socialism" or "household responsibility system." Thus, began the fourth stage. The state still owned the land, but the farmer under contract, would produce stated amounts of basic commodities for sale to the state, and sell off surplus produce for personal profit in the free market. After 1979 some new features were added such as the expansion of the size of private plots for farm families on which they can exercise complete control of production. The state set up a sort of farm credit system, but the system has been gradually creating a small class of "wealthy" peasants, rich by Chinese standards, who have, by dint of competition, been able to rise above the common farmers. The farmers were also encouraged to plant commercial crops such as cotton, jute, tea, sugarcane, and vegetables at larger levels than before. Priority was also accorded to the production and distribution of farm machinery.

Nearly half a million communes now occupy 90 percent of China's agricultural land. The production teams, composed of 80 to 160 persons are responsible for the cultivation on the collectives of 8 to 38 hectares farm size. In addition to the collectives another unit of farm organization was developed---the State Farm (in the late 1970s and early>80s)---which is state-owned and operated enterprise. Most State Farms are located in the sparsely populated regions, such as Manchuria, Xinjiang (and Xizang to a lesser extent).

Additionally, the administration has large investments of capital and of agricultural modernization. These efforts may be grouped in four major categories: water conservancy and increase in irrigation; intensification and modernization of agricultural techniques including mechanization, improved crop varieties, and pest control; reclamations of land and increase in the cultivated area. Floods and droughts have always periodically destroyed crops in several parts of the country especially in the plains of Huanghe and the Chang Jiang. Until recently a major objective of the administration' planning has been to adopt water control and conservancy measures, such as the construction of small-scale dams, ponds, and canals. Most of these are located in the Huanghe basin, where drought and flood problems are the most frequent and serious. However, currently the administration is building the world' largest dam on the Chiang Jiang River in the hilly region near Yichang, the so-called *Three Gorges Dam*. This is China' most ambitious project since the Great Wall. It will be 607 feet (185 meters) high, 1.3 miles (nearly 2 km) wide, and displace nearly 2 million people and hundreds of villages to be built at a cost of billions of dollars the projected is likely to be completed by 2009. When completed the multipurpose dam will be used for generation of electricity, flood control, irrigation, and water transport.

The administration has placed significant emphasis on the further intensification and mechanization of agriculture. The application of scientific knowledge and experimentation to develop better seeds, equipment, and farming techniques have been undertaken, together with the development of administration and propaganda machinery to diffuse their use among the farmers, the communes and collective farms. An important aspect of application of scientific techniques has been a greater utilization of fertilizers on state farms as opposed to the use of traditional fertilizers such as animal manures, pond mud, plant residues and ashes, oil seed cakes, although most farmers continue to use traditional fertilizers on private farms. During the last two decades China's relatively small chemical fertilizer industry has received a considerable boost and expanded rapidly by the purchase of plants from abroad, mostly from Japan and western Europe.

Mechanization of Chinese agriculture has been slow, and uneven, primarily due to such factors as the small size of the farms and topographic conditions. The use of tractors is growing especially in Manchuria and Xinjiang regions where extensive farming techniques are more feasible. In recent years some priority has been afforded to improved farming techniques. Although some progress has been made, by and large the rural scene still is dominated by human and animal power and traditional hand tools. Rapid mechanization has, however, released a flood of surplus labor onto both the countryside and the cities and created some problems of unemployment.

Currently nearly one-tenth of China's land surface is listed as farmland as compared to nearly one-half of India's and about one-third of the United States' territory used for farming. Although a large part of the country is in form of "negative land" for the purposes of farming, expansion in cultivated acreage to increase food production is possible in areas which may be called second-rate land. Immediately gaining control of the country the government placed emphasis upon land reclamation and on assessing the amount of potential land capable of future cultivation, but gains in the land reclamation have been modest, and restricted primarily to the desert areas of Xinjiang and Inner Mongolia where agricultural extension has been made possible by providing greater irrigation facilities. Further extension undoubtedly exist in Manchuria, Xinjiang, and some scattered parts elsewhere, but these areas are agriculturally marginal, as they suffer from short growing season, lack of water, or poor drainage that make them uneconomic to be used for cultivation under present conditions. Hill lands of Southern China hold some promise for expansion, but reclamation would require mechanized equipment and coordination with overall water and soil conservancy schemes in order to avoid flooding and silting problems for the intensively used valleys below.

Grain crops occupy a large part of the agricultural land, since these form the basis of national diet. The favored grains have been rice and wheat, with sorghum, millets, barley, and maize as secondary crops. Rice and wheat are the leading crops—nearly 190 million metric tons of rice and 125 million metric tons of wheat are raised annually. Maize production approximates a little less than that of wheat. China produces nearly one-third of world's rice, and one-sixth of its wheat output in normal years. Besides it is a major world producer of several other grains such as millets, sorghum, barley, and soybeans. While most of these grain crops are raised universally, country's most significant cropping boundary is the crest of Qinling Mountains in the west, and Huai River in the east, which forms the northern limit of rice cultivation. Although rice is grown as far north as Manchuria, but this line limits its continuous cultivation. South of this line, China is a green, humid and subtropical rice-citrus fruits-palms, tea-land, while the northern part of the country is a dry brown, wheat-millet-barley-white potato-land. Such crops as maize, soybeans, kaoliang, tobacco, and broad beans are to be found in both regions. North of the Qinling-line the growing season is shorter (five to seven months) as against ten months to a year in the south where double cropping is more widespread. The Chang Jiang Valley has nearly half of its land double-cropped. Rainfall is about twice as heavy in the south. The grass area of the south is larger, but the existence of extensive plains in the north gives it a larger area for cultivation. Nearly half the cropland of China is irrigated, and most of this is in rice.

Forestry Through centuries of deforestation much of the good timber that once covered China's landscape has been lost, and nearly one-seventh of the land surface now is under forest. Most of the Tibetan highlands lie above the timberline. Inner Mongolia and Xinjiang are too dry for the growth of natural forest, except for their high mountains. Moreover, it is quite likely that bamboo, grasses, shrubs and scrubby plants have taken over the areas that once were covered with forest. Apparently, for its size, China is not well endowed with forest resources. Since 1955 there have been concerted efforts to increase forested areas by afforestation, but have not always been particularly successful. Despite increasing demand for wood and paper products and the administration's efforts to develop forest resources, during the recent years forestry still lags and it China has been a timber importer.

Livestock and Fishing Animal husbandry constitutes an important component of agricultural production, although China has not been renowned for its livestock wealth. Of special importance are the hogs, and it has the world's largest swine herd. Hogs are valuable for their meat, and high quality of manure they produce, and are raised by most farm families. In addition, cattle, sheep, and goats are raised. It also produces a large quantity of poultry, consisting mostly of chicken.

Fishing In terms of total fish-catch, China ranks third among the fishing nations in the world. Fish consumption provides an important nutritive element in the meat-poor country. Its seacoast is relatively long, indented and irregular, the continental shelf is extensive, and the offshore ocean currents provide desirable conditions for fishing grounds. However, its coastal waters and numerous inland waters are only partially developed. The primary fishing grounds are along the southern coast from Hangzhou Bay (near Shanghai) southward to Hainan Island, a distance of over 3,000 miles. The fishing fleet comprises non-motorized and some motorized boats with small catch capacity. Shanghai and Guangzhou (Canton) are the leading fishing ports. Freshwater inland fisheries are also important. China has a long tradition of aquaculture and now leads in the production of freshwater catch in the world. The most productive inland fish farming area is in the Xi Basin and Delta between Guangzhou and Macau. Pond-culture is a significant sideline occupation for most farm families in a broad area of the southeast, although increased pressure by the administration to convert fishponds, lakes, and drainage channels into land for crop production is resulting in the decline in freshwater fish production. Despite the large fish consuming market at home, fish exports to Japan, Hong Kong and other Asian nations generally provides badly needed foreign exchange to many farmers engaged in fishing.

Mineral Resources Generally speaking, minerals did not play an essential role in the evolution of China's distinctive culture, although it is well-endowed with several minerals, chief of which is coal, and substantial quantities of other minerals such as tin, lead, zinc, iron-ore, copper, asbestos, antimony, manganese, tungsten, pyrites and oil also exist. But exploitation of these has remained marginal, except that of coal, and oil during the last few decades. The country ranks third in coal production in the world (after United States and Russia); mining nearly one-sixth of the world's output. The deposits are widely scattered, but most of these are located in the northern part of the country. The provinces of Shanxi and Shaanxi produce over one-half of China's production. Other major coal-producing areas are: Heilingjiang, Shandong, Hubei, and Liaoning. In addition to these northern provinces significant amounts of coal exist in Sichuan, Yugui, Guangxi, and Guizhou. Most of nation's coal is bituminous of good quality; lignite and anthracite deposits are also present in several areas. Coal is the major source of energy requirements for the country's drive to industrialize. The production has gone up dramatically during the last forty years. A major problem is that south of the Chang Jiang the reserves are widely scattered, and are comparatively meager to cope with the nation's industrialization programs.

Before 1950 China was considered to be very poor in petroleum reserves and the production of oil was very modest. Over

Map 20.2

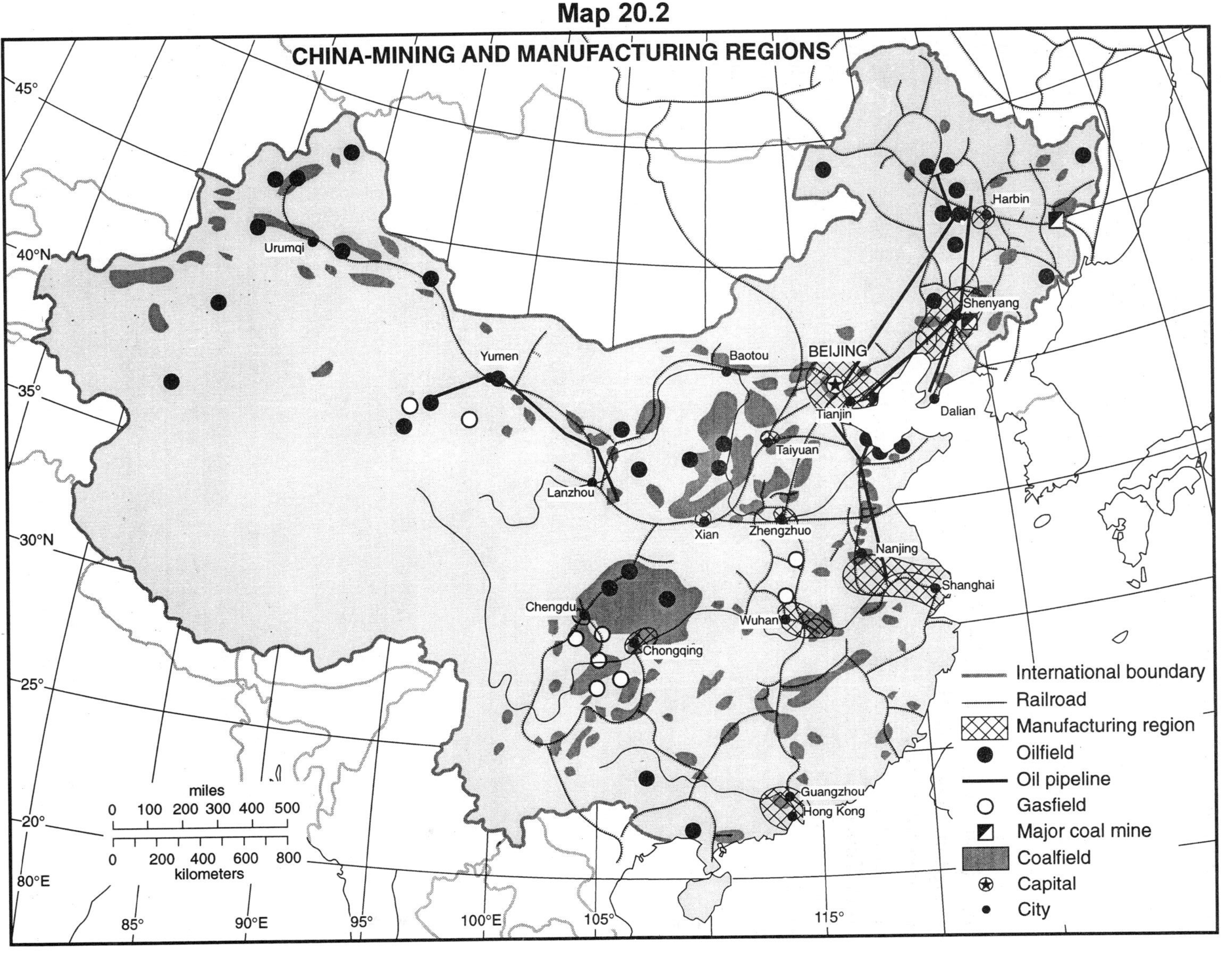
CHINA-MINING AND MANUFACTURING REGIONS
Harbin
Shenyang
Dalian
BEIJING
Tianjin
Baotou
Taiyuan
Xian
Zhengzhuo
Nanjing
Shanghai
Wuhan
Chengdu
Chongqing
Guangzhou
Hong Kong
Lanzhou
Yumen
Urumqi
International boundary
Railroad
Manufacturing region
Oilfield
Oil pipeline
Gasfield
Major coal mine
Coalfield
Capital
City
miles
0 100 200 300 400 500
0 200 400 600 800
kilometers
45°
40°N
35°
30°N
25°
20°
80°E
85°
90°E
95°
100°E
105°
115°

85 percent of the nation's needs had to be met with imports. Since 1960, however, it has become an exporter and the country is rapidly becoming a major oil producer. The current production amounts to nearly 5 percent of the world. Major fields are located in Xinjiang, Gansu, Qinghai, Sichuan, Shandong and Henan provinces. Some shale oil is found at Fushun in Liaoning, (near the coal fields) and in Guangdong. Light oil of good quality has also been found in the Xi River estuary of the South China Sea, and the Qaidam Basin in Xinjiang. China has contracted with western oil companies to jointly explore and develop oil deposits in the China Sea, the Yellow Sea, the Gulf of Tomkin and the Po Hai. Most of the country's production is consumed at home; only a small amount is exported. Mineral fuels and lubricants account for about 8 percent of value of country's exports.

Iron ore is found in almost all the provinces; the reserves are located mostly in Hainan Island, Gansu, Guizhou, southern Sichuan, and Guangdong. The largest mines are located north of the Chang Jiang River that supply the ore to neighboring iron and steel plants. China is well supplied with ferro alloys, and manganese. Reserves of tungsten are also known to be quite large. The tungsten production is nearly one-third of the world's total output. Reserves of antimony are perhaps the largest in the world. Tin reserves are also plentiful and production is about one-sixth of the world. About one-fifth of the world's production pyrites is derived from China. The nation is also rich in copper, gypsum, asbestos, tungsten, lead, zinc, cement and in gold deposits (though not much extraction of gold has taken place due to inaccessible sites).

The country has rich overall energy potential, and the production from various sources is estimated to be about 7-8 percent of the world. It has an extensive river network and mountainous terrain capable of generating a large amount of hydroelectric power, most of which is generated in the southwest, where the coal supplies are poor but demand for energy is growing. In contrast, in the north and the northeast the potential is smaller, but these contain the sites of the early hydroelectric stations built by the Japanese. Here in the northern regions water flow is given to considerable seasonal fluctuations because there is considerable drop in water level in the rivers and flooding during summers.

In sum, while China possesses a rich overall energy potential, most of which remains to be developed. The locations of most energy resources are distant from their industrial users. For example, the southwestern Tibet in the area of the sharp bend of the Brahmaputra River, is potentially very rich for the generation of hydroelectric power, but is remote and inaccessible for the industrial centers. Generally speaking, China is rich in coal and oil; Central China contains abundant coal, and southwest has great hydroelectric potential. But the industrialized regions around Guangzhou (Canton) and the lower Chang Jiang Basin around Shanghai cannot obtain much of the energy from the major sites where energy generation is feasible or actually produced.

Industry From the early times China has been primarily an agricultural country. Historically, neither trade nor manufacturing has dominated its economy. The state traditionally placed numerous restrictions to the growth of large-scale manufacturing until the later part of the 20th century. Only the handicraft industry was accorded special encouragement and allowed to develop to its fullest. Each local region had to be self-sufficient resulting in a duplication of handicraft industry for the manufacture of tools, and a variety of handicraft objects were produced in different regions. Several regions became well known for the production of metal and wood products, others in textiles, or items in lacquer or ivory. These products and silk fabrics were even exported to the European markets via the fabled Silk Road centuries before the beginning of modern industrial revolution in Europe. Only by the third quarter of the 19th century, when industrial revolution began churning out manufactured goods on a large scale in Europe, manufactured goods of the West began to flow into China.

Traditional handicraft industries remained strong until the 18th century, although these have not lost popularity entirely until the present time. By the 19th century the westerners had obtained concessions from China in mining, trade, and industry, and the European settlements had taken foothold along the coastal locations. China began importing western goods as well as modern means of production to produce goods in the country by affording such advantages to the foreign-owned factories as beneficial tarriffs rates and concessions, as well as by providing cheap Chinese labor to such factories. A considerable start was made in several types of industries in the cities along the coastal seaboard.

Foreign industrial enterprises were attracted by the huge Chinese markets for cheap goods, by cheap labor, and to some extent by China's natural resources. The first large-scale factory industry to be developed was the manufacture of cotton textiles in such coastal cities as Shanghai, Tianjin and Guangzhou. Besides textiles the Westerners established such other industries as flour milling, cigarette making and pressing of vegetables for oil in several cities. In the long run, the native handicrafts suffered greatly in competition with the modern, Westernized, mass-produced goods. However, an intensive, modernized, heavy industry was initiated and developed first by the Japanese during the 1930s when they occupied northwestern part of the county (Manchuria) in Liaoning and Jilin provinces, rather than by the Europeans. The Japanese capitalized on the local resources, such as proximity to the major iron-ore and coal fields, cheap, captive Chinese labor and the prospect of gaining control over vast territories that could provide a haven for the resettlement of the burgeoning population in Japan. The major industrial manufacturing plants included the production of heavy iron and steel equipment (metallurgical machine building materials and heavy transport equipment).

Since the Communist takeover, China has gone through several phases of industrial development. Between 1949 and 1952, the economy was badly hurt by civil war and the administration embarked on a period of rehabilitation and recovery. They introduced some fundamental changes, and took control of all sectors of economy. Factories, transport lines, and raw material bases were restored, private industries were nationalized and industrial output was brought to the pre-revolutionary levels.

As a result of considerable attention that the Communist regime has accorded to the development of industry, the overall industrial output grew at a rate of over 10 percent per year during the later 1980s and 1990s. Clearly, this has been a direct response to the adoption of the partial privatization and western techniques to bolster the industrial output. Although the capitalist system is still considered as unplanned and chaotic, the system's individual terms are considered highly efficient in terms of both production and distribution as well as profit making. The capitalist countries are no longer viewed with contempt or hostility.

China's industrial work force, estimated to be about 100 million, is the largest in the world, and probably exceeds the combined total of all other developing countries. The nation's industrial development has surpassed all other sectors in economic growth and degree of modernization. The highest priority has been given to the metallurgical and machine building industries, which now account for nearly a third of the total gross value of industrial output. In such endeavors, however, technological innovations have suffered at the hands of a system that prizes growth in output rather than improvements in quality. That has resulted in the importation of items such as specialized steels. In the chemical industry, the main preoccupation of the authorities is to expand the output of chemical fertilizers, plastics, and synthetic fibers. China is now the world's leader in the production of nitrogenous fertilizers (producing nearly one-fifth of the world's output). Textiles, the traditional manufacturing item accounting for one-seventh of the gross industrial output, a growing proportion of which consists of synthetics, is widely distributed throughout the country, but there are a number of important textile centers, such as Shanghai, Guangzhou (Canton) and Harbin.

Seventy percent of China's energy consumption is provided by coal. Overall, production has gone up significantly during the last twenty years, but it still falls short of demand. Petroleum output also increased rapidly from the very low levels in the early 1960s but remains stationary at the levels of 1970s. The northwestern region and offshore deposits are potentially rich in reserves, but nearly half of the country's oil production is derived from the oilfields in the Northwest (Manchuria). Moreover, China has vast, largely untapped, hydroelectric power potentials and natural gas reserves. The administration is planning to develop nuclear power plants in the Shanghai and Guangzhou areas.

Overall, the distribution of industry remains uneven, despite serious attempts by the government since the mid-1950s to develop industry in the interior at the cost of the major coastal centers. Although the percentage growth of industry in the interior regions, in general, exceeded that of the coastal locations, the far larger initial manufacturing base of the coastal cities has resulted in coastal cities' dominance in China's industrial economy.

In the northeast, the most important manufacturing region is the *Shenyang-Fushun-Anshan* belt which lies close to the fields of iron-ore and coal, and is well-connected by a system of railroads in the Liaoning peninsula leading to the important ports of Luda and Lushun. Industry here consists of heavy metallurgical items like shipbuilding, and cement and chemicals. The highest in gross value of industrial output, in the country, however, is the manufacturing belt along the east coast that contains the industrial and commercial centers of *Shanghai, Hangzhou, and Nanjing*. This manufacturing region predates the Communist rule and developed and modernized greatly since the 1950s. The chief industries are: textiles, synthetic fibers, consumer goods, and iron and steel. Good transport facilities (including canal transport), cheap abundant labor, and early start are the major reasons for the industrial development of the region. The third important manufacturing region is that of *Tianjin-Tangshan-Beijing*. It is in fact one of the earliest industrial regions of China, expanded and modernized during the Communist rule. Heavy iron and steel industries predominate capitalizing on the proximity to the mineral rich provinces of Shanxi and Jehol (important iron-ore, and coal region in the country), good transportation facilities, outlet to the outside world through the port of Tianjin, and nearness to the country's administrative center of Beijing. Textiles, chemicals, and consumer goods are some other manufactured items of the region. Fourth region lies in the delta of Xi River with the main focus on *Guangzhou* and *Kowloon* (in Hong Kong). Industry here consists of light manufacturing dealing with items as textiles (cotton, silk and synthetic), ceramics, rubber goods, electrical goods and consumer goods. The former British colony of Hong Kong that reverted to China in 1997, has a great number of manufacturing plants of light industry (particularly that of textiles and consumer goods). Shipbuilding is also an important industry of the region.

In addition to these industrial regions, a number of old industrial units, especially specializing in handicraft and cottage industry are dispersed in the country which the government is attempting to modernize. Among the more recent in heavy industry undertakings by the Communist government was to develop a large complex of metallurgical and iron and steel industry in the Middle Chang Jiang basin near and around the cities of *Wuhan, Wuchang* and *Hanyang*.

Transportation and Trade China places a great emphasis on the development of transport, because it is closely related to the development of the national economy, the strengthening of political unification and the consolidation of the national defense system. Nevertheless the nation's domestic transport system remains underdeveloped, and constitutes a major constraint on country's economic growth. Only 2.5 percent of the labor force is engaged in it, generating less than one-sixteenth of the value of nation's gross product. From 1949 to 1952 the primary concern was to repair the existing lines of communication, and to strengthen political control. Initially, China's railroads and highways were concentrated in the coastal regions, and the communications to the interior were poor. The situation has been largely rectified, and all of China except for certain areas in Xizang is accessible by rail, road, water, or air. Railroad track consists of 47,600 miles (76,600 km), and paved roads. Most of the railroad network is concentrated in the northeast, the lower Huanghe, middle and lower Chang Jiang and lower Xi river basins. Xinjiang is also connected to the rest of China by a railroad.

China's trade, both domestic and foreign, is modest; accounting for a small portion of its overall economy and 1.5 to 2 percent of the world trade. The direction of foreign trade has marked changes since the 1950s. In 1954 more than three-quarters of the total foreign trade was with the Communist world, but by 1965 it fell to only one-third of its total. By the 1980s foreign trade with all Communist countries was less than 10 percent of the total trade and more than 70 percent with the non-Communist developed countries and the remainder with non-Communist developing nations. In general, China has a sizable trade surplus, most of it with the developing countries. A large majority of its imports consists of industrial and capital goods such as machinery, transport equipment, textile industry products, metal products, chemicals; the remainder is made up of foodstuffs and consumer goods. These goods come from Japan, the U.S.A., Hong Kong, and West Germany. Nearly half of China's imports come from East Asia and Southeast Asia; and a little over half its exports go to the same countries. About 70 percent of the exports consist of manufacturing goods, of which textiles and clothing are the most important. Agricultural products (food, and inedible raw materials) are sent to Hong Kong. Exports also include coal and oil.

TAIWAN

The small, thriving island of Taiwan lies 100 miles (161 km) off the southeast coast of China mainland. In the north the East China Sea separates it from the Ryukyu Islands, and Japan. To its east lies the Pacific Ocean, and to the south the South China Sea and the Philippine Seas that separate it from the Philippines. Though linked by geography, culture, and history to China mainland, it is politically under the jurisdiction of the Nationalist government since 1940 and controls over 22 islands (including the Quemoy, Matsu, and Pescadore groups) west of the main island of Taiwan, which collectively contain a total area of nearly 14,000 sq miles (36,260 sq km).

Located strategically close to the southern half of China's coast at a distance of only 400 miles from Shanghai, China's largest city, and 360 miles from Hong Kong it lies within the easy air reach of China's military establishment. The air and naval forces on the island give Taiwan potential control of the west-Pacific sea laws, whereas the Taiwan Straits have provided an obstacle to Communist conquest that enabled Nationalist China to establish here an independent state. In sum, accessible location on the major sea routes in East Asia, and contemporary international politics have given Taiwan highly distinctive position in the world.

Taiwan belongs to the great island system that rims the western Pacific Ocean. The island is largely mountainous with mostly north-south trending ridges in the interior that slope gently westward to a small basin and plains bordering the shallow Taiwan Strait.

Broadly speaking, the island can be divided into two sections: in the east are the folded and faulted mountain ranges sloping steeply to the Pacific; and a low and gently sloping foothill coastal plain on the west. The coastline on the west is straight and bordered by sand dunes and lagoons. Several of the ridges in the interior rise to over 10,000 feet in elevation. The drainage of the island is mostly radial.

The island has a humid, subtropical climate. The lowlands have a warm,

Map 20.3

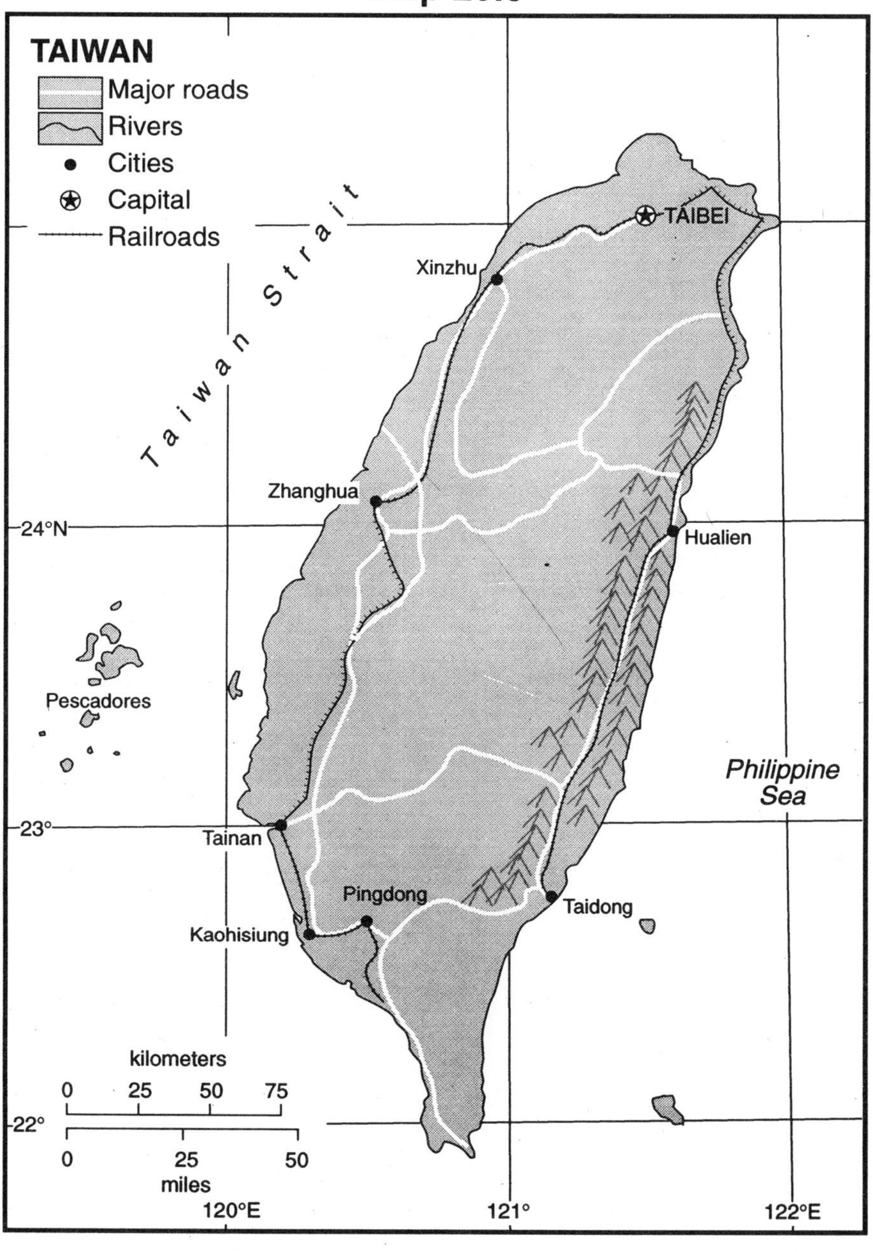
TAIWAN
Major roads
Rivers
Cities
Capital
Railroads
Taiwan Strait
TAIBEI
Xinzhu
Zhanghua
24°N
Hualien
Pescadores
Philippine Sea
23°
Tainan
Pingdong
Taidong
Kaohisiung
kilometers
0 25 50 75
22°
0 25 50
miles
120°E
121°
122°E

year-long growing season. In general, summers are long (April to November), and warm on the plains, while the high mountains in the interior receive snow during winter. The island lies on the northeast margin of the tropics in the path of prevailing winds. The weather is influenced by the northeasterly winds during the winter and southwesterly winds during the summer, which bring abundant rainfall in both seasons averaging 100 inches (2,500 mm). Rainfall is heavier in the north than in the south. During late summer, typhoons bring destruction and in their wake disastrous floods in parts of the fertile plains.

Taiwan has abundant timber and forestry resources in the mountains. Nearly 70 percent of the lands are covered with forest. Commercially productive trees are those of camphor, bamboo, and cypress. Taiwan has in the past provided three-fourths of world's natural camphor, but its importance has diminished with the development of synthetic camphor. Inaccessibility, low quality of timber, and costs of cutting trees have been responsible for the declining importance of forestry. With a population of close to 22 million in the mid-1990s Taiwan is a very crowded land. The density of population is over 1,560 people to a square km. Population per sq km tripled after Japan occupied the island in 1895. At that time it contained only 2 million inhabitants. By 1956 population swelled to over 9 million, registering an unprecedented rate of growth of 4 percent a year during the mid-1950s resulting primarily from the inflow of 2 million refugees from the mainland. The rate steadily declined, dropping to less than 1 percent a year level. During the late 1960s the government introduced a serious family program. Meanwhile, urban development had intensified and in response to growing urban opportunities during the Japanese and later the Nationalist governments, the traditionally extended and large families began concentrating more resources on a few children. Currently three-fourths of the population lives in urban areas.

Official records identify two major groups living in the island: Taiwanese and Mainlanders, forming 84 and 14 percent respectively of the population. Two percent of the population is listed as that of the aborigines who live in isolated mountain areas. Most of the Mainlanders and their parents were involved with the Nationalist army and were refugees from the Communist-led takeover of the mainland. They occupy key positions in the administration, and speak Mandarin Chinese. The Taiwanese are the descendants of the Chinese immigrants largely from Fujian and Guangdong provinces in southeast China and speak Hakka and other southeast China dialects. Because of the prominence of the Mainlanders in the administration, Mandarin has become the official language of Taiwan. The principal religions, in addition to the forms of worship of the aborigines are Daoism (the Chinese folk-religion) and Buddhism, which account for over 43 and 43 percent of the population respectively. Confucianism has greatly influenced these two religions and colored the people's ethics, morality and academic thinking. But religion, in general, remains a passive factor and the people tend to be eclectic regarding their religious beliefs, many practicing a little of several faiths. Christians constitute a small, but significant minority; and most of them are Protestants. Muslims number about a million, most of whom live in larger cities.

Starting off initially with a large agricultural base, and specializing in producing rice and sugar during the Japanese occupation, Taiwan has in the mid 1990s only 12 percent of its labor force engaged in agricultural production. Its two dominant crops are still rice and sugarcane, but has since established a more balanced economy and has become a model of high agricultural and industrial productivity. Despite its large population growth it is not only self-sufficient in foodstuffs but can export a modest surplus of rice, sugar, tea, and tropical fruits. A number of crops, such as sweet potatoes, corn and a variety of fruits including pineapples, bananas, soybeans, and various types of vegetables are raised.

Nearly a quarter of Taiwan's total area is arable and all available land is cultivated to the maximum, including sloping areas, dry river beds, and reclaimed marshland. The Japanese introduced improved strains of rice, chemical fertilizers, modern irrigation methods, and the Nationalists continued modernization plans, particularly in rice planting. Now Taiwanese agriculture is an intensive semi-mechanized gardening system with a highly integrated cropping system that provides two crops a year. And crop yields compare with the highest in Asia. Fishing, particularly pond fishing, prevalent in China mainland is also important. Commercial fishing fleet since has greatly expanded.

Taiwan's mineral production is modest, although a number of minerals have been found. Coal, copper, gold and silver, sulphur, and manganese are produced in small quantities. Coal is mined in the north, but is rapidly becoming exhausted, and petroleum has replaced it as a major energy source. Petroleum and natural gas are obtained in small quantities on shore, but the continental shelf may contain larger reserves. A petroleum refinery at Gaoxiang imports crude oil; and the aluminum refinery also at Gaoxiang depends on the imports from Southeast Asia. But the rising oil costs and national defense needs during the 1980s were responsible for the development of nuclear power plants which now account for one-third of Taiwan's installed capacity and nearly one-half of actual energy production.

Manufacturing emerged as the largest sector of economy during the 1970s and utilizes 30 percent of all labor force and produces over a third of Taiwan's gross domestic product.

The island's economic advantages lie in its abundant cheap labor supply, generation of capital formation essentially from foreign sources (particularly from the United States), and willingness to adopt technological innovations. During the 1950s and 1960s labor-intensive light industry predominated producing such consumer goods as foodstuffs and textiles aimed at both domestic and foreign markets. In the 1960s and 1970s investment shifted to more capital-intensive heavy industry items such as appliances, vehicles, steel, petrochemicals, machinery, and ships. During that time labor supply became scarce and wages increased and greater effort was directed to skill-intensive, high technology industries (chemicals, pharmaceuticals, precision instruments, and electronics, etc.). Over three-fourths of Taiwan's population resides in the cities. **Taibeh (Taipeh)** in the north is the seat of administration and has a population close to 2.6 million. **Gaoxiang** (population 1.5 million) in the south on the western coast is the major port, and is an industrial center.

Prospects Despite its relatively small size and limited resources, Taiwan has experienced one of the most spectacular economic developments of any developing country during the 20th century. The per capita product and personal income quintupled since 1949. The economy has been transformed from agricultural to industrial. But it faces a troubling political future as the shadow of Mainland China lengthens across the Taiwan Strait.

Like Japan, Taiwan has achieved its modernization and recent economic growth based on foreign trade. Its success has derived primarily from the discipline, industry and skills of its labor force, coupled with the advantage of a central location, and limited resources, and intermediate technology, its manufacturing has depended heavily on imported materials, equipment, and technology that was obtained mostly from Japan and the United States, which are also its largest trading partners.

Because its domestic market is limited and manufactures, it depends heavily on exports directed mostly at the United States. Its main exports are sophisticated electronic equipment, textiles, garments, synthetic fibers, and plastic goods. Besides the United States, Hong Kong, Japan, Germany, and Singapore are other importers. The island imports electronic machinery, chemicals, iron and steel, and crude petroleum. The major import sources are Japan, the U.S., Germany, Hong Kong, and Australia.

Further Readings

Buultjens, R. "China's Development Efforts and Quest for the Future," *Focus*, April, 1977, pp. 1-8.

Fureng, D. *Industrialization and China's Rural Modernization*. New York, 1992.

Institute of International Economics, *China in the World Economy*. Washington, D.C., 1994.

Smith, C. *China: People and Places in the Land of One Billion*. Boulder, Colorado, 1990.

21

China: A Regional Survey

China is so vast and diverse a country, that it may be more meaningful to understand the infinite variety that characterizes its lands by identifying its various regions that contain broad aspects of internal unity and coherence. This chapter is devoted to China's regions, identified in terms of their overall physical, and cultural characteristics.

Northeastern China The western name for northeastern China is Manchuria, the historic home of the Manchus. The Manchus invaded the area south of the Great Wall in 1644 and established the Manchu dynasty which ruled it until 1911. During the last fifty years huge numbers of Chinese settled in the Manchu-land, and Manchuria ceased to be a land of the Manchus. It is now overwhelmingly Chinese in race and culture. Although a part of greater China, during most of the past 2,000 years, Manchuria has been a zone of conflict between the USSR, Japan, Britain and China. Japan occupied it in 1931 and set up a government headed by a Manchurian ruler. Japan's primary interests were commercial and strategic. The area has been the meeting ground of many cultures: Manchus, Mongols, Koreans, and Chinese have, at earlier times, struggled for dominance in the area. In more recent times the region has been exposed to the Russian and Japanese cultural influences.

In the middle portion of the region are the Manchurian Plains consisting of the river plains of the Sung-hua and the Liao, measuring some 600 miles (965 km) from north to south, and 400 miles from east to west. The plain is surrounded by mountains on the east and west, and the valley of the Liao in the south fronts the Gulf of Liaodong. A narrow strip of coastal lowland leads to the Huanghe where the Great Wall reaches the coast. In the north and northeast, the Sung hua enters the Amur river. To the west the low mountains afford easier access to Inner Mongolia, the Gobi Desert and the People's Republic of Mongolia. Although the Manchurian Plain has long, cold, continental winters, similar to those of North Dakota in the United States.

Manchuria is China's prairie land and an important agricultural area. The growing season is long enough for one crop, and rainfall though not abundant (15 inches in the south to 28 inches annual in the north) is sufficient for dry-farming. In the southern part of the plains crops such as kaoliang, millet, soybeans, corn and wheat are grown, and considerable surplus is exported to other regions of China; in the north farming is mostly restricted to wheat and soybeans. The large expanses of the plain are well adapted to mechanized farming, and China's larger "state farms" and "collectives" are located here.

Manchuria contains China's larger industrial complexes, based in part on its considerable mineral reserves. Over one-half of the country's iron-ore and magnesite deposits are located in *Eastern Manchuria*, as well as aluminum ores, and a good amount of coal. Thus, the basis for both heavy-and-metals industries exists. Proximity to other fuel and raw materials has added to Manchuria's industrial potentials. An excellent network of railroads with connecting links to Russia, Korea, as well as to China south of the Great Wall serves the industries.

Manchuria experienced dramatic industrial and urban development since the 1930s, first under the Japanese, and later on during the Communist regime, primarily due to the substantial mineral wealth, and agricultural potential of the region. There are six "millionaire" urban centers: **Shenyang** (3.6 million), is close to the mineral belt in the uplands east of the Manchurian Plains adjoining the North Korean border. Together with **Fushun** and **Anshan** (both containing a little over a million people each) it is a major industrial center of China specializing in iron and steel products, and heavy metallurgical items. Iron-ore and coalfields are close to both Fushun and Anshan. **Harbin** (population 2.4 million) lies in the basin of the Sunghua, and is well connected by railroad links with Russia and North Korea, and by a north-south to **Dalian** (population 1.2 million). The latter is the main naval port and an exit station for the products of the region. **Changchun** (1.7 million) is also a major industrial and commercial center. It lies on the north-south railroad from Harbin to Dalian as well as is linked eastward with North Korea.

Manchuria has been a crossroads to northeast Asia. A railroad from Harbin connecting the region with Vladivostok, Russia's major port on the Sea of Japan, is 300 miles shorter than the all-Russian route beyond the Amur River. Russia's interest in this line, therefore, has always been extremely keen. During the mid-1930s Japan's interest, apart from Manchurian food and industrial resources, was to counterbalance any threat that Russia might apply against Japan.

West and South of the Manchurian Plains, and the Greater Hinggaling Mountains are the *Mongolian Desert Fringelands* of Inner Mongolia.

Inner Mongolia covers nearly one-third of a million sq miles territory between the Great Wall and the Gobi Desert. Much of it is a semi-arid grassland having a fairly good grazing potential. In 1949 the Communist administration rearranged the territory administratively by creating the Inner Mongolian Autonomous Region as one of the provinces of the nation. The southern section of Inner Mongolia consists of desert territory which is a part of the Gobi Desert to the west that primarily spans the southern part of the nation of Mongolia (the Mongolian People's Republic). The southern boundary of the Gobi Desert is defined by the Great Wall of China. Inner Mongolia is also conveniently characterized as "Chinese Mongolia."

Inner Mongolia is a cultural and environmental transition zone, a grassland where dry farming is possible. Culturally, it contains several tribal groupings, each occupying traditional pasturelands and speaking their distinctive dialects. In the east where rainfall rises to 12 inches or more Chinese farmers have been settled in the region who eke out a precarious living by cultivating land. Westward of the great northern bend of the Huanghe rainfall is scanty, and the grassland gives way to the desert. Along the river are some irrigated oases. The Mongolian fringelands are the "dust bowl" of China. Most of the region, particularly the western desert section, receives less than 5 inches of annual rainfall. Only 3 percent of Inner Mongolia is given to cultivation. In addition to millet, kaoliang, and wheat, the Chinese have introduced new crops such as soybeans, and linseed in the few farming tracts.

Historically, the region has been a zone of conflict between the nomad Mongols and the sedentary Chinese. To protect themselves from the nomads who occasionally descended on China for plunder, the ancient Chinese constructed the Great Wall that "marks the desert from the sown". During the last century, however, the sedentary farmer gradually pushed back the nomads to put under plough the dry sodded territory to more than a hundred miles beyond the Great Wall in the Mongolian fringelands.

Surrounding the Manchurian Plains to the north and west, and in part to the south are mountains—the Hinggaling Mountains, the Greater Hinggaling are in the north, and the Lesser to its west and southwest. The Lesser Hinggaling Range parallels the Amur in the north, and the Greater Hinggaling comes close to the Mongolian border.

The Lower Huanghe Region is the largest, unbroken level land in the country, covering some quarter of a million sq miles and contains at least 180 million people (nearly one-sixth of China's total). The region epitomizes the Chinese political and cultural heritage. No other part of China has played such an important role in Chinese history. With a density of over 1,000 persons per sq mile, and twice that figure for the cultivated area, the region is one of the world's most crowded regions. More popularly known as the "North China Plain" and the Yellow Plain, the region stretches southward from the Great Wall in the north to the hills north of the Chang Jiang basin in the south, and east of the hills of the Loess Region.

The largest section of the plain consists of the lower basin and the delta of the Huanghe or the Yellow River, and the southern part of the lower Huai River. The term "Yellow Plain" loosely equates with the area of the north China plain because of the color of its soil. The plain is an extensive alluvial stretch of land built by the deposition of the excess sediment of the rivers as they enter the plain. The deposition of the excess silt has raised the bed of the stream and has resulted in periodic flooding of the adjoining fertile farmland. Since at least the 1st century, the Chinese have constructed confining dikes: Periodic flooding during periods of heavy rains covers more than 10,000 sq miles and affects millions of people. The construction of Sanmen Dam has now reduced the periodic danger of floods that have occasionally caused great devastation to the farmland.

The climate of the region is as unpredictable as the flow of the rivers. The cold winds from Mongolia grip the area. Summers can be very hot. Rainfall is on the average increasing from 20 inches annually in the north to 30 inches in the south, and occurring mostly in summer. When they arrive a few weeks later, spring planting may be delayed and the early fall frost may precede the harvest. Occasionally, there is famine through drought, just as heavier rainfall may cause flooding. The term "land of famine" has been applied to the region. The chief causes of famine are attributed to the climate vagaries associated with the region.

Wheat, kaoliang, millets, and corn are the major grain crops, and a large variety of other crops are grown such as soybeans, cotton, tobacco, peanuts and a large number of garden vegetables. In the last half century, commercial crops (tobacco, cotton, and peanuts) have been increasingly produced. Nearly two-thirds of the land is in cultivation, and at least a third of the cultivated area produces two crops a year.

The Lower Huanghe basin is seething with humanity, and covered by an intricate mosaic of small fields. Although the countryside is dotted with rural villages, urbanization is higher than for the country. Two of China's "millionaire" cities—Beijing and Tianjin (populations: nearly 6 and 4.5 million respectively) are located at the northern end of the plain. **Beijing** is an interior metropolis and pre-eminent among the cities of the region, and after Shanghai the second largest city in China. It is the political, cultural, educational and transportation center of the country, and its development traces back nearly 1,000 years of long and venerable history. Time and again it has served as the political capital of the nation, and its glorious past has accorded it a cultural and sentimental significance far exceeding its size. Its regional situation in the northwestern corner of the plain at point where the mountain barrier between agricultural China and nomadic Mongolia is narrowest, gives Beijing an excellent strategic military and political advantage. Historically, Beijing occupied a logical position for the capital of an invading Mongol or Manchu armies, and was equally well situated for the Han Chinese rulers who were concerned with relations to the north. The present Communist rulers kept it as a capital for its traditional and cultural qualifications. Thus, in recent years it has succeeded in retaining much of its historic and political importance. The city has now acquired modest industrial development based upon the availability of coal from nearby uplands to the west. Its commercial significance derives from its status as the transportation hub. **Tianjin** is China's second important port after Shanghai. Although it has a poor harbor because its site is nearly 40 miles inland on a river, it possesses geographical advantages of situation that make it the port of entry and exit for a good part of the plain. It is the commercial center for north China, and second to Shanghai in cotton and wool textiles manufacturing. As a gateway to the land route to the northeastern provinces and to Beijing, it has enormous military significance.

The Shandong Region The administrative province of Shandong is equally divided between lowland and upland; its peninsula projects northeastward into the Yellow Sea to stand opposite the hilly Liaodong peninsula of Manchuria, forming pillars of the marine gate leading to the Bo Hai Bay. Southwest of the peninsula, an upland rises to 5,600 feet altitude which is separated from the peninsula by a low, alluvial corridor that provides a route between the lower Huanghe to the port of Qindao on the Yellow Sea.

A large percentage of the area is under such crops as wheat, barley, corn, millets, kaoliang, peanuts, cotton, and sweet potatoes. Shandong is well known for its silk. Population densities are generally high, but less than those of the Huanghe area due to unfavorable topography of a part of the region.

The region is rich in minerals including excellent bituminous coals. The coastline is mostly rocky, and contains excellent harbors, and is important for fishing. Mineral resources, good communications with the lower Huanghe hinterland, and the presence of strategically located harbors such as Qindao on the southern coast form the back of industrial activities. **Qingdao** (population 1.5 million) is the major port and ranks third in the nation among the largest cotton-textile industrial centers. It is also a major naval station for China, and is the rail terminus for a large hinterland of the middle and lower Huanghe plain.

The Loess Plateau One of the distinctive regions of hills in North China, it lies to the west of the lower Huanghe basin and its dominant characteristic is the widespread occurrence of yellow wind-lain silt, known as *loess*. The thickness of this fine windblown ranges from very little to a maximum of 300 feet (91 meters). The source of *loess* is the Ordos Desert within the Great Loop of the Huanghe, outside the Great Wall, where the supply of lake and river sediments is wind-blown and deposited over primarily in this region (some of which is also carried eastward to the lower Huanghe, and Sandong regions).

Cultivated land amounts to about one-sixth of the region, as contrasted to a much larger percentage in the lower Huanghe basin. Millets, wheat, kaoliang, occupy three-fourths of cultivated land; other crops include potatoes, cotton, tobacco, and opium. A major resource of the region is its high-grade coal. Industrial development in the region has, however, lagged because of poor communications. China's 80 percent coal deposits and modest amount of iron-ore are potentially valuable for future expansion in industrial development. Iron and steel, cotton mills, machine plants, and chemical industries are located at **Taiyuan**, and **Lanzhou**; the latter also has an oil refinery.

The Wei River plain was the cradle of Chinese civilization. Some of the earliest traces of which are found around the city of **Xian**. For the first millennium of recorded Chinese history the Wei plain was the heartland of China, from where Chinese culture and the state spread eastward and southward. **Xian** (population close to 2 million) is the gateway to the region controlling the communications northwestward and southwestward. The passage of the railroad through it has given it added importance. Taiyuan in Shaanxi (population: 1.6 million) is a major industrial center. From Xian to Lanzhou extends the famous ancient highway which connected ancient Beijing to Europe over which moved silk and porcelain in early times.

Central Mountain Region extends from the Tibetan Highlands almost to the Pacific into the heart of the densely settled agricultural China. From east to west the region extends some 850 miles (1,360 km); from north to south the distance is quarter of that figure. The west-east 11,000-foot Qinling, and its southeastward extension, Daba Shan, along the Chang Jiang and the Han rivers, separate north from south China. The Qinling often formed the boundary zone of independent rival states in the past. The southern section of the mountain block also locked in the Sichuan region from eastern China. After the Sino-Japanese War in 1938, the strategic protection afforded by the mountains, both in the north and in the east make Sichuan an ideal base for the Chinese government's resistance to Japan.

The Qinling mountains make a formidable barrier between the north and south China not only as a topographic obstacle to communication and as a cultural divide, but its greater significance lies in its more enduring role as a climatic divide. It shields the northern section from the southeast maritime air of most of its moisture, leaving northern China dry, while blocking the northwest, cold, continental airmasses from entering southern China. North of the Qinling system, therefore, the climate is characterized by low precipitation and cold, long winters. North of the Qinling only 10 to 15 percent of farmland is irrigated, and yields dry crops. South of it, climate is warm, precipitation is ample, the growing season is long and nearly 40 percent of the cropland is under irrigation.

The southern section of the region is remarkable for its rocky, steep gorges, canyons, and rapids on the Chang Jiang River which courses its way through gorges of 1,000 to 2,000 feet deep (608 meters). The rough topography makes the Chang Jiang the sole gateway to the secluded Sichuan basin. An important attribute of the gorges is their hydroelectric potential. An ambitious multipurpose dam near Yichang is being constructed to supply a large amount of hydroelectric power, besides providing irrigation and communication facilities.

This is a region of difficult communications. A few automobile roads and two railroads from Sichuan basin in the west to Shaanxi and to Beijing in the north cross the western section of the mountains.

Sichuan Basin is the most distinctive and isolated of all regions of China, enclosed as it is on all sides by rugged mountains, and is among the most productive agriculturally. Within its nearly 80,000 sq miles territory virtually all types of topography, climates, vegetation, crops and human activities are found and over 110 million people. Rich in resources, large in size and isolated enough, the region can constitute an independent state. In fact, it was the core of one of the political dynasties ruling China for nearly half a century after the breakup of the Han Empire in A.D. 220.

The basin is a land of hills, and low mountains, cut by rivers flowing through steep-sided valleys only 5 percent of the land is level, and the only plain is the alluvial Chengdu basin. Sichuan literally means four streams and refers to the chief four streams of the province. Within it most of the lower parts of the basin are navigable for most of the length of the Chang Jiang River. Until recently, there were few motorable highways, and until 1952 there was no railroad within the region. Despite the construction of few railroads, and highways within the region, the Chang Jiang remains the primary link with eastern China.

Although Sichuan is located far in the interior and surrounded by high mountains, precipitation is about 40 inches. The region is sheltered by the Qinling Mountain in the north from the cold winds in winter, and rarely experiences snowfall or frost except on higher altitudes. Humid, warm climate almost throughout the year is characteristic; the growing season may last up to 350 days. Precipitation occurs in winter as well as in summer, although heavy summer rains, and melting of snows in the high western mountains may cause flooding down stream in the Central Lake Region.

Despite hilly terrain, nearly 40 percent of the region is cultivated. A greater variety of crops than any other region of China is grown here. Both the wheat, millet, and corn of northern China, and the rice, and sugarcane of the south are grown. Rice is the usual summer crop occupying 60 to 80 percent of the cultivated land. Sweet potatoes are important on hilltops. Silk and tea are also widely produced together with some cotton, and tobacco. Tung or wood oil is a major cash crop. Citrus fruits are grown south of the Chang Jiang. The Chengdu plain in the western part of the region is particularly important for agriculture; a large part of it is occupied by cultivated fields. China's model irrigated system, developed nearly 2,200 years ago and operating with little change since, is located in the Chengdu plain.

Historically, Sichuan has been famous for two products—sugar and salt. The area to the west of Neijiang, the third largest city of the region, is well known for sugarcane and for manufacture of salt from the nearby salt wells. Mineral resources of the region are modest. Besides salt, some coal, iron ore, natural gas, petroleum, copper and gold are known to exist, but exploitation is slow, for Sichuan region remains primarily an agricultural region, a "rice bowl" of the nation. Urbanization and industrialization are at lower than the national level. The two leading cities, **Chongqin** (population: about 2.5 million), and **Chengdu** (population: 1.8 million) have traditionally acted as rivals for the political supremacy of the region. In modern times the former has asserted its dominance on the basis of its commercial and industrial leadership, and latter has derived its importance from its position as a seat of administration and learning.

Central Lakes Basins East of Yichang the Chang Jiang widens and flows eastward through the basins of China's two great lakes, the Dongting and the Poyang. The latter is much smaller of the two but the hills that separate them do not form serious obstacles to land communications, and the two are linked by highways and a railroad; thus the floodplain of the Chang Jiang forms a connecting link between them. The southern and northern parts of the basis, however, exhibit some marked topographic and climatic differences.

Climatically, the difference lies in the length of growing season. Although the southern portions are hot and humid like the northern parts, the growing period is much longer. This difference is reflected in the agriculture of the two sections, the wheat-rice transition zone of the north gives place to the predominantly paddy fields of the south. The intensity of cultivation also increases southwards. The cities of **Changsha** and **Nanshang** are among the leading rice markets of China. North of Dongting Lake is one of the major cotton regions of China, supplying raw material to the textile industries of Hankou.

The southern hill lands of the region are important for their mineral wealth. Nearly 90 percent of China's tungsten and antimony, and substantial amounts of coal, iron ore, gypsum, sulfur, lead, zinc and silver are obtained in this section. Iron ore and coal from the mines close by provide raw material and fuel for an important center of heavy iron-and-steel manufacturing at **Hanyang**. Dominating the commercial, economic, and political life of the region is the **Wuhan** metropolitan district, which has a population of nearly 3.5 million, and includes the triplet-cities of **Wuchang**, Hanyang and Hankau. The largest manufacturing complex of heavy iron and steel metallurgical industries is located in this metropolitan district, which is served by an extensive network of water-transport, strategic highways and railroads.

Lower Chang Jiang Plains Several thousand miles of waterways and canals crisscross the area between Wuhu and Shanghai so that within a 100-mile radius of Shanghai it is possible to travel by small boats from one settlement to another. The river, along with a network of canals, and lakes may well form one of the busiest transport webs in the world. The most famous of the canals is the Grand Canal, constructed between the 7th and the 13th centuries, to link Beijing with Hangzhou----a distance of nearly 1,000 miles (1609 km) long. It was used to carry tribute rice to the imperial capital at Beijing during the medieval times.

Historically, the Chiang Jiang has been a cultural divide between the northern and southern parts of China, separating the "Mandarin" speaking north from the Wu-speaking peoples of the south. But the plains of Chiang Jiang constitute a transition zone. Like the North China Plains it has predominantly level topography but its plains have isolated small hills and ridges, reminiscent of the south. It enjoys relatively abundant rainfall that characterizes the south, but experiences, occasional drought conditions like the "dry" lands of the north in contrast to the tropical, rainy south. Winters are relatively mild, and of brief duration. The growing season is sufficiently long to permit two successive harvests, including a rice harvest in summer, and wheat in the comparatively dry and cooler winters. Hence, the dominant crops of both north (wheat, soybeans, and barley, and south (rice) achieve equal status in this intermediate area of diversified agriculture.

Agriculturally, the north-bank lands form a rice-wheat transition zone, with local cotton specialization near the mouth of the river. To the south of the river, rice, silk and tea gain prominence. Near lakes, streams and coastal waters thousands of household depend on fishing for their livelihoods. Pond-fish culture and duck raising in the wet-paddy fields are universally employed on the farms. Cotton is important in the delta and processed in the large mills in Shanghai. A third of China's cotton is grown in the region. Some of the world's finest silk is produced in the immediate hinterland of Shanghai. Every acre of land has been intensively cultivated. Tiny fields still dot the landscape, despite efforts by the Communist administration to combine holdings into larger cooperative farms. Crop yields are among the highest in China reflecting the meticulous care (hand cultivation, fertilization and multiple cropping) given to farming.

This region is one of the most "humanized" portions of the earth, the density of population rises to between 2,500 to 3,500 persons per sq mile. The density is reflective of the productiveness of the well-watered plain, although its nearly level topography leads to frequent flooding. About 60 million people live in an area of about 20,000 sq miles. The high population densities are partly a result of the degree of urbanization of the region which is nearly twice that of China, and higher than any other part of China of comparable size. This is one of China's prominent industrial regions and contains several large cities. The three "millionaire" cities of **Hangzhou**, **Suzhou**, and **Nanjing** all were imperial capitals in the past, the last one was also a national capital before World War II, and the former two are famous for silk-reeling and weaving. The largest metropolis in China, **Shanghai**, with a population close to 7.5 million, is among the great cities of the world, and endows extraordinary commercial significance to the region. The city and most of its port facilities are not on the Chang Jiang, but on a small tributary 14 miles from the mouth of the Chang Jiang. Although the site of the port is a poor one, and silting is a major concern, the situational advantages are unrivaled in China. It serves a vast hinterland of three-quarters a million sq miles in the productive market of nearly 300 million people that extends far beyond the lower Chang Jiang basin. The port handles more than a half of China's international and inland trade. Approximately one-half of China's manufacturing labor is concentrated within its metro region. Textiles are the leading industries, followed by food processing, and light industries, printing, and manufacture of ships, steel and machinery plants. The industry can draw upon fuel and raw materials of a large tributary area at reasonable cost, because of cheap water transport and plentiful cheap labor.

The Xi River Delta and the Guangzhou Hinterland lies between the coastal southeast and the Southwestern Highland region (Guizhou-Yungui region). Physically it includes two different sections, one, the coastal delta section backed by a comparatively small river plain, and the other, the primarily hill country which contains narrow river valleys and farmlands. The Xi and two other rivers carry a large volume of sediment and have built a compound delta in the vicinity of **Guangzhou** (formerly Canton). Though the delta plain is relatively small, it is the largest plain south of the Central Lakes region. The port city of Guangzhou, however, is the political, and cultural focus of the region.

Climatically, it is tropical China, for most of the Guangzhou hinterland lies within 25 degrees of the equator. There is no winter. The coastlands bear the main brunt of the summer typhoons from the tropical Pacific. Rainfall averages the highest in the country averaging over 65 inches spread evenly throughout the year with a pronounced summer maxima.

The growing season is year-round; yielding two or more crops (primarily of rice) generally on the same land. Even a third crop of rice can be raised or a third one may consist of a dry crop, such as wheat, sweet potatoes, or maize. Cultivated land in the entire region amounts to one-seventh of the total area, and a large portion of it is irrigated. But in the plain, more than half of the land is given to agriculture. The Xi delta is one of the most densely populated agricultural lands in China, averaging over 4,000 persons of cultivated land. Abundant tropical fruits are also produced. Many farmers raise fish, devoting a good part of their farm to ponds. China's most productive fisheries are located along the coast. Silk is important in the delta around Guangzhou.

Guangzhou (population: 3 million) dominates all of China south of the Chang Jiang, and is commercially second only to Shanghai. Commanding a hinterland larger than that of Shanghai, which can be easily reached by navigable streams. Nearness to Hong Kong has been both an advantage as well as fraught with competition, especially since the recent takeover of Hong Kong by the Communist administration.

Hong Kong (population 6.6 million) is a special case. It is unique in historical development, economic structure and political orientation. It occupies a strategic position some 115 km (70 miles) south-east of Guangzhou. The island of Hong Kong was ceded to Great Britain in 1842, and additional territory was leased on the mainland for 99 years which officially reverted to Chinese sovereignty in July 1997. In an unprecedented decision China agreed to respect Hong Kong's political and economic autonomy for a period of 50 years under the rubric "One Country—Two Systems".

Hong Kong experienced phenomenal growth between 1960-80. Its deep water harbour, freeport status, and nodal location along the major shipping lanes to Europe and America favored its development as a major commercial center, and made it one of the busiest ports in the world. Excellent warehousing and banking facilities together with a large pool of skilled Chinese labor helped the city to grow into a chief entrepot between Shanghai and Singapore. Trade with China, that accounts for nearly one-half of the total (a good deal of which was illegal before 1997) helped enormously in the city's commercial importance. A variety of light industries, such as electrical and electronic equipment, and more recently ship building contribute to its solid economic base.

The closing days of the 2nd millennium witnessed the closure of another foreign colony, when **Macau** (population: 300,000) a Portuguese by China in 1998. Macau remained in the shadow of Hong Kong, and suffered as a trading post after the British acquisition of Hong Kong. Some light industry exists here. Scarcity of land, however, made both the cities among the most densely populated areas in the world—both contain over 16,000 persons per sq km.

The region is characterized by ethnic diversity. Nearly a third of the ethnic groups are non-Chinese of Thai background, most of them are highly sinicized, and intermarry freely with the Chinese. Of the remaining ethnic groups, almost half are the Yao mountaineers. Four or five mutually unintelligible Chinese dialects are spoken by the people of the region. In the northern part Mandarin dialect predominates; whereas, Cantonese prevails in the eastern and southern sections; Hakka is mostly used in the northeast part of the region. Around Shantou, another local dialect is preferred.

Southeast Coastal Hill Lands Located along the southern coast, it is one of the least developed, and isolated region of China. The embayed, irregular coastline sheltering numerous harbors reflects the rugged topography of the area. Interior isolation, external contacts and interest in marine activities mark the geography of this region. The mostly hilly terrain leaves out a small coastal plain and a few short rivers reach inland from an irregular coastline. No other region of China is so oriented to the sea or so detached from the interior. Isolation from other regions and the difficulty of internal communication have produced a variety of linguistic differences within the region. For example, people living in the coastal cities of

Ningbo, Fuzhou, Xuamen, that lie at comparatively short distances from one another use mutually unintelligible, spoken dialects; in the interior there are other linguistic groups. In the province of Fujian alone over 100 dialects are spoken.

This region has very hot subtropical climate with 50 inches of rainfall on the lower coastal areas and over 80 inches on the interior mountains. Typhoons, most common during the late summer, bring heavy rains, when their destructive force bring havoc to a good part of the region. Winters are frost-free near the coast.

A great variety of semitropical fruits are grown and exported. Wet-rice is the predominant crop, and two crops a year may be raised on the same land. Sweet potato is an important staple crop, because tea is commercially significant. The areas of arable land are restricted, and the soils are, in general, poor. Population of the region is relatively sparse. As a direct consequence of limitations on agricultural activities, home industries are important, particularly in village settlements, although nationally known handicrafts are produced mainly in the large cities.

Several minerals are obtained in the region although exploitation still remains limited. Fujian area is the important for iron-ore reserves. Flourspar and alurite are also produced. Salt evaporation from sea is an important activity along the shores of Hangzhou bay. Communication is limited, except for the few railroads that have opened up parts of the interior sand contact with other parts of China as well as within the region has primarily been by sea.

Hundreds of islands along the coast contain rich fishing grounds. Harbors such as Fuzhou, Ningbo, Xiamen are the important fishing markets. Each port has large number of boat people. In 1979 the Chinese government initiated a new "open-door policy" in order to attract technology transfer and foreign investment by developing *Special Economic Zones* (SEZ) as manufacturing and export centers. Many of these centers are to be located on the southeastern coast. Of these only **Shenzhen**, which adjoins Hong Kong has elicited healthy response, primarily because of its location. With a large inflow of foreign money it has become a burgeoning metropolis of over 1 million people.

Southwestern Highland Region lies to the southwest of the Central Lakes Region and includes the two remote provinces of China: Guizhou and Yungui (Yunnan). Southwest of the Central Lake basin, the hills begin to rise to merge into an eroded plateau area of Guizhou at altitudes averaging 4,000 feet (1,206 meters). Further west at a higher level averaging 7,000 feet is the Yungui plateau, sloping land comprising nearly three-fourths of the land surface. In Yungui the topography is more rugged, containing deep narrow chasms and jagged ranges that break the overall plateau character of topography. Virtually no alluvial plains have been built by the upper Salween and Mekong River, and they are not navigable and serve mainly as barriers to communications between China and Myanmar. Yungui plateau contains several lakes.

A variety of subtropical fruits are grown and exported. Wet-rice is the universal summer crop cultivated mostly by the Chinese farmers. Sweet potato is another important crop which is also a staple food. Arable land is restricted. The Chinese have not assimilated the aborigines who have pushed to the steep, comparatively less productive steeper hillsides on which they grow less productive cereals such as corn, barley and wheat.

The southwestern highlands is a transition zone, containing a strange mixture of the topography of eastern Tibet, and the demography of Thailand, Vietnam and Laos. The region has retained its frontier character because of its rough topography, poor communication, isolation from the heartland of Chinese civilization, as well as because of the hostility of the non-Chinese tribesmen. The region, as a whole, forms a great ethnic frontier. In 1990 the total population of the two provinces, Guizhou and Yungui was close to 70 million, of which the non-Chinese amounted to about 30 million. In Guizhou, the most numerous group is that of the Miao mountaineers. In Yungui, the Chinese form the largest single ethnic group, but are in a minority in a total population divided into 60 different ethnic groups. Foremost among the tribal groups are the Thai groups, followed by Lolo, Miao, and Muslims. Although the Guizou tribesmen have been suppressed ruthlessly during the last two centuries, Chinese control in the Yungui-Myanmar frontier has been less than complete. The Communist regime has recognized the status of these minority ethnic groups by organizing their territories into so-called "autonomous areas."

The region contains a variety of minerals. Yungui is the chief tin-producing area. In addition modest amounts of copper, silver, and coal are produced. Guizhou produces small amounts of coal and mercury. **Kunming** (with a population of a million and a half), is the major commercial center, and a railroad junction linking China with Hanoi in Vietnam. Another avenue to the outside world is provided by the WWII Burma Road through the difficult country in the gorges of the Mekong and the Salween rivers.

The Far West or Outer China South of the People's Republic of Mongolia and west of the Great Wall are vast stretches of territory covering some 630,000 sq miles (1.6 million sq km), and home to nearly 18 million people which can be appropriately described as "Outer China" or China's "Far West." It includes the two provinces: Xinjiang in the north and Xizang (Tibet) in the south. Xinjing, means the "New Territory," and often referred to as "The New Frontier," although historically it is far from being new. Xinjiang has been a frontier region of China for over 2,000 years and the Chinese garrisons have been in the chief cities for a long time. Traditionally, it has been popularly known as Chinese Turkestan, a term that provides a better ethnic characterization for the region. Except for Tibet, Xinjiang is the only minority province that contains non-Chinese population. Nearly 70 percent of an estimated 18 million that inhabit the region are the Uygurs, who are Turkic in origin and language, and Islamic in religion. Their chief occupation is cultivation, and most are sedentary. Since the late 1950s there has been a large-scale in-migration of the Han Chinese into the area that has occasionally caused bitter economic and cultural animosity between the Muslim natives and the Chinese settlers. The Uygurs' anti-Chinese sentiment has arisen by their fear of being displaced in their own land by the Chinese settlers. Adjoining the former Soviet nations in Central Asia and containing several Muslim groups of Uygurs, Kazaks, Uzbeks, and Tajiks that inhabit both sides of the international border, who dream of a territorial union in Central Asia, is the politically sensitive region of Xinjiang. Isolated cases of rebellion by the Kazaks against the government took place in the early 1960s at a time when the relations between China and the Soviet Union were particularly strained.

Xinjiang commands the historic gateway to Europe. Highways have crossed its territory since the dawn of history, and provided a link between ancient China and the Roman world. These highways were the lines of communication for the early Europeans, and of the monks that brought Buddhism from India. The fabled Silk Road that linked China to the Mediterranean lands during the medieval times takes a southerly route, south of Tarim Basin, and passes through oases sheltering such medieval cities as Hotien (Khotan) and Shache (Yarkand), and made famous by Marco Polo and by the silk caravans. Parallel to the silk road now runs a railroad that goes from Xian to Lanzhou, and beyond through the Gansu Corridor westward to Urumqui.

Physically, Xinjiang is divided into two great basins by the mighty Tien Shan: the Junggar Basin in the north, and Tarim Basin in the south. The Tien Shan is among the great mountain chains of Asia, and rises to 23,000 feet in the west and 17,000 feet in its eastern extension. Most of Xinjiang receives less than 5 inches of rain in a year, and only the higher reaches of the Tien Shan may catch as much as 30 inches. The two divisions present some contrasting physical and economic characteristics. The Junggar basin is predominantly pastoral and the Tarim predominantly agricultural. The former is populated largely by Mongols professing Lamaistic Buddhism, although some Muslim Kazak pastoralists are also distributed on the Tien Shan. In the Tarim oases the inhabitants are chiefly Uygurs of Turkic origin. The Chinese are concentrated mainly in the oasis-towns as merchants, administrators, artisans, garrison troops and in recent years as technicians. Besides the Uygurs, Kazaks and Chinese and a dozen of other ethnic groups inhibit Xinjiang. Prominent among them are: Mongols, Krygyz, Russians, Uzbeks, Tajiks, and Manchus.

Xinjiang occupies the deep interior of the Asian landmass, and is inaccessible to maritime influences. The Arctic Ocean is the nearest ocean nearly 2,000 miles to the north with barrier mountains along the Siberian frontier. The conspicuous climatic characteristics are extreme temperature ranges (daily and annual) and extreme aridity.

Cultivated land, limited to the oases, accounts for less than 15 percent of the land. Although wheat is the leading crop, other grain crops include corn, kaoliang, rice and barley. Excellent fruits, tobacco, and cotton are also important crops in the oases. In recent decades notable improvements have taken place in Xinjiang agriculture and animal husbandry. The administration has also opened up some new areas for cultivation through land reclamation programs.

Xinjiang is considered potentially rich in some minerals such as coal, iron-ore, and oil, although the mining activity is limited. The proven reserves of oil are considered to be quite large; but the production is only 3 to 4 million metric tons annually, confined primarily to the oilfield at Hami (the much larger and productive Yumen field is in the adjacent province of Gansu). Industrial development in the region is still in infancy consisting mainly of handicraft items. At **Urumqi** (population: half a million), the Communist regime had established a large modern cotton mill and a large electric power plant. At the southern edge of the Tarim basin on the ancient Silk Road lie some of the historic trade centers such as Hotien, Yarkand, and Kashgar, which are major service centers in southern Xinjiang. In the middle of the Tarim Basin lies the nearly rainless desert of Takla Makan. China's nuclear testing sites and deployment of nuclear arsenal is located in the Tarim basin.

Xizang (Tibet) In China's Far West and among its "outer regions" are the high and extensive plateaus of Tibet that are rimmed in by the mighty snowbound mountain chains rising over 20,000 feet (6,096 meters) the Karakoram in the west, the Kunlun in the north, the Amne Machin in the east, and the Himalaya in the south. Covering an area of half a million sq miles this region forms the roof of Asia and is the least accessible parts of the continent of Asia. Tibet has constituted one of the world's great barriers and isolating factors in history. For thousands of years, two of the world's greatest civilizations, the Indian and the Chinese, evolved on opposite sides of this great highland barrier.

Tibet's high plateau-like topography, particularly that of its western portion, its Central Asiatic location, its inland drainage and lack of fresh water and vegetation make its climate dry and desert-like. Most of it is cut off from the summer Indian monsoons by the Himalayan barrier. In the southeast, moisture bearing winds occasionally blow up the valleys of the Brahmaputra, and Mekong, and bring some summer rain to the Zangbo (Tsangpo or Brahmaputra). A large part of Tibet, lies at 12,000-15,000 feet altitude. In the vast middle section are hundreds of lakes, both fresh and salt, where salt, potash and borax are found around their margins. For eight months the ground lies frozen; and in summer large areas become swampy. Tibet is too cold and dry for grass, trees or cultivation. Eastern Tibet is a land of great valleys, and high mountain ranges, and canyons, where the mighty rivers of East Asia---Huanghe, the Chang Jiang, Mekong, and Salween----flow in their deep gorges. Here there is good deal of precipitation, and extensive forests cover the lower slopes of valleys.

Much of Tibet is unattractive and hazardous for travel. Vegetation, animal life, and human settlements are confined primarily to the lower eastern and southern plateaus and valleys. In the south the main population centers are located in the 12,000 foot high valley of Zangpo River in which **Lhasa**, the monastery city and the capital, and the traditional seat of the Dalai Lamas, is situated. Lhasa, has estimated population of over 150,000 (up from 69,000 in 1952). Ancient roadways are focused on it connecting it to China, and India. A motorable highway across the high ranges links it with Sichuan and Qinghai provinces of China.

The Tibetans profess Lamaistic Buddhism faith. The church and state have functioned together administered by the spiritual and temporal rule of the Dalai Lama before the Communist take-over. Thousands of monasteries that dotted the settlements possessed most of the best agricultural land. Approximately one-third of male population in 1950 consisted of monks. After Tibet's occupation in 1951, the Communist administration expropriated large monastic landholdings, demolished the monastic order, drove the monks from their monasteries, and introduced land reform and other changes designed to implement socialism similar to the other parts of China.

Large numbers of Han Chinese (half a million by 1956 and 1.5 million by 1963) were brought into and settled on the confiscated Tibet's monastic lands. Others came as administrators, traders, garrison soldiers, and technical personnel drastically altering the demographic and racial character of the land. From a virtually nonexistent minority in 1951, the Chinese became a formidable force by the mid-1990s. In early 1950s, Tibet's population was close to 1.5 million in which Chinese numbered a few thousand; by the mid-1990s the estimated population of Tibet had risen to over 4 million of which approximately over 2.5 million were Chinese.

The only extensive agricultural region is the Zangbo valley, where barley, wheat, potatoes, millets and turnips are grown. In it are also located the capital city of Lhasa and nearly all other larger settlements such as Shigatse. The yak is Tibet's typical draft and pack animal, which provides hides and skin for clothing and shelter; meat and milk for food; dung for fuel, and bones as implements. It is well acclimated to high altitudes. Sheep, cattle and horses also thrive in some regions. Although much of the population is still engaged in pastoral activities, the Chinese regime has opened up several stockbreeding cooperatives. The recent advances made by irrigation and the growing of forage crops is decreasing the amount of nomadic pastoralism. In addition to the vast salt reserves, there are deposits of gold, and copper. Traditionally, foreign trade was carried by pack animals such as yak, mules and horses and consisted of imports of tea and silk from China and manufactured foods from India in exchange for hides, wool and salt from Tibet.

Hainan China possesses several islands, especially along the south and southeast China coast including the two clusters of small islands, and the two larger ones, Hainan and Taiwan. Hainan geologically is an extension of the southeast coastal hills, parted by a shallow 20-mile channel. Much of the island is mountainous, rising to over 100 feet in elevation. A coastal plain occupies a northern quarter of the island, and extends around the western and eastern shores, that contain several mangrove swamps along the coast.

The island of Hainan is the only sizable tropical section of China where annual temperatures average about 77°F (25°C) for the most part. The summer monsoon brings heavy rainfall, especially on the southern half of the island. Rice is the leading crop, but coffee, cocoa and rubber are also produced, the latter three reflect the tropical location. High quality iron ore is developed in the southwestern section. Industry is relatively insignificant. About six and a half million people live on about a 13,000 sq mile territory of the island. Occupying a commanding situation of the Gulf of Tonkin, there is a fine naval base strategically located at the southern end of the island. The Chinese administration has selected Hanian as one of the 20 Special Economic Zones (SEZ) to be developed for a free market economy, containing a new city, a high-reach industrial complex, a duty-free port, and world class resort facilities to attract foreign investment. Beyond Hainan China's territorial claims include a number of small islands scattered through the South China Sea as far as 40°N latitude, including the Spratly and Parcel groups, which are a bone of contention between China and several countries in Southeast Asia (claimed by China, Malaysia, Philippines, Taiwan and Vietnam).

Prospects China enters the 21st century in a period of transition. By virtue of its size, and extensive resource base, productive capacity and military strength, it is potentially a world power. The present administration has had remarkable success in eradicating poverty, and in providing food, health, services, clothing and housing to its large and burgeoning population. This transformation of the nation is nothing short of miraculous by comparison with the turbulent years preceding the Communist take-over. But three-fourths of its population is largely rural and remains relatively poor. The per capita income is low (that approximates 860 dollars a year). Urbanization and industrialization are on the increase, and per capita income is growing but the low-level development of the rural masses leave hundreds of millions still in relative poverty. Millions in the cities, as well, experience shortages of housing whereas many millions keep on moving to the larger cities such as Shanghai, Beijing, and Guangzhou in search of employment.

Currently China faces three basic problems: population growth, industrial progress and the defense of her borders. Population remains a continuing problem, despite strict measures to check its growth rate and the dramatic drops in birth rates, the addition of 13-14 million peoples annually remains a serious threat. Industrial progress is also limited by infrastructural bottlenecks and deficiency of capital formation.

Despite these problems China's economy has been in a transition since the early 1980s as the country has moved away from a Soviet-type economic system. Agriculture has been decollectivized, the small non-agricultural private sector has grown rapidly, and government priorities have shifted toward light, rather than heavy industry. She has decided to increase her interaction with the international economy. As a result, her foreign trade has since grown faster than her gross national product. It is quite likely that the government's decision to permit China to be used by western firms as an export platform may eventually make the country economically strong enough to be a competitive force to her neighbors such as South Korea and Malaysia, and perhaps Japan.

China has been able to resolve its border disputes with the neighbors, but the existence of Taiwan, long considered as an errant Chinese province, poses a real problem and a serious challenge to the Chinese government.

Further Readings

Cressey, G.B. *Asia's Lands and Peoples*. New York, 1963.

Ginsburg, N. *The Pattern of Asia*. Englewood Cliffs, N.J., 1978.

Victor, F.S. *The Chinese Cities -- The Growth of the Metropolis Since, 1949*. Hong Kong, 1985.

22

East Asia: Mongolia and Korea

MONGOLIA

Sandwiched between China and Russia, Mongolia is a vast highland country occupying an area of about 600,000 sq miles (1.5 million sq km) in the heart of eastern Asia. Referred to as Outer Mongolia in the past, it became an independent republic called the People's Republic of Mongolia in 1924, and since 1992 has been known simply as Mongolia.

Mongolia is one of the most sparsely populated nations in the world, containing only 2.5 million people. According to a 1996 estimate, and has a population density of 4 persons to a sq mile (1.6 to a sq km). Racially it is quite homogeneous, belonging to the classic type to which the people lent their name. Ninety-five percent of the population belongs to the various Mongol groups; the Khalka-speaking Mongols constitute nearly four-fifths of the entire population. Other Mongolian groups include Dorbed, Bayad, Buryat, and Dariganga. The rest of the population consists of Turkic-speaking Kazaks, who have been traditionally Muslims, and are located mostly in the western part of the country. Russians form a small but important section of the population and are concentrated primarily in the cities. The Chinese, who were formerly important in cities, trade and finance, have largely left the country.

After the establishment of the vast Mongolian empire in the 13th century, Buddhism derived mostly from Tibetan sources was introduced in the country. However, Buddhism was swept away during the 1930s when the government greatly influenced by Russia dismantled the country's existing religious structures. Buddhist monasteries were closed, and monastic lands confiscated. The number of Buddhist monks dropped from 100,000 in 1924 to 110 in 1990. Historically, Shamanism has had considerable influence among the several nomadic tribes and remained, along with Buddhism, one of Mongolia's main religions until the 16th century. In their push for empire, the Mongols came in contact with Islam during the 13th to 16th centuries, which established its ascendancy over the conquerors. Although the Mongol Yuan dynasty of China had adopted Buddhism, Lamaism (Tibetan Buddhism) that incorporated Shamanistic traits remained the predominant religion. Shamanism is now a marginal cult, though some of the superstitions surrounding it are still widespread. The constitution of 1992 allows freedom of worship, and all traditional forms of religious practices—Lamaistic Buddhism, Shamanism, Islam, and others can be followed, but the formal practice of religion has been greatly reduced during the 20th century, and reliable data on the religious affiliations of

Map 22.1

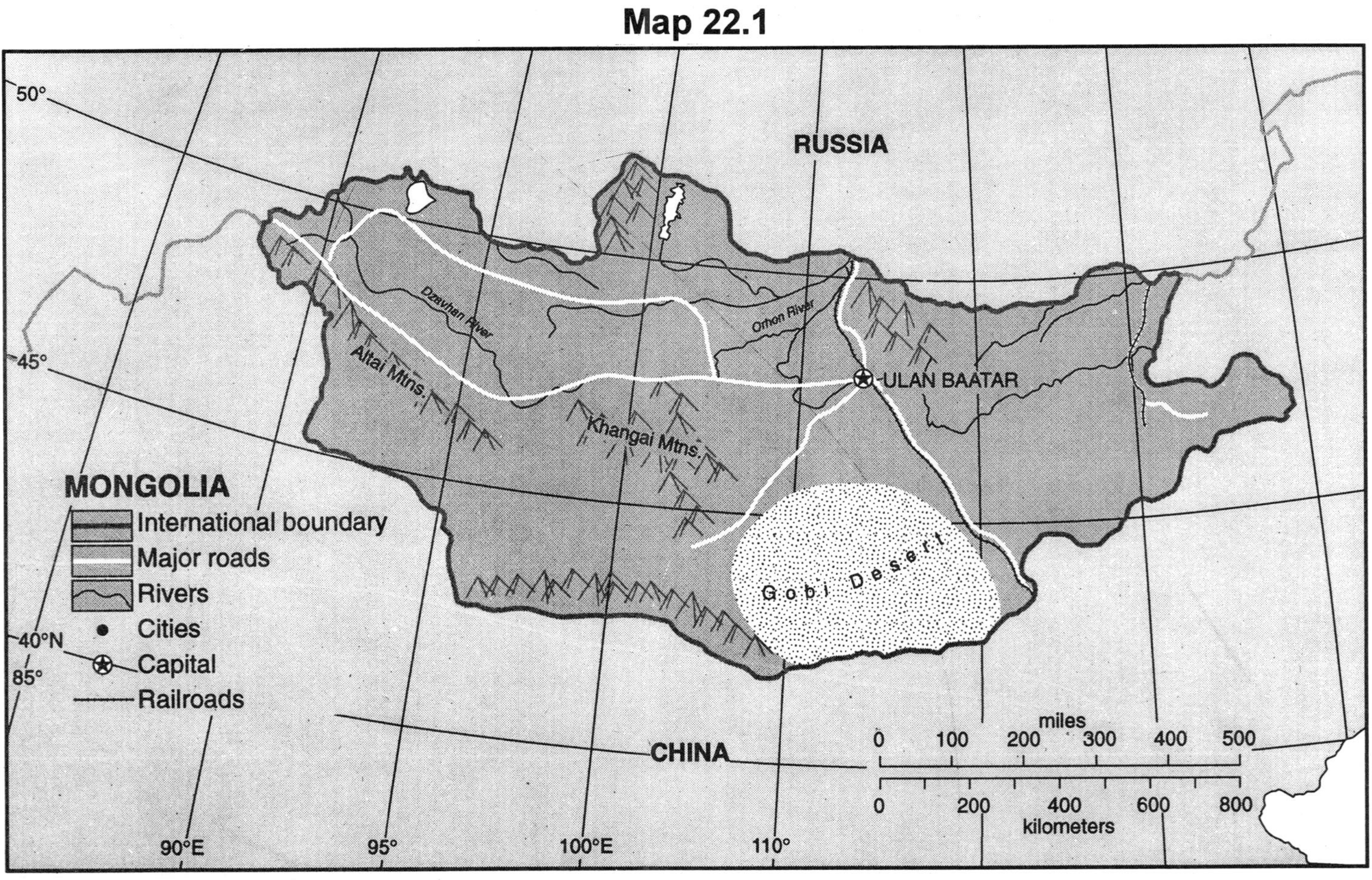
RUSSIA
CHINA
ULAN BAATAR
Gobi Desert
Khangai Mtns.
Altai Mtns.
Dzavhan River
Orhon River
MONGOLIA
International boundary
Major roads
Rivers
Cities
Capital
Railroads
miles
0 100 200 300 400 500
0 200 400 600 800
kilometers
50°
45°
40°N
85°
90°E
95°
100°E
110°

the people are not available.

The Mongols have a long and a remarkable history. Their ancestors, the Huns, lived in Central Asia from the 3rd to the 1st century B.C. in several nomadic tribal groupings. Not until the late 12th and early 13th centuries a single Mongolian feudal state was formed by Genghis Khan and his successors who controlled a vast empire that stretched from Bulgaria to Manchuria and included much of China, Russia, and Central Asia. During the declining days of the Mongol Empire in the 16th and 17th centuries Russia and China struggled to gain control of the Mongol territory, and China succeeded in controlling a large part of the empire. Outer Mongolia finally gained independence in 1921. It seceded an eastern part of its territory to China. The Chinese part is now known as "Inner Mongolia", and the remaining part became an independent nation, known as the Mongolian People's Republic. Because of its location between Russia and China, the Mongol territory had been dominated by its more powerful neighbors, China and Russia between 1691-1921. After independence it became closely tied to the Soviet Union, and remained so until the end of the 1980s.

Today there are more Mongolians living outside the country than in it; some three and a half million live in China's Inner Mongolia Autonomous Region, and another one million live in the Tuva and Buryat regions of Russia that adjoin Mongolia to the north.

Physical Features Most of Mongolia is a vast plateau area of rolling to level topography with elevations between 3,000 and 6,000 feet (912 meters and 1824 meters). Only one-third of the country's surface is made up of hills and mountains located in the extreme west, and the southwestern regions. Two-thirds of the surface lies in the nearly flat Gobi Desert which is approached over a mountain rim from all sides, in which the surface gradually descends. Across the surface of the Gobi one may drive an automobile for miles on a featureless landscape, the monotony of which is relieved by only a few residual monadnocks. This nearly level plain lies at an elevation of about 5,400 feet (1,642 meters) in the southern part of the country to over 6,000 feet in the north, with lower elevations in the middle of the Gobi Desert.

Many of such depressions in the Gobi are located between the Inner Mongolia (in China) and Ulaanbataar, the capital of Mongolia in the north, which are parted by low hills. In several cases they contain intermittent playa lakes.

The highest and the longest mountain is the westernmost Mongolian Altai Mountains that rises to over 13,000 feet, sweeps in from the northwestern tip of the country and extends toward the southeast for nearly 1,000 miles with a lesser range splitting off to the southeast as the Gobi Altai Mountains. There are several in its eastern part of the Altai that provide gateways to the Junggar and Tarim basins of western China.

Near the center of the country are located the Hangayn (Khangai) Mountains with peaks rising to over 12,000 feet (3,648 meters). Unlike the Altai, however, the Hangayn contains gentler slopes and crests and is often covered with fine pastures. In the far north there are also the mountain ranges that join the Tannu Ola mountains along the Russian border.

The entire country is prone to seismic

movements and severe earthquakes. There are thousands of lakes in the country, mostly in the northern and western regions. A large number of those are salty, transitory, and variable in size, and without any outlets. The largest fresh-water lake is the Hövsgöi that occupies a structural depression in the northwestern region of the country. Many of the lakes are of volcanic origin.

Tucked far away from the moderating influences of the oceans, Mongolia experiences an extreme continental climate, resembling that of Montana in the United States and Alberta. Its location at a high latitude and high altitude makes the winters more severe for its latitudes. The difference between the mean temperatures of January and July can reach 80°F (44°C). January averages --7° and --10°F (--21C and –23.3°C). The growing period in most of Mongolia is about 100 days.

Moisture is of great importance in this dry country, most of which is received from the northerly winds. Because the high altitude areas are located in the north and west, precipitation is highest in the north and diminishes toward the south, from an average of 12 inches (30.5 centimeters) in the northern mountains to less than 2 inches (5.1 cms) in the southwest Gobi Desert.

Most of the country is nearly barren, excepting the higher elevations of mountains which catch some moisture and are covered with vegetation of larches, aspens, and birches. The steppe lands predominate covering 75 percent of the surface with large, tall grasses, which in the northern half of the country turn southward into short grassland and further south into desert steppes, and wasteland.

Between 1921 and 1990 Mongolia came under the Soviet hegemony, and received technical, economic, and military assistance from the Soviet Union and mostly followed the Soviet guidance in political and cultural matters, both domestic and international. Until 1950 the only contact with the rest of the world was through the Soviet territory. A recognition by the People's Republic of China came in that year, and the reestablishment of trade southward. Foreign contacts increased with the construction of the Trans-Siberian Railway in 1956, and the trip from Moscow to Beijing was shortened by over 700 miles (1,183 km).

Historically, the Mongols have led a nomadic existence, and cultivation has not been significant in national economy, although this has dramatically changed after Mongolia adopted a communist pattern of government in the early 1920s. Herding has been the chief source of livelihood in the past, and from it the people derived eighty to 90 percent of their food. In the early 1990s nearly 30 percent of the labor force was engaged in agriculture but generated only 16 percent of the gross national income. Mining and manufacturing have become important during the past few decades and now contribute to nearly one-quarter of nation's domestic product.

Livestock herding is still a significant element of economy. Nearly 14.5 million sheep, 9 million goats, 3.5 million cattle, 2.5 million horses, a million goats, and half a million camels contribute food and clothing for the people, and provide raw materials for the expanding manufacturing economy and trade. Mutton and cheese for food; dung for fuel, and wool for the manufacture of the characteristic felt *gers* or *yurts* is provided by this enormous population of animals. Traditionally, horses, camels and yaks have provided means of transport. The Mongol grazing economy has thus been quite self-sufficient.

In the early 1990s nearly 60 percent of the population was listed as urban, in striking contrast to the primarily, nomadic life that traditionally favored the *yurt* or *ger* population. A distinctive feature of the countryside is still the *yurt*. The *ger* is a cone-shaped tent-like structure of felt (nowadays of canvas laid over felt) stretched over light, collapsible wooden frames. Building materials are scarce on the steppe, and the *ger* structure is light, strong, and easy to assemble, transport, and reerect. Besides being portable, it is ideally suited to Mongolia's rigorous climate. Its round, squat shape can withstand strong winds unlike the "modern" one-man camping tents. The felt is impenetrable by rain or snow, and insulates well. It is still being used by herdsmen moving from pasture to pasture.

The urban centers have experienced a phenomenal growth in the second half of the 20th century. In 1950 only a quarter of the population lived in the cities. The capital, **Ulaanbataar** (population nearly 630,000) is the largest and most important urban center containing a quarter of the country's population. It lies at the strategic railroad that connects Beijing with the Russian territory. Ulaanbataar houses, one-half of nation's industrial labor force which is primarily engaged in light industry specializing in such items as processed food and meat, beverages, dairy products, and textiles. Two other important cities, **Darhan** and **Suhbaatar**, are located close to the country's northern border. The former was founded in 1961 as a planned city and contains a population of 90,000. Second only to Ulaanbataar it is nation's major industrial center. Suhbaatar, founded in 1937, is a transport and trade center, on the Trans-Mongol Railway, near the Russian border.

Economic Development Nearly eighty percent of Mongolia's land is in meadows and pastures, and nine percent in forests. Less than one percent of the land, limited mostly to the northern parts of country is under permanent cultivation. Because of the long, cold winters, only a single crop in a year is possible. Eighty percent of the cropland is in grains---mostly in spring wheat but with some in barley or oats---all the rest is in fodder crops (hay). Yields are generally low. Potatoes and other vegetables occupy a tiny fraction of the cropland. Three-fourths of the cropland is in state farms, the remaining in cooperatives. The state farms are large, averaging 700 sq miles in size and typically include some livestock production as well as crops.

Livestock raising---based on millions of head of sheep, goats, cattle, horses, and camels---account for about seventy percent of the value of agriculture. The livestock are widely distributed throughout the entire country. The horses and cattle reach their greatest concentration in the wetter north-central regions, whereas the goats and camels are proportionately more numerous in the drier west and south. Since the 1930s most of the livestock have been raised in agricultural cooperatives, although the administration has been making efforts to slowly privatize these since the early 1990s.

Mongolia possesses several minerals, chief of which are coal, iron, tin, tungsten, copper, gold, and silver, but just a few of these (coal and copper and some gold and tungsten) are extracted at a commercial scale.

Although the administration has devoted a great amount of effort to industrialize the nation, much of the country's industrial production is limited to processing raw materials or meeting basic consumer needs. The main manufactured products are: processed foods (meat, beverages, dairy products, and flour), woolen textiles, footwear, hides, skins, furs, lumber, paper, matches, and furniture. About half of the industrial employment resides in or near Ulaanbataar. Heavy industry based mainly on iron ore and copper is concentrated in Darhan, and forests products are processed in Suhbaatar.

Minerals, and metals are the principal exports, (40 to 45 percent of all exports). These surpassed the traditional exports of livestock and their by-products during the 1980s. Raw material goods, food products, consumer and construction materials are the other export items. The main imports are machinery and transport equipment, followed by fuels, minerals, and consumer goods. Until the early 1990s three-fourths of Mongolia's trade was with the former Soviet Union and much of the rest with the former Communist block countries of Eastern Europe. Since then, it has sought to increase its trade with other nations. However, the development of a broad-based international trade is hampered by several factors, notably the absence of direct access to the sea, the need to use Russian and Chinese ports and the high transport costs.

Prospects Mongolia is passing through a period of administrative, political, economic and social transition. With the collapse of the Soviet Union in 1990, the political and economic systems, thus far shackled to the Soviets, have been freed. The new constitution of 1992 calls for a liberalization of economic and social forces. Mongolia experienced great economic hardship as it moved from a command economy to a system of market economy. One-third to one-half of the budget was previously obtained from the now-defunct Soviet Union. Its international debt has become excessively high. With the escalation of national debt, and rapidly growing population, its per capita income is dropped to among the lowest in the world.

The worsening economic situation has undoubtedly been exacerbated by Mongolia's landlocked position between the two powerful neighbors on which it had to depend for passage to the outside world for its international trade. Mongolia is now gradually increasing its trade with nations other than these two neighbors.

KOREA

Divided politically into North and South by an artificial barrier roughly along the 38th parallel at the end of World War II, Korea has been historically and culturally a single nation. Because of its strategic location between China, Japan, and the Russian Far East, it has suffered from the aggressive inroads of its powerful neighbors. The peninsular shape has also been a critical factor in its history, as it has acted historically as a bridge between Asian mainland and the maritime powers. The political division into North and South reflects the key significance of Korea's situation, for it has become now, as in the past, a meeting place for different and

Map 22.2

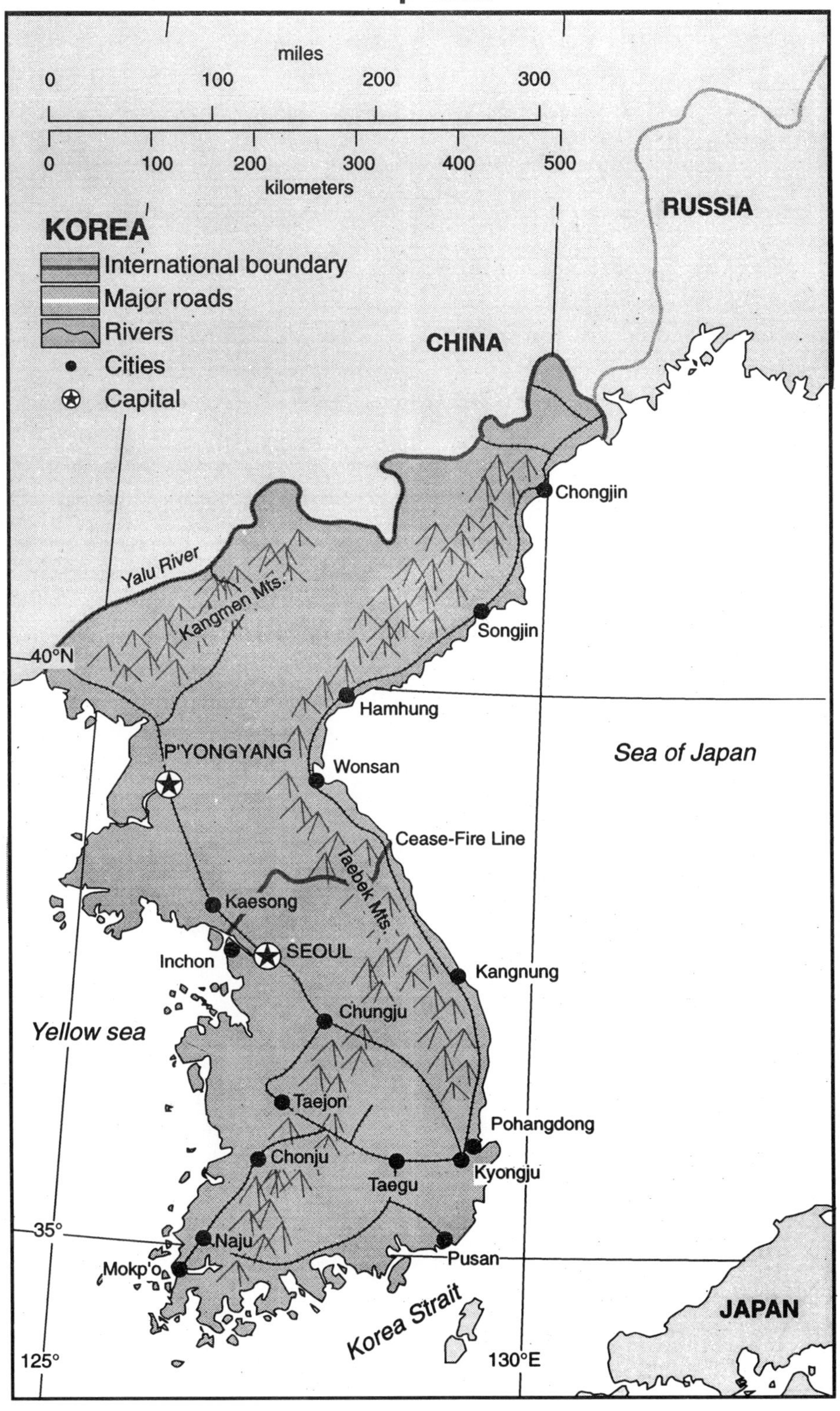
miles
0
100
200
300
0
100
200
300
400
500
kilometers
KOREA
International boundary
Major roads
Rivers
Cities
Capital
RUSSIA
CHINA
Chongjin
Yalu River
Kangmen Mts.
Songjin
40°N
Hamhung
Sea of Japan
P'YONGYANG
Wonsan
Cease-Fire Line
Taebek Mts.
Kaesong
SEOUL
Inchon
Kangnung
Chungju
Yellow sea
Taejon
Pohangdong
Chonju
Taegu
Kyongju
35°
Naju
Pusan
Mokp'o
Korea Strait
JAPAN
125°
130°E

often hostile ideologies. In 1948, the Republic of Korea was formed in the south, and the Democratic People's Republic of Korea came into existence in the north. The peninsula is still divided by a demilitarized zone that cuts across the land in which North Korea is ruled by what may be described as one of the few remaining outposts of Communist totalitarian ideology, and South Korea by an administration that owes allegiance to market economy, and liberal democratic forces.

North and South Koreas are small countries---the peninsula stretches some 600 miles (970 km) in length and 125 to 200 miles (200 to 320 km) in width. It reaches to within 120 miles of the island of Honshu in Japan, separated by the Sea of Japan and approximately the same distance of the Shandong Peninsula of China off the Yellow Sea. There are about 3,500 islands, mainly along the southern and western coastline.

The Two Koreas: North and South

Korea split into two countries disregarded the essential cultural and historical unity of the peninsula. The demilitarized zone originally created as a temporary zone separating the northern from the American zone for the purpose of accepting Japanese surrender has hardened into political division of the nation. Following the division in 1945, the government of North Korea was strongly influenced by Russia and China, and established a repressive, highly secretive, and authoritarian regime, and South Korea by an equally repressive anti-Communist government. The split not only ignored the basic cultural and economic interests and desire of the Korean people, it was against the logic of geography.

In 1950, North Korea sought to reunite the country by force and invaded South Korea across the 38th parallel precipitating a U.N. military response led by the United States, and the devastating Korean War (1950-1953). The Red Chinese army entered the peninsula to assist the North Korean army. Eventually, the ceasefire was arranged in 1953 after a good deal of devastation of the Korean peninsula.

Geographically, there are a number of regional differences between the North and South that have created a corresponding interdependence. The north is far more mountainous, and contains much less arable land than the south. Climatically, the north is more continental. Food production is heavily concentrated in South Korea, and North Korea is food-deficit. Formerly food deficiency in north was made up by shipments of food from the south. Given the concentration of population and trade centered in south, there is need for regional interchange.

The distribution of several raw materials for industry also reveals regional contrasts. There is much greater concentration of mineral resources in the north. North Korea produces most of Korea's bituminous coal, nearly all of its iron ore, and over three-quarters of its mineral output. Hydro power is also concentrated in the north, which produces twenty times as much as the south. North Korea specializes in heavy industry and also contains Korea's largest plants producing fertilizers. South Korea's farms can well utilize chemical fertilizers and industrial machinery of the north. Thus, two Koreas are potentially interdependent.

At the end of the Russo-Japanese war (1904-5), Japan occupied the country, that lasted until the Second World War (1939-

45). The Japanese during their occupation introduced changes in Korean development that had direct impact on the Japanese nation, rather than on the amelioration of the Koreas. Food production was nearly doubled, but the new output went largely to Japan to feed the expanding Japanese population, the Korean farmer was no better off. The fishing industry was greatly expanded, but eighty percent of the catch went to Japan to feed the Japanese people there. Japan's demands on the Korean timber, added to the traditional practice of shifting cultivation in many regions of Korea led to widespread deforestation of Korea's hills. That greatly worsened erosion and seriously aggravated the fuel shortage. Meanwhile population of Korea began to increase resulting from eradication of famines with the construction of an extensive transport network by the Japanese, and from the public health measures. These measures were essential to the Japanese so that they could exploit the Korean economy. The net result of such developments, was a big increase in population but a corresponding deterioration of Korean standard of living. Japan also developed Korea's rich mineral wealth, most of the output went to Japan, although it did create some local manufacturing but the finished products were largely exported.

Japan had to relinquish control of Korea after it surrendered to the American forces at the end of World War II. Korea at that time was drained of its economic resources, had no political leadership and was poorly prepared to the demanding task of establishing an independent existence in the troubled postwar world. The Japanese surrender was followed by the unfortunate division of the country into two states---a Communist-dominated north, and an anti-Communist south. This division was agreed upon by the American forces and Chinese-backed Korean army, the two sides which were controlling Korea's territory. The actual dividing line, that closely followed the 38th parallel along a demilitarized zone, dates from 1953. The direct result of the rivalry between the two sides completely overlooked the Korean interests and national desire for unity and independence.

The Physical Environment Korea is largely mountainous, and contains only small valleys and narrow coastal plains. The Taebaek-San Maek (Taebaek Mountains) are the largest chain that runs along the eastern coastline in a generally north-south direction, curving from northwest in North Korea to southeast in South Korea. From this chain branch of several ranges, and the principal rivers have sources in it. Of the other rivers---Yalu that make boundary with China, is the most important. The highest point---Paiton---rises to 6,822 feet (1,950 meters) in the extreme north, close to China's border.

In general, the northern interior, adjacent to China's Manchurian region is more rugged, but most relief is between 2,000 and 5,000 feet (610-1525 meters). In the southern part of Korea, the topography is of lesser altitude. But most of the topography contains a profusion of ranges of hills in the interior, often lapped by small encircling alluvial plains. Other lowlands that fringe the coastal areas are not extensive; they are narrow on the eastern side of the peninsula, and relatively broader though irregular on the west. The eastern coast is relatively straight; the western is indented, and contains many islands.

The climatic characteristics of Korea resemble those of central and northern China than of Japan. In winter it comes under the influence of a high-pressure system, and drifts of cold, dry air from north and west. During fall and spring, conditions are tempered by the passage of cyclonic storms which bring more precipitation in the south than in the north. Snow also stays on the ground longer on the ground in the north, but melts quickly in the mild temperatures of the south. The northern interior has bitterly cold winters. Only in the extreme southern fringe the mean January temperature is above freezing. The frost-free period lasts from 150 days in the northern interiors to 226 days in the southern parts.

Summers are hot and humid, and receive a larger amount of precipitation. The annual range of temperature is much greater in the north and in the interior regions than in the south. The average monthly temperature in January at Seoul (South Korea) is about 23°F (-5°C) and about 78°F (25°C) in August. In Poyangyang January temperatures average 17°F (-8°C) and summer temperatures average a few degrees short of Seoul.

The annual rainfall ranges from 25 to 50 inches (625 to 1,524 mm), and occasionally, late summer typhoons and storms cause heavy showers along the southern coast. Precipitation varies with orographic (topographic) position. The highest amounts are in the south with nearly 60 inches in the hills in South Korea, and least in the sheltered upper Yalu Basin in North Korea (about 25 inches).

Terrain and climatic conditions favor the growth of extensive forests in Korea. Nearly three-quarters of North Korea and close to two-thirds of South Korea is forest land. The best timbers lie along the China border in North Korea which contains large strands of birch, spruce, and pine trees. However, most of the commercial quality timber has largely been removed in the densely settled areas and coastal slopes. Forestry has considerably declined since the World War II, although recent reforestation programs have stressed economic forestry particularly in North Korea.

Cultural Characteristics and Historical Development The Koreans share a common racial origin with peoples of North Asia and the society is homogeneous, with only small percentage (less than one percent) of foreigners, most of them of urban Chinese, located mostly in South Korea. Cultural and racial characteristics reflect typically Mongoloid features and vary only slightly from one region to another. All Koreans speak the Korean language, which belongs to the Altaic family and is related to Japanese but contains Chinese loan words. Although Koreans have borrowed cultural elements from their powerful neighbors----China and Japan----culturally, they are quite distinct and have developed their own language, literature, and national mores.

There is little uniformity of religious beliefs within Korea. In a typical family in South Korea, for example, the women may follow Buddhism, while the men may adhere to the Confucian ethical system. In North Korea nearly 68 percent of the people were listed as atheists or non-religious in 1980. In South Korea, however, majority of the population (52 percent) declares themselves as religious who actively participate in religious functions regularly, and 48 percent as non-religious. One-sixth of North Korea's population practice traditional belief-systems that appear to be an amalgam of Confucianism, Buddhism, and Chondogyo ("Society of the Heavenly Way"). The latter was founded by a Confucian teacher in 1860, and combines elements of Buddhism, Confucianism, and Christianity and even Taoism. Basically, the way of life and value system Koreans are grounded in Confucian teachings, including ancestor worship, that is particularly strong in the rural areas. Buddhism is more popular with women, and old Buddhist temples are a part of the Korean landscape, although less than 2 percent of the population in North Korea, and a quarter of the declared "religion-practicing" population in South Korea are adherents of Buddhism.

Christianity is new in Korea, but claims a large following, numbering 5 to 6 million of which 4 million are Protestants and has a profound effect on the modernization of Korean society. Shamanism—a religious belief in gods, demons, and ancestral spirits responsive to a priest or shaman remains strong in rural areas, especially among women. The Communist regime in North Korea has constitutionally allowed freedom of religion but does not practice it, allegedly for the fear that it will weaken the Communist party and government. After the Korean War (from 1950 and 1953), churches and Buddhist temples were confiscated and many were converted to other purposes.

North Korea: Resources and Development At the time of partitioning of the peninsula in 1948, North Korea inherited a much larger share of the resources---80 to 90 percent of all mineral deposits on the peninsula, but it was essentially a peasant economy. The means of production under the Communist rule since then have been socialized, and priorities and emphasis in economic development set by the government. As in other Communist countries special emphasis was placed on capital goods rather than on consumer goods. North Korea contained a much larger territory (48,000 sq miles as compared to less than 40,000 sq miles of South Korea) but a much smaller population base. But it was relatively less developed economically than its southern neighbor—a contrast that intensified since the partition of the peninsula.

Of the 200 or so known minerals of economic value, iron ore, coal, tungsten, sulphur, and zinc are especially plentiful in North Korea. Iron ore reserves are estimated at over 2 billion tons although production is less than 10 million tons annually. Rich deposits of coal both bituminous and anthracite also found in the southern part of the country, not far from Pyongyang. The Japanese paid special attention to the development of the hydro power along the Yalu River and its tributaries during their occupation. Power production is therefore based primarily on hydroelectricity, but thermal electricity is becoming increasingly important because of growing demands of industrialization.

Since World War II Korea changed from an agricultural to an industrial nation. Mining and manufacturing together now account for nearly one-third of the labor force and represent an important segment of the economy. Major industrial development during and since the 1950s has been concerned with the production of iron and steel particularly of heavy metallurgical machinery centered at Chongju; agricultural machinery near Pyongyang; and of textiles located mostly at Pyongyang.

Production of electricity, both coal-derived thermal and from hydro sources, also increased during this period, although it has not kept pace with industry.

Agriculture makes a much smaller contribution to economy, although more than 40 percent of the labor force is engaged in it. Because of the chronic labor shortage agricultural productivity remains low. There has been an increase in cultivated land, and irrigation facilities, as also in the increased use of chemical fertilization and mechanization through the government programs of socialization. The country remains chronically deficient in food grain production. Foodgrains is one of the important items of imports. By 1958 all farms were incorporated into 3,000 cooperatives, controlled by management committees which establish production quotas, and prescribe type and amount of seed and fertilizer to be used on farms. There are also state and model farms, and livestock husbandry is concentrated on the state farms.

The main foodcrops are grains, such as rice, corn, barley, millets and wheat. The production of grains has increased since the 1950s, but has not kept pace with the growth of population which in the mid 1990s stood at 22.5 million (in 1945 it was less than 9 million). Sweet potatoes, soybeans, and fruit trees are grown extensively. Industrial crops include cotton, tobacco, flax and rapeseed. Forestry has declined since World War II, as the administration has been paying greater attention to manufacturing, and fishing which have become a greater source of national income.

Railroads are the principal means of transport, and highway transportation is relatively less important. River transport plays an important role in transporting agricultural goods and minerals. **Pyongyang**, the capital, and a major city, has a population of over 2.5 million, and connected by railroad with the Manchurian province of China and with South Korea.

In the early 1990s more than two-thirds of North Korea's trade was conducted with the Soviet Union, and China. During the last two decades, trade has been permitted with the non-Communist countries, including Japan, Hong Kong, Netherlands, United Kingdom and Australia. Imports mainly consist of machinery, oil and fuel, chemical and rubber products. Major exports are minerals, cement, fish products and textiles.

Economic Development in South Korea As compared to North Korea, natural resources in South Korea are meager. Coal, iron ore, gold, lead, tungsten and zinc, comprise two-thirds of the total value of mineral resources, but in quantity not even one-tenth of the North Korean productions. Iron ore is exported to Japan, and tungsten to the United States. Anthracite coal is the major exploited energy resource. Nearly three-fourths of the hydro power is derived from Han River, not far from Seoul. In contrast to North Korea, thermal electric power is much more important than hydroelectric power. The first oil refinery started to produce petroleum products in 1964, and power stations have since switched from coal to oil.

Relatively speaking, South Korea has a larger amount of cultivated land (nearly one-fifth of the land is under cultivation as compared to one-sixth in North Korea). It is much more peninsular and maritime in character. Though it too is largely mountainous but its plains are more extensive than of North Korea. Nearly 18 percent of the labor force is engaged in agricultural activities as compared to less than one percent in mining and over 26 percent in manufacturing. In terms of contribution to gross domestic product agriculture accounts for about 9 percent of nation's total production. Both the farm population and the proportion of national income from agriculture have been decreasing. Rice is the principal crop. Barley, wheat, soyabeans, are the other crops that provide staple foods for many of the farmers, along with rice, rice being the commercial crop sold to the wealthy and the urban dwellers. During the period of Japanese control agriculture became increasingly commercialized and most riverine plains were made to yield greater productions by large-scale irrigation projects, the application of commercial fertilizers, and the practice of more intensive methods of cultivation. Export food production kept pace with population growth, more so in South Korea, which is now self-sufficient in its food requirement, although the share of agriculture to nation's domestic product proportionately has decreased.

Since partition manufacturing has become the most important segment of nation's economy, producing close to 30 percent of nation's wealth. The government stresses importance of manufacturing. In the 1960s and 1970s, the government emphasized the growth of large establishments such as fertilizer plants and oil refineries by giving financial assistance; as a result small and middle industries producing consumer goods did not grow much due to financial difficulties. The trend began to be reversed in the 1980s and 1990s when the administration withdrew itself of direct involvement, although heavy industry including metals, chemical and fertilizer industries continued to be developed.

In terms of value and employment, however, textile industry has been the most important single industry. South Korea has now become a major market and dominant center for consumer goods industries; besides textiles, food processing has also become an important industry.

On the whole, South Korea is remarkably well-served with transportation facilities. The Japanese had developed the ports of Inchon, Makyo and Pusan into major rice-exporting centers, with sizable shipping facilities during their occupation. These were connected with the agricultural hinterlands by modern highways and railroad network. In the postwar period when the country, with foreign aid, particularly from the United States, started to rebuild the economy, the demand for the expansion of the transportation system became a necessity. The need for further development was, in part, necessitated by a phenomenal increase in population. In 1944 South Korea had a population of about 19 million, which has grown to 47 million by 1999. The transportation system has since then, expanded considerably. With the introduction of modern highway and air services, and with the rapid growth of population, especially of cities, urban transportation facilities have become increasingly congested. Since 1960 road transport has become more important, and now accounts for nearly 90 percent of passenger travel.

City growth has been even more rapid than the population in general. The Korean War (1950-53) had a damaging effect on the Korean economy. The influx of refugees from North Korea crowded into the southern cities. South Korea has now five "millionaire" cities: **Seoul**, **Pusan**, **Taegu**, **Inchon** and **Kwangju**, as compared just one (**Seoul**) in 1950. Only one city in North Korea, **Pyongyang**, the capital, contains over a million people (2.3 million). The South Korean cities experienced phenomenal increases bolstered by a rapid industrialization and a great surge in the economy. The capital city is **Seoul** which, with a population of nearly 11 million, is among the most populous cities of the world. Its large hinterland encompassing before partition the whole of Korea, its central location, west-coast orientation toward China and the mainland, make it an excellent capital and commercial center. It is also an important industrial center and connected with **Inchon**, the nation's major port on the west coast. **Pusan** on the east coast, with about 3.7 million people, grew especially after 1910 as the principal port for trade with Japan, and has been the chief point of entry for American aid and foreign commerce.

South Korea's principal trading partners are the United States, Japan, Hong Kong, and Southeast Asia.

Prospects North Korea inherited a larger resource-base and a more industrialized section of the peninsula at the time of partition but it pursued a rigid, inward-looking economic policy. During the last few years a substantial part of its expenditure was devoted to the development of nuclear technology and weapons. Economic development in general, and agricultural development in particular lagged because the emphasis was placed on the production of capital goods and heavy manufacturing items rather than on consumer goods and food production. The pursuance of lopsided economic policies resulted in its population facing starvation in the mid-1990s. South Korea, on the other hand, adopted a more liberal and free-market economic system, and the results have been spectacular. Its economic growth averaged between 6 and 10 percent annually for the last three decades. It is, in small measure, duplicating the Japanese role as an exporter of consumer goods including automobiles and electronic materials.

Given the complementary nature of the two Koreas' natural resources, and the desire for a unified state by most Koreans, suggestions for the reunification of Korea have been made in recent years. A united Korea can undoubtedly expect to enjoy considerable economic prosperity, and may have no need to spend large sums of money on military forces, which both North and South Koreas are currently doing. But the political designs by the rival governments to retain their distinct regimes offers resistance to the reunification efforts. Recent trade negotiations between the two governments have demonstrated greater promise. While the artificial division of Korea lasts, its economy and its people will suffer.

Further Readings

Ginsburg, N. *The Pattern of Asia*. New York, 1978.

Spencer, J.E. *East by South*. New York, 1971.

Warden, R.L. *Mongolia: A Country Study*. Washington, D. C., 1991.

23

East Asia: Japan

The island nation of Japan lies off the east coast of Asia mainland at a strategic location close to Russia, China and South Korea on the western rim of the Pacific Ocean. The country consists of nearly 3,400 islands that stretch some 1,200 miles (1,931 km) from northeast to southwest. The four larger ones: Hokkaido, Honshu, Kyushu and Shikoku make up 97 percent of the country. To the west, the Sea of Japan separates it from the eastern shores of Russia and North and South Korea. To the north, the Soviet island of Sakhalin lies barely 26 miles away, and to the northeast the Russian-controlled Kuril island is less than 10 miles, both these territories have historically been Japanese possessions or claimed by Japan. To the east and south lies the Pacific Ocean, the gateway to the United States, and to the southwest is East China Sea and on the mainland the People's Republic of China.

The physical proximity of Russia, China and Koreas is of great strategic significance and important in understanding Japan's political geography. Japan's location along the shortest sea route between eastern Asia and western North America, particularly in regard to the United States is also significant. Japan is not just far to the west of North America, but lies close to the main great circle route between the continents of Asia and North America. The Pacific Ocean acts as a great commercial link between the two continents; most of the world's major ocean routes pass through Japan. Thus location has imparted commercial advantage to Japan, for here meet two of the world's great trade routes; one originating in Europe, crossing the Indian Ocean to Singapore, moving north past Hong Kong and Shanghai to Kobe and Yokohama, and the other starting from North America going westward via Japan to China and the Philippines.

In political and military terms, the islands provide a major outpost of the non-Communist world on the eastern margin of Asia restricting Chinese, and Russian access to the Pacific. On the other hand, if Japan were to become allied with Chinese or Russian interests, it could act as a barrier to western encroachments into eastern Asia. The islands have the advantage of accessibility and at the same time of relative security. This has allowed the outside ideas to penetrate easily—from China in ancient times and from the West after Japan's self-imposed isolation during modern times. The insular position has brought safety. Not until the World War II in 1945, any invader ever landed on Japanese shores. The Mongols under Kublai Khan tried in 1281, but a typhoon thwarted that attempt.

Japan is a "miracle" of modern times. With astonishing speed during the last half a century or so it has emerged as an economic super-power, an industrial giant, and one of the most thriving and vibrant nations. Its presence is being increasingly felt throughout the world as we move into the 21st century, for there is hardly anyone in contemporary world who remains untouched by decisions made in Tokyo. The Japanese manufactures flood the world markets with technologically advanced equipment ranging from optical goods and cameras to microwave ovens, VCRs, stereo systems, cars, huge oceangoing vessels----materials that reflect the precision, skills, and innovative designs and functional ease---characteristic of the Japanese society. Capitalizing on its economic might, it has become the only major counterbalance to the United States and Europe in a post-Soviet world.

It would be, however, inappropriate to consider Japan as a mere economic giant. The so-called "miracle" of Japan's rise to great power status is indeed astonishing, but it is no miracle; it is based on a tradition of impressive achievements, grounded in extreme devotion to artistic and cultural values, combined with a historic sense of national mission and identity. However, the fierce Japanese sense of nationalism has been preserved not only by the retention of historic values but also by selective adaptability, and an eager searching for the kind of change based on direct western models that the Japanese considered essential for the preservation of their independence. The Japanese were ready to reject only those aspects of their past that they thought interfered with newly acquired changes. Their objective was the preservation of Japan, and not its Westernization. Though they borrowed heavily from western models, and techniques in pursuit of economic prosperity, they specifically safeguarded Japan and preserved its identity. Preservation of centuries-old Japanese art, literature, sculpture, architecture, ceramics, and the careful humanization of the land has given a distinctive quality to the Japanese culture.

Cultural and Historical Background

Ethnically, the Japanese are akin to the peoples of eastern Asia, the Chinese and the Koreans. The original natives, the indigenous Ainus have nearly been assimilated into the general population who in their pure form are on the verge of extinction, although small groups survive in Hokkaido. Koreans form a minuscule minority for less than 1 percent of the population most are descendants of those who migrated to Japan in the first half of the 20th century, when Korea was a Japanese colony. Japan, thus, contains one of the most cohesive ethnic societies.

Japanese, the national language belongs to the Altaic linguistic group and is largely akin to Korean and Chinese. Until the 4th century it had no written form, when the system of writing in characters was borrowed from China, and by 9th century redesigned retaining the Chinese ideographic forms. Like Chinese, the language is highly complex, containing some 50,000 characters, involving 40 brushstrokes, which has earned it the notoriety of being the "devil's language." Its standardized and somewhat simplified form was established through the creation of a national educational system in the late 19th century. The unique cultural identity of the Japanese people is based primarily on the fact that the Japanese islands have been inhabited by the same racial and linguistic group, with virtually no admixture from outside, for nearly 1,500 years. Although

Japanese culture contains many similarities to the culture of China, some of which trace back to pre-Chrisitan era, the Japanese assimilated and transformed the Chinese cultural borrowing into a uniquely Japanese mold.

Regarding the religious inclinations of the Japanese, Shinto, the indigenous religion, various sects of Buddhism, and Christianity together with a number of "new religions" that have emerged since the 19th century coexist in relative peace. Shintoism and related religions claim their adherents of about 40 percent of the population. Practicing Buddhists form another 38 percent and Christians nearly 4 percent. Most of 18 percent of the remaining population practices an eclectic form of several religions or are non-believers in formalized religions. Most of Japanese are, however, affected by the practices of a number of religious beliefs. For example, in a family a person may believe in several Shinto gods and at the same time belong to a Buddhist sect.

Shintoism is a polytheistic religion, and its followers worship natural objects, major historical or mythical figures, or Hindu gods and Chinese spirits in a uniquely Japanized form. Each rural settlement has at least one shrine of its own. After the Meiji Restoration in 1868 Shinto became a state-supported religion, but this institution was abolished after World War II. Buddhism arrived from Korea in the mid 6th century. By the next century contacts with China had been firmly established, Buddhism also traveled from there reinforcing the prevailing Buddhist hold. Several Buddhist monasteries were built throughout the country. A century later (8th century) it had acquired the status of a national religion. From China too, came Zen Buddhism in the 12th century, which has since claimed a large following. Christianity was brought in the 16th century by the Roman Catholic missionaries and was initially well received both as a religion and as a symbol of European culture. The "new religions" were formed after the mid 19th century. Most have their roots in Shintoism, but have also been influenced by Buddhism, neo-Confucianism, and Christianity, and some like the Soka Gakkai ("Value Creation Society") have powerful political organizations.

The aboriginal Ainu population is proto-caucasoid in origin, which originally arrived from the mainland in the southwestern parts of present-day Japan probably during the postglacial period of earth's history (even before the present territorial configuration of the country had taken shape). The earliest man may well have been able to walk overland from the mainland. Gradually the Ainus partly contributed to the racial composition of the modern Japanese. From the mainland the Chinese later on carried their culture into Korea and brought it in southern Japan whose peoples were themselves in contact with Korea by the 3rd century B.C.

By the mid 6th century Buddhism had been officially introduced and the Chinese culture had formally made inroads into Japan. The Japanese leaders dispatched commissions to the court of China, as there was a growing realization that their knowledge was incomplete, and they should study all possible aspects of the Chinese culture. Thus began a program of "modernization" that would in due course permeate throughout southern Japan. All types of subjects were studied----ranging from Buddhist theology to city planning, from painting to manufacturing. This enormous task of culture transfer was integrated into a distinctive, evolving Japanese milieu, and blended with the elements of native existing culture.

The centuries between 900 and 1400 A.D. were the most important periods of cultural activity transformation, and maturity in Japanese history. Although the Japanese culture was modeled upon that of China, but slavish copying was never permitted. For example, the Japanese religion grew into a mixture of Buddhism and native animism. The Chinese ideographic writing underwent substantial transformation as to be distinctively Japanese. The house types using movable panel walls, designed to let in light and sun, clean floors and mats were clearly Japanese innovations blended into Chinese housing styles. The Japanese agriculture developed its own methods, tools and customs retaining the basic Chinese modes.

All through these times the leaders of Japan had carefully discarded those elements of Chinese culture that would have minimized their control over the various clans, but retained and modified the elements of the utilitarian arts, agriculture and settlements that they had learned from the Chinese. They were also anxious that the early patterns of social stratification remain strong. Thus a sharp social cleavage between the masses and the small group of leaders, holders of wealth and large owners of land eventually developed. In China, on the other hand, a centralized system had historically been the pattern. In Japan provincial administration on the other hand, remained powerful, the emperor remained a figurehead. This situation often led to a competitive struggle between the leading classes. The struggles for power between the leaders of class in turn were responsible for the development of a professional soldiery, which gradually became the **samurai** class, enjoying social and legal status, and deriving income from taxation of the peasantry. The military tradition of the soldiery in modern Japanese history prior to and including WWII stems largely from Japan's military tradition.

Throughout the feudal era, the peasant class which had been relegated to a lower social status in comparison to the **samurai** class had been protesting for better treatment. Thus, internal stresses had been gathering in the society. The arrival of the occidental culture through trade or contacts with the outside world was perceived as a threat to the society which the Japanese thought would aggravate the internal stresses. For example, the arrival of Christianity was considered a threat as it carried allegiance to a God who lay beyond the sea outside Japan. Therefore, the Japanese decided to close Japan to the outside world. The Tokugawa rulers shut off the Japanese ports to Asian and European traders, the Spanish and Portuguese were expelled, and Japan sealed itself in a hermit-like state. The significant political result of this was that Japan escaped the fate of colonial status, so abhorrent to the several Asian nations. The net outcome of this policy of a long period of hermitage was to solidify the Japanese culture and gave the nation a distinctive identity.

Commodore Perry arrived in 1858, and called upon the Japanese to end their two and a half centuries of isolation. Although earlier similar overtures had been made and overturned, the American overture was made through a show of force. Reluctantly Japan started a new period in its history. Within a few years trade

Map 23.1

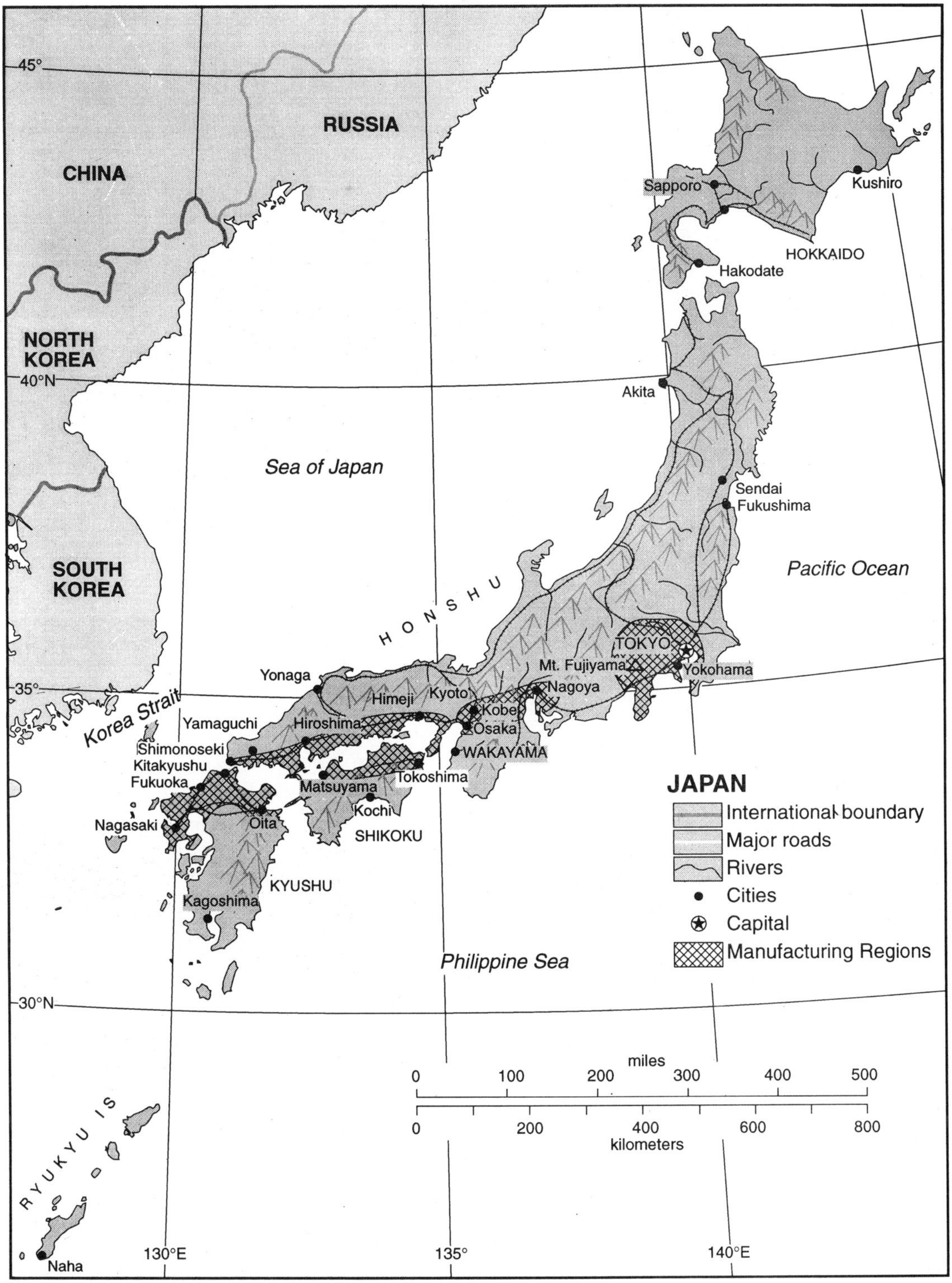

RUSSIA
CHINA
NORTH KOREA
SOUTH KOREA
45°
40°N
35°
30°N
130°E
135°
140°E
Sea of Japan
Pacific Ocean
Philippine Sea
Korea Strait
HONSHU
HOKKAIDO
SHIKOKU
KYUSHU
RYUKYU IS
Sapporo
Kushiro
Hakodate
Akita
Sendai
Fukushima
TOKYO
Yokohama
Mt. Fujiyama
Nagoya
Yonaga
Kyoto
Himeji
Kobe
Osaka
WAKAYAMA
Hiroshima
Yamaguchi
Shimonoseki
Kitakyushu
Fukuoka
Nagasaki
Oita
Matsuyama
Tokoshima
Kochi
Kagoshima
Naha
JAPAN
International boundary
Major roads
Rivers
Cities
Capital
Manufacturing Regions
miles
0 100 200 300 400 500
0 200 400 600 800
kilometers

agreements with several occidental countries were signed heralding a new phase of modernization. The abolition of landed feudalism and renovation of administrative system was the first step in the modernization process. Railway lines were constructed in the 1870s. and focus in economy was diverted from agriculture to industrialization and trade.

With the beginning of the Meiji rule in 1868, the country took a new direction in economic and military policies. In 1872 compulsory elementary education was instituted. Soon after, the old privileged warrior class of *samurais* was eliminated, and a reformed bicameral parliament *Diet* was set up. Realizing that Japanese culture was out of date compared to the foreign trade, and industrial and military might of the West, the new leaders started developing contacts with the West by sending students to study abroad to learn the western sciences, technology, and cultures. Because many of these new leaders were derived from the old military class, their immediate concern was the building of a strong military force that could deter the colonial imperialism of the occident.

Partly as a consequence of the burgeoning population and partly due to the military tradition of the society Japan embarked upon a period of conquests between 1894 and 1931. It quickly gained control of Taiwan, the Pescadores, and the Liaodang Peninsula (China) in 1894-95; Korea in 1910; and in 1931 Manchuria. The Japanese invasion of China came in 1937 which alienated the western democracies. In 1941 Japan launched a surprise attack on Pearl Harbor, and in 1942-43 the Japanese forces pressed their conquest of Southeast Asia. Finally, after atom bombs were dropped on Hiroshima and Nagasaki, it surrendered to the U.S. forces. Horrified at the massive destruction that the American bombs had caused, the nation decided to forsake the military tradition. Even before the World War II the Japanese had shown great appreciation for western industrial technology and started to undertake a thorough-going industrialization program. For nearly half a century the Japanese government sent commissions, observers, and students to the west to study the material economy of the west in a concerted effort to bring Japan up to date as quickly as possible. The pace quickened during peaceful times since the end of World War II. Thus, the industrial revolution arrived in Japan during the last 50 years.

Modern Japan has now become a unique blend of the ancient oriental and modern occidental cultures in all their forms. The American dime stores full of artifacts labeled "Made in Japan", the American style farmstead in Hokkaido, the cultured pearl, the Japanese Army, Mitsubishi, Toyota and Honda automobile companies, the high-tech cameras and photo-equipment material, the religious shrine and tourist centers are the expressions of the industrial modernization of Japan harmoniously coexisting with the guarded, time-honored Japanese cultural traditions.

Physical Framework Physically, the two overwhelming aspects of Japan are its mountains and its insularity. Land and water are so close to each other that one is always within sight of mountains and seas. The Japanese word for landscape or scenery is **sansui** from the two characters for "mountain" and "water". Two-thirds of the country is mountainous, and less than one-fourth is level enough for agriculture, and only 12.5 percent of the total land surface is under cultivation. The mountains are for the most part fairly "young", in an early stage of dissection in which steep slopes predominate, and there is a dense network of steep-sided river valleys.

Within the rugged mountainous framework there are 200 volcanoes, one-third of which have been active within historic times. Fujiyama is the most famous of the active peaks, although it has not erupted since the early 18th century. It is 12,388 feet (3,776 meters) high, and has a symmetrical cone which reaches 37 degrees toward the summit. Many of the Japanese volcanoes are conical in shape and contain *calderas* or circular-shaped depressions occupied by lakes. Several volcanoes are active and given to frequent, and violent eruptions. More than 1,000 hot springs in the country are associated to volcanic intrusions, although some owe their temperatures to friction along fault zones. Health resorts are developed around several hot springs.

Earthquakes are common throughout the country which receive about 1,500 shocks of varying intensity a year. The destructive ones occur once in three years on the average, affecting only smaller areas.

Much of Japan's coastline is indented, more on the Pacific side than on the Sea of Japan. Much of the Pacific coast thus contains embayments and coves, providing a number of harbors. The Inland Sea between the southwestern coast of Honshu Island and the northern coast of Kyushu and Shikoku, itself may be considered one large harbor containing several lesser harbors lining its shores. The coasts of southern Honshu and western Kyushu are dotted with natural harbors on which hundreds of fishing villages are located. The coastline along the Sea of Japan is relatively smooth with fewer areas of natural harbors. The coasts of Hokkaido are also mostly regular.

The mountains are divided into many small land blocks, separated by generally small lowlands and deep valleys; there is no long or continuous mountain range. These land blocks have resulted from intense faulting (movement of adjacent rock masses along a fracture) and warping (folding). Most streams are short, swift, and torrential, and are in their youthful stage. They have not had time to erode very far to build up extensive flood plains or deltas, but when the swift-flowing streams reach the sea, they carry with them fine material that might otherwise be deposited on the land.

The country has no large plains. Level land is limited to a few alluvial flood plains and deltas, and the dissected terraces of streams. Most of the flat land is located around the coast, and is broken up by local mountain barriers into isolated plains. The four main islands contain three dozen areas large enough to be identified as lowlands. The largest of these, and by far the most important in economic terms, is the Kanto Plain with an area of 2,500 sq miles (6,475 sq km) in eastern Honshu. It accommodates a quarter of the population of Japan and two of its largest cities, Tokyo and Yokohama. Other plains to the west of the Kanto include the Nobi Plain in east central Honshu, and the Kinki Plain at the eastern end of the Inland Sea. To the north the larger plains are located around Sendai (northeastern Honshu) and on Hokkaido the Ishikari lowland. It is noteworthy that each of these major plains is located on the Pacific side of the country; by comparison, larger plains are few in number on the regular coast along the Japan Sea, the most prominent being the Niigata Plain.

A good deal of Japan's surface is covered by volcanic ash and lava, and river alluvium, and hence the soils of many of the coastal plains as well as the uplands, are derived from this source. Unfortunately most soils tend to be thin, immature, and are not very fertile, except those of the thicker alluvial deposits on the larger few lowlands identified above. Their light porous structure also makes them to be vulnerable to serious erosion, and makes irrigation more difficult.

Climate In general, the climate of Japan is distinguished by its humidity and marked seasonal changes, and is mostly free from the disastrous fluctuations characteristic of India or northern China. For most part the entire country receives ample rainfall and generally enjoys a favorable climate. No part of the country is too cold for agriculture except the northernmost part in Hokkaido which is climatically marginal for cultivation primarily reflecting its very cold winters where the frost-free season lasts 140 days as compared to 300 days in most of southern Japan. Japan is warmer than parts of northern China to the west in the same latitudes. It is located off the east coast of a large landmass, and continental influences are tempered by marine conditions.

Broadly speaking, Japan's climate is monsoonal and governed by seasonal winds. The main influences affecting its climate are the country's latitudinal extent, the surrounding oceans, its proximity to the Asian landmass while local variations are related to variations in altitude and aspect, and to the distribution of ocean currents in the waters surrounding Japan. The broad features of Japan's climate are consistent from year to year, but there can be marked variations over comparatively short distances.

The key climatic determinant is the interaction between the sea and the land, resulting in the monsoonal pattern of air streams. In winter the country is under the influence of the predominantly westerly and northwesterly winds that blow from the northern interior of Asia, and bring cold, dry, polar continental air over the archipelago. Thus, the Japanese winters are comparatively much drier than the summers. Precipitation in Osaka, for example, is three times in the three summer months than in the winter. No part of Japan is dry, however, and there is no true dry season. The northwestern monsoon winds that traverse the Japan Sea before reaching the islands pick up moisture and bring precipitation in the form of rain or snowfall to the area facing the Sea of Japan.

In summer, there is a reversal of wind direction. They are southeasterly and bring warm, moist air from the adjoining Pacific Ocean. The summer rainy season brings copious precipitation. Although the precipitation varies with topography, but most of the country receives more than 40 inches (1,020 millimeters) annually mainly during the summer. The smallest amounts of precipitation occur on eastern Hokkaido (30 to 40 inches) which is near the northern limits of the maritime air; the eastern Inland Sea and north-central highlands of Honshu, both areas are protected in all directions by coast mountains.

Japan's insularity also markedly affects the climatic pattern. The southern and southeastern coasts are washed by a warm

Map 23.2

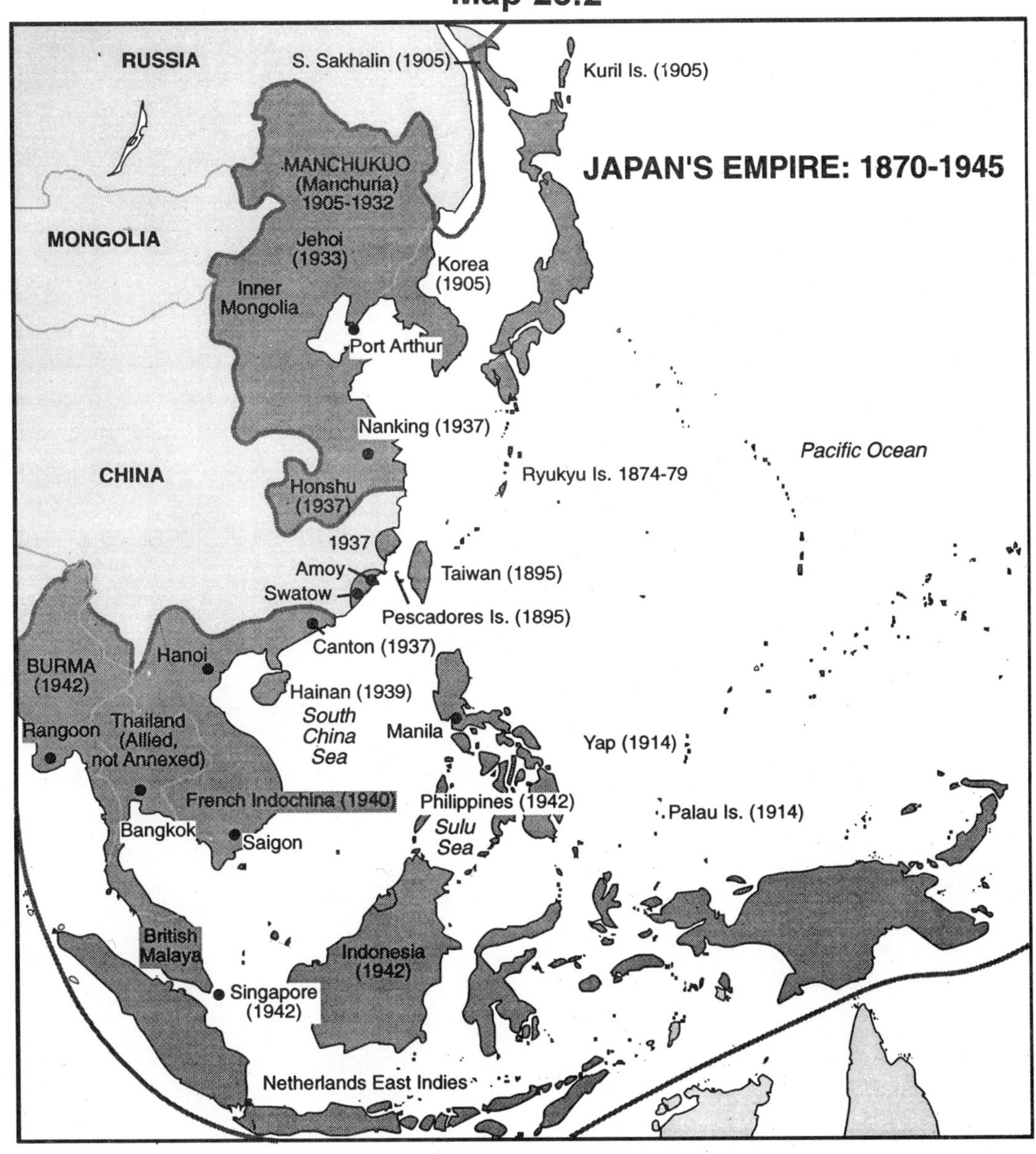
RUSSIA
S. Sakhalin (1905)
Kuril Is. (1905)
JAPAN'S EMPIRE: 1870-1945
MANCHUKUO
(Manchuria)
1905-1932
MONGOLIA
Jehoi
(1933)
Korea
(1905)
Inner
Mongolia
Port Arthur
Nanking (1937)
Pacific Ocean
CHINA
Ryukyu Is. 1874-79
Honshu
(1937)
1937
Amoy
Taiwan (1895)
Swatow
Pescadores Is. (1895)
Canton (1937)
BURMA
(1942)
Hanoi
Hainan (1939)
Rangoon
Thailand
(Allied,
not Annexed)
South
China
Sea
Manila
Yap (1914)
French Indochina (1940)
Philippines (1942)
Palau Is. (1914)
Bangkok
Saigon
Sulu
Sea
British
Malaya
Indonesia
(1942)
Singapore
(1942)
Netherlands East Indies

Kuroshio ocean current moving northeastward from the South China Sea to as far north as latitude 35° which has an average minimum winter temperature of 68°F (20°C), and warms the coast of southern Kyushu, Shikoku, and Honshu to about the latitude of Tokyo. A branch of the Kuroshio, the Tsushima current, moves westward along the Japan Sea, but its influence is less than that of the parent stream. Conversely, a cold current, known as the Oyashio, moves southward from the Bering Sea, cooling the northeastern coast of Honshu until it meets the waters of the Kuroshio, causing dense fogs in summers, particularly off Hokkaido, as noted earlier, the Sea of Japan acts an important climatic influence, especially in winter. The Sea of Japan side of Japan, therefore, is an area of copious precipitation all the year round, but with a marked winter, rather than a summer maximum.

Another characteristic feature of Japan's climate is the occurrence of violent storms or typhoons that invade southern sections of the country, averaging three to four each year during summer, especially in late August and September. These violent storms originate throughout the year in tropical latitudes of the Pacific, but by August their tracks swing to the south of the Japanese archipelago, striking the west coast of Kyushu, southern coast of Shikoku and southern part of the Kanto Plain. Often high winds and heavy downpours associated with the storms cause enormous damage whenever they strike land.

As regards temperature conditions, clearly the major determinants are the latitudes, and the local relief within the country. Obviously the southern parts of the country are generally warmer than the northern ones. The transitional seasons of spring and fall are shorter in the north. At Asahikawa in the central part of the northern island of Hokkaido the mean temperature in January, the coldest month, is 16°F (--9°C) and in August, the hottest month is 70°F (21°C), with an average temperature of 43°F (6°C). At Tokyo the mean temperature in January is 39°F (4°C), the mean temperature in August 81°F (27°C) and the average 59°F (15°C). The warmest temperatures occur on Kyushu, and the southern islands. At Kagoshima, the mean for January is 45°F (7°C), and 81°F (27°C) for August, and the average is 63°F (17°C). Variations in temperatures (and climate in general) are also caused by the disposition of local relief features. In general, the mountainous backbone of the country interrupts air flow from northwest and southeast that cause heavy snows during winter along the Sea of Japan. The western mountain slopes of Honshu also act as a major topographic barrier for the cold, winter winds they emanate from the Asian landmass, after these have crossed the Sea of Japan, protecting the eastern coast of Honshu from the cold winds from the west.

Vegetation and Forests Vegetation in Japan reflects the major climatic pattern, particularly in the latitudinal and altitudinal variations in temperatures. Since most of the country receives abundant precipitation, there are few places where natural grassland may be found. Forests occupy nearly two-thirds of the land surface and meadows and pastures a little less than 2 percent of the land. However, the original vegetation cover, has been extensively modified through centuries of human activity. Bulk of the present forest cover is either secondary growth which has been intensively used. Since population has grown rapidly since the 1870s, the forests have been utilized on a large scale. More trees have been cut down than were replaced by natural growth. This is because Japan has traditionally used wood for building houses. Although modern buildings have increasingly used concrete, steel, and synthetic plywood, Japanese paper, furniture, and other industries consume enormous quantities of wood.

In many areas the original forest cover has been almost entirely removed, and replaced by scrubby second growth in the form of wild grass and brush known as *genya*. Japan has been called a "wood civilization," and Japan must import large quantities of timber and timber products despite the fact that large amounts of roundwood is produced (nearly 30 million cubic feet a year) domestically in order to meet the large demand at home which is estimated to be over two and a half billion cubic feet a year. Japan is unable to satisfy their wood reeds, especially for its pulp and paper industry.

Although creditable progress has been made in reforestation and forest management, present exploitation exceeds the domestic demand. Nearly half of the forests are in public domain where modern forest management practices are in use, the remaining half of the private forests utilize practices of clean-cut timber without proper thinning.

The forests of Japan may be grouped into three main categories. First, boreal forests cover the summits of the higher mountains in northern Honshu and the lowlands in eastern Hokkaido; the predominant trees in these forests are spruce, fir, and birch. Second, mixed forests, largely deciduous that cover most of central Honshu down to the Inland Sea region. Oak, beech, and maple are intermixed with firs, pine and cedar (Japanese **sugi**) appears at higher altitudes. This subtropical category covers lower elevations as far north as central Honshu near latitude 37°N. The camphor tree is one of the many subtropical species found in southern Honshu and Kyushu and Shikoku islands at lower elevations.

Population and Urbanization Pattern

Japan has a population of a little over 126.7 million, making it the seventh most populous nation after China, India, Russia, the United States, Indonesia and Brazil. The overall population density that approaches one thousand persons per sq mile is exceeded by only a few countries. In terms of agricultural density---that is in terms of its cultivated area---it is nearly four and a half times that of the overall density figure, that necessitates the need for a part of the food supply to be derived from the fishery products, and supplemented by grain and food imports.

The density figures, however, mask the fact that a large part of the Japanese population resides on a very small area of land within a small country, particularly in the three great metropolitan regions on Tokyo-Yokohama, Nagoya, and Osaka-Kobe-Kyoto, in which a little less than half of Japan's entire population resides. In striking contrast to the urban congestion, almost one-third of Japan has a population density of one person per sq mile. Most of this includes areas of steep slopes, volcanic rocks, or unusable soils.

Traditionally, Japanese lived in clustered villages, and only few in dispersed settlements. Only since the 1950s when Japan shifted her cultural focus from agriculture to industrialization there has been a massive redistribution of population from the rural areas, which peaked in the late 1960s and stabilized in the 1970s and 1980s. The process drained the rural areas of young people in their productive and most fertile age groups. In the late 1960s the urban share of the population was close to fifty percent.

In general, Japan has experienced spectacular growth since 1868, reflecting the steadily increasing pace of urbanization. Population numbered close to 33 million, and urbanization was, in all probability, less than 30 percent around 1870. An official count of the population in 1920 yielded the figure of 57 million. Between 1920 and 1939, the country's major industrial centers grew substantially and both total population and its urban component gained in massive numbers. During the WWII there was a temporary but marked migration to the rural areas to avoid aerial bombing. But since 1945 there has been a renewal of the population expansion along with the rapidly growing pace of urbanization.

Japan is one of the most urbanized countries in the world and an estimated 78 percent of her population resides in cities or their suburbs. The largest urban area is Tokyo, which with its port of Yokohama and adjacent metro area contains nearly 15 million inhabitants to make it probably the biggest urban sprawl in the world. Another ten cities exceed one million residents. Most of the larger cities are coastally industrial and commercial centers connected to international sources of supply and markets by modern port facilities. Because of the rapid growth of manufacturing since Second World War, cities face congestion and pollution, but economic development that accompanied Japanese urbanization has brought material prosperity and the pull of economic opportunities in the cities keep pumping more people into the cities.

A striking feature in the modern growth of the Japanese population has been the decline of birth and death rates particularly since World War II. Largely in response to improved birth control measures and health conditions. In 1950 only 5 percent of the Japanese population was age 65 or over. At the present time, 14 percent of Japan's population or close to 17.5 million people are over age 65 years and the figure is likely to trend upward. By the year 2000 it may reach 23 percent. The rapidly aging of Japan's population is obviously fraught with grave social and economic consequences. A major concern it faces in the near future is a shortage of population in the working age groups. In addition, its top-heavy age structure will inevitably create enormous burdens in pension, and medical costs for its future taxpayers. Low birth (and death) rates have resulted in low population increases, the net reproduction rate is now below one (i.e., fewer than one female child born per woman during her childbearing years), the long-term prospect is for continued decline. Not enough children are born in the country or likely to be born in the foreseeable future to ensure the long-term stability of the population. The prospect of a declining pool of future taxpayers for sustaining the aging population is thus a major concern for the country.

By the mid 20th century, Japan faced problems of rural overpopulation. It had lost its overseas empire, built with astonishing speed, and had failed to provide an adequate base of supplies or a rich market for its products. It thus turned to industry utilizing and honing the manufacturing techniques and skills it had painstakingly learned from the West. The rapid rehabilitation of industry which was sheltered during the WWI resulted in the concentration and growth of population in the nation's principal urban and industrial districts, which lie mostly in the Pacific coastal areas and around the Inland Sea. Consequent to this urbanization trend has been the marked decline of the rural areas. Most of the adult males migrate to the Pacific coast, and many of those who remain at home periodically leave as temporary laborers in the cities, leaving behind residual populations consisting primarily of the elderly.

During the 1950s Japan's leadership also took steps to make possible the limitation of population. Legalized abortion and other measures of control, some of which appear harsh by contemporary standards, including the practice of *mabiki* (or thinning out), a euphemism for female infanticide, produced annual declines in the birthrates. It was quite clear that Japan could control her rate of population growth. It was, undoubtedly, made possible with the demographic consequences of urbanization and modernization. The real demographic challenge that Japan now faces is its burgeoning welfare system because the cost to the nation of carrying such large numbers of old people appears incalculable. Already some cuts in the welfare system have been made in order to minimize the enormity of a growing problem.

Mineral Resources Mining is a relatively unimportant branch of the Japanese economy. Although endowed with an inadequate mineral base in general, Japan is fortunate in having more plentiful coal, though it is of relatively poor quality, and much of it is expensive to mine. Coal accounts for about half of all mineral production, and one-sixth of its energy requirements. There is also a marked shortage of coking coal, and Japan has to depend on both coal and oil imports to a growing extent for its industries. The two major coal producing regions are northern Kyushu and western Hokkaido. The largest developed deposits are in northwestern Kyushu near Kitakyushu, close to the sea, where it can be easily distributed to the consuming areas by water transport. Population and manufacturing are particularly concentrated around the head of the Inland Sea, where Kyushu coal is delivered at low cost. But Kyushu fields have lost much of their importance because of the gradual exhaustion of the richer seams and the poor quality of the coal mined. The Hokkaido fields also lie near the sea; its coal seams are thicker than those of Kyushu, and the use of mechanized large-scale mining methods have resulted in higher output per miner.

Japan's oil deposits are very meager; the annual domestic production of crude oil is less than 4 million barrels which amounts for a negligible fraction of the country's oil consumption. The oil-bearing belt extends from northern Honshu on Sea of Japan to Ishikari lowlands in Hokkaido. Most of the wells are shallow, and the oil is generally of poor quality. Natural gas is produced in Honshu; a major area is in the Kanto region. Japan controlled the southern half of Sakhalin Island from 1905 to 1945 and developed and produced petroleum for its domestic market, but the supply was shut off when Sakhalin returned to Russia at the end of World War II. In order to feed a variety of its modern industrial and transport operations, Japan has to import oil which has steeply risen especially since 1950. Oil is increasingly used as fuel in thermal power stations. Nearly 60 percent of energy consumed in 1984 was derived from oil; the contribution of other sources such as coal, natural gas, nuclear power and hydroelectric power collectively is 40 percent.

Although Japan is one of the world's major steel makers, domestic resources and production of iron ore are insufficient; the fields are widely scattered, and the quality of ore is poor. The major field is at Kamaishi in northeastern Honshu. To meet the deficiency Japan must import iron ore from mainland Asia, (South China, Korea, Malaysia, the Philippines, India), and Australia. Most of the key lesser metals associated with the iron and steel industry are available domestically in small quantities. Manganese, nickel, tungsten, cobalt, tin, mercury, antimony, lead, bauxite reserves are all negligible or inadequate, amounting to a small fraction of the domestic demands.

Among the metallic ores, only copper is more important. Domestic production was sufficient for Japanese pre-World War II requirements. With the large-scale industrialization since 1950, domestic production is far from sufficient to meet the demand, and a good deal of it must be imported. Lead and zinc are found in association with copper. Other metallic ores mined include silver, gold, molybdenum, and titanium. The most noteworthy nonmetallic resource is sulfur, one of the basic materials needed for industry.

Because of its climate and topographic configuration Japan's hydroelectric power resources are considerable. A central mountain spine, steep slopes, ample and well-distributed rainfall, and the concentration of major markets at the foot of the mountains create an overall situation conducive to hydro development. The extensive network of small, torrential rivers can also be utilized for irrigation, but flooding is a serious problem in many parts of the country. As a result of the rugged topography, the hydroelectric potential is unevenly distributed. The variable flow of the most rivers make it necessary to build small standby thermal electric plants. Most canyons are narrow and offer little storage room. Although hydroelectric development is largely concentrated in central Honshu, and some parts of Kyushu, the distribution is well located in relation to the important industrial regions. The scattered location of the sites on a number of small rivers and mountain streams has helped to make power widely available making it possible that industrial development was not limited to a few sites where coal was cheap. Most of the installations are small, the larger ones are in central Honshu, and the smaller ones are widely distributed. Because of higher costs involved in the construction and maintenance of hydro plants, Japan has turned to oil-fired thermal or nuclear thermal plants, distributed primarily on the coast where refineries, and markets are concentrated. Hydroelectric power accounts for less than 5 percent of energy consumed in the country, as compared to nearly 60 percent obtained through oil-fired energy, and nearly 9 percent supplied by nuclear sources. Despite possession of relatively vast hydroelectric resource, three-fourths of which has actually been developed Japan, the country has turned to cheaper and more efficient sources for the production of electricity to offset the periodic shortages of power that were associated with hydropower.

Agriculture Japan by no means provides the most suitable natural conditions for agricultural development. The steep and densely wooded mountainous terrain, and generally poor soils leave only around one-eighth of the land fit for cultivation. But by most standards, Japanese agriculture can be called one of most efficient in the world particularly in terms of the use it makes of its available resources. Japanese national average yields are exceeded only in a few small favored areas in the world. And it is undoubtedly a tribute to the ingenuity and hard labor of generation of Japanese farmers that a near self-sufficiency in most food crops has been achieved despite a substantial population growth since the late 19th century.

As the rapid growth of urban factory industry during the World War I and since then offered large numbers of rural agrarian workers full time industrial work away from their villages, the proportion of the labor force employed in agriculture began to decline. On the other hand, population in general had been growing, and demand for food products had been increasing which in return accelerated the steady increase in agricultural prices. The administration, on its part, adopted a policy of providing incentives to farmers to grow more food, vegetables, produce, meat, and dairy products. This policy was particularly successful after World War II when the production of meat, dairy products, citrus fruits and a variety of vegetables increased spectacularly.

As industrialization and urbanization increased, Japanese consumers steadily switched from rice to bread, and western-style breakfasts continued to grow in popularity. Despite attaining self-sufficiency in rice, a variety of citrus fruits, and vegetables, reliance on imported wheat, beef, pork, chicken, and dairy products continued to grow. Overall national self-sufficiency in 1980 was around 70 percent. By the early 1990s, food, beverages (and tobacco) accounted for about 15 percent of Japan's total imports, and she had become one of the world's largest importers of a number of major crops. The rapid growth of industry during the post-World War II had already weaned away the workers from the villages at an unprecedented rate. By the end of 1980s, less than 10 percent of the labor force was employed in agriculture; in the early 1990s this figure fell to less than 7 percent. Agriculture's share to the nation's total output has also declined sharply during the last century—from about 40 percent in 1890 to less than 3 percent in 1990.

Japanese agriculture is land-intensive; the emphasis is on high production per unit of land. The intensity of pressure on the cultivated land is reflected in a number of agricultural practices that are best associated with its agriculture, and much of the Asian agriculture. Through the country, except in most level areas, *terracing* is practiced. Over the ages, terraces have been laboriously carved of hillsides and are supported by retaining walls of earth, brush, masonry or boulders, some of which are several feet in height. As in several countries in Southeast Asia, terracing of the sloping hillsides is particularly significant for raising wet crops such as rice. *Multiple cropping and intercropping* are also widely practiced in making the most of the limited area fit for cultivation. One-third of Japan's cultivated land is double-cropped. As in many wet rice lands in Asia, *transplantation*, a land-conserving practice is widely practiced for paddy. After a period of several weeks, when the rice plant is a foot high, the seedlings are gently pulled from the flooded seedbeds and transplanted to larger paddy fields.

Perhaps more than any other nation, a characteristic feature of Japanese agriculture is an intensive use of organic and inorganic fertilizers—such as bean and oil cake, bone meal, seaweed, and fish cake. During the last few decades chemical fertilizers have been increasingly used at an enormous scale.

Nearly 90 percent of the cropped land is devoted to food crops. The single most important crop in terms of crop area, production, value, and social significance is rice. Rice is grown on the best land, distributed primarily on the productive lowlands and the terraced lower slopes of the hillsides. Abundant rain, and a generally mild climate helps the farmers to raise two to three crops annually in Kyushu and Shikoku, although the cultivation of a single crop in a year extends to far north up to even Hokkaido where wheat gains in importance.

During the last four decades the farmers have with government support, spent heavily to mechanize rice cultivation within the limits of their small fields. Today, farmers rent or buy motorized planters that neatly set out trays of seedlings. Virtually all aspects of planting have been changed by technology. A lightweight tilling machine has replaced the ox-driven plow; herbicides have eliminated weeding; chemical fertilizers have increased production; and mechanized harvesters and vehicles haul the grain to the market. The use of early-maturing and high-yielding varieties have extended the usable growing season and widened the area of growth to regions earlier considered less suitable. Japanese crop yields are among the highest in the world—for rice it is 80-85 bushels per acre for a single harvest as compared with 50 to 60 bushels in China, 50 in the United States, and 30 or so bushels in India and in Southeast Asia. Japan is now nearly self-sufficient in rice, and even exports small quantities.

Wheat, barley, rye, oats are other food crops which collectively occupy half of the area devoted to rice. Wheat is usually grown as a subsidiary crop or between rice harvests without irrigation. Barley, rye and oats make use of higher, unirrigated or less fertile land. Like wheat their cropped area lies mostly in Hokkaido. With the introduction of western habits of diet, the consumption of bread has increased. Sweet potatoes form a big crop in the south. Vegetables and beans are generally rotated with rice or grown on field borders.

Apart from timber and forest, the two most important tree crops in Japan are tea and mulberry (used to feed silkworms). Tea gardens occupy a fraction of the cropland, occupying most of the country's unrewarding uplands. Diluvial uplands and steep, terraced hill slopes not suited for most food crops are usually given to tea growth. Shizonka, Prefacture is the nation's most important area for the tea cultivation. The crop has suffered from competition from cheaper, plantaion-grown tea in India and Sri Lanka. Japan produces roughly 4 percent of the world's tea production, and imports some.

The importance of silk as cash crop and as an important source of foreign exchange, has greatly diminished recently as a result of the development of rayon, and nylon. Japan now produces almost a sixth of world's artificial fibers including rayon and nylon. Silk culture, however, remains almost a monopoly of Japan and China and an important item of exports from both. It is concentrated in areas of dense population or uplands, or on land that can be spared. The major area of mulberry trees lies in central Honshu. The raising of silkworms until they spin their cocoons is an intensely laborious job, requiring prodigious amounts of hand labor, and has become a household undertaking. Japan exports most of silk to the United States; China is now emerging as its solid competitor.

Fishing Japan is the leading fishing country in the world; her production approximates one-sixth of all the fish caught commercially in the world. Next to rice, fish forms an essential part of Japanese diet, and is eaten in a variety of ways: salted, smoked, pickled, raw, and cooked. The national taste for seafood is well known. Virtually everything caught is consumed or used. A part of the fish catch, especially the scraps and entrails, is used as fertilizer or turned into animal feed or fish oils.

The traditional importance of fishing in Japan is derived from a combination of need and opportunity. A small, crowded country with little arable land has fortunately been blessed with substantial aquatic resource. The three main islands possess a long, indented coastline, and no part of the country is far from the sea. Fishing along the shores of Inland Sea was one the basic elements of economy since the beginning of the Christian era. As the Japanese moved northward along the coast of Honshu, fishing villages were among the early settlements. By the 17th century most of the nearest shore waters were already fished. The growth of population in the last century, and the modern techniques learned, have led to a remarkable development of the fishery industry in modern Japan with the result that the Japanese eat more fish per capita and harvest a greater variety of fish products than the people of any other nation. In 1990, the total Japanese fish catch of all types amounted to a little over 10 million metric tones, the highest in the world.

Although the fishing industry is distributed over 3,000 fishing villages distributed along all the coasts of Japan major concentrations are on the coasts of the Inland Sea and eastern coast of central Honshu. The coastline is long and very indented, which maximizes contact with the sea and provides numerous good harbors, particularly for fishing boats. A propitious convergence of cold and warm currents, especially off the coast of Hokkaido where the north-flowing warm Kuroshio Ocean current from equatorial waters and the cold south-flowing Oyashio meet and mix, yield a rich harvest of plankton and fish.

The Japanese fishery industry is essentially a marine fishery, although inland pond fishing has gained recently. Pond raising of fish started in the early 19th century and the culture of pearl industry dates only from 1915. Rivers and streams, lakes, even mandated paddy fields in season, also provide freshwater fishing. Ninety percent of freshwater fish is consumed as food. The remainder is made into fertilizer and oil.

A major portion of Japan's fish catch is derived from waters within 20 miles (34 km) of the Japanese coast, although Japanese fishing vessels have long scoured the western and northern Pacific for their catch. In fact, Japanese fishermen have been active in every ocean in the world, including the Antarctic, where they account for a large share of the world whale catch. The Second World War had serious consequences for Japanese fishing, particularly the loss of rich fishing rights off the coasts of the Kurils, Sakhalin, Korea, and Kamchatka, from which she obtained about one-sixth of her total prewar catch. Sakhalin and the Kurils were absorbed by the Soviet Union in 1945. The government of South Korea has barred the Japanese fishermen from all waters around Korea. In spite of these restrictions, the fish resources available to Japan are extensive, but as in the case of forests, they are being used up rapidly. Despite its dominant international position, Japanese industry now faces a number of serious problems, partly resulting from structural weaknesses within the industry and partly due to the restrictions imposed upon it by nations that have claimed a 200-mile economic zone in their coastal waters. Until the 1960s fish products remained an important export item that helped pay for her what it had to import. But in recent years imports of fishery products have exceeded exports.

Patterns of Manufacturing Japan's industrial vitality is one of the wonders of the modern world. She has had the highest economic growth rate of any of the industrial nations since 1960. Fully 10 percent of the world's economic output is now generated by Japan; nearly half of which is derived from manufacturing and trade. The most remarkable aspect of its post-World War II economic growth has been the rapid development of manufacturing. Over a quarter of the working population is employed in industry, and a large number of cities are devoted primarily to manufacturing. By 1970, just 15 years after the World War II, she overtook West Germany in steel production, and emerged as the third largest industrial nation and now exceeds production of steel in Latin America, Africa, and the rest of Asia combined. Japan has become a greatly feared competitor whose products are in increasing demand.

As the world's second largest producer of motor vehicles (and is a major exporter), Japan builds over half of the world's ships. She is second in petroleum refining, leads in building of television sets and radios, and is a major producer of crude steel, synthetic rubber, aluminum, sulfuric acid, plastics, cement, pulp and paper, and cotton yarn. she is second only to the United States in electrical machinery manufacture, and the production of synthetic fibers. The Japanese gross national product is among the highest in the world, and her per capita income has passed that of the United States and nearly all the large, European industrial countries. Japan's phenomenal development contrasts especially with the slow and difficult progress of industrialization and commercialization of the two larger nations of Asia, India and China. Between 1870 and 1900 Japan laid the foundation of a major industrial-commercial structure. By 1939 it was among the world's leaders in commercial and industrial production, competing with European traders in the markets of Asia, although the products initially were shoddy and of low quality. Even after the devastation of her industrial cities, and merchant marine during the Second World

War, she recovered its factory output by 1951. During the two decades between 1960 and 1980, Japan's gross national product rose at an average of nearly 7.5 percent each year, outstripping the growth rates of all other industrial nations. But the most notable feature of this remarkable economic resurgence was the improvement in quality and variety of the manufactured items which made her a fierce competitor in international trade.

Japan's most spectacular growth was in the production of automobiles, iron and steel, machinery (including robots, petrochemicals, precision equipment---notably cameras), and advanced electronic products (including computers, and telecommunication equipment). All this was accomplished despite the paucity of natural resources at home and the loss of the former Japanese possessions in Korea, Manchuria, China, and Taiwan which provided most of the necessary coal, iron, and food imports. Japan accomplished in fifty years the economic progress that took three or four times as long in Britain, West Europe and the United States.

Although an industrial latecomer, Japan had certain advantages in her ability to utilize the technological and organizational methods of the West developed through trial and error over a long period. Initially, the natural advantage was her small size and easy internal circulation conducive for movement of raw materials and goods, and for external trade in contrast to the several large developing countries such as China, India and Brazil that are struggling for industrialization and economic development. No part of Japan is far from the sea. Most of the population and production areas are on or near the coast where they can be served easily by water transport. In the early stage of industrialization, this natural circumstance was of crucial importance, and now continues to play a significant role in industries when much of the raw materials for manufacturing must be imported. A highly developed and an efficient railroad and an extensive highway system supplement the encircling sea routes. The small gauge (one meter instead of the standard 1.4 meter) railroad lines are particularly well suited to the sharp curves and grades required for Japan's mountains. A large number of relatively easily accessible and well-distributed hydro power sites also helped Japan's rapid industrialization.

Japan's location has been a valuable asset. The islands are suited to assemble, process, and distribute goods to and from most of the Pacific rim area—an area of varied resources and large populations for whom Japan is a convenient supplier of manufactured goods. Convenient sea routes put it within easy reach of more than a billion people of Asia and the Americas. These natural advantages remained untapped for several centuries during the time of Japan's insularity and isolation, but put to effective use as industrialization and commercialization turned insularity and maritime position into a valuable resource during the last five decades. Japan has a dominant shipping power in the western Pacific and competes strongly for leadership in the trans-pacific trade.

The Japanese have long demonstrated a remarkable willingness to use foreign ideas and techniques and adopt the technologies of other countries. Pottery making came from China by way of ancient Korea, as did silk growing. In the late 19th century Japanese emissaries went to

Europe and the United States to learn the fundamentals of modern science, medicine, engineering and finance. Today, Japanese devour western technical journals. The overseas offices of large Japanese companies spend so much of their time collecting information that they have been accused of operating like an industrial intelligence network. There has been nothing slavish in Japanese utilization of foreign expertise. The insularity of the islands helped to stimulate early growth of Japanese nationalism. Whatever has been borrowed has been transformed, often improved upon, and molded into something uniquely Japanese. Borrowing from western models was specifically designed to safeguard Japan and preserve her identity. Modernization and Westernization became a national crusade. In order to stay ahead, Japanese companies have bought the rights (patents) to a good deal of foreign technology, and also have spent large sums on their own research and development.

Still another important reason for worldwide appeal for Japanese exports has been the demand for excellence in the nation's domestic marketplace. More than 85 percent of all goods made in Japan are marketed for domestic consumption. An intense competition for excellence in the production of goods for the choosy home market has been the hallmark of Japanese manufacturers.

It is also remarkable that Japan has been able to combine high quality of its goods with one of the world's highest production rates. A Japanese factory hand on the average turns out more goods per hour than his American or European counterpart. Devotion to duty partly accounts for some of this, but the final factor is the reliability and sophistication of the plants. Japanese manufacturers invest heavily in the most advanced equipment available. Most manufacturing plants have highly automated assembly lines, often producing several times the production rates of the most efficient western manufacturing units of such industries as automobile making. By using computer-controlled production system and robot equipment, Japanese manufacturing steel companies, for example, have been known to ship the finished steel priced 20 percent less than their foreign competitors. High labor productivity relative to other major industrial countries and the extensive use of technological innovation have indeed been important to Japan's strong industrial position in the world.

A principal reason for Japan's economic success has been the high level and rapid growth of capital investment. The Japanese over the years have been the world's most conscientious savers, hoarding away nearly one-fifth of their disposable incomes into the savings schemes offered by the government, and commercial banks. A boom in capital investment acted as a powerful engine for industrial development in the years following Second World War. Moreover, Japanese economic growth has been carefully monitored, and fostered by the policies of the government. A singularly healthy cooperation between the administration, business community and industrial workers has helped enormously the growth of Japanese industry. A highly dedicated labor force, efficient, and mostly automated plants, and cheap assembly costs for raw materials enable the Japanese to undersell their competitors, even though most of the industrial raw materials have to be imported.

The industry is highly monopolistic, despite the presence of large number of small and relatively inefficient manufacturing enterprises and most of Japanese industries are controlled by a few great families or such as Mitsui, Mitsubishi, Honda, Sumimoto, Yawata, Fuji and combines derived essentially from the old *zaibatsu* cartel that was supposedly dissolved at the end of World War II. Among the larger corporations, there have been mergers, and tie-ups resulting in huge enterprises.

After World War II the character of the Japanese industry underwent radical change not only in the quality of production but in the type of goods it produced. Japan switched from lighter industry and the production of smaller, everyday items of consumer goods, to heavy industry (from food, textiles, clothing, and other consumer goods principally for a growing domestic market to heavy industries) including chemicals, metals, heavy machinery, and transport equipment. In early 1990s transport equipment, heavy machinery, chemicals and metals accounted for nearly one-half of the total value of industrial production. Japan can now compete better with the newly emerging Asian producers on the Asian market such as China and India in heavy machine and precision goods than in textiles or light consumer goods. Textiles and consumer-goods production is more widely spread. The wartime destruction of her many factories meant that she could build newer plants geared to more profitable heavy industry by taking advantage of advanced modern technology. Japan's steel, for example, undersells American steel on the U.S. west coast despite Japan's need to import most of its ore and despite high U.S. tariffs. Japanese shipbuilding is booming and ships have become a major export. Since 1956 Japan has become the largest shipbuilding nation.

Manufacturing Belt and the Cities

Most of Japan's manufacturing activity is located in an industrial belt that closely corresponds to the highly urbanized area that stretches from the Kanto Plain and hugs the northern Kyushu along the coast, with extreme concentrations at four nodal centers: Tokyo-Yokohama, Kobe-Osaka, Nagoya, and Kitakyushu. Collectively, the four industrial nodes account for nearly 80 percent of the country's total industrial output. Very few manufacturing establishments, except for those engaged in silk weaving, rice milling, fish canning, and food processing are found elsewhere, and those are widely scattered throughout the country. Steel installations in Kamaishi on the northeast coast of Honshu and Muroran in Hokkaido, based on local supplies of ore are also the major exceptions, which lie outside the industrial belt. These secondary manufacturing districts are now growing larger.

The largest of Japan's industrial nodes centers around Tokyo-Yokohama in the Kanto Plain but includes the adjoining cities of Kawasaki, Yokosuka and Chiba. This region is the primary concentration of almost every kind of manufacturing activity, and produces close to 30 percent of Japan's total industrial output. The principal manufacturers range from iron-and-steel fabrications, to beer, to high-quality optical equipment and electronic materials. Tokyo is also the center of printing and publishing industry, and Yokohama is an important center of shipbuilding. Much of the coal for power, and heating comes by rail from the coal fields nearly 100 miles to the north, and some from Hokkaido. Electric power is also generated by the nearby hydro installations in the mountains to the west. Labor is concentrated in the cities and in the surrounding Kanto Plain and workers from considerable distances have to commute daily via a highly-developed system of electric interurban transport. The centrally located Tokyo Bay provides excellent communications by sea. The location of the national capital in **Tokyo** is responsible for business, and manufacturing advantages. With a population of nearly 15 million, metropolitan Tokyo is probably the world's largest urbanized district. Although Tokyo and Yokohama are important ports, **Yokohama** (Population: 3.3 million) has an excellent harbor, and has grown as Tokyo's shipping outpost for the trans-pacific market. Tokyo is also the major financial and banking center and the largest railroad hub. It lies in the most densely populated, and Japan's largest lowland. Excellent port facilities provide for the importation of raw materials, and the export of manufactured goods, and the dense population provides ample labor supply and a concentrated domestic market.

Second to the Kanto Plain's industrial district is the Kobe-Osaka-Kyoto manufacturing node that includes the cities of Nara, Sakai, and by extension Wakayama on the south and Himeji on the west. **Osaka** (population: 2.7 million) at the head of the Inland Sea specializes in the country's traditional products, such as cotton and synthetic textiles. It acted as the major commercial center for Japan during the Tokugawa rule (1600-1867). Central location, transport facilities and level site were the major advantages in its rise as an important manufacturing center. **Kobe** (population close to 1.5 million) is a deep-water harbor, and uses a larger volume of oversea shipping traffic. Inland **Kyoto,** with its magnificent gardens, and temples, and almost equal in size to Kobe, is the old imperial capital, a historic center of arts and crafts, and Japanese culture. It is not a modern industrial city; instead it specializes in artistic crafts, such as silk weaving, pottery, lacquer, bronze and toys. The key situation of Osaka-Kobe at the head of the Inland Sea gives both these ports a westward orientation in foreign trade and shipping, as reflected in lesser importance of transpacific trade as compared with Tokyo-Yokohama. The Osaka-Kobe-Kyoto region lacks both cheap power and raw materials but has a large skilled hinterland and a central location. Steel and machinery are made in all of these manufacturing centers, but large-scale shipbuilding is concentrated primarily in Kobe. Coal from Kyushu, and coal and iron from abroad are collected at Kobe for steel making for ships and machinery. Osaka and Kobe have become a huge urban sprawl, much like Tokyo and Yokohama.

A third major industrial node is the city and environs **of Nagoya,** in a densely populated area centered on alluvial lowland, much like Osaka and Tokyo. Nagoya is a replica of Osaka in bay-head location and is important for textiles, automobiles, and machinery. It is an old castle town and administrative center located on the main rail and road lines connecting Kyoto and Tokyo. It is relatively distant, by Japanese standards, from coal sources, but is better placed for hydro power and draws upon stations in the nearby mountains. Nagoya is much less diversified in its industries, and the value of products is nearly one-third that of the previous two industrial regions. Of the major industrial nodes, Nagoya is the only one in which textiles are the most important product, but a few other light industries, especially pottery, are also significant. During World War II it became important as an aircraft manufacturing center, and since then it has become a major automotive center for transport equipment (cars, airplanes) and oil-refining.

North Kyushu, and the adjacent sections of Honshu along the Inland Sea make up the fourth industrial region, and unlike other three, the region is not dominated by any one city, but contains several middle-sized industrial centers, chief of which are Fukuoka, Kitakyushu, Yamaguchi, Hiroshima and Shimonoseki. Kitakyushu is composed of the group of the former cities (Yawata, Moji, Kokura, Tobata and Wakamatsu). Kitakyushu, right next to coal area, contains one of Japan's largest steel plants. The region specializes in heavy metallurgical industries; other industries include cement, glass, chemicals, sugar-refining, rice milling, and rice brewing. The region is blessed with rice milling, coal fields, and cheap means of transport. Some of the coal seams are ideally placed near the sea for assembling seaborne import of iron and coking coal. Hydroelectric power is of little importance. **Kitakyushu and Shimonoseki** are important ports for mainland trade. **Nagasaki,** further down the Kyushu coast facing East China Sea, and Japan's westernmost large city was Japan's "window on the world" during the self-imposed seclusion policy of the 17th and 18th centuries. Nagasaki became later on an important fueling station and terminus for the shipping routes between Asia and Europe. Nearby, some 150 miles (254 km) to the north is Japan's large naval base. Nagasaki is still a major shipping, shipbuilding, and ship repair center. Fukuoka produces primarily consumer items. **Hiroshima,** at the eastern edge of the Kyushu-southern Honshu industrial part of the region, is a center of diversified heavy industry. It was completely destroyed by the first atomic bomb, and has now rebuilt its heavy manufacturing complex. Since the core of the region lies at the western maritime entrance to Japan, it has become the center for considerable import trade in food stuffs and raw materials. Fukuoka, Hiroshima, Kitakyushu each has over one million population.

Very few of Japan's major industries are located outside the four industrial districts. A steel plant in Muroran on the northern island of Hokkaido, formerly used imported ores from Korea, and now gets most of its coal supply from the nearby fields but it is relatively far from the Japanese main markets.

Railroads play an extremely important role in transport, primarily in the carrying of passengers despite a stiff competition from road and air transport. Starting in 1891 using electricity from the nation's first power station, Japan rapidly expanded its intraurban and suburban railroad systems, particularly during the two world wars. The first subway was built in Tokyo's downtown district in 1927. Besides Tokyo subway, construction also progressed in major cities such as Osaka, Nagoya, Sapporo, Yokohama, Fukuoka, Kyoto, and Kobe since then. In 1964 a rail line providing high-speed passenger service on an electrified double-track route from Tokyo and Osaka started.

Japan's railway network is the most extensive in East Asia, and except for India-Pakistanis, is the most highly developed in Asia as a whole. In the early

1990s there were nearly 24,000 miles (39,000 km) of mainline railways, a third of which were electrified. The Japanese railways are mostly of meter gauge (3 feet, 6 inches), instead of standard gauge in most of the rest of the world. This gauge is particularly suited to the sharp curves and grades required in a mountainous country, and their rolling stock is light weight. The main lines of the railroad system paralleled the shoreline, with cross-island connections from Pacific coast to Sea of Japan on the four major islands. Since the Second World War Japanese railroads have greatly increased their operations, and are increasingly facing serious traffic problems caused by congestion within the large cities. Most commuter trains are very crowded during rush hours. There are several monorails in major cities which are also used for commuting. The length of daily commuting time has become a serious problem in large cities, for example in Tokyo in some instances commuters have to spend two hours each way.

By comparison to the railroad and coastal shipping services, road development in Japan has been relatively limited. Although the road track over a million kilometers (approximately 603,000 miles) long, 70 percent of which is paved. Roads have never been fully modernized, and the construction of a nationwide network of expressways is several years away. Because of the mountainous terrain, most roads, both built-up and in rural areas, are narrow, winding, and of steep gradient; only the larger cities are served by modernized roads and streets.

During the last four decades the number of four-wheeled motor vehicles and three-wheeled trucks has increased at a phenomenal rate. Japan has a very light density of motor vehicles per unit land, a phenomenon that has considerably aggravated the traffic problems within the cities.

Japan is one of the world's premier seafaring nations and has the largest merchant marine fleet in the world, grossing over 10 million tons. Most of its vessels are huge (especially the oil tankers) and are engaged in the carriage of both Japanese and foreign cargoes over the oceans all over the world. Many carry the so-called "third-country" trade, or between two countries other than Japan. Kobe and Yokohama are the leading trading ports; the former is the outport for Osaka and Kyoto, and the latter for Tokyo. Among the other ports Nagoya, Kawasaki, Chiba, Kitakyushu and Sakai are the prominent ones. Although seafaring is Japan's ancient commercial activity, most of the ports were modernized after the nation reopened its doors to foreign trade in 1859 following a self-imposed isolation of 250 years. Prior to Second World War shipping was dominated by a handful of companies linked to the giant manufacturing companies, the so-called **zaibatsu.** The postwar ownership pattern has a greater distribution of control and ownership. Government subsidies have been greatly helping in the expansion of fleet by new construction.

Air transport was limited before Second World War, but has gained in importance considerably as Japan became industrially a premier nation. Despite competition by railways, the volume of domestic air transport has grown significantly during the last three decades. Tokyo is the nation's largest single focus of air transport, followed by Osaka, Nagoya, Sapporo, and Fukuoka. All the large metropolitan areas are connected by air routes. Because of the presence of several islands, the country is covered by a denser network of inter-island air transport system.

Foreign Trade Because the natural resource base of Japan is inadequate to support her large population, she must turn to the outside world for the means by which national livelihood could be supported. Clearly the health of her economy must depend upon foreign trade, both upon imported raw materials and upon foreign markets for its manufactured goods. Foreign trade normally accounts for about a quarter of her national income. Before the late 19th century Japan had very little foreign trade, even the domestic trade was restricted. Exports were limited to such items as raw silk and tea, silk textiles, porcelains, and some other manufactured products of cottage industry; and most imports were minerals, raw cotton, foodgrains and lumber. As industrial production began to increases exports of such items as cotton textiles, chemicals, glass, cement and simpler items of machinery also went up. A significant change in the composition of exports occurred in the late 20th century. The share of textiles and food products decreased considerably, while export of machinery and transport equipment grew dramatically, accounting for the largest component of exports. Motor vehicles, office machinery, iron and steel products, power-generating machinery, metal-working machinery accounted for some 29-30 percent of all exports in the mid 1990s. High value specialties and precision items like optical goods, electronic equipment, radios, camera, tape recorders, receivers, televisions and highly automated scientific instruments were also significant items of export. Food items, which were the leading items of imports now and accounted for 14 percent of imports. Mineral fuels, lubricants and related materials were the leading import items accounting for nearly a quarter of all imports (among those, crude petroleum products accounted for 16-17 percent of all imports).

The United States is Japan's largest single customer, normally comprising from about a quarter to a third of all Japanese trade. Japan's most important pre-Second World War trading areas were in East Asia. The close of the World War I found Japan stripped of her major pre-war possessions in Asia. Korea and Taiwan and Manchuria were no longer Japanese dependencies and her major customers; instead they were themselves industrializing and trying to compete Japan in the world markets. Japan still gets a sizable portion of her food and mineral imports from them. In the recent years trade with China, and Indonesia has grown measurably, currently account for over 5 percent each of the total value of Japanese imports. But more importantly Japan's trade with the Middle East grew enormously during the decades of the 1980s and 1990s due to oil imports. Furthermore, Japan has been able to enlarge her market range during the period that now includes North America and Europe. Germany, France and Canada are her other trading partners.

Prospects Spurred by an unprecedented expansion of industrial production and an aggressive export policy, Japan has had an extraordinarily rapid rate of economic growth in the 20th century, particularly after the Second World War. This spectacular growth has brought the country to the forefront of the world economy and made it the second largest economic power, ranking only behind the United States. Interestingly, because of her trading skills and the quality of her products, Japan now exports watches to Switzerland, beer and cameras to Germany, shoes to Italy, and automobiles to the United States.

Japan's economic future depends, however, on a number of factors. The pace of economic expansion, which had reached the highest levels of any industrialized nation in the 1960s and 1970s, slackened during the last two decades. Strong competition from the United States, the industrialized western Europe, and the newly-industrializing Asian nations exerted new pressures on Japan's economic growth. And she is discovering that markets for her goods may reach their limit with the traditional partners than her productive capacity in the decades to come. The real potential lies her making inroads into the markets of the developing countries in Asia, and Africa. Japan has already penetrated into the markets of such nations as Thailand, Indonesia, India, Vietnam and South Africa at a time when these nations themselves are trying to build their own economies, and would offer restrictions on Japan's increasing pressures.

At home, Japan faces a number of problems that would constrain her future development. Continued, and possibly increasing dependence on imports of essential items such as oil, iron ore, and food stuffs is likely to adversely affect the nation's development and social welfare programs, particularly in view of the rising standards and increased consumption of imported goods. Price-volatility has already had a negative impact on her development plans.

As Japanese cities become more crowded and polluted, environmental hazards would require stricter regulations on pollution control, and would certainly demand increased expenditures, thereby reducing the capital available for investment. Investments on the social welfare front have also been on the increase, and are likely to escalate in the coming decades as the elderly population would grow (the age-group of 65 and above was only 5 percent in 1950 and is likely to grow to 23 percent in 2025). This will create an enormous burden in pension, medical, and infra-structural costs. It is quite likely that the emphasis on technological innovations will shift away from the reliance on imported bulky natural resources and heavy manufacturing to activities requiring still greater knowledge and skills. Given Japan's sales promotion techniques and the quality of her manufacturing goods, there does exist potential for the extension of markets for the country in the newly developing nations in Asia, Africa and Latin America. In sum, Japan may possibly continue to enjoy economic prosperity well into the 21st century.

RYUKYU ISLANDS

The Ryukyuan archipelago extends 400 miles (650 km) from southern Japan to Taiwan and contains over 140 islands covering a total land area of nearly 1,291 sq miles (2078 km). Okinawa Island with an area of 485 sq miles and a population of over a million is the largest of the islands. The islands have long been inhabited, possibly since the Neolithic Age. The people appear to be descendants of Japanese and Southeast Asians who migrated to the Ryukyus during prehistoric times. There must have been some contacts with both China and Japan, and the Ryukyuans' socioeconomic systems reflect both Japanese and Chinese influences. In ancient times the islands formed an independent kingdom. By the 14th century a faction of the Ryukyuan leadership, located in Okinawa, accepted Chinese sovereignty which imitated a period of trading contact with China. This resulted in an assimilation of

Chinese culture by the Okinawans. In time, trade contacts with the Asian mainland from Korea to Vietnam included Japan on the north and Indonesia on the south, which also affected Ryukyuan culture systems. Japanese influence had been growing in the 17th century over northern and central Ryukyu Islands, and in the late 19th century. As the European contacts in the east exerted political pressures during this period, Japan assumed administrative control over the whole of the Ryukyu Islands. In 1879 the Ryukyus became an integral part of Japan.

At the end of World War II, the United States took control of the islands, and the military government was replaced in 1951 by a civil administration based in **Naha** (on Okinawa), the capital and largest city of the islands, controlled primarily by Japan. Following Japanese agitation for the return of the islands, Japan took over the Ryukyu Islands, from the United States, but the U.S. continued to maintain military installations and troops on Okinawa Island.

Less than half of the islands are permanently inhabited; many of the smaller islands are just coral reefs. The larger ones are volcanic in nature, and some are big enough to provide some agricultural land. The population is primarily rural, and agriculture is the dominant occupation of the people. Sweet potatoes and rice are the staple crops. The traditional economy was badly upset during the World War II. An artificial economy significantly affected by the American military presence has since developed. There is some pineapple production now and tuna fishing has become important. Three-fourths of the Okinawan population consists of smallholding sugarcane growers. Sugar refining and pineapple canning are the important manufacturing activities, with the products going to Japan. Tea and tobacco are raised for home consumption. The Ryukus remain, however, a food-deficit area for the local population, most of the food products are consumed by the American military installations. Okinawa has experienced a surge in urbanization since World War II. Naha's population estimated to be close to half a million has doubled since 1970. The islands' proximity to Japan, Taiwan and China reflects the strategic importance of the Ryukyus, and was an important factor in the establishment of American military bases in Okinawa. The local population, however, recently looked upon the military presence with increasing disfavor.

Further Readings

Buckley, R. *Japan Today.* Cambridge, 1985.

Friedman, D. *The Misunderstood Miracle: Industrial Development and Political Change in Japan.* Ithaca, 1988.

Kornhauser, D. *Japan.* New York, 1982.

MacDonald, D. *A Geography of Modern Japan.* Ashfork, UK, 1985.

24

Central Asia

The five nations of Central Asia—Kazakstan, Kyrgyzstan, Tajikistan, Turkmenistan and Uzbekistan---are fairly distinctive in their physical setting, cultural and historical characteristics, and levels of social and economic development, to make this region a well defined and cohesive part of the Asian mosaic. Physically, their landscapes are dominated by deserts. Settlement is chiefly in oases and along the two main rivers in the south as well as in the Piedmont zone between the mountains and the desert plain. Culturally, all are Turkic in ethnicity and language, except for the Persian Tajiks, and all, except the Slavic and other minorities identify with Islam. Politically, all of these were constituent units of the Soviet Union before its break-up in 1991. Although the Soviets pursued a policy of ethnic autonomy in the region throughout the period of their control, but exercised a constant pressure of linguistic "Russification," particularly in the urban centers and in Kazakstan (where Russians and Ukrainians comprised a substantial minority), these nations remained relatively unscathed by Russian cultural influence.

THE PHYSICAL SETTING

Covering an area of over 1.5 million sq miles (nearly 4,000 sq km) the Central Asian nations impinge on the frontiers of Iran, Afghanistan, China and Mongolia, and before 1991 formed the southernmost part of the USSR. Tucked in the middle of the world's largest continent and remote from the open waters of the world's great oceans, the region lies at the historic crossroads between Europe and Asia, and was subjected to repeated conquests from outside and contained little political integration from within.

Kazakstan is the largest among the Central Asian nations, possessing a territory larger than all other Central Asian republics combined. A third of its vast expanse consists of lowlands stretching from the Kirghiz Steppe lands in the north, and flat, Caspian-Aral Depressions in the west to hilly plateaus towards the east, and high mountains along China-Krygyz borders in the southeast where the highest points of over 20,000 feet (6250 meters) are reached. The western and southwestern parts of the republics are dominated by the low-lying Caspian Depression, south of which (and west of the Aral Sea) lies the Ustyurt Plateau. The world's largest inland body of water, the Caspian makes Kazakstan's border for 1,450 miles, 2,345 km). Other large bodies of water include Lake Balkhash in the eastern part of the country and the shrinking Aral Sea. The Aral Sea is heavily

Table 24.1 *Central Asia: Basic Data, 1999*

Country	*Area (Thousand sq km/sq mi)*	*Population (Millions)*	*GNP ($/capita)*	*Capital*	*Urban Population Percent (est.)*
Kazakstan	2724.9/1,052.1	15.3	1,350	Astana	56.0
Kyrgystan	198.5/76.6	4.7	480	Bishkek	34.7
Tajikistan	143.1/55.3	6.2	330	Dushanbe	28.3
Turkmenistan	488.1/188.5	4.7	940	Ashgabat (Ashkabad)	44.8
Uzbekistan	447.4/172.7	24.5	1,010	Tashkent	38.4

Source: Encyclopedia Britannica, Book of the Year, 2000.

contaminated by the salty dust and chemical deposits. The entire area around it is in terrible environmental decay. The northern part of the country faces Russia, and is drained by Irtysh, Ishim, and Tobol rivers which run northwards, draining to the Arctic waters. In the west the Ural River flows into the Caspian Sea. In the southern section of Kazakhstan the waters of the Syr Darya flow into the Aral Sea. In the extreme northeast, are the Tarbagatay and the Alatau ranges. The Tien Shan rises along the Kyrgyzstan border.

The largely mountainous republics of Kyrgyzstan and Tajikistan border China, and contain the highest peaks in Central Asia: Victory Peak at 24,406 feet (7,439 meters), Lenin Peak at 23,405 feet (7,134 meters) and Communism Peak at 24,590 feet (7,495 meters) in the Tien Shan and the Pamir Mountain systems. In the southwestern part of the Tien Shan is the fertile Valley of Fergana and Lake Issyk-Kul. Though less than 10 percent of Tajikistan and a large portion of Kyrgyzstan are lowlands, the fertile and productive valleys such as Fergana, Vaksh and Pauj are important for the human geography of these mountainous nations. The upper courses of Central Asia's two major rivers----the Syr Darya and the Amu Darya---drain these two republics. The entire area is prone to seismic activity occasionally frequented by severe earthquakes.

Eighty percent of the territory of Uzbekistan and Turkmenistan consists of two desert lowlands: the Kyzyl Kum in Uzbekistan and the Kara Kum in Turkmenistan. Both rank among the world's largest sand deserts. Southeast of the Aral Sea, small hills break the flatness of the Kyzyl Kum Desert in Uzbekistan whereas the western ranges of the Tien Shan Mountains spill over the eastern part of Uzbekistan rise to nearly 14,500 feet (4,300 meters). Between the mountain ranges, the heavily populated oases contain an extensive network of canals.

The two historic rivers----the Syr Darya and the Amu Darya----that drain the two republics of Uzbekistan and Turkmenistan have been depleted during the past several decades due to overuse in both agriculture and industry. As most of the waters in the delta of the Amu River and the lower course of the Syr River have dried up, the Aral Sea has been shrinking. In some places the shoreline has receded more than 75 miles (120 km). All this has created problems of salinization of the Aral. The pollution from the overuse of insecticide, and chemical fertilization for irrigation, the Syr and the Amu basins have further contaminated wide areas and

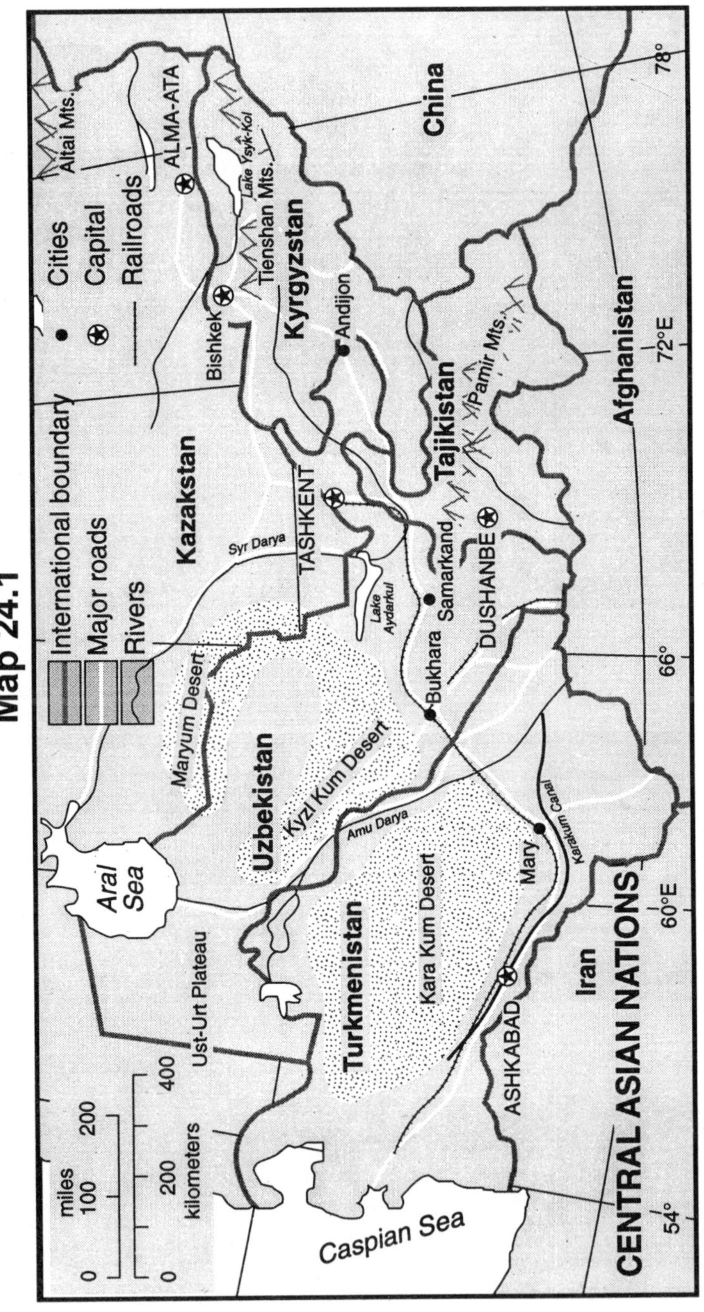
Map 24.1
International boundary
Major roads
Rivers
Cities
Capital
Railroads
miles
0
100
200
0
200
400
kilometers
Kazakstan
Uzbekistan
Turkmenistan
Kyrgyzstan
Tajikistan
Afghanistan
Iran
China
Aral Sea
Caspian Sea
Ust-Urt Plateau
Maryum Desert
Kyzl Kum Desert
Kara Kum Desert
Syr Darya
Amu Darya
Karakum Canal
Lake Aydarkul
Lake Ysyk-Kol
Altai Mts.
Tienshan Mts.
Pamir Mts.
ALMA-ATA
Bishkek
Andijon
TASHKENT
Samarkand
DUSHANBE
Bukhara
Mary
ASHKABAD
CENTRAL ASIAN NATIONS
54°
60°E
66°
72°E
78°

have had disastrous effects on the environment and on the health and livelihood of the human population around the Aral littoral.

The Kopet Mountains lie along the Turkmenistan-Iran border.

CLIMATE

The conspicuous features of Central Asia's climates are marked dryness and large temperature ranges. Generally speaking, this sharply continental climate of most parts is a reflection of its deeply interior location in the heart of the Eurasian landmass, a largely desert territory and a mountainous terrain that covers most of the Central Asian territory. The great heights of the Altai, the Tien Shan and the Pamirs in the southeast prevent moisture from entering Central Asia and the vast distances to larger and open water bodies in any other direction accentuate dryness.

Average annual precipitation rises as one moves from the dry western desert margins to the foothills of the Altai, and Tien Shan from only 3 inches (75 millimeters) annually to the south of the Aral Sea to nearly 14 inches at Tashkent that lies close to the base of the mountains in the east. Further east, precipitation is again low in the valleys which are surrounded by high mountains that block rain-bearing winds, but a little higher in the open, Piedmont area. Inadequate precipitation has led to a dependence on irrigation almost everywhere. Most of the precipitation falls in winter, occurring from the winds emanating in the Mediterranean or is *orographic* in character (i.e., produced by the presence of mountains).

Summers are hot; temperatures in south Kazakhstan reach 68°F (20°C) but rise much higher in the southern republics of Turkmenistan to 95°F (35°C). Winter temperatures are low, ranging from -2°F to 3°F (-19° to -16°C) in the north to 23° to 29°F (-5° to 1.4°C) in the south. Continental climate, however, exhibits not only great seasonal fluctuations but large daily ranges in temperatures as well. The seasonal averages thus mask an extremely wide daily range of temperatures. For example, in most areas, temperatures are seldom lower than 95°F (35°C) during summer days except in the mountains and an absolute maximum high temperature in the Kara Kum Desert may reach 122°F (50°C) in shade. By contrast, winter temperatures in the extreme south on the border with Afghanistan, drop to -27°F (-33°C). In the largely mountainous republics of Kyrgystan and Tajikistan temperatures are tempered particularly during the brief summer by altitude, the disposition of the terrain and the exposure to the sun rays.

HISTORICAL AND CULTURAL PATTERNS

The early history of Central Asia is largely obscure and only faint glimpses of its cultural patterns prior to Alexander the Great's conquests in the 4th century B.C. can be had through ritual songs and legends of the nomadic tribes. The ancient Greeks and later, the Arabs were familiar with the area and its inhabitants. The territory between the Amu Darya (the Oxus of the Greeks) and the Syr Darya (the Jaxartes of the Greeks) was known to the Arabs as the Jahun-Sayhun region which contained the sites of the famed urban centers such as Bukhara and Samarkand during the medieval times.

Islam came to the region with the Arab conquest in the mid-7th century and became firmly rooted by the 11th century when the Turks were ruling over most of Central Asia. The area from Kazakhstan to Turkmenistan then came to be known as Turkistan. Except for the Tajiks, who are the descendants of the Iranian peoples since the first-millennium B.C. and were ruled by them when the Turks had established over Central Asia, all the groups had adopted Turkic languages. Only the Tajiks remained linguistically distinct as speakers of Persian-related tongues.

The most notable event of the medieval times was the conquests of the Mongols who established an extensive empire under Genghiz Khan in the 13th century. The present-day distribution of ethnic groups were probably formed during and after the collapse of the Mongols empire.

The Russian encroachment in Central Asia started in the mid-19th century, first in the Kazakh territory and later in the area that lay south of it, gradually annexing all of Central Asia between 1840 and 1870.

The post-revolutionary period of the Soviet occupation after 1921 was marked by large-scale economic transformation of the land. Railroad links were forged with other parts of the Soviet Union; land reforms were enacted; and modernization of agriculture introduced, all accomplished with Russian technology and finances. In an exuberant burst of agricultural developments of their "Virgin Lands," however, the focus appeared misplaced. The whole of Central Asian lands were galvanized to produce primarily cotton, an emphasis that proved ecologically disastrous.

The Russian decision to propagate agriculture in the "Virgin Lands" was also accompanied by a large-scale migration of Russian farmers into these new agricultural colonies, many of them later on moved to the newly emerging urban centers. The native tribal groups meanwhile took to sedentary occupation in the agricultural lands. The indigenous tribal population became slowly a minority in the urban centers. Most Central Asian cities now contain non-local population in large numbers; the Russians obviously dominate and form disproportionately high percentages in industry, and transport sectors.

The five largest ethnic groups in Central Asia, in descending order of size, are the Uzbeks, Kazaks, Tajiks, Kyrgyz, and Turkmen, each residing within the republic that bears its name. (For example, Tajiks in Tajikistan; Uzbeks in Uzbekistan, etc.) The sixth group of the Karakalpakstans occupies a territory in the Amu Darya plain and has a subordinate political status within Uzbekistan. The incorporation of Central Asia into Fascist Russia in the late nineteenth and early 20th centuries and later the Soviet Union, and the immigration of large numbers of several groups---including Russians, Ukrainians, Jews, Germans and Koreans—gave a distinctive multi-ethnic character to the cultural geography of Central Asia.

Predictably, the Russians rank second to the indigenous groups within each country, heir proportion of the total population varying greatly—from 35 percent in Kazakstan and 16 percent in Krygyzstan to a significantly lower level of 6 percent in Uzbekistan. Prior to 1980, they formed a majority in Kazakstan, but have

Map 24.2

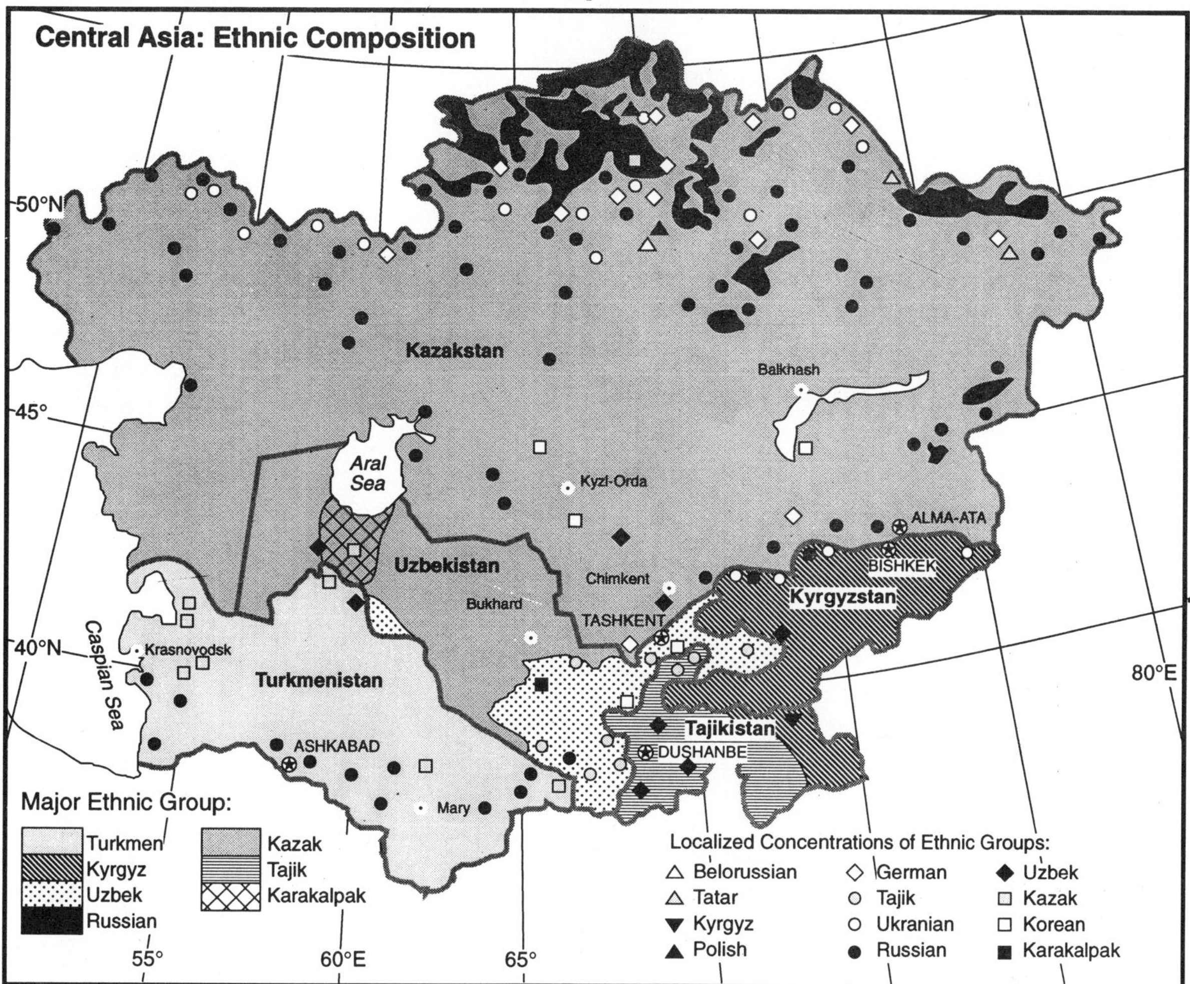

since experienced a substantial decline as did other non-indigenous groups such as the Ukrainians, Germans and Koreans. For the last three decades the indigenous groups have experienced proportionate gains in numbers, arising primarily out of their much higher rates of natural increase, but also from an increased out-migration of Russians.

All of the Central Asian nations contain substantial ethnic minorities, consisting largely of other indigenous ethnic groups reflecting their shared cultural and historic development, and also of the several non-indigenous groups, such as the Russians, Ukrainians, Germans, etc. In the case of Kazakstan, which contains the largest proportion of non-indigenous population, for example, the Russians account for over one-third of its population, the Germans, Tartars and others nearly a quarter, leaving the Kazaks in a position of a narrow majority. Similarly, nearly a quarter of Tajikistan is composed of the Uzbeks, and so closely are the Tajiks and the Uzbeks intermixed that the Soviet partition of the area in 1924 failed to segregate them. The political boundaries demarcated by the

Soviets in the mid-and-late 1920s, therefore, could not produce ethnic homogeneity.

Map 24.2 shows the ethnic composition in the five Central Asian nations.

Linguistically, all of Central Asia, with the exception of the Indo-European Tajikistan, belongs to the Turkic family, and speak languages that are closely related to Turkish. The Tajik language is related to Persian or Farsi. The language of Turkmenistan is closer to the Turkish of Turkey than other Turkic languages. The Karakalpas, living along the southern shores of the Aral Sea are linguistically closer to the Kazakhs, although culturally they share common traditions with the Uzbeks.

In view of the multi-ethnic character of the Central Asian republics, Russian, Ukrainian, and German are spoken as a second language primarily by the ethnic Russians, Ukrainians or Germans. The introduction of Russian as an "official language" during the Soviet control remained primarily a language of the administration, and of the urban elite. As Russian was made compulsory in schools, it was considered essential for advancement in workplace and was adopted by those living in large cities who acquired a working knowledge of it but a vast majority did not accept it. Recent trends indicate that the gains made through linguistic "Russification" in Central Asia may be reversed in the near future as the nationalist sentiments are likely to become more strident.

To the linguistic homogeneity, Islam provides another element of cultural affinity. All the major Central Asian nationalities identify with it. The Kazakhs, Kyrgyz, Uzbeks, and Turkmen practice the Sunni brand of Islam, as do a majority of the Tajiks. Rural communities generally identify more strongly with religion. However, in general, it is more permissive than elsewhere in the world. Ethnic Russians and Ukrainians, who live mostly in the larger urban centers, associate with Orthodox Christian churches. By and large, religion remained a weak force politically during the Soviet occupation, but is likely to grow in importance as the Central Asian nations begin to establish stronger regional connection with the neighboring Islamic countries such as Iran, Turkey, and Saudi Arabia.

ECONOMIC ACTIVITY

Central Asia is relatively rich in raw materials for agricultural and industrial production although the resources are unevenly distributed. Most industrial raw materials such as coal, iron ore, copper, lead, and zinc are concentrated in the foothills and intermountain basins in the eastern, mountainous regions. Heavy industry is mostly located in northeastern Kazakhstan, and in the Tashkent area in the foothills of the Tien Shan Mountains. Oil extraction and natural gas are confined to Fergana Valley, western Turkmenistan, and the Caspian shore areas. Hydroelectric power potential is considerable in the mountains of the Tien Shan and developed to a considerable extent. In agriculture, irrigated cotton in the plains of the Amu and Syr rivers; and a variety of food crops in the foothills of the mountains in the east and southeast are important.

A brief review of the economic and political characteristics of the Central Asian countries follows.

KAZAKSTAN

The most northerly among the Central Asian republics, Kazakstan has a long border with Russia on the west and north, and was the first to be annexed to the Soviet Union. To the south are the republics of Turkmenistan and Kyrgyzstan and in the southwest is the Caspian Sea. China lies to the east. Farming occupies some 40 percent of the population consisting largely of the Kazaks and the Slavic settlers. The Virgin and the "Idle Lands" program launched by the Soviet Union in 1953 opened up the vast grasslands of northern Kazakstan to wheat farming by the Slavic settlers. Livestock breeding is also important. The Kazakhs raise goats, sheep, cattle and poultry as their traditional occupation. Kazakhstan is a significant producer of Karakul and Astarkhan wools. The principal crops produced are wheat, barley, potatoes, vegetables, sugar beets, melons and other fruits, grown primarily in the southern part of the country.

The republic possesses abundant mineral wealth, and several mining and processing industries have been developed. There are major coal fields in the Karaganda, Turgay, and Ekibastuz basins and important deposits of iron-ore, lead, tungsten, manganese, copper, silver, nickel, and antimony exist. The Caspian Depression and the Mangyshlak Peninsula contain large reserves of oil and gas, and are considered among the world's largest unexploited oil-bearing areas. In 1992, Kazakstan contracted with Chevron Corporation to exploit these potential fuel resources. Currently country's fuel production is mainly derived from coal of the Karaganda field.

Manufacturing industry employs nearly one-fifth of the nation's labor force, composed primarily of Russian and Ukrainian settlers. It represents a significant sector of the economy, contributing a respectable 45 percent to the nation's domestic product. Parts of the country are highly industrialized. Major manufacturing industries include metal-processing, the production of such items as cement, chemical fertilizers, machines, synthetic fibers, canned vegetables and fruits, sugar, meats, and pharmaceuticals.

Railroads carry most of the freight; the two trunk lines, the Trans-Siberian, and the Kazak (Turk-Sib) traverse the country east to west. Given the country's large size, and relatively less developed roadways, air transport is employed to cover most of the passenger traffic. The republic has an extensive network of oil pipelines.

The settlements have distinct regional patterns reflecting the varied ethnic make up of the population. The Russians, Ukrainians, and Belorussians that collectively account for close to one-half the population are settled on large collective and state farms in villages of the northern plains. These populated areas are separated by the arid area to the south where the nomadic Kazak sheep breeders live in temporary quarters, known as *yurts* (the round tents with sturdy pole frames covered by heavy felt). This extremely wide dispersion of population is reflected in a large number of small settlements. The density of population is the lowest in Central Asia----nearly 6 persons to a kilometer. Fewer than 100 settlements are designated as cities or towns in a population of over 17 million.

By the late 1980s Slavic population outnumbered the indigenous Central Asian ethnic groups, although they were concentrated mostly in the urban areas.

Map 24.3

RUSSIA
RUSSIA
CHINA
UZBEKISTAN
KAZAKSTAN
miles
0 100 200 300
0 100 200 300 400 500
kilometers
50°N
Ural'sk
Ural R.
Aktyubinsk
Kustanay
Petropavlovsk
Koktchetav
Pavlodar
Irkysh R.
Semipalitinsk
Zyryanovsk
Tobol R.
Karaganda
Chu R.
Gurjev
45°
Aralsk
Baykonur
Ayaguz
Tanabagatay Mts.
Balkash
Lake Balkash
Aral Sea
Ili R.
Kzyl-Orda
Caspian Sea
Kzyl Kum Desert
Syr R.
Chu R.
ALMA-ATA
Tien Shan
International boundary
Major roads
Rivers
Cities
Capital
40°N
45°
55°
60°E
75°
80°E
85°

The Kazakhs and other non-Slavic groups make up for less than a quarter of the urban population.

Kazakstan contains only a few large urban centers. Fewer than two dozen cities have populations over 1,000,000. Chief among these are: Alma Ata (1.1 million), Karaganda (608,680), Chimkent (438,800), and Semey (344,700). With the exception of Karaganda which was founded by the Russians as a mining and manufacturing town in 1857, these urban centers are located at the ancient sites, and contain "old" sections to which were added newer and planned areas during the Soviet period.

Alma-Ata (Almaty) is the former capital, located in the east and close to China's Junggar province at the foothills of the Tien Shan Mountains. One of the most beautiful places in Central Asia, the present city was founded by the Russians in 1855 and experienced phenomenal growth during the 1930s and 1940s, primarily on account of the location of metallurgical and machine building industry near the city. Besides containing heavy industry, the city is a center of food-processing and light industries as well. Alma-Ata lies on the Turk-Sib railroad that connects it with Russia and the other republics of Central Asia. The new capital replaced Almaty in November, 1997, and was renamed Astana in 1998.

Karaganda, the major industrial center and the second largest city of Kazakhstan, lies in the major coal-mining area. It specializes in iron and steel industry, and a few other industries such as food-processing, cement, and footwear. **Chimkent**, the third largest city is located in the south in the Syr River basin, on the site of ancient caravan routes that connected medieval China with Central Asia. An important industrial, and railroad junction, its major industries are: food processing, and manufacture of chemicals, cement and automobile parts. **Semey** (formerly called Semipalatinsk) was founded as a Russian frontier fortress in 1718 near the site of medieval caravan routes from Mongolia to Russia. The city grew enormously after the completion of this Turk-Sib railroad in 1931, and became an important manufacturing center of textiles, food-processing and footwear.

As the largest and most influential of the five Central Asian republics of the former USSR, Kazakstan is likely to play an important role in Central Asia. However, the issue of the former Soviet, and currently Russian-controlled, nuclear warheads deployed in the vicinity of the Aral Sea, making the country one of the four largest nuclear powers in the world, has been the focus of international concern. The country has joined NATO's "Partnership for Peace," and the U.S. has offered military assistance in case of need. It signed the Treaty on the Non-Proliferation of Nuclear Weapons with the USSR and the U.S.A. in 1993 and pledged to remove all nuclear weapons by 1999.

KRYGYZSTAN

Bounded by Kazakhstan on the north and northwest, by China and the east and south, and by Uzbekistan and Tajikistan to the west and south, the mountainous country of Kyrgyzstan is a much smaller nation (76,600 sq miles or 198,500 sq km in area) than its larger northern neighbor. Its eastern and southern borders run along the crests of Tien Shan and Pamirs Mountains.

The Kyrgyz are traditionally a pastoral nomadic people. Two-thirds of its estimated population of 4.7 million resides in rural areas. Agriculture provides employment to one-third of the country's labor-force and contributed nearly 38 percent to the nation's domestic product in 1990. Less than one-third of the country's area is arable, 70 percent of which depends on irrigation. Agricultural output, especially in the Fergana Valley, has substantially increased during the last five decades through canal irrigation. The principal crops raised are: cereal grains, potatoes, vegetables, cotton, tobacco, fruits, and sugar beets. Most of the country's cotton is produced in the Fergana Valley. Mulberry trees in the Fergana Valley support the raising of silkworms. Livestock raising is an important part of agricultural activity. Sheep constitute nearly two-thirds of the livestock numbers, followed by cattle, goats and horses, raised notably in the Chu Valley and the littoral of Issyk-Kul lake. Horses are used as draft animals as well as a source of meat, and *koumiss* (mare's fermented milk) is widely used.

Kyrgyzstan has considerable mineral deposits, including coal, gold, tin, antimony, zinc, uranium, and tungsten, but unlike its neighbors, is deficient in petroleum and natural gas for which it has to rely heavily on imports from other former Soviet republics. However, hydroelectric power development is substantial which provides nearly one-third of the domestic energy requirements of the nation. The remainder is derived from coal and imported oil.

Modern industrial development took place later than in other Central Asian nations. However, by the late 20th century the republic had become a producer of machinery, electronics and instruments, textiles, and food-processing (in particular sugar-refining). Light industries are also fairly developed now utilizing local agricultural products such as meat, fruits and vegetables. Wool is an important exportable product, and mills produce cotton and silk fabrics, cloth, and leather goods.

Kyrgystan is predominantly rural; only 35 percent of the population is designated as urban, which is composed largely of the non-Kyrgyz groups. Urban development took place largely after the 1930s, but the Kyrgyz remained an urban minority **Bishkek** (formerly Frunze) is the capital, and the largest city (population 631,000). It lies on Chu River, at an altitude of 2600 feet (790 meters) in a picturesque setting, close to the Issyk-Kul lake and the Tien Shan Mountains. Most of the city's development dates since the WWII. Machine building and metal industries are important.

Several highways connect the Kyrgyzstan to the neighboring countries primarily to offset the problems of difficult terrain. Before 1924 there were only two narrow-gauge lines joining the northern parts of the country to Kazakhstan and the Turk-Sib (Turkistan-Siberian) railroad. In 1948 railroad extensions linked the capital city of Bishkek to Issyk-Kul.

Kyrgystan depended entirely on the USSR for defense and its forces were fully integrated into the Soviet armed forces. After it became independent, it started raising a national army. In 1992 it entered into a collective defense with the other Central Asian nations and with Russia a collective security treaty.

TAJIKISTAN

Smallest and the most mountainous of all the Central Asian nations, Tajikistan is bordered by Afghanistan on the south, by China on the east, by Kyrgyzstan on the north, and by Uzbekistan on the north and west. Culturally, its national identity is complicated by its ethnic ties and outside influences. Despite the political, economic and cultural links with other nations of Central Asia, the Tajiks share close ethnic relationship with the Iranian people and other Tajik groups living in northeastern Afghanistan. The Tajik and Persian languages are closely related and are mutually intelligible.

Farming is the major economic activity, with cotton growing as the leading crop produced. Raising of livestock---including long-horned cattle, Gissar sheep and goats---and the cultivation of fruits, grains (primarily wheat, rice, and barley) and vegetables are important. Horticulture has been traditionally important, and apricots, pears, apples, cherries, grapes, pomegranates, figs and nuts are widely grown. Almonds and dried apricots are the country's important exports. Only 7 percent of the land is arable, three-fourths of which is under irrigation. Agriculture in Tajikistan depends to a large extent on irrigation, developed during the 1930s by the Soviet government and extended after World War II. Modern agriculture has relied on the heavy use of chemical fertilizers and pesticides, and the rivers descending from the Tajikistan have been carrying large amounts of pollutants into and through Uzbekistan and Kazakstan.

Tajikistan is rich in mineral wealth, such as coal, iron, lead, antimony, zinc, gold, tin, and tungsten. Large reserves of natural gas and petroleum also exist, fair amounts of which are produced. Natural gas is exported after meeting meager domestic needs whereas most of the petroleum produced is consumed domestically. Besides gas and petroleum, other energy resources include sizable coal deposits and hydroelectric power, but most of the electric power produced in the country is hydroelectric, obtained by the utilization of waters of the fast-flowing mountain streams. Major power stations operate on the Syr Darya and the Vaksh River. A new hydroelectric power station was built in 1994 with assistance from Russia and Pakistan. The chief mining areas are in the north. Coal mining and oil extraction began earlier than the extraction of natural gas which was started in the mid 1960s.

Tajikistan's industry is based mainly on its agricultural production and consists of a dozen cotton mills and a few silk factories, and on the extraction and processing of the country's mineral resources (predominantly natural gas and petroleum). The biggest cotton-milling center is at the capital, Dushanbe. Other light industries include the manufacture of knitted goods, leather tanning, and food-processing industries (including the processing of fruits, vegetable oils and tobacco). Metalworking consists of the manufacture of power equipment, cables, agricultural and household implements.

As most of Tajikistan is mountainous, nearly two-thirds of its surface is covered with bare rock, sand, permanent snow and trees. It is the least urbanized state in Central Asia. Settlements are mostly strung in the irrigated farming areas along the foothill regions which are turned into oases, with cotton plantations, gardens and vineyards. The capital city of **Dushanbe** (592,400) is in the Gissar Valley in the southwestern part of the republic, on the medieval caravan route from China to the fabled cities of Samarkand and Bukhara (Uzbekistan). It is an important regional railroad junction. A leading cotton textile center of the state, it also produces silk, textile machinery, clothing, leather goods, tractor parts, and foodstuffs.

On account of the predominant mountainous terrain of the country, railroads are limited to only 554 miles (891 km) of track that carries nearly half of the country's freight. The Soviet government had constructed several roadways, most of which contain paved surfaces. Air connections to the outside world include flights to Pakistan, Iran, Russia, and other Central Asian republics via Dushanbe. The principal imports are machinery, and industrial products, and exports cotton yarn, natural gas and petroleum.

After the collapse of the USSR, Tajikistan suffered seriously resulting from the disruptions of the trading system that depended heavily on the Soviet Union. Civil war broke out in 1992, and high inflation caused large-scale disaffection. Although the country joined the International Monetary Fund and the World Bank in 1993, Russia continued to assist heavily. Foreign aid by Iran and Pakistan, and internal restructuring of the economy continued in 1994 and 1995, but the pace of "privatization" and movement toward market economy remained cautiously slow. The future economic development appeared to depend mostly on the country's as yet largely unexploited mineral reserves, as well as the established cotton industry.

TURKMENISTAN

Turkmenistan is the second largest and the least densely populated nation in Central Asia. It is the southernmost of Central Asia's five republics and is bounded in the northwest by Kazakhstan, and by Uzbekistan on the east and northeast, by Iran on the south and by Afghanistan on the southeast. The Caspian Sea lies along its western boundary. Most of the country is a desert wasteland, except for few, occasional oases dotted along the foothills of the Kopet Mountain in the south and along the Amu Darya in the north.

The economy is predominantly agricultural. In 1992 agriculture contributed an estimated 31 percent of gross domestic product, and utilized 44-45 percent of the working population. Nearly 90 percent of the country's territory is covered by the Karakum desert, but widespread irrigation since the 1940s has enabled substantial agricultural development. The cultivation of fine-staple cotton, and raising of Karakul sheep are the principal agricultural activities. In addition to the cultivation of cotton, other important crops are grains, vegetables and fruits (in particular grapes and melons) which are grown primarily in the Amu Darya delta, and the oases along the river. However, the extensive use of chemical fertilizers and pesticides, and irrigation has led to pollution from the chemical runoff. The soil and water salinization of the Amu Darya delta and of the Aral Sea has further caused severe environmental hazards. Livestock breeding of the Karakul sheep is important for the production of wool. Silkworm breeding, and the raising of horses, and camels (as a source of wool, milk and meat) occupy a significant position in the agricultural economy. In less populated western section of the nation, people raise sheep, goats, and camels, and cultivate grains and fruits.

Turkmenistan is rich in mineral resources, particularly in natural gas and petroleum, potassium, and salt. Petroleum deposits and the associated industry are located in the Caspian plain and in the off-shore fields to the west of the Cheleken Peninsula in the Caspian Sea. The oil is high grade, and pipelines link the producing region to other areas of the nation. Petroleum and natural gas are used for the modest domestic needs, but most of the production is exported.

Industry contributes nearly 40 percent of the natural domestic production when only 20 percent of working population are employed in it. Industrial activity is chiefly associated with the extraction and processing of the country's oil and natural gas resources and cotton-processing. These activities account for nearly two-thirds of the value of total industrial production. During the Soviet control, a complete reorganization of the economy was completed by 1930. While retaining the old branches (cotton-ginning, oil-pressing, and carpet making), new ones (heavy industry and light industry such as food processing) were also developed that resulted in a big increase in industrial production. The republic's engineering and metal-processing enterprises are associated with the repair of diesel locomotives, rail cars, and agricultural machinery. Textiles and cotton-ginning, silk-weaving and woolen fabrics, carpet and rug making are other important industries. Turkmen carpets are internationally known and form one of the major exports of the nation. The food processing industry includes the production of vegetable oil, processing of fish and meat, flour milling, and wine making.

The settlement patterns may be divided into two broad categories: (a) an oasis region of Kopet Mountain and other oases along the Amu Darya characterized by adequate water supply, and agricultural land, and (b) a desert region of the Karakum. Cattle raising is the chief occupation. The economic and cultural center of an important oasis in the south is **Ashkabad** (416,000), which is also the national capital. Located at the northern foot of Kopet Mountains at the edge of the Karakum desert, it is an important junction on the Trans-Caspian railroad, just 19 miles (30 km) from the Iranian border. The city has textile, carpet-weaving, glasswork, and machine building industries.

The country's railroad track is limited. A new railroad and oil pipeline project are being developed to connect Turkmenistan with Iran to provide the country with direct outlet for nation's exports to the Middle East and the West. Before independence, its trade was exclusively oriented to the Soviet Union. The principal imports are machinery, light industrial products and processed foods, and exports are cotton yarn, and natural gas and petroleum. Water transport includes a merchant fleet in the Caspian Sea connecting the nation with Azerbaijan.

UZBEKISTAN

Lying south and southeast of Kazakhstan, west of Kyrgyzstan and Tajikistan, and north of Turkmenistan, and Afghanistan, Uzbekistan's territory falls mainly between Central Asia's two major rivers—Syr Darya and Amu Darya, though they only partly form its boundaries. Although the territory does reach the Caspian Sea, it includes the southern part of the Aral Sea. The Soviet government created the Uzbek Soviet Socialist Republic as a constituent unit of the USSR in 1924, but in 1991 after the dissolution of the Soviet Union it declared its independence.

With a population of 24.5 million, the largest in Central Asia, the nation is not only the most populous, but contains the largest number of the native Turkic peoples, overwhelmingly outnumbering the other ethnic groups, and have registered the highest growth rate in the region. The cultural and historic roots of the Uzbeks date back to the ancient times. Several cities such as Bukhara, Khiva, Kokand, and Samarkand were cultural, political and trade centers for centuries.

Although a large part of the country is occupied by the extensive desert of Kyzl Kum, some of the world's richest irrigated oases exist in a narrow band of densely populated area in the east. The most important of the oases is the Fergana Valley, drained by the Syr Darya, and divided primarily between Uzbekistan and Tajikistan. This and other oases such as Tashkent, Andizhan, Bukhara, Khiva, Kokand and Mary (Merv), Samarkand, and Zeravshan form the rich agricultural tracts that fulfill the nation's major produce of high-grade cotton. Most of the major oases are located where mountain streams descend on to the lowland except such oases or Zeravshan that are located in the desert lowlands.

Uzbekistan is Central Asia's largest agricultural producer. In cotton production it ranks third in the world (China and India rank higher). Besides cotton, rice, sugar beets, vegetables, fruits (particularly grapes, pomegranates, figs and melons are grown. Known for its orchards and vineyards, Uzbekistan is also important for raising Karakul sheep and silkworms. Plenty of sunlight, mild winters, fertile irrigated soil, and good pastureland make conditions suitable for the cultivation of these crops and for cattle raising. Raising silkworms is a traditional occupation among farmers, dating to the 4th century. The Fergana Valley is especially known for silk production.

The country's mineral resources include metallic ores such as copper, zinc, lead, tungsten and gold. Uzbekistan possesses substantial reserves of natural gas, oil, and coal. Most of the natural gas is consumed domestically, and gas pipelines link the important cities and stretch from Bukhara to the Ural Mountains in Russia. Petroleum fields exist in the Fergana Valley, in the vicinity of Bukhara, and in Karakalpakstan. Dams on the Syr Darya and its tributaries are utilized to produce hydroelectricity. The country is deficient in water resources. The existing canals—the Great Fergana, Northern Fergana, Southern Fergana and Tashkent---now face shortfalls in irrigation waters due to depletion of the rivers.

In manufacturing, Uzbekistan is Central Asia's major producer of machinery and heavy equipment. The manufactured items include machines and equipment for cotton cultivation, harvesting and processing, and for use in textile industry, irrigation, and road construction. Cement, textile, chemical fertilizers, and tea packing are some other industries. The country predictably exports cotton, natural gas, oil, silk and fruits, as well as manufactured goods such as machines, cement, textiles, and fertilizers.

The country is nearly self-sufficient in energy sources, and agricultural products development. The disruption of the Soviet trading system caused by the collapse of the USSR in 1991 did not particularly affect the nation. Since independence, the nation has followed a slow and cautious path of privatization. Economic development in the future would depend largely on overcoming the current infrastructural handicaps such as the antiquated means of distribution and processing of raw materials. Neither surface nor air transport now available is adequate to handle the transport of the produce such as fruits and vegetables. There are few well-developed highways. Most of the country's trade was with the Soviet Union; the nation is considering plans to enlarge the trading area to include the developing countries.

The majority of Uzbekistan's population lives in rural areas. In the early 1990s just over 40 percent of the population was registered as urban, and only 16 of the cities contain population over 100,000. Population comprised mostly of the 'non-Uzbeks.

Tashkent With a population of a little over 2 million, is the largest city of Central Asia, and the capital and the "primate" city of Uzbekistan. It lies in a large oasis along the Chirchik River on the Trans-Caspian railroad in the foothills of the Tien Shan Mountains. Dating back to the 1st century B.C., the city has been a historic trade and handicraft center on the historic caravan routes from the Orient to Europe during medieval times. When the Russians occupied it in 1865, the walled city had a population of 70,000. Several old buildings, mausoleums, and religious shrines survive in the "older" section. Tashkent lies in the most industrialized part of Uzbekistan, and contains one of the largest textile mills in Asia. Other industries include food-and tobacco-processing plants, and factories that manufacture machinery, electrical equipment, chemicals and furniture. The modern, planned section built during the Russian period, co-exists with the old Oriental quarters with its narrow, winding streets, numerous mosques, and bazaars.

Other major cities of Uzbekistan—**Samarkand** (370,000), **Audizham** (293,000), **Bukhara** (238,000), **Fergana** (198,000), and **Kokand** (176,000) are considerably smaller than Tashkent, but are just as important from the cultural and historic standpoint. Their histories extend back to ancient times, and they have served as political, and trade centers for centuries. With the exception of Bukhara these cities are located in the industrial heartland of the nation; each, including Bukhara, on the historic routeway (between China and the Middle East) has a significant industrial component. It specializes primarily in the manufacture of textiles, processed food, and machinery. The "old" sections of these cities were partially reconstructed during the Soviet period.

PROSPECTS

As Central Asia is hurtled into the "new" post-Soviet era, it lurches towards a profoundly different world. The monolithic political structure of the Soviet rule is collapsed and all the five republics are now independent, sovereign nations. Among the possible scenarios, the most likely one to emerge is the movement towards the introduction of a western-style democracy and market economy.

Since the medieval times, Central Asia has remained a backwater in Asian history.

Isolated, and economically backward, it has been away from the world's main currents of political and economic history during the last several decades. Though relatively rich in mineral and agricultural resources, the economic development of the region was essentially geared to the needs of the Soviet Union.

Politically, the boundaries of the five present-day republics created during the Soviet regime do not reflect the historic ethnic and cultural patterns of the people, and all the republics contain substantial minority populations, a situation which is fraught with the prospect of future conflicts. Culturally, the peoples of Central Asia managed to retain their traditional ethnic heritage under difficult circumstances. They are now attempting to use the recently-acquired democratic freedoms to assert their identity. This heretofore suppressed nationalistic movements of the various minority groups during the Soviet rule are beginning to emerge. The central administrations of these new republics are likely to experience difficulties in holding the ethnic groups together as their nationalistic pressures show signs of increasing the risk of political instability and even fragmentation.

An emerging factor of increasing importance is the growing strength of the Islamic religious tradition that was suppressed during the Soviet rule, which is now likely to affect the future political configuration of the region. The religious aspirations that lay dormant over long periods of authoritarian Soviet rule are now beginning to assert themselves and these nations are now attempting to forge new regional connections with neighboring Islamic countries such as Turkey, Iran, and Saudi Arabia (without renouncing their basic secular political structures).

The vast lands of Kazakstan had provided the Soviets with testing sites, and installation of nuclear armaments in the area east of the Aral Sea and the full extent of the environmental pollution resulting from such testing was kept a guarded secret. As the region emerges from the shadows of secrecy, Central Asia faces the challenge of cleaning up the mess that the Soviet testing over the last three decades had created. The enormous costs of environmental cleanup added to those of rebuilding of the backward economies, as well as these nations' own efforts to become economically competitive in world markets would require substantial financial assistance from the international community including the United States and Russia.

In perspective, whatever course the Central Asian nations adopt---the continuation of keeping economic links with Russia and other former Soviet republics, as the formation of some kind of common market or the formation of a united Turkestan---the future destinies of these republics will be of more than regional significance.

Further Readings

Akiner, S. (ed.) *Political and Economic Trends in Central Asia.* New York, 1994.

Malik, H. (ed.) *Central Asia: Its Strategie Importance and Future Prospects.* New York, 1994.

Mandelbaum, M. (ed.) *Central Asia, and the World.* New York, 1994.

25

Conclusion: Asia Enters the New Millennium

Although millennial dates are arbitrary marks on the calendar, humanity's fascination with such dates remains strong as it throws off certain metaphysical sparks. The approach of the year 2000 (by the western calendar) produces an unusual interest for it reflects humanity's perennial fascination with zeroes. People have an irresistible curiosity regarding the shape of future beyond this magical date. Futuristic predictions in several books and articles have focused on the year 2000 for nearly half a century now, and the volume of such predictions keeps on growing.

In line with this tradition, therefore, it may be permissible to speculate on the likely trends of development and change that might shape the Asian scene as the new millennium approaches. As a broad initial generalization, however, it is safe to assume that whatever might be locked in the womb of future, it is very likely that the century to come (as indeed the centuries to follow) will be complex, fast-paced and turbulent for most Asian nations. The accelerating march of technology is clearly responsible for much of this, but the pace of change is international. As it gathers further momentum, the nations of Asia are bound to be caught up in the web of this fast-paced global change. Although the unseen will figure prominently in shaping the texture of Asia's future, there are number of clearly discernible trends that are on the horizon.

Increasing Population Pressures

Population pressures have been building up nearly everywhere in Asia, especially during the last century, more so in the two larger Asian nations of India and China, which collectively account for over 2 billion of Asia's 3.6 billion people. Although the rate of population growth in China has gone down during the last few decades and is almost one-half that of India (population growth rates for India and China are 1.8 and 0.9 percent per year respectively) both nations together are adding more than 30 million human beings to the world's population annually. Even if the rates of population increase were to fall in future the increased pressure of additional human beings will undoubtedly create enormous problems of overcrowding, environmental pollution and in general, lowering of the standard of living. Japan, where population growth is among the lowest in the world, the problem is of a different type, that of "graying society" which will surely tax the adult segment that will have to shoulder the burden of the "unproductive" (or less productive) aged population.

Many futurists have predicted that the

rich nations will generally get richer and the poor poorer in the 21st century. In their view in order to rise to a level of prosperity, a developing country must achieve decades of economic growth while simultaneously holding its population stable. According to them few Asian nations will manage that trick successfully. Japan, and Singapore (the developing nations in the 1930s) have been able to become prosperous after decades of sustained efforts. Thailand, South Korea, Taiwan, and Hong Kong have been, for some time trying to emulate and duplicate Japan's example. Applying the futurist's logic to China and India, the two largest of Asia's nations may take a much longer time to be industrial giants like Japan. Their populations are too large and the gross domestic product too small (China's per capita gross product is $860 and is expected to reach $1,500 by the year 2010). In 2000, India's population has also crossed the one billion mark. Its gross domestic per capita is expected to grow from the present $370 to $820. Therefore, the dynamics of the 21st century presents a gloomy outlook for developing countries like India. The Middle East nations do not face economic problems yet, since their population base is too small. But their economies depend on a single resource (oil). The volatility of oil market in the world and the non-renewable nature of this resource make the oil-rich nations highly vulnerable to unrestricted future population growth.

Environmental Concerns One of the trends most certain to explode dramatically in the years immediately ahead is the deterioration of the natural environment, and the need to clean up the mess that decades of neglect has created. Increasing urbanization and industrialization have blanketed the skies of larger cities with atmospheric smog. Chemical pollutants emitted by an ever increasing number of automobiles and industries in and around such cities as Beijing, Delhi, Tehran, and Tokyo-Yokohama (and many others) have created massive environmental problems threatening health and safety of the people.

Centuries of indiscriminate cutting of trees in the densely populated areas to clear land for cultivation has been universal all over the continent. With most of Asia's good land already under plow, a population of close to 5 billion in the latter part of the next century would probably have to make do with less than half the arable land per capita that exists today. That would set the stage for environmental disaster.

The realization of the fragility and vulnerability of the continent's most familiar and cherished ecosystems----the Ganga, the Caspian, the Himalaya---and the drive to reverse or even halt the environmental degradation caused by disturbance to the ecosystem by the removal of tree cover is not strong as yet in Asia, but is likely to grow in the years ahead.

As the nations of Central Asia emerge from shadows of secrecy after the collapse of the Soviet Union, the full extent of environmental mischief in which economic goals has been a predominant objective is gradually becoming known. The pollution from the overuse of insecticide, and chemical fertilization for irrigation have contaminated wide areas in the basins of the Syr and Amu rivers, as well as the environmental pollution created by the nuclear testing and of dumping of nuclear "waste" have had disastrous effects on the environment on the health and livelihood of the human population around Aral and Caspian littorals. The challenge to western nations of assisting in the rebuilding of Central Asia's impoverished economies, as well as these nations' own efforts to become competitive in the 21st century, will be compounded by the enormous costs of environmental cleanup, and by the burden of building their industrial sector.

At any rate, it can be safely predicted that environment will be one of the troubling issues in the years ahead.

Growth of Nationalism Nearly all Asian countries contain numerous ethnic and linguistic minority groups that are becoming increasingly assertive in claiming their cultural, political, economic and territorial rights. Many of the countries in South and Southeast Asia attained independence in the mid and late 1940s and early 1950s. At the same time they were trying to bring tranquility and territorial consolidation, they will run a serious risk of becoming more fragmented as nationalist pressures mount.

While pressure groups are actively organized to influence government policy everywhere, they have exhibited militancy and a violent streak. In India the Sikhs and the Kashmiri Muslims; in Indonesia the Timorese; the "Tamil Tigers" in Sri Lanka; and the Kurds in Iraq have been engaged in bloody confrontation with the central governments for the creation of independent "homelands" for their cultures, and ethnicity. India has been able to defuse the problem of the separatists, at least for the time being, but pacification in other countries is a long way off. The nationalist movements in China, in the provinces of Xinjiang, Inner Mongolia and Tibet (Xizang) have been put down by force, although the central government has attempted to defuse the nationalist threats by devolving substantial power from central to regional administration and by recognizing the distinctive character of various minority groups.

With the collapse of the monolithic political structures of the Soviet Union in Central Asia (Kazakstan, Kyrgystan, Tajikistan, Turkmanistan and Uzbekistan) the strength of many "hidden" nationalistic forces has emerged. Each central administration in this region has under its control a number of distinct and vocal minorities, many of which have been deprived of their linguistic and cultural identity over long periods of authoritarian rule. Besides the indigenous ethnic groups, the central Asian nations contain several non-indigenous groups, such as the Russians, Ukrainians, Germans, etc. The introduction of Russian as an "official language" during the Soviet control was never accepted by a vast majority of the population. Linguistic "Russification" will surely be reversed in favor of indigenous languages as the nationalist sentiments become more strident in the near future.

An emerging factor of increasing importance is the growing strength of the Islamic religious tradition that was suppressed during the Soviet rule, and that is now likely to affect social and political configuration of the nations of Central Asia. The religious aspirations that lay dormant over long periods of authoritarian Soviet rule are now beginning to assert themselves, and these nations are trying to forge new regional connections with neighboring Islamic countries.

Growing Regional Integration The push toward increasing economic and political integration within Asian countries started much later than on the other continents but seems by now to have gained momentum. The earliest regional association was the League of Arab States (more generally known as the Arab League) founded in 1945. It is a loose federation designed to strengthen ties linking the Arab States, and to coordinate their policies (supposedly political). Having failed to bring harmony among the mutually suspecting or hostile nations, it has never functioned smoothly. It has no political or economic agenda except to create some kind of unity among the Arab States. The Organization of the Petroleum Exporting Countries (OPEC) which was established in 1960 has clearly defined goal to link nations whose main source of export earnings is petroleum; it aims to unify and coordinate members' petroleum policies and safeguard their interests. The organization cuts across regional boundaries; seven of its eleven members are Asian countries mostly from the Middle East, the remaining are from Africa and South America, such as Algeria, Libya, Nigeria, Indonesia, and Venezuela. The Arab nation-members of the OPEC separately formed an association in 1968, named the Organization of Arab Petroleum Exporting Countries (OAPEC), which is merely duplication of OPEC, designed only for the Arab countries.

In other regions of Asia, the Association of South East Asian Nations (ASEAN) which started in 1989 is the most important formal grouping of countries within the Pacific Basin. Its core members until recently were Indonesia, Thailand, Malaysia, the Philippines, Singapore and Brunei. Vietnam, Myanmar and Laos were admitted in the mid and late 1990s. The Pacific Basin is, in population size and territorial extent, larger and more heterogeneous than Europe. The main purpose of the association is to enhance trading and economic links within the Pacific community. On much the same lines, the nations of South Asia also formed an economic association in 1985 in order to improve regional cooperation. This union, known as the South Asian Association for Regional Cooperation (SAARC), aimed primarily at developing improved trading links within the region and forging closer economic cooperation. The SAARC is at a nascent stage of its development.

In recent years, contradictory forces within South Asia and the Pacific Basin have both helped and hindered progress towards increased integration, although, on balance, they would seem to indicate slow, but steady move towards greater interdependence. Among Southeast Asian nations, Singapore, Malaysia, and Thailand have shown a consistent economic growth and political stability. The problem posed by countries such as Indonesia is the uncertainty that is undoubtedly created by the absence for a process for achieving the replacement of one regime by another. In South Asia, democracy is relatively stable in India, but political turmoil in the other nations have already weakened the economic links which these nations are struggling to develop.

Likely Trends in Southwest and Central Asia The existence of oil, a vital resource on which much of the future development of the world depends—has invested the Persian Gulf and Caspian Basin nations in Central Asia, with enormous economic and political power. The collapse of the Soviet Union, and the growing links between the Middle East and Central Asia and South Asia have catapulted the region into a globally strategic area and a source of conflict. Oil has transformed much of the character of the region, and has made it a focus of aggressive interests of major external powers as well as enormously exacerbated tensions within the region. The fact that demand for oil has been rising, especially from the industrializing Asian nations, makes the region vitally significant not only for the major world powers but for the rest of the world as well.

Another developing scenario, unrelated to oil, is the growing militant Islamic nationalism in the Gulf States and in Central Asia which is likely to affect the political texture of the region with implications beyond the Gulf region.

In South Asia By virtue of her size, and historical importance, India has played a vital role in shaping the lives and fortunes of the region, and is likely to remain a dominance force in South Asia. Pakistan and Bangladesh were a part of British India prior to 1947, and the kingdoms of Nepal and Bhutan have remained within India's hegemony. The region encompasses diverse religions, languages, castes, and cultures. Pluralism within it has often resulted in ethnic conflicts and civil wars. The democratic structures designed by the founding fathers of these nations (excepting the kingdoms of Nepal and Bhutan) have come under heavy stresses and strains, often threatened by fragmented polity. India's democracy has been a "functioning democracy," elsewhere authoritarian rules have been the rule.

Except for Pakistan, South Asian nations initially adopted semi-socialistic form of economic policies, but gradually moved away from central planning. India, and Pakistan moved to economic liberalization in the mid-1990s and replaced central planning with market economy. Although compared to East Asian and Southeast Asian economies, the region's growth has been anemic, the path of economic liberalization adopted lately promises higher growth. Population pressures have been a major problem, but thanks to the successful green revolution, India, Pakistan and Bangladesh can feed their people now.

The social, demographic, economic and political problems, however, will continue to affect South Asia's development, and the economic outlook appears modestly promising.

In Southeast Asia Historically, Southeast Asian nations are relatively young; most are still struggling towards peace and stability. With the exception of Thailand, none has had the sustained experience of an established, dominant, indigenous political order. There has historically been no regionally dominant state, such as India or China, or in another setting, Brazil. With the passing of colonial rule, and with a newer order trying to emerge, Southeast Asia remains in a state of political, social and economic flux. An exception is the tiny city-state of Singapore, where the experiment of a democratic government, parliamentary traditions, and rigorous economic reconstruction based on private enterprise have brought political stability and prosperity.

Economic integration within the region developed by such groupings as ASEAN, and SAARC is still at a nascent state and remains feeble. However, Southeast Asian economies, particularly those of Malaysia, Indonesia, and the Philippines adopted capitalism as the organizing basis of economic life and became integrated into the global economies. The results were spectacular—the economies of each of these nations grew at rates between 8 and 12 percent a year during the 1980s, earning them the nickname of "Asian Tigers". The sharp currency crises that gripped these economies in 1997 can be attributed to underlying institutional weaknesses, and can hopefully be addressed.

In East Asia The Pacific Rim Asian nations, Korea and Taiwan in particular, have shown remarkable growth during the past two decades, and to a lesser degree China has tried to duplicate their performance. Their growth rates have ranged between 6-10 during this period. This was made possible as Korea and Taiwan adopted essentially capitalistic development strategy in the 1960s and 1970s, and in a much-modified form by China more recently. Japan's extraordinary growth in the 20th century was of course the precursor of the other three nations, although its productive capacity may well be reaching a saturation point. As the competition from the U.S., industrialized western Europe and the newly-industrializing Asian nations intensify, this scarce-resource nation faces increased pressure on its growth. However, Japan's international economic power is well established, and the nation is expected to play an increasingly important role within the Pacific Basin.

The sharp currency crises that gripped Korea, Hong Kong, and even Japan has sent shock waves in these nations. The economic salvation of these nations lies in eradicating institutional and structural weaknesses (such as corruption, insufficient banking regulations, populist policies, etc.) that they are prone to.

Further Readings

Wilber, C.O. (ed.) *The Political Economy of Development and Underdevelopment*. New York, 1973.

World Bank Atlas. Washington, D.C., 1998.

World Bank. *Key Indicators of Developing Asian Countries*. New York, 1991.

—. *World Development Report*. New York, 1997.

Index